Introduction to Optical Electronics

Second Edition

AMNON YARIV

California Institute of Technology

HOLT, RINEHART AND WINSTON

New York Chicago San Francisco Atlanta
Dallas Montreal Toronto London Sydney

Library of Congress Cataloging in Publication Data

Yariv, Amnon.
Introduction to optical electronics.
Includes index.
1. Lasers. 2. Electro-optics. 3. Quantum electronics. I. Title.
TA1675.Y37 1976 537.5'6 76-11773
ISBN 0-03-089892-7

Printed in the United States of America
234567890 038 0987

Preface

The rapidly growing field of quantum electronics can be divided roughly into two broad categories. The first is concerned chiefly with the atomic aspects of the problem. These include the study of energy levels, lifetimes, and transition rates in laser media, and also the mechanisms and the physical origins of phenomena such as Raman and Rayleigh scattering and second-harmonic generation. This branch of the field leans heavily on the formalism of quantum mechanics and has consequently become the domain of the physicist and, to a lesser extent, the physical chemist.

The second category deals with the *coherent interactions of optical radiation fields with various atomic media*. Here we tend to accept the existence of certain physical phenomena and concern ourselves with their implications and applications. The physical properties may now be represented by parameters characteristic of the material. Two typical examples are: (1) the analysis of power output and frequency pulling in laser oscillators in which the physical phenomena of spontaneous emission and atomic dispersion are important, and (2) the problem of optical second-harmonic generation and phase matching in which the complicated quantum mechanical considerations involved in understanding the optical nonlinearity are lumped into the nonlinear constant.

This second aspect of quantum electronics is more closely linked to applications and has consequently attracted the attention of the applied physicist and the electrical engineer. In this area, a good deal of the emphasis is on optics rather than on quantum physics and many of the concepts encountered here have their counterparts in radio and microwave electronics. For this reason I have decided to refer to the subject matter as optical electronics and to choose the same name for the book's title.

Since the appearance of the first edition, a number of topics have become important. These topics have been incorporated into this new edition and involve the following:

Expanded treatment of Gaussian beam propagation in homogeneous and in focusing media
Unstable optical resonators
Heterojunction injection lasers
High pressure CO_2 lasers
Noise and detection error probability in binary communication channels
Mode locking in homogeneously broadened laser systems
Propagation in symmetric and asymmetric dielectric waveguides
Mode coupling and directional coupling in dielectric waveguides
Distributed feedback lasers

Although the first edition was aimed at students in the senior year or in the first year of graduate studies, it was used mostly by graduate students. To encourage this trend, I have augmented the level of mathematical sophistication used in some of the discussions. Nevertheless, I still believe that the ever-increasing role of coherent optics in science and technology will require an early exposure to this area on the part of most electrical engineering and applied physics students. With this in mind, I have undertaken to present the material without the use of quantum mechanics. Instead of inventing quasi-classical substitutes for quantum mechanical concepts, I decided to ask the student to accept on faith certain statements whose justification can only be provided by quantum mechanics. Somewhat to my own surprise, I found that this was necessary only when introducing the concepts of stimulated and spontaneous transitions. The rest of the material can then be treated using classical formalism. An introductory knowledge of atomic physics and of electromagnetic theory would be helpful, although the basic results are derived in the text.

I am grateful to Drs. U. Ganiel, R. MacAnally, and S. Kurtin for critical comments and suggestions and to Ruth Stratton and Dian Rapchak for their patient and competent typing of the manuscript.

Pasadena, California
June, 1976

AMNON YARIV

Contents

CHAPTER

1

Electromagnetic Theory

1.0 Introduction

In this chapter we derive some of the basic results concerning the propagation of plane, single-frequency, electromagnetic waves in homogeneous isotropic media, as well as in anisotropic crystal media. Starting with Maxwell's equations we obtain expressions for the dissipation, storage, and transport of energy resulting from the propagation of waves in material media. We consider in some detail the phenomenon of birefringence, in which the phase velocity of a plane wave in a crystal depends on its direction of polarization. The two allowed modes of propagation in uniaxial crystals—the "ordinary" and "extraordinary" rays—are discussed using the formalism of the index ellipsoid.

1.1 Complex-Function Formalism

In problems that involve sinusoidally varying time functions we can save a great deal of manipulation and space by using the complex function formalism. As an example consider the function

$$a(t) = |A| \cos (\omega t + \phi_a) \tag{1.1-1}$$

where ω is the circular (radian) frequency[1] and ϕ_a is the phase. Defining the complex amplitude of $a(t)$ by

$$A = |A|e^{i\phi_a} \tag{1.1-2}$$

we can rewrite (1.1-1) as

$$a(t) = \text{Re}\,[Ae^{i\omega t}] \tag{1.1-3}$$

We will often represent $a(t)$ by

$$a(t) = Ae^{i\omega t} \tag{1.1-4}$$

instead of by (1.1-1) or (1.1-3). This of course is not strictly correct so that when this happens *it is always understood* that what is meant by (1.1-4) is the *real part* of $A \exp(i\omega t)$. In most situations the replacement of (1.1-3) by the complex form (1.1-4) poses no problems. The exceptions are cases that involve the product (or powers) of sinusoidal functions. In these cases we must use the real form of the function (1.1-3). To illustrate the case where the distinction between the real and complex form is not necessary, consider the problem of taking the derivative of $a(t)$. Using (1.1-1) we obtain

$$\frac{da(t)}{dt} = \frac{d}{dt}[|A| \cos(\omega t + \phi_a)] = -\omega|A| \sin(\omega t + \phi_a) \tag{1.1-5}$$

If we use instead the complex form (1.1-4), we get

$$\frac{da(t)}{dt} = \frac{d}{dt}(Ae^{i\omega t}) = i\omega Ae^{i\omega t}$$

Taking, as agreed, the real part of the last expression and using (1.1-2), we obtain (1.1-5).

As an example of a case in which we have to use the real form of the function, consider the product of two sinusoidal functions $a(t)$ and $b(t)$, where

$$\begin{aligned} a(t) &= |A| \cos(\omega t + \phi_a) \\ &= \frac{|A|}{2}[e^{i(\omega t+\phi_a)} + e^{-i(\omega t+\phi_a)}] \\ &= \text{Re}\,[Ae^{i\omega t}] \end{aligned} \tag{1.1-6}$$

and

$$\begin{aligned} b(t) &= |B| \cos(\omega t + \phi_b) \\ &= \frac{|B|}{2}[e^{i(\omega t+\phi_b)} + e^{-i(\omega t+\phi_b)}] \\ &= \text{Re}\,[Be^{i\omega t}] \end{aligned} \tag{1.1-7}$$

[1] The radian frequency ω is to be distinguished from the real frequency $\nu = \omega/2\pi$.

with $A = |A| \exp(i\phi_a)$ and $B = |B| \exp(i\phi_b)$. Using the real functions, we get

$$a(t)b(t) = \frac{|A|\,|B|}{2}[\cos(2\omega t + \phi_a + \phi_b) + \cos(\phi_a - \phi_b)] \quad \textbf{(1.1-8)}$$

Were we to evaluate the product $a(t)b(t)$ using the complex form of the functions, we would get

$$a(t)b(t) = ABe^{i2\omega t} = |A|\,|B|e^{i(2\omega t+\phi_a+\phi_b)} \quad \textbf{(1.1-9)}$$

Comparing the last result to (1.1-8) shows that the time-independent (dc) term $\frac{1}{2}|A|\,|B| \cos(\phi_a - \phi_b)$ is missing, and thus the use of the complex form led to an error.

Time-averaging of sinusoidal products.[2] Another problem often encountered is that of finding the time average of the product of two sinusoidal functions of the same frequency

$$\overline{a(t)b(t)} = \frac{1}{T}\int_0^T |A| \cos(\omega t + \phi_a)|B| \cos(\omega t + \phi_b)\,dt \quad \textbf{(1.1-10)}$$

where $a(t)$ and $b(t)$ are given by (1.1-6) and (1.1-7) and the horizontal bar denotes time-averaging. $T = 2\pi/\omega$ is the period of the oscillation. Since the integrand in (1.1-10) is periodic in T, the averaging can be performed over a time T. Using (1.1-8) we obtain directly

$$\overline{a(t)b(t)} = \frac{|A|\,|B|}{2} \cos(\phi_a - \phi_b) \quad \textbf{(1.1-11)}$$

This last result can be written in terms of the complex amplitudes A and B, defined immediately following (1.1-7), as

$$\overline{a(t)b(t)} = \tfrac{1}{2}\,\mathrm{Re}\,(AB^*) \quad \textbf{(1.1-12)}$$

This important result will find frequent use throughout the book.

1.2 Considerations of Energy and Power in Electromagnetic Fields

In this section we derive the formal expressions for the power transport, power dissipation, and energy storage that accompany the propagation of electromagnetic radiation in material media. The starting point is

---

[2] The problem of the time average of two nearly sinusoidal functions is considered in Problems 1.1 and 1.2.

Maxwell's equations (in MKS units)

$$\nabla \times \mathbf{h} = \mathbf{i} + \frac{\partial \mathbf{d}}{\partial t} \tag{1.2-1}$$

$$\nabla \times \mathbf{e} = -\frac{\partial \mathbf{b}}{\partial t} \tag{1.2-2}$$

and the constitutive equations relating the polarization of the medium to the displacement vectors

$$\mathbf{d} = \epsilon_0 \mathbf{e} + \mathbf{p} \tag{1.2-3}$$

$$\mathbf{b} = \mu_0(\mathbf{h} + \mathbf{m}) \tag{1.2-4}$$

where $\mathbf{i}$ is the current density (amperes per square meter); $\mathbf{e}(\mathbf{r}, t)$ and $\mathbf{h}(\mathbf{r}, t)$ are the electric and magnetic field vectors, respectively; $\mathbf{d}(\mathbf{r}, t)$ and $\mathbf{b}(\mathbf{r}, t)$ are the electric and magnetic displacement vectors; $\mathbf{p}(\mathbf{r}, t)$ and $\mathbf{m}(\mathbf{r}, t)$ are the electric and magnetic polarizations (dipole moment per unit volume) of the medium; and ϵ_0 and μ_0 are the electric and magnetic permeabilities of vacuum, respectively. We adopt the convention of using lowercase letters to denote the time-varying functions, reserving capital letters for the amplitudes of the sinusoidal time functions. For a detailed discussion of Maxwell's equations, the reader is referred to any standard text on electromagnetic theory such as, for example, Reference [1].

Using (1.2-3) and (1.2-4) in (1.2-1) and (1.2-2) leads to

$$\nabla \times \mathbf{h} = \mathbf{i} + \frac{\partial}{\partial t}(\epsilon_0 \mathbf{e} + \mathbf{p}) \tag{1.2-5}$$

$$\nabla \times \mathbf{e} = -\frac{\partial}{\partial t}\mu_0(\mathbf{h} + \mathbf{m}) \tag{1.2-6}$$

Taking the scalar (dot) product of (1.2-5) and $\mathbf{e}$ gives

$$\mathbf{e} \cdot \nabla \times \mathbf{h} = \mathbf{e} \cdot \mathbf{i} + \frac{\epsilon_0}{2}\frac{\partial}{\partial t}(\mathbf{e} \cdot \mathbf{e}) + \mathbf{e} \cdot \frac{\partial \mathbf{p}}{\partial t} \tag{1.2-7}$$

where we used the relation

$$\frac{1}{2}\frac{\partial}{\partial t}(\mathbf{e} \cdot \mathbf{e}) = \mathbf{e} \cdot \frac{\partial \mathbf{e}}{\partial t}$$

Next we take the scalar product of (1.2-6) and $\mathbf{h}$:

$$\mathbf{h} \cdot \nabla \times \mathbf{e} = -\frac{\mu_0}{2}\frac{\partial}{\partial t}(\mathbf{h} \cdot \mathbf{h}) - \mu_0 \mathbf{h} \cdot \frac{\partial \mathbf{m}}{\partial t} \tag{1.2-8}$$

Subtracting (1.2-8) from (1.2-7) and using the vector identity

$$\nabla \cdot (\mathbf{A} \times \mathbf{B}) = \mathbf{B} \cdot \nabla \times \mathbf{A} - \mathbf{A} \cdot \nabla \times \mathbf{B} \tag{1.2-9}$$

results in

$$-\nabla\cdot(\mathbf{e}\times\mathbf{h}) = \mathbf{e}\cdot\mathbf{i} + \frac{\partial}{\partial t}\left(\frac{\epsilon_0}{2}\mathbf{e}\cdot\mathbf{e} + \frac{\mu_0}{2}\mathbf{h}\cdot\mathbf{h}\right) + \mathbf{e}\cdot\frac{\partial\mathbf{p}}{\partial t} + \mu_0\mathbf{h}\cdot\frac{\partial\mathbf{m}}{\partial t} \tag{1.2-10}$$

We integrate the last equation over an arbitrary volume V and use the Gauss theorem [1]

$$\int_V (\nabla\cdot\mathbf{A})\,dv = \int_S \mathbf{A}\cdot\mathbf{n}\,da$$

where $\mathbf{A}$ is any vector function, $\mathbf{n}$ is the unit vector normal to the surface S enclosing V, and dv and da are the differential volume and surface elements, respectively. The result is

$$-\int_V \nabla\cdot(\mathbf{e}\times\mathbf{h})\,dv = -\int_S (\mathbf{e}\times\mathbf{h})\cdot\mathbf{n}\,da = \int_V\left[\mathbf{e}\cdot\mathbf{i} + \frac{\partial}{\partial t}\left(\frac{\epsilon_0}{2}\mathbf{e}\cdot\mathbf{e}\right) + \frac{\partial}{\partial t}\left(\frac{\mu_0}{2}\mathbf{h}\cdot\mathbf{h}\right) + \mathbf{e}\cdot\frac{\partial\mathbf{p}}{\partial t} + \mu_0\mathbf{h}\cdot\frac{\partial\mathbf{m}}{\partial t}\right]dv \tag{1.2-11}$$

According to the conventional interpretation of electromagnetic theory, the left side of (1.2-11), that is,

$$-\int_S (\mathbf{e}\times\mathbf{h})\cdot\mathbf{n}\,da$$

gives the total power flowing *into* the volume bounded by S. The first term on the right side is the power expended by the field on the moving charges, the sum of the second and third terms corresponds to the rate of increase of the vacuum electromagnetic stored energy $\mathcal{E}_{\text{vac}}$ where

$$\mathcal{E}_{\text{vac}} = \int_V\left[\frac{\epsilon_0}{2}\mathbf{e}\cdot\mathbf{e} + \frac{\mu_0}{2}\mathbf{h}\cdot\mathbf{h}\right]dv \tag{1.2-12}$$

Of special interest in this book is the next-to-last term

$$\mathbf{e}\cdot\frac{\partial\mathbf{p}}{\partial t}$$

which represents the power per unit volume expended by the field *on* the electric dipoles. This power goes into an increase in the potential energy stored by the dipoles as well as into supplying the dissipation that may accompany the change in $\mathbf{p}$. We will return to it again in Chapter 5, where we treat the interaction of radiation and atomic systems.

Dipolar dissipation in harmonic fields. According to the discussion in the preceding paragraph, the average power per unit volume expended by the field on the medium electric polarization is

$$\overline{\frac{\text{Power}}{\text{Volume}}} = \overline{\mathbf{e} \cdot \frac{\partial \mathbf{p}}{\partial t}} \tag{1.2-13}$$

where the horizontal bar denotes time-averaging. Let us assume for the sake of simplicity that $\mathbf{e}(t)$ and $\mathbf{p}(t)$ are parallel to each other and take their sinusoidally varying magnitudes as

$$e(t) = \text{Re}\,[Ee^{i\omega t}] \tag{1.2-14}$$

$$p(t) = \text{Re}\,[Pe^{i\omega t}] \tag{1.2-15}$$

where E and P are the complex amplitudes. The electric susceptibility χ_e is defined by

$$P = \epsilon_0 \chi_e E \tag{1.2-16}$$

and is thus a complex number. Substituting (1.2-14) and (1.2-15) in (1.2-13) and using (1.2-16) gives

$$\begin{aligned} \overline{\frac{\text{Power}}{\text{Volume}}} &= \overline{\text{Re}\,[Ee^{i\omega t}]\,\text{Re}\,[i\omega Pe^{i\omega t}]} \\ &= \tfrac{1}{2}\,\text{Re}\,[i\omega\epsilon_0\chi_e EE^*] \\ &= \frac{\omega}{2}\,\epsilon_0|E|^2\,\text{Re}\,(i\chi_e) \end{aligned} \tag{1.2-17}$$

where in going from the first to the second equality we used (1.1-12). Since χ_e is complex we can write it in terms of its real and imaginary parts as

$$\chi_e = \chi_e' - i\chi_e'' \tag{1.2-18}$$

which, when used in (1.2-17), gives

$$\overline{\frac{\text{Power}}{\text{Volume}}} = \frac{\omega\epsilon_0\chi_e''}{2}\,|E|^2 \tag{1.2-19}$$

which is the desired result.

We leave it as an exercise (Problem 1-3) to show that in anisotropic media in which the complex field components are related by

$$P_i = \epsilon_0 \sum_j \chi_{ij} E_j \tag{1.2-20}$$

the application of (1.2-13) yields

$$\overline{\frac{\text{Power}}{\text{Volume}}} = \frac{\omega}{2}\,\epsilon_0 \sum_{i,j} \text{Re}\,(i\chi_{ij}E_i^*E_j) \tag{1.2-21}$$

1.3 Wave Propagation in Isotropic Media

Here we consider the propagation of electromagnetic plane waves in homogeneous and isotropic media so that ϵ and μ are scalar constants. Vacuum is, of course, the best example of such a "medium." Liquids and glasses are material media that, to a first approximation, can be treated as homogeneous and isotropic.[3] We choose the direction of propagation as z and, taking the plane wave to be uniform in the x-y plane, put $\partial/\partial x = \partial/\partial y = 0$ in (1.2-1) and (1.2-2). Assuming a lossless ($\sigma = 0$) medium, (1.2-1) and (1.2-2) become

$$\nabla \times \mathbf{e} = -\mu \frac{\partial \mathbf{h}}{\partial t} \tag{1.3-1}$$

$$\nabla \times \mathbf{h} = \epsilon \frac{\partial \mathbf{e}}{\partial t} \tag{1.3-2}$$

$$\frac{\partial e_y}{\partial z} = \mu \frac{\partial h_x}{\partial t} \tag{1.3-3}$$

$$\frac{\partial h_y}{\partial z} = -\epsilon \frac{\partial e_x}{\partial t} \tag{1.3-4}$$

$$\frac{\partial e_x}{\partial z} = -\mu \frac{\partial h_y}{\partial t} \tag{1.3-5}$$

$$\frac{\partial h_x}{\partial z} = \epsilon \frac{\partial e_y}{\partial t} \tag{1.3-6}$$

$$0 = \mu \frac{\partial h_z}{\partial t} \tag{1.3-7}$$

$$0 = \epsilon \frac{\partial e_z}{\partial t} \tag{1.3-8}$$

From (1.3-7) and (1.3-8) it follows that h_z and e_z are both zero; therefore, a uniform plane wave in a homogeneous isotropic medium can have no longitudinal field components. We can obtain a self-consistent set of equations from (1.3-3) through (1.3-8) by taking e_y and h_x (or e_x and h_y) to be zero.[4] In this case the last set of equations reduces to Equations (1.3-4) and (1.3-5). Taking the derivative of (1.3-5) with respect to z and

---

[3] The individual molecules making up the liquid or glass are, of course, anisotropic. This anisotropy, however, is averaged out because of the very large number of molecules with random orientations present inside a volume $\sim\lambda^3$.

[4] More fundamentally it can be easily shown from (1.3-1) and (1.3-2) (see Problem 1.4) that, for uniform plane harmonic waves, **e** and **h** are normal to each other as well as to the direction of propagation. Thus, **x** and **y** can simply be chosen to coincide with the directions of **e** and **h**.

using (1.3-4), we obtain

$$\frac{\partial^2 e_x}{\partial z^2} = \mu\epsilon \frac{\partial^2 e_x}{\partial t^2} \tag{1.3-9}$$

A reversal of the procedure will yield a similar equation for h_y. Since our main interest is in harmonic (sinusoidal) time variation we postulate a solution in the form of

$$e_x^{\pm} = E_x^{\pm} e^{i(\omega t \mp kz)} \tag{1.3-10}$$

where $E_x^{\pm} \exp(\mp ikz)$ are the complex field amplitudes at z. Before substituting (1.3-10) into the wave equation (1.3-9) we may consider the nature of the two functions $e_x^{\pm}$. Taking first e_x^{+}: if an observer were to travel in such a way as to always exercise the same field value, he would have to satisfy the condition

$$\omega t - kz = \text{constant}$$

where the constant is arbitrary and determines the field value "seen" by the observer. By differentiation of the last result, it follows that the observer must travel in the $+z$ direction with a velocity

$$c = \frac{dz}{dt} = \frac{\omega}{k} \tag{1.3-11}$$

This is the *phase velocity* of the wave. If the wave were frozen in time, the separation between two neighboring field peaks—that is, the wavelength—is

$$\lambda = \frac{2\pi}{k} = 2\pi \frac{c}{\omega} \tag{1.3-12}$$

The e_x^{-} solution differs only in the sign of k, and thus, according to (1.3-11), it corresponds to a wave traveling with a phase velocity c in the $-z$ direction.

The value of c can be obtained by substituting the assumed solution (1.3-10) into (1.3-9), which results in

$$c = \frac{\omega}{k} = \frac{1}{\sqrt{\mu\epsilon}} \tag{1.3-13}$$

or

$$k = \omega\sqrt{\mu\epsilon}$$

The phase velocity in vacuum is

$$c_0 = \frac{1}{\sqrt{\mu_0\epsilon_0}} = 3 \times 10^8 \text{ m/s}$$

whereas in material media it has the value

$$c = \frac{c_0}{n}$$

where $n \equiv \sqrt{\epsilon/\epsilon_0}$ is the *index of refraction.*

Turning our attention next to the magnetic field h_y, we can express it, in a manner similar to (1.3-10), in the form of

$$h_y^{\pm} = H_y^{\pm} e^{i(\omega t \mp kz)} \tag{1.3-14}$$

Substitution of this equation into (1.3-4) and using (1.3-10) gives

$$-ikH_y^+ e^{i(\omega t - kz)} = -i\omega\epsilon E_x^+ e^{i(\omega t - kz)}$$

Therefore, from (1.3-13),

$$H_y^+ = \frac{E_x^+}{\eta} \qquad \eta = \sqrt{\frac{\mu}{\epsilon}} \tag{1.3-15}$$

In vacuum $\eta_0 = \sqrt{\mu_0/\epsilon_0} \simeq 377$ ohms. Repeating the same steps with H_y^- and E_x^- gives

$$H_y^- = -\frac{E_x^-}{\eta} \tag{1.3-16}$$

so that in the case of negative ($-z$) traveling waves the relative phase of the electric and magnetic fields is reversed with respect to the wave traveling in the $+z$ direction. Since the wave equation (1.3-9) is a linear differential equation, we can take the solution for the harmonic case as a linear superposition of e_x^+ and e_x^-

$$e_x(z, t) = E_x^+ e^{i(\omega t - kz)} + E_x^- e^{i(\omega t + kz)} \tag{1.3-17}$$

and, similarly,

$$h_y(z, t) = \frac{1}{\eta}\left[E_x^+ e^{i(\omega t - kz)} - E_x^- e^{i(\omega t + kz)}\right]$$

where E_x^+ and E_x^- are arbitrary complex constants.

Power flow in harmonic fields. The average power per unit area—that is, the intensity (W/m^2)—carried in the direction of propagation by a uniform plane wave is given by (1.2-11) as

$$I = \overline{|\mathbf{e} \times \mathbf{h}|} \tag{1.3-18}$$

where the horizontal bar denotes time averaging. Since $\mathbf{e} \parallel x$ and $\mathbf{h} \parallel y$, we can write (1.3-18) as

$$I = \overline{e_x h_y}$$

Taking advantage of the harmonic nature of e_x and h_y, we use (1.3-17) and (1.1-12) to obtain

$$\begin{aligned} I = \tfrac{1}{2}\,\mathrm{Re}\,[E_x H_y^*] &= \frac{1}{2\eta}\,\mathrm{Re}\,[(E_x^+ e^{-ikz} + E_x^- e^{ikz})] \\ &\quad \times [(E_x^+)^* e^{ikz} - (E_x^-)^* e^{-ikz}] \\ &= \frac{|E_x^+|^2}{2\eta} - \frac{|E_x^-|^2}{2\eta} \end{aligned} \tag{1.3-19}$$

The first term on the right side of (1.3-19) gives the intensity associated with the positive ($+z$) traveling wave, whereas the second term represents the negative traveling wave, with the minus sign accounting for the opposite direction of power flow.

An important relation that will be used in a number of later chapters relates the intensity of the plane wave to the stored electromagnetic energy density. We start by considering the second and fourth terms on the right of (1.2-11)

$$\frac{\partial}{\partial t}\left(\frac{\epsilon_0}{2}\mathbf{e}\cdot\mathbf{e}\right)+\mathbf{e}\cdot\frac{\partial \mathbf{p}}{\partial t}$$

Using the relations

$$\begin{aligned}\mathbf{p} &= \epsilon_0 \chi_e \mathbf{e}\\ \epsilon &= \epsilon_0(1+\chi_e)\end{aligned} \tag{1.3-20}$$

we obtain

$$\frac{\partial}{\partial t}\left(\frac{\epsilon_0}{2}\mathbf{e}\cdot\mathbf{e}\right)+\mathbf{e}\cdot\frac{\partial \mathbf{p}}{\partial t}=\frac{\partial}{\partial t}\left(\frac{\epsilon}{2}\mathbf{e}\cdot\mathbf{e}\right) \tag{1.3-21}$$

Since we assumed the medium to be lossless, the last term must represent the rate of change of electric energy density stored in the vacuum as well as in the electric dipoles; that is,

$$\frac{\mathcal{E}_{\text{electric}}}{\text{Volume}}=\frac{\epsilon}{2}\mathbf{e}\cdot\mathbf{e} \tag{1.3-22}$$

The magnetic energy density is derived in a similar fashion using the relations

$$\begin{aligned}\mathbf{m} &= \chi_m \mathbf{h}\\ \mu &= \mu_0(1+\chi_m)\end{aligned}$$

resulting in

$$\frac{\mathcal{E}_{\text{magnetic}}}{\text{Volume}}=\frac{\mu}{2}\mathbf{h}\cdot\mathbf{h} \tag{1.3-23}$$

Considering only the positive traveling wave in (1.3-17) we obtain from (1.3-22) and (1.3-23)

$$\begin{aligned}\frac{\overline{\mathcal{E}_{\text{magnetic}}+\mathcal{E}_{\text{electric}}}}{\text{Volume}} &= \left(\frac{\epsilon}{2}\right)\overline{(e_x^+)^2}+\left(\frac{\mu}{2}\right)\overline{(h_y^+)^2}\\ &= \frac{\epsilon}{4}|E_x^+|^2+\frac{\mu}{4}|H_y^+|^2\\ &= \frac{\epsilon}{4}|E_x^+|^2+\frac{\mu}{4}\frac{|E_x^+|^2}{\eta^2}\\ &= \tfrac{1}{2}\,\epsilon|E_x^+|^2\end{aligned} \tag{1.3-24}$$

where the second equality is based on (1.1-12), and the third and fourth use (1.3-15). Comparing (1.3-24) to (1.3-19), we get

$$\frac{I}{\overline{\mathcal{E}}/\text{Volume}} = \frac{1}{\sqrt{\mu\epsilon}} = c \tag{1.3-25}$$

where $\overline{\mathcal{E}} = \overline{\mathcal{E}_{\text{magnetic}}} + \overline{\mathcal{E}_{\text{electric}}}$ is the total field energy density. In terms of the electric field we get

$$I = \frac{c\epsilon|E|^2}{2} \tag{1.3-26}$$

1.4 Wave Propagation in Crystals—The Index Ellipsoid

In the discussion of electromagnetic wave propagation up to this point, we have assumed that the medium was isotropic. This causes the induced polarization to be parallel to the electric field and to be related to it by a (scalar) factor that is independent of the direction along which the field is applied. This situation does not apply in the case of dielectric crystals. Since the crystal is made up of a regular periodic array of atoms (or ions), we may expect that the induced polarization will depend, both in its magnitude and direction, on the direction of the applied field. Instead of the simple relation (1.3-20) linking **p** and **e**, we have

$$\begin{aligned} P_x &= \epsilon_0(\chi_{11}E_x + \chi_{12}E_y + \chi_{13}E_z) \\ P_y &= \epsilon_0(\chi_{21}E_x + \chi_{22}E_y + \chi_{23}E_z) \\ P_z &= \epsilon_0(\chi_{31}E_x + \chi_{32}E_y + \chi_{33}E_z) \end{aligned} \tag{1.4-1}$$

where the capital letters denote the complex amplitudes of the corresponding time-harmonic quantities. The 3×3 array of the χ_{ij} coefficients is called the electric susceptibility tensor. The magnitude of the χ_{ij} coefficients depends, of course, on the choice of the x, y, and z axes relative to that of the crystal structure. It is always possible to choose x, y, and z in such a way that the off-diagonal elements vanish, leaving

$$\begin{aligned} P_x &= \epsilon_0\chi_{11}E_x \\ P_y &= \epsilon_0\chi_{22}E_y \\ P_z &= \epsilon_0\chi_{33}E_z \end{aligned} \tag{1.4-2}$$

These directions are called the *principal dielectric axes of the crystal.* In this book we will use only the principal coordinate system. We can, instead of using (1.4-2), describe the dielectric response of the crystal by means of the electric permeability tensor ϵ_{ij}, defined by

$$\begin{aligned} D_x &= \epsilon_{11}E_x \\ D_y &= \epsilon_{22}E_y \\ D_z &= \epsilon_{33}E_z \end{aligned} \tag{1.4-3}$$

From (1.4-2) and the relation

$$\mathbf{D} = \epsilon_0 \mathbf{E} + \mathbf{P}$$

we have

$$\begin{aligned} \epsilon_{11} &= \epsilon_0(1 + \chi_{11}) \\ \epsilon_{22} &= \epsilon_0(1 + \chi_{22}) \\ \epsilon_{33} &= \epsilon_0(1 + \chi_{33}) \end{aligned} \tag{1.4-4}$$

Birefringence. One of the most important consequences of the dielectric anisotropy of crystals is the phenomenon of birefringence in which the phase velocity of an optical beam propagating in the crystal depends on the direction of polarization of its **e** vector. Before treating this problem mathematically we may pause and ponder its physical origin. In an isotropic medium the induced polarization is independent of the field direction so that $\chi_{11} = \chi_{22} = \chi_{33}$ and, using (1.4-4), $\epsilon_{11} = \epsilon_{22} = \epsilon_{33} = \epsilon$. Since $c = (\mu\epsilon)^{-1/2}$, the phase velocity is independent of the direction of polarization. In an anisotropic medium the situation is different. Consider, for example, a wave propagating along z. If its electric field is parallel to x, it will induce, according to (1.4-2), only P_x and will consequently "see" an electric permeability ϵ_{11}. Its phase velocity will thus be $c_x = (\mu\epsilon_{11})^{-1/2}$. If, on the other hand, the wave is polarized parallel to y it will propagate with a phase velocity $c_y = (\mu\epsilon_{22})^{-1/2}$.

Birefringence has some interesting consequences. Consider, as an example, a wave propagating along the crystal z direction and having at some plane, say $z = 0$, a linearly polarized field with equal components along x and y. Since $k_x \neq k_y$, as the wave propagates into the crystal the x and y components get out of phase and the wave becomes elliptically polarized. This phenomenon is discussed in detail in Section 9.2 and forms the basis of the electrooptic modulation of light.

Returning to the example of a wave propagating along the crystal z direction, let us assume, as in Section 1.3, that the only nonvanishing field components are e_x and h_y. Maxwell's curl equations (1.3-5) and (1.3-4) reduce, in a self-consistent manner, to

$$\begin{aligned} \frac{\partial e_x}{\partial z} &= -\mu \frac{\partial h_y}{\partial t} \\ \frac{\partial h_y}{\partial z} &= -\epsilon_{11} \frac{\partial e_x}{\partial t} \end{aligned} \tag{1.4-5}$$

Taking the derivative of the first of Equations (1.4-5) with respect to z and then substituting the second equation for $\partial h_y/\partial z$ gives

$$\frac{\partial^2 e_x}{\partial z^2} = \mu\epsilon_{11} \frac{\partial^2 e_x}{\partial t^2} \tag{1.4-6}$$

If we postulate, as in (1.3-10), a solution in the form

$$e_x = E_x e^{i(\omega t - k_x z)} \tag{1.4-7}$$

then Equation (1.4-6) becomes

$$k_x^2 E_x = \omega^2 \mu \epsilon_{11} E_x$$

Therefore, the propagation constant of a wave polarized along x and traveling along z is

$$k_x = \omega\sqrt{\mu\epsilon_{11}} \tag{1.4-8}$$

Repeating the derivation but with a wave polarized along the y axis, instead of the x axis, yields $k_y = \omega\sqrt{\mu\epsilon_{22}}$.

Index ellipsoid. As shown above, in a crystal the phase velocity of a wave propagating along a given direction depends on the direction of its polarization. For propagation along z, as an example, we found that Maxwell's equations admitted two solutions: one with its linear polarizations along x and the second along y. If we consider the propagation along some arbitrary direction in the crystal, the problem becomes more difficult. We have to determine the directions of polarization of the two allowed waves, as well as their phase velocities. This is done most conveniently using the so-called index ellipsoid

$$\frac{x^2}{(\epsilon_{11}/\epsilon_0)} + \frac{y^2}{(\epsilon_{22}/\epsilon_0)} + \frac{z^2}{(\epsilon_{33}/\epsilon_0)} = 1 \tag{1.4-9}$$

This is the equation of a generalized ellipsoid with major axes parallel to x, y, and z whose respective lengths are $2\sqrt{\epsilon_{11}/\epsilon_0}$, $2\sqrt{\epsilon_{22}/\epsilon_0}$, and $2\sqrt{\epsilon_{33}/\epsilon_0}$. The procedure for finding the polarization directions and the corresponding phase velocities for a *given* direction of propagation is as follows: Determine the ellipse formed by the intersection of a plane through the origin and normal to the direction of propagation and the index ellipsoid (1.4-9). The directions of the major and minor axes of this ellipse are those of the two allowed polarizations[5] and the lengths of these axes are $2n_1$ and $2n_2$, where n_1 and n_2 are the indices of the refraction of the two allowed solutions. The two waves propagate, thus, with phase velocities c_0/n_1 and c_0/n_2, respectively, where $c_0 = (\mu_0\epsilon_0)^{-1/2}$ is the phase velocity in vacuum. A formal proof of this procedure is given in References [2] and [3].

To illustrate the use of the index ellipsoid, consider the case of a uniaxial crystal (that is, a crystal which possesses a single axis of threefold, fourfold, or sixfold symmetry). Taking the direction of this axis as z,

---

[5] These are actually the directions of the **D**, not of the **E**, vectors. In a crystal these two are separated, in general, by a small angle; see References [2] and [3].

symmetry considerations[6] dictate that $\epsilon_{11} = \epsilon_{22}$. Defining the principal indices of refraction n_o and n_e by

$$n_o^2 \equiv \frac{\epsilon_{11}}{\epsilon_0} = \frac{\epsilon_{22}}{\epsilon_0} \qquad n_e^2 \equiv \frac{\epsilon_{33}}{\epsilon_0} \tag{1.4-10}$$

the equation of the index ellipsoid (1.4-9) becomes

$$\frac{x^2}{n_o^2} + \frac{y^2}{n_o^2} + \frac{z^2}{n_e^2} = 1 \tag{1.4-11}$$

This is an ellipsoid of revolution with the circular symmetry axis parallel to z. The z major axis of the ellipsoid is of length $2n_e$, whereas that of the x and y axes is $2n_o$. The procedure of using the index ellipsoid is illustrated by Figure 1-1.

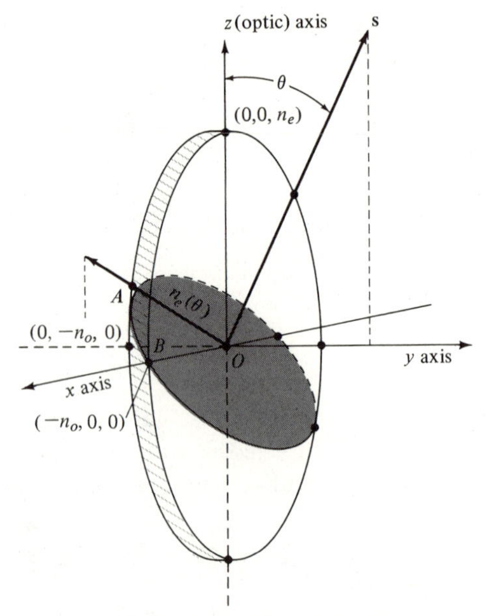

Figure 1-1 Construction for finding indices of refraction and allowed polarization for a given direction of propagation **s**. The figure shown is for a uniaxial crystal with $n_x = n_y = n_o$.

The direction of propagation is along **s** and is at an angle θ to the (optic) z axis. Because of the circular symmetry of (1.4-11) about z we can choose, without any loss of generality, the y axis to coincide with the projection of **s** on the x-y plane. The intersection ellipse of the plane normal to **s** with the ellipsoid is shaded in the figure. The two allowed polarization directions are parallel to the axes of the ellipse and thus correspond to

---

[6] See, for example, J. F. Nye, *Physical Properties of Crystals* (Oxford University Press, New York, 1957).

the line segments OA and OB. They are consequently perpendicular to **s** as well as to each other. The two waves polarized along these directions have, respectively, indices of refraction given by $n_e(\theta) = |OA|$ and $n_o = |OB|$. The first of these two waves, which is polarized along OA, is called the *extraordinary wave*. Its direction of polarization varies with θ following the intersection point A. Its index of refraction is given by the length of OA. It can be determined using Figure 1-2, which shows the intersection of the index ellipsoid with the y-z plane.

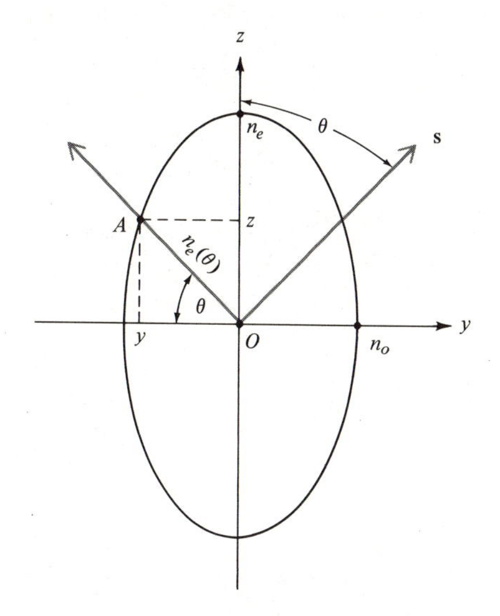

Figure 1-2 Intersection of the index ellipsoid with the z-y plane. $|OA| = n_e(\theta)$ is the index of refraction of the extraordinary wave propagating in the direction **s**.

Using the relations

$$n_e^2(\theta) = z^2 + y^2$$

$$\frac{z}{n_e(\theta)} = \sin\theta$$

and the equation of the ellipse

$$\frac{y^2}{n_o^2} + \frac{z^2}{n_e^2} = 1$$

we obtain

$$\frac{1}{n_e^2(\theta)} = \frac{\cos^2\theta}{n_o^2} + \frac{\sin^2\theta}{n_e^2} \tag{1.4-12}$$

Thus, for $\theta = 0°$, $n_e(0°) = n_o$ and for $\theta = 90°$, $n_e(90°) = n_e$.

The ordinary wave remains, according to Figure 1-1, polarized along the same direction OB independent of θ. It has an index of refraction n_o. The amount of birefringence $n_e(\theta) - n_o$ thus varies from zero for $\theta = 0°$ (that is, propagation along the optic axis) to $n_e - n_o$ for $\theta = 90°$.

Normal (index) surfaces. Consider the surface in which the distance of a given point from the origin is equal to the index of refraction of a wave propagating along this direction. The surface is called the normal (index) surface. The normal surface of the ordinary wave is a sphere, since the index of refraction is n_o and is independent of the direction of propagation. The normal surface of the extraordinary wave is an ellipsoid. In a uniaxial crystal it becomes an ellipsoid of revolution about the optic (z) axis in which the distance $n_e(\theta)$ to the origin is given by (1.4-12). The intersection of the normal surfaces of a positive ($n_e > n_o$) uniaxial crystal with the **s-z** plane is shown in Figure 1-3.

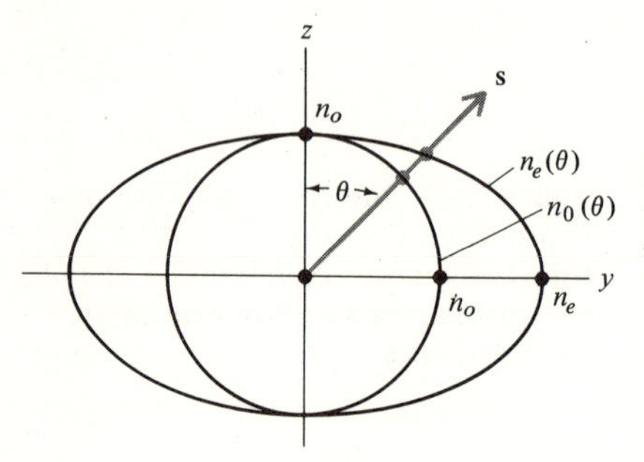

Figure 1-3 Intersection of s-z plane with normal surfaces of a positive uniaxial crystal ($n_e > n_o$).

PROBLEMS

1-1 Consider the problem of finding the time average

$$\overline{a^2(t)} = \frac{1}{T}\int_0^T a^2(t)\,dt$$

of

$$\begin{aligned} a(t) &= |A_1|\cos(\omega_1 t + \phi_1) + |A_2|\cos(\omega_2 t + \phi_2) \\ &= \text{Re}\,[V_a(t)] \end{aligned}$$

where

$$V_a(t) = A_1 e^{i\omega_1 t} + A_2 e^{i\omega_2 t}$$

and $A_{1,2} = |A_{1,2}|e^{i\phi_{1,2}}$. $V_a(t)$ is called the *analytic signal* of $a(t)$. Assume that $(\omega_1 - \omega_2) \ll \omega_1$ and integrate over a time T, which is long compared to the period $2\pi/\omega_{1,2}$ but short compared to the beat period $2\pi/(\omega_1 - \omega_2)$.[7] Show that

$$\overline{a^2(t)} = \tfrac{1}{2}[V_a(t)V_a^*(t)]$$

[7] When this condition is fulfilled, $a(t)$ consists of a sinusoidal function with a "slowly" varying amplitude and is often called a quasi-sinusoid.

1-2 Show how we can use the analytic functions as defined by Problem 1-1 to find the time average

$$\overline{a(t)b(t)} = \frac{1}{T}\int_0^T a(t)b(t)\,dt$$

where $a(t)$ is the same as in Problem 1-1, and the analytic function of $b(t)$ is

$$V_b(t) = [A_3 e^{i\omega_3 t} + A_4 e^{i\omega_4 t}]$$

so that $b(t) = \text{Re}\,[V_b(t)]$. Assume that the difference between any two of the frequencies ω_1, ω_2, ω_3, and ω_4 is small compared to the frequencies themselves. *Answer:* $\overline{a(t)b(t)} = \frac{1}{2}\,\text{Re}\,[V_a(t)V_b^*(t)]$.

1-3 Derive Equation (1.2-21).

1-4 Starting with Maxwell's curl equations [(1.2-1), (1.2-2)] and taking $\mathbf{i} = 0$, show that in the case of a harmonic (sinusoidal) uniform plane wave the field vectors $\mathbf{e}$ and $\mathbf{h}$ are normal to each other as well as to the direction of propagation. [*Hint:* Assume the wave to have the form $e^{i(\omega t - \mathbf{k}\cdot\mathbf{r})}$ and show by actual differentiation that we can formally replace the operator ∇ in Maxwell's equations by $-i\mathbf{k}$.]

1-5 Derive Equation (1.3-19).

1-6 A linearly polarized electromagnetic wave is incident normally at $z = 0$ on the x-y face of a crystal so that it propagates along its z axis. The crystal electric permeability tensor referred to x, y, and z is diagonal with elements ϵ_{11}, ϵ_{22}, and ϵ_{33}. If the wave is polarized initially so that it has equal components along x and y, what is the state of its polarization at the plane z, where

$$(k_x - k_y)z = \frac{\pi}{2}$$

Plot the position of the electric field vector in this plane at times $t = 0$, $\pi/6\omega$, $\pi/3\omega$, $\pi/2\omega$, $2\pi/3\omega$, $5\pi/6\omega$.

REFERENCES

[1] Ramo, S., J. R. Whinnery, and T. Van Duzer, *Fields and Waves in Communication Electronics*. New York: Wiley, 1965.

[2] Born, M., and E. Wolf, *Principles of Optics*. New York: Macmillan, 1964.

[3] Yariv, A., *Quantum Electronics*, 2d Ed. New York: Wiley, 1975.

CHAPTER

2

Propagation of Rays

2.0 Introduction

In this chapter we take up the subject of optical ray propagation through a variety of optical media. These include homogeneous and isotropic materials, thin lenses, dielectric interfaces, and curved mirrors. Since a ray is, by definition, normal to the optical wavefront, an understanding of the ray behavior makes it possible to trace the evolution of optical waves when they are passing through various optical elements. We find that the passage of a ray (or its reflection) through these elements can be described by simple 2×2 matrices. Furthermore, these matrices will be found to describe the propagation of spherical waves and, in the next chapter, of Gaussian beams such as those which are characteristic of the output of lasers.

2.1 Lens Waveguide

Consider a paraxial ray[1] passing through a thin lens of focal length f as shown in Figure 2-1. Taking the cylindrical axis of symmetry as z, denoting the ray distance from the axis by r and its slope dr/dz as r', we can relate the

[1] By paraxial ray we mean a ray whose angular deviation from the cylindrical (z) axis is small enough that the sine and tangent of the angle can be approximated by the angle itself.

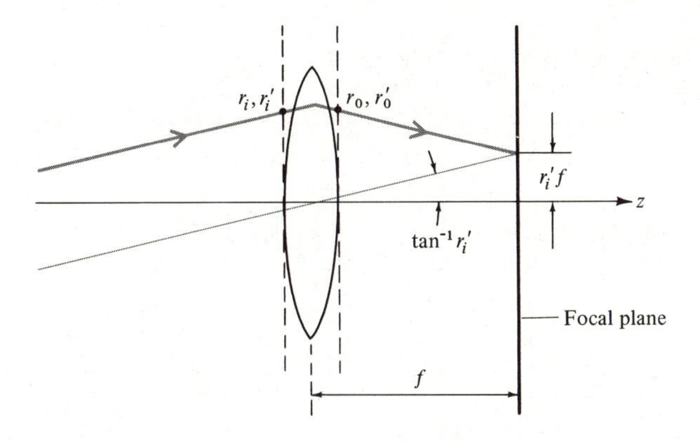

Figure 2-1 Deflection of a ray by a thin lens.

output ray (r_{out}, r'_{out}) to the input ray (r_{in}, r'_{in}) by means of

$$\begin{aligned} r_{out} &= r_{in} \\ r'_{out} &= r'_{in} - \frac{r_{out}}{f} \end{aligned} \tag{2.1-1}$$

where the first of (2.1-1) follows from the definition of a thin lens and the second can be derived from a consideration of the behavior of the undeflected central ray with a slope equal to r'_{in}, as shown in Figure 2-1.

Representing a ray at any position z as a column matrix

$$\mathbf{r}(z) = \begin{vmatrix} r(z) \\ r'(z) \end{vmatrix}$$

we can rewrite (2.1-1) using the rules for matrix multiplication (see References [1]–[3]) as

$$\begin{vmatrix} r_{out} \\ r'_{out} \end{vmatrix} = \begin{vmatrix} 1 & 0 \\ -1/f & 1 \end{vmatrix} \begin{vmatrix} r_{in} \\ r'_{in} \end{vmatrix} \tag{2.1-2}$$

where $f > 0$ for a converging lens and is negative for a diverging one.

The ray matrices for a number of other optical elements are shown in Table 2-1.

Consider as an example the propagation of a ray through a straight section of a homogeneous medium of length d followed by a thin lens of focal length f. This corresponds to propagation between planes a and b in Figure 2-2. Since the effect of the straight section is merely that of increasing r by dr', using (2.1-1) we can relate the output b and input (at a) rays by:

$$\begin{vmatrix} r_{out} \\ r'_{out} \end{vmatrix} = \begin{vmatrix} 1 & d \\ -1/f & (1 - d/f) \end{vmatrix} \begin{vmatrix} r_{in} \\ r'_{in} \end{vmatrix} \tag{2.1-3}$$

Notice also that the matrix corresponds to the product of the thin lens matrix times the straight section matrix as given in Table 2-1.

Table 2-1 RAY MATRICES FOR SOME COMMON OPTICAL ELEMENTS AND MEDIA

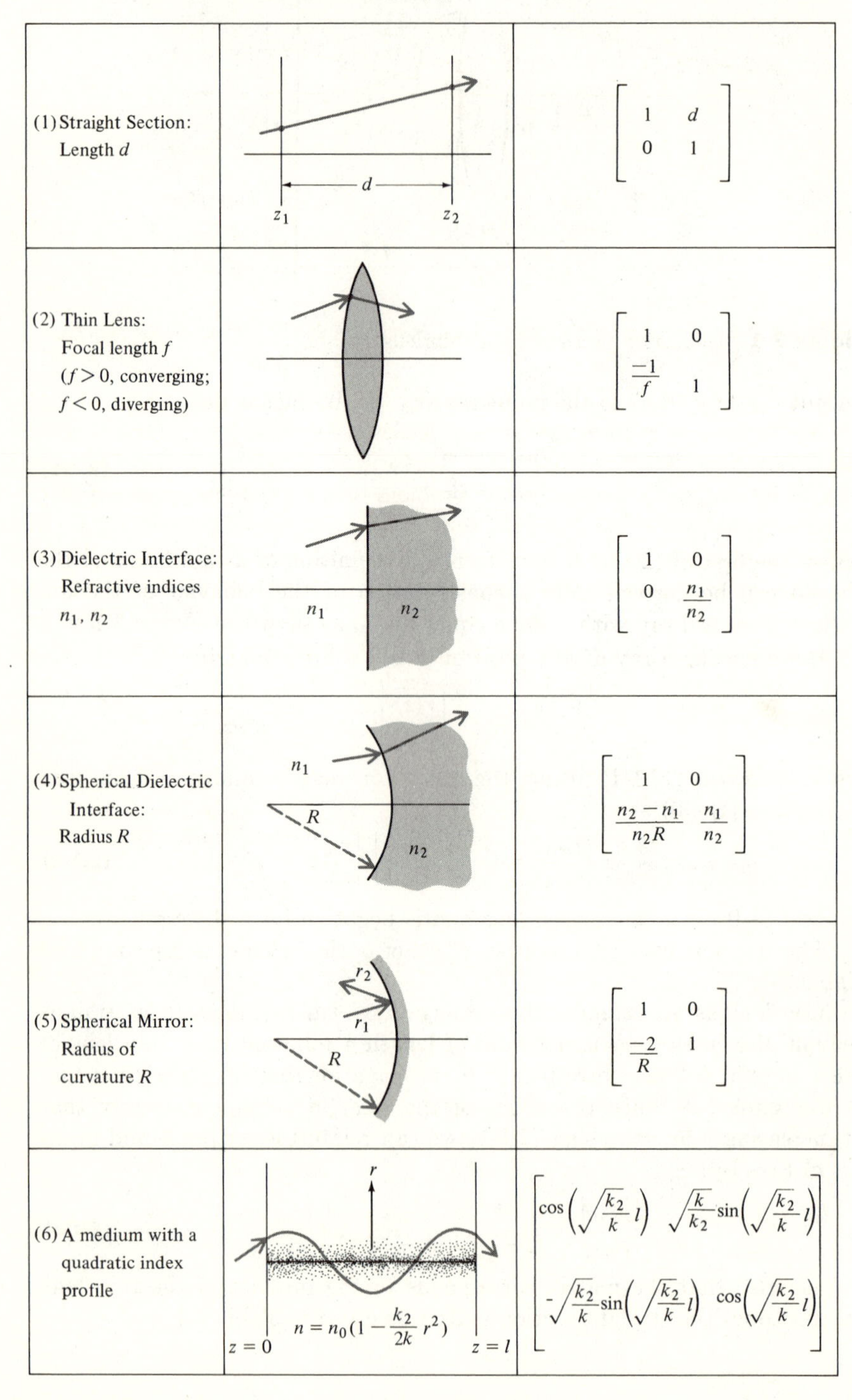

Element	Diagram	Ray matrix
(1) Straight Section: Length d	d, z_1, z_2	$\begin{bmatrix} 1 & d \\ 0 & 1 \end{bmatrix}$
(2) Thin Lens: Focal length f ($f>0$, converging; $f<0$, diverging)		$\begin{bmatrix} 1 & 0 \\ \frac{-1}{f} & 1 \end{bmatrix}$
(3) Dielectric Interface: Refractive indices n_1, n_2	n_1, n_2	$\begin{bmatrix} 1 & 0 \\ 0 & \frac{n_1}{n_2} \end{bmatrix}$
(4) Spherical Dielectric Interface: Radius R	n_1, R, n_2	$\begin{bmatrix} 1 & 0 \\ \frac{n_2 - n_1}{n_2 R} & \frac{n_1}{n_2} \end{bmatrix}$
(5) Spherical Mirror: Radius of curvature R	r_2, r_1, R	$\begin{bmatrix} 1 & 0 \\ \frac{-2}{R} & 1 \end{bmatrix}$
(6) A medium with a quadratic index profile	r, $n = n_0(1 - \frac{k_2}{2k} r^2)$, $z = 0$, $z = l$	$\begin{bmatrix} \cos\left(\sqrt{\frac{k_2}{k}}\, l\right) & \sqrt{\frac{k}{k_2}} \sin\left(\sqrt{\frac{k_2}{k}}\, l\right) \\ -\sqrt{\frac{k_2}{k}} \sin\left(\sqrt{\frac{k_2}{k}}\, l\right) & \cos\left(\sqrt{\frac{k_2}{k}}\, l\right) \end{bmatrix}$

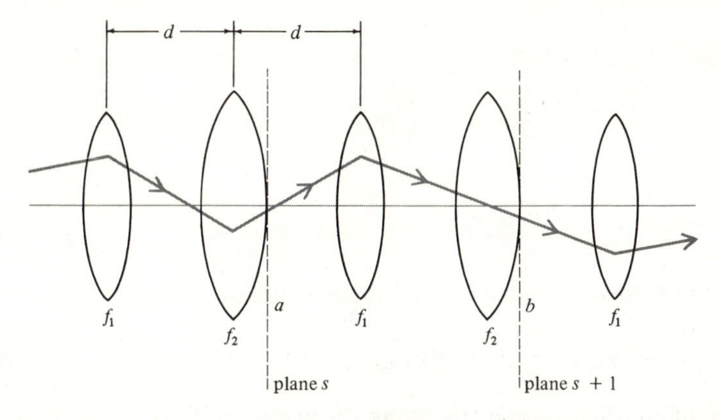

Figure 2-2 Propagation of an optical ray through a biperiodic lens sequence.

We are now in a position to consider the propagation of a ray through a biperiodic lens system made up of lenses of focal lengths f_1 and f_2 separated by d as shown in Figure 2.2. This will be shown in the next chapter to be formally equivalent to the problem of Gaussian-beam propagation inside an optical resonator with mirrors of radii $R_1 = 2f_1$ and $R_2 = 2f_2$ which are separated by d.

The section between the planes s and $s + 1$ can be considered as the basic unit cell of the periodic lens sequence. The matrix relating the ray parameters at the output of a unit cell to those at the input is the product of two matrices, one for each lens, each of which is of the form of the matrix in (2.1-3).

$$\begin{vmatrix} r_{s+1} \\ r'_{s+1} \end{vmatrix} = \begin{vmatrix} 1 & d \\ -\frac{1}{f_1} & \left(1 - \frac{d}{f_1}\right) \end{vmatrix} \begin{vmatrix} 1 & d \\ -\frac{1}{f_2} & \left(1 - \frac{d}{f_2}\right) \end{vmatrix} \begin{vmatrix} r_s \\ r'_s \end{vmatrix} \tag{2.1-4}$$

or, in equation form,

$$\begin{aligned} r_{s+1} &= Ar_s + Br'_s \\ r'_{s+1} &= Cr_s + Dr'_s \end{aligned} \tag{2.1-5}$$

where A, B, C, and D are the elements of the matrix resulting from multiplying the two square matrices in (2.1-4) and are given by

$$\begin{aligned} A &= 1 - \frac{d}{f_2} \\ B &= d\left(2 - \frac{d}{f_2}\right) \\ C &= -\left[\frac{1}{f_1} + \frac{1}{f_2}\left(1 - \frac{d}{f_1}\right)\right] \\ D &= -\left[\frac{d}{f_1} - \left(1 - \frac{d}{f_1}\right)\left(1 - \frac{d}{f_2}\right)\right] \end{aligned} \tag{2.1-6}$$

From the first of (2.1-5) we get

$$r_s' = \frac{1}{B}(r_{s+1} - Ar_s) \tag{2.1-7}$$

and thus

$$r_{s+1}' = \frac{1}{B}(r_{s+2} - Ar_{s+1}) \tag{2.1-8}$$

Using the second of (2.1-5) in (2.1-8) and substituting for r_s' from (2.1-7) gives

$$r_{s+2} - (A + D)r_{s+1} + (AD - BC)r_s = 0 \tag{2.1-9}$$

for the difference equation governing the evolution through the lens waveguide. Using (2.1-6) we can show that $AD - BC = 1$. We can consequently rewrite (2.1-9) as

$$r_{s+2} - 2br_{s+1} + r_s = 0 \tag{2.1-10}$$

where

$$b = \tfrac{1}{2}(A + D) = \left(1 - \frac{d}{f_2} - \frac{d}{f_1} + \frac{d^2}{2f_1f_2}\right) \tag{2.1-11}$$

Equation (2.1-10) is the equivalent, in terms of difference equations, of the differential equation $r'' + Gr = 0$, whose solution is $r(z) = r(0) \exp[\pm i\sqrt{G}z]$. We are thus led to try a solution in the form of

$$r_s = r_0 e^{isq}$$

which, when substituted in (2.1-10), leads to

$$e^{2iq} - 2be^{iq} + 1 = 0 \tag{2.1-12}$$

and therefore

$$e^{iq} = b \pm i\sqrt{1 - b^2} = e^{\pm i\theta} \tag{2.1-13}$$

where $\cos\theta = b$.

The general solution can be taken as a linear superposition of $\exp(is\theta)$ and $\exp(-is\theta)$ solutions or equivalently as

$$r_s = r_{\max} \sin(s\theta + \alpha) \tag{2.1-14}$$

where $r_{\max} = r_0/\sin\alpha$ and α can be expressed using (2.1-8) in terms of r_0 and r_0'.

The condition for a stable—that is, confined—ray is that θ be a real number, since in this case the ray radius r_s oscillates as a function of the cell number s between $r_{\max}$ and $-r_{\max}$. According to (2.1-13), the necessary and sufficient condition for θ to be real is that [5]

$$|b| \leqslant 1 \tag{2.1-15}$$

In terms of the system parameters we can use (2.1-11) to reexpress (2.1-15) as

$$-1 \leqslant 1 - \frac{d}{f_2} - \frac{d}{f_1} + \frac{d^2}{2f_1f_2} \leqslant 1$$

or **(2.1-16)**

$$0 \leqslant \left(1 - \frac{d}{2f_1}\right)\left(1 - \frac{d}{2f_2}\right) \leqslant 1$$

If, on the other hand, the stability condition $|b| \leqslant 1$ is violated, we obtain, according to (2.1-10), a solution in the form of

$$r_s = C_1 e^{(\alpha_+)s} + C_2 e^{(\alpha_-)s} \tag{2.1-17}$$

where $e^{\alpha\pm} = b \pm \sqrt{b^2 - 1}$, and since the magnitude of either $\exp(\alpha_+)$ or $\exp(\alpha_-)$ exceeds unity, the beam radius will increase as a function of (distance) s.

Identical-lens waveguide. The simplest case of a lens waveguide is one in which $f_1 = f_2 = f$; that is, all the lenses are identical.

The analysis of this situation is considerably simpler than that used for a biperiodic lens sequence. The reason is that the periodic unit cell (the smallest part of the sequence that can, upon translation, recreate the whole sequence) contains a single lens only. The (A, B, C, D) matrix for the unit cell is given by the square matrix in (2.1-3). Following exactly the steps leading to (2.1-11) through (2.1-14), the stability condition becomes

$$0 \leqslant d \leqslant 4f \tag{2.1-18}$$

and the beam radius at the nth lens is given by

$$r_n = r_{\max} \sin(n\theta + \alpha)$$

$$\cos\theta = \left(1 - \frac{d}{2f}\right) \tag{2.1-19}$$

Because of the algebraic simplicity of this problem we can easily express $r_{\max}$ and α in (2.1-19) in terms of the initial conditions r_0 and r_0', obtaining

$$(r_{\max})^2 = \frac{4f}{4f - d}\left(r_0^{\,2} + d r_0 r_0' + d f r_0'^{\,2}\right) \tag{2.1-20}$$

$$\tan\alpha = \sqrt{\frac{4f}{d} - 1} \bigg/ \left(1 + 2f\frac{r_0'}{r_0}\right) \tag{2.1-21}$$

where n corresponds to the plane immediately to the right of the nth lens. The derivation of the last two equations is left as an exercise (Problem 2-1).

The stability criteria can be demonstrated experimentally by tracing the behavior of a laser beam as it propagates down a sequence of lenses spaced uniformly. One can easily notice the rapid "escape" of the beam once condition (2.1-18) is violated.

2.2 Propagation of Rays between Mirrors [6]

Another important application of the formalism just developed concerns the bouncing of a ray between two curved mirrors. Since the reflection at a mirror with a radius of curvature R is equivalent, except for the folding of the path, to passage through a lens with a focal length $f = R/2$, we can use the formalism of the preceding section to describe the propagation of a ray between two curved reflectors with radii of curvature R_1 and R_2, which are separated by d. Let us consider the simple case of a ray which is injected into a symmetric two-mirror system as shown in Figure 2-3(a). Since the x and y coordinates of the ray are independent variables, we can take them according to (2.1-19) in the form of

$$\begin{aligned} x_n &= x_{\text{max}} \sin (n\theta + \alpha_x) \\ y_n &= y_{\text{max}} \sin (n\theta + \alpha_y) \end{aligned} \tag{2.2-1}$$

where n refers to the ray parameter immediately following the nth reflection. According to (2.2-1), the locus of the points x_n, y_n on a given mirror lies on an ellipse.

Reentrant rays. If θ in (2.2-1) satisfies the condition

$$2\nu\theta = 2l\pi \tag{2.2-2}$$

where ν and l are any two integers, a ray will return to its starting point following ν round trips and will thus continuously retrace the same pattern on the mirrors. If we consider as an example the simple case of $l = 1$, $\nu = 2$, so that $\theta = \pi/2$, from (2.1-19) we obtain $d = 2f = R$; that is, if the mirrors are separated by a distance equal to their radius of curvature R, the trapped ray will retrace its pattern after two round trips ($\nu = 2$). This situation ($d = R$) is referred to as symmetric confocal, since the two mirrors have a common focal point $f = R/2$. It will be discussed in detail in Chapter 4. The ray pattern corresponding to $\nu = 2$ is illustrated in Figure 2-3(b).

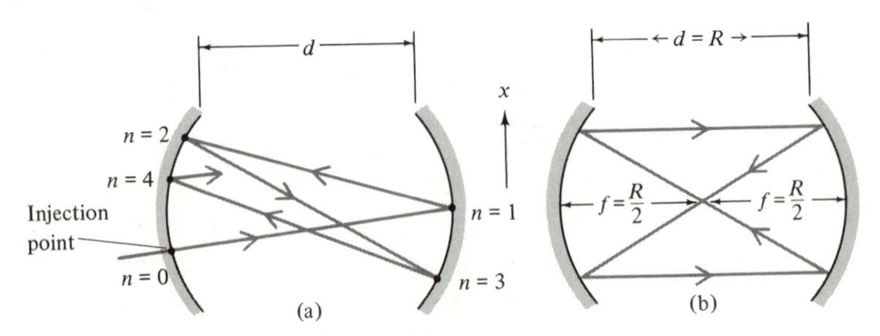

Figure 2-3 **(a)** Path of a ray injected in plane of figure into the space between two mirrors. **(b)** Reentrant ray in confocal ($d = R$) mirror configuration repeating its pattern after two round trips.

2.3 Rays in Lenslike Media [7]

The basic physical property of lenses that is responsible for their focusing action is the fact that the optical path across them $\int n(r, z)\, dz$ (where n is the index of refraction of the medium) is a quadratic function of the distance r from the z axis. Using ray optics, we account for this fact by a change in the ray's slope as in (2.1-1). This same property can be represented by relating the complex field amplitude of the incident optical field $E_{\mathrm{R}}(x, y)$ immediately to the right of an ideal thin lens to that immediately to the left $E_{\mathrm{L}}(x, y)$ by

$$E_{\mathrm{R}}(x, y) = E_{\mathrm{L}}(x, y) \exp\left[+ik \frac{x^2 + y^2}{2f}\right] \tag{2.3-1}$$

where f is the focal length and $k = 2\pi n/\lambda_0$.

The effect of the lens, therefore, is to cause a phase shift $k(x^2 + y^2)/2f$, which increases quadratically with the distance from the axis. We consider next the closely related case of a medium whose index of refraction n varies according to[2]

$$n(x, y) = n_o\left[1 - \frac{k_2}{2k}(x^2 + y^2)\right] \tag{2.3-2}$$

where k_2 is a constant. Since the phase delay of a wave propagating through a section dz of a medium with an index of refraction n is $(2\pi\, dz/\lambda_0)n$, it follows directly that a thin slab of the medium described by (2.3-2) will act as a thin lens, introducing [as in (2.3-1)] a phase shift proportional to $(x^2 + y^2)$. The behavior of a ray in this case is described by the differential equation that applies to ray propagation in an optically inhomogeneous medium [8],

$$\frac{d}{ds}\left(n \frac{d\mathbf{r}}{ds}\right) = \nabla n \tag{2.3-3}$$

where s is the distance along the ray measured from some fixed position on it and $\mathbf{r}$ is the position vector of the point at s. For paraxial rays we may replace d/ds by d/dz and, using (2.3-2), obtain

$$\frac{d^2r}{dz^2} + \left(\frac{k_2}{k}\right) r = 0 \tag{2.3-4}$$

If at the input plane $z = 0$ the ray has a radius r_0 and slope r_0', we can write the solution of (2.3-4) directly as

$$\begin{aligned} r(z) &= \cos\left(\sqrt{\frac{k_2}{k}}\, z\right) r_0 + \sqrt{\frac{k}{k_2}} \sin\left(\sqrt{\frac{k_2}{k}}\, z\right) r_0' \\ r'(z) &= -\sqrt{\frac{k_2}{k}} \sin\left(\sqrt{\frac{k_2}{k}}\, z\right) r_0 + \cos\left(\sqrt{\frac{k_2}{k}}\, z\right) r_0' \end{aligned} \tag{2.3-5}$$

---

[2] Equation (2.3-2) can be viewed as consisting of the first two terms in the Taylor-series expansion of $n(x, y)$ for the radial symmetric case.

That is, the ray oscillates back and forth across the axis, as shown in Figure 2-4. A section of the quadratic index medium acts as a lens. This can be proved by showing, using (2.3-5), that a family of parallel rays entering at $z = 0$ at different radii will converge upon emerging at $z = l$ to a common focus at a distance

$$h = \frac{1}{n_0}\sqrt{\frac{k}{k_2}} \cot\left(\sqrt{\frac{k_2}{k}}\, l\right) \tag{2.3-6}$$

from the exit plane. The factor n_0 accounts for the refraction at the boundary, assuming the medium at $z > l$ to possess an index $n = 1$ and a small angle of incidence. The derivation of (2.3-6) is left as an exercise (Problem 2-3).

Equations (2.3-5) apply to a focusing medium with $k_2 > 0$. In a medium where $k_2 < 0$—that is, where the index increases with the distance from the axis—the solutions for $r(z)$ and $r'(z)$ become

$$\begin{aligned} r(z) &= \cosh\left(\sqrt{\frac{|k_2|}{k}}\, z\right) r_0 + \sqrt{\frac{k}{|k_2|}} \sinh\left(\sqrt{\frac{|k_2|}{k}}\, z\right) r_0' \\ r'(z) &= \sqrt{\frac{|k_2|}{k}} \sinh\left(\sqrt{\frac{|k_2|}{k}}\, z\right) r_0 + \cosh\left(\sqrt{\frac{|k_2|}{k}}\, z\right) r_0' \end{aligned} \tag{2.3-7}$$

so that $r(z)$ increases with distance and eventually escapes. A section of such a medium acts as a negative lens.

Physical situations giving rise to quadratic index variation include:

1. Propagation of laser beams with Gaussian-like intensity profile in a slightly absorbing medium. The absorption heating gives rise, because of the dependence of n on the temperature T, to an index profile [9]. If $dn/dT < 0$, as is the case for most materials, the index is smallest on the axis where the absorption heating is highest. This corresponds to a $k_2 < 0$ in (2.3-2) and the beam spreads with the distance z. If $dn/dT > 0$, as in certain lead glasses [10], the beams are focused.

2. In optically pumped solid state laser rods the portion of the absorbed pump power which is not converted to laser radiation is conducted as heat to the rod surface. This heat conduction requires

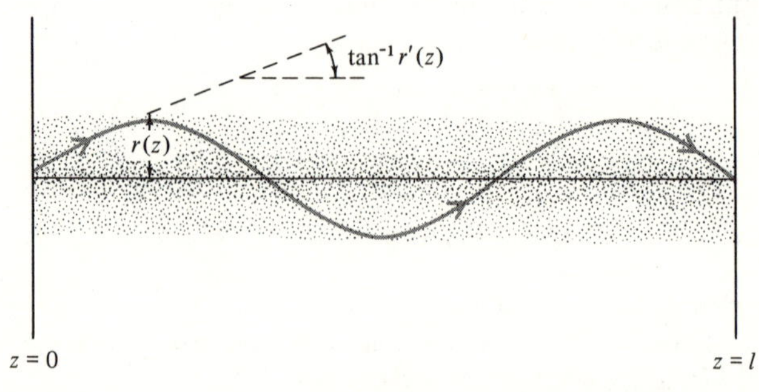

Figure 2-4 Path of a ray in a medium with a quadratic index variation.

a temperature gradient in which T is maximum on axis. The dependence of n on T then gives rise to a positive lens effect for $dn/dT > 0$ and a negative lens for $dn/dT < 0$.

3. Dielectric waveguides made by sandwiching a layer of index n_1 between two layers with index $n_2 < n_1$. This situation will be discussed further in Chapter 7 in connection with injection lasers.

4. Optical fibers produced by cladding a thin optical fiber (whose radius is comparable to λ) of an index n_1 with a sheath of index $n_2 < n_1$. Such fibers are used as light pipes.

5. Optical waveguides consisting of glasslike rods or filaments, with radii large compared to λ, whose index decreases with increasing r. Such waveguides can be used for the simultaneous transmission of a number of laser beams, which are injected into the waveguide at different angles. It follows from (2.3-5) that the beams will emerge, each along a unique direction, and consequently can be easily separated. Furthermore, in view of its previously discussed lens properties, the waveguide can be used to transmit optical image information in much the same way as images are transmitted by a multielement lens system to the image plane of a camera [11].

PROBLEMS

2-1 Derive Equations (2.1-19) through (2.1-21).

2-2 Show that the eigenvalues λ of the equation

$$\begin{vmatrix} A & B \\ C & D \end{vmatrix} \begin{vmatrix} r_s \\ r_s' \end{vmatrix} = \lambda \begin{vmatrix} r_s \\ r_s' \end{vmatrix}$$

are $\lambda = e^{\pm i\theta}$ with exp $(\pm i\theta)$ given by Equation (2.1-12). Note that, according to Equation (2.1-5), the foregoing matrix equation can also be written as

$$\begin{vmatrix} r_{s+1} \\ r_{s+1}' \end{vmatrix} = \lambda \begin{vmatrix} r_s \\ r_s' \end{vmatrix}$$

2-3 Derive Equation (2.3-6).

2-4 Make a plausibility argument to justify Equation (2.3-1) by showing that it holds for a plane wave incident on a lens.

2-5 Show that a lenslike medium occupying the region $0 \leqslant z \leqslant l$ will image a point on the axis at $z < 0$ onto a single point. (If the image point occurs at $z < l$, the image is virtual.)

2-6 Derive the ray matrices of Table 2-1.

REFERENCES

[1] Pierce, J. R., *Theory and Design of Electron Beams*, 2d Ed. Princeton, N.J.: Van Nostrand, 1954, Chap. 11.

[2] Ramo, S., J. R. Whinnery, and T. Van Duzer, *Fields and Waves in Communication Electronics*. New York: Wiley, 1965, p. 576.

[3] Yariv, A., *Quantum Electronics*, 2d Ed. New York: Wiley, 1975.

[4] Siegman, A. E., *An Introduction to Lasers and Masers*. New York: McGraw-Hill, 1968.

[5] Kogelnik, H., and T. Li, "Laser beams and resonators," *Proc. IEEE*, vol. 54, p. 1312, 1966.

[6] Herriot, D., H. Kogelnik, and R. Kompfner, "Off-axis paths in spherical mirror interferometers," *Appl. Opt.*, vol. 3, p. 523, 1964.

[7] Kogelnik, H., "On the propagation of Gaussian beams of light through lenslike media including those with a loss and gain variation," *Appl. Opt.*, vol. 4, p. 1562, 1965.

[8] Born, M., and E. Wolf, *Principles of Optics*, 3d Ed. New York: Pergamon, 1965, p. 121.

[9] Gordon, J. P., R. C. C. Leite, R. S. Moore, S. P. S. Porto, and J. R. Whinnery, "Long-transient effects in lasers with inserted liquid samples," *J. Appl. Phys.*, vol. 36, p. 3, 1965.

[10] Dabby, F. W., and J. R. Whinnery, "Thermal self-focusing of laser beams in lead glasses," *Appl. Phys. Letters*, vol. 13, p. 284, 1968.

[11] Yariv, A., "Three dimensional pictorial transmission in optical fibers," *Appl. Phys. Lett.*, vol. 2, p. 88, 1976.

CHAPTER

3

Propagation of Optical Beams in Homogeneous and in Guiding Media

3.0 Introduction

The propagation of rays through lenses and lenslike media was discussed in the preceding chapter. A closely related topic of fundamental importance in quantum electronics is the propagation of optical beams. These beams usually take the form of planelike waves whose energy density is localized, for reasonable propagation distances, near the propagation axis. The output of laser oscillators will be found to consist of one or more of such beams. This is also the form of the fields set up by feeding electromagnetic energy into a resonator formed by two curved reflectors. The understanding of the characteristics of these modes is thus a prerequisite to the study of many laser-related phenomena.

3.1 Wave Equation in Quadratic Index Media

The most widely encountered optical beam in quantum electronics is one where the intensity distribution at planes normal to the propagation direction is Gaussian. To derive its characteristics we start with the Maxwell equations in an isotropic charge-free medium.

$$\nabla \times \mathbf{H} = \epsilon \frac{\partial \mathbf{E}}{\partial t}$$

$$\nabla \times \mathbf{E} = -\mu \frac{\partial \mathbf{H}}{\partial t} \tag{3.1-1}$$

$$\nabla \cdot (\epsilon \mathbf{E}) = 0$$

Taking the curl of the second of (3.1-1) and substituting the first results in

$$\nabla^2 \mathbf{E} - \mu\epsilon \frac{\partial^2 \mathbf{E}}{\partial t^2} = -\nabla \left(\frac{1}{\epsilon} \mathbf{E} \cdot \nabla \epsilon \right) \tag{3.1-2}$$

where we used $\nabla \times \nabla \times \mathbf{E} \equiv \nabla(\nabla \cdot \mathbf{E}) - \nabla^2 \mathbf{E}$. If we assume the field quantities to vary as $\mathbf{E}(x, y, z, t) = \text{Re}\,[\mathbf{E}(x, y, z)e^{i\omega t}]$ and neglect the right side of (3.1-2),[1] we obtain

$$\nabla^2 \mathbf{E} + k^2(\mathbf{r})\mathbf{E} = 0 \tag{3.1-3}$$

where

$$k^2(\mathbf{r}) = \omega^2 \mu \epsilon(\mathbf{r}) \left[\frac{1 - i\sigma(\mathbf{r})}{\omega\epsilon} \right] \tag{3.1-4}$$

thus allowing for a possible dependence of ϵ on position $\mathbf{r}$. We have also taken k as a complex number to allow for the possibility of losses ($\sigma > 0$) or gain ($\sigma < 0$) in the medium.[2]

We limit our derivation to the case in which $k^2(\mathbf{r})$ is given by

$$k^2(r, \phi, z) = k^2 - kk_2 r^2 \qquad r = \sqrt{x^2 + y^2} \tag{3.1-5}$$

where, according to (3.1-4),

$$k^2 = k^2(0) = \omega^2 \mu \epsilon(0) \left(l - i \frac{\sigma(0)}{\omega\epsilon(0)} \right)$$

so that k_2 is some constant characteristic of the medium. Furthermore, we assume a solution whose transverse dependence is on $r = \sqrt{x^2 + y^2}$ only so that in (3.1-3) we can replace ∇^2 by

$$\nabla^2 = \nabla_t^2 + \frac{\partial^2}{\partial z^2} = \frac{\partial^2}{\partial r^2} + \frac{1}{r}\frac{\partial}{\partial r} + \frac{\partial^2}{\partial z^2} \tag{3.1-6}$$

The kind of propagation we are considering is that of a nearly plane wave in which the flow of energy is predominantly along a single (for example, $\mathbf{z}$) direction so that we may limit our derivation to a single

---

[1] This neglect is justified if the fractional change of ϵ in one optical wavelength is slight.

[2] If k is complex (for example, $k_r + ik_i$), then a traveling electromagnetic plane wave has the form of $\exp[i(\omega t - kz)] = \exp[+k_i z + i(\omega t - k_r z)]$.

transverse field component E. Taking E as

$$E = \psi(x, y, z)e^{-ikz} \tag{3.1-7}$$

we obtain from (3.1-3) and (3.1-5) in a few simple steps,

$$\nabla_t^2\psi - 2ik\psi' - kk_2r^2\psi = 0 \tag{3.1-8}$$

where $\psi' \equiv \partial\psi/\partial z$ and where we assume that the variation is slow enough that $k\psi' \gg \psi'' \ll k^2\psi$.

Next we take ψ in the form of

$$\psi = \exp\left\{-i\left[P(z) + \frac{k}{2q(z)}r^2\right]\right\} \tag{3.1-9}$$

that, when substituted into (3.1-8) and after using (3.1-6), gives

$$-\left(\frac{k}{q}\right)^2 r^2 - 2i\left(\frac{k}{q}\right) - k^2r^2\left(\frac{1}{q}\right)' - 2kP' - kk_2r^2 = 0 \tag{3.1-10}$$

If (3.1-10) is to hold for all r, the coefficients of the different powers of r must be equal to zero. This leads to [1]

$$\left(\frac{1}{q}\right)^2 + \left(\frac{1}{q}\right)' + \frac{k_2}{k} = 0 \qquad P' = -\frac{i}{q} \tag{3.1-11}$$

The wave equation (3.1-3) is thus reduced to (3.1-11).

3.2 Gaussian Beams in a Homogeneous Medium

If the propagation medium is homogeneous, we can, according to (3.1-5), put $k_2 = 0$, and (3.1-11) becomes

$$\frac{1}{q^2} + \left(\frac{1}{q}\right)' = 0 \tag{3.2-1}$$

Introducing the function $s(z)$ by the relation

$$\frac{1}{q} = \frac{s'}{s} \tag{3.2-2}$$

we obtain directly from (3.2-1)

$$s'' = 0$$

so that

$$s' = a \qquad s = az + b$$

or, using (3.2-2),

$$\frac{1}{q(z)} = \frac{a}{az + b} \tag{3.2-3}$$

where a and b are arbitrary constants, we will find it more convenient to deal with a parameter q, so that we may rewrite (3.2-3) in the form

$$q = z + q_0 \tag{3.2-4}$$

From (3.1-11) and (3.2-4) we have

$$P' = -\frac{i}{q} = -\frac{i}{z + q_0} \tag{3.2-5}$$

so that

$$P(z) = -i \ln\left(1 + \frac{z}{q_0}\right) \tag{3.2-6}$$

where the arbitrary constant of integration is chosen as zero:[3]

Combining (3.2-5) and (3.2-6) in (3.1-9) we obtain

$$\psi = \exp\left\{-i\left[-i\ln\left(1 + \frac{z}{q_0}\right) + \frac{k}{2(q_0 + z)} r^2\right]\right\} \tag{3.2-7}$$

We take q_0 to be purely imaginary and reexpress it in terms of a new constant ω_0 as

$$q_0 = i\frac{\pi\omega_0^2 n}{\lambda} \qquad \lambda = \frac{2\pi n}{k} \tag{3.2-8}$$

The choice of imaginary q_0 will be found to lead to physically meaningful waves whose energy density is confined near the z axis. With this last substitution, let us consider, one at a time, the two factors in (3.2-7). The first one becomes

$$\exp\left[-\ln\left(1 - i\frac{\lambda z}{\pi\omega_0^2 n}\right)\right] = \frac{1}{\sqrt{1 + (\lambda^2 z^2/\pi^2\omega_0^4 n^2)}} \exp\left[i \tan^{-1}\left(\frac{\lambda z}{\pi\omega_0^2 n}\right)\right] \tag{3.2-9}$$

where we used $\ln(a + ib) = \ln\sqrt{a^2 + b^2} + i\tan^{-1}(b/a)$. Substituting (3.2-8) in the second term of (3.2-7) and separating the exponent into its real and imaginary parts, we obtain

$$\exp\left[\frac{-ikr^2}{2(q_0 + z)}\right] = \exp\left\{\frac{-r^2}{\omega_0^2[1 + (\lambda z/\pi\omega_0^2 n)^2]} - \frac{ikr^2}{2z[1 + (\pi\omega_0^2 n/\lambda z)^2]}\right\} \tag{3.2-10}$$

[3] The integration constant will merely modify the phase of the field solution (3.1-7). This is equivalent to a mere shift of the time origin.

If we define the following parameters

$$\omega^2(z) = \omega_0^2\left[1 + \left(\frac{\lambda z}{\pi\omega_0^2 n}\right)^2\right] = \omega_0^2\left(1 + \frac{z^2}{z_0^2}\right) \qquad \text{(3.2-11)}$$

$$R = z\left[1 + \left(\frac{\pi\omega_0^2 n}{\lambda z}\right)^2\right] = z\left(1 + \frac{z_0^2}{z^2}\right) \qquad \text{(3.2-12)}$$

$$\eta(z) = \tan^{-1}\left(\frac{z}{\pi\omega_0^2 n}\right) = \tan^{-1}\left(\frac{z}{z_0}\right) \qquad \text{(3.2-13)}$$

$$z_0 \equiv \frac{\pi\omega_0^2 n}{\lambda}$$

we can combine (3.2-9) and (3.2-10) in (3.2-7) and, recalling that $E(x, y, z) = \psi(x, y, z)\exp(-ikz)$, obtain

$$E(x, y, z) = E_0\frac{\omega_0}{\omega(z)}\exp\left\{-i[kz - \eta(z)] - i\frac{k^2}{2q(z)}\right\}$$

$$= E_0\frac{\omega_0}{\omega(z)}\left\{\exp -i[kz - \eta(z)] - r^2\left(\frac{1}{\omega^2(z)} + \frac{ik}{2R(z)}\right)\right\} \qquad \text{(3.2-14)}$$

$$k = \frac{2\pi\eta}{\lambda}$$

This is our basic result. We refer to it as the fundamental Gaussian-beam solution, since we have excluded the more complicated solutions of (3.1-3) (that is, those with azimuthal variation) by limiting ourselves to transverse dependence involving $r = (x^2 + y^2)^{1/2}$ only. These higher order modes will be discussed separately.

From (3.2-14) the parameter $\omega(z)$, which evolves according to (3.2-11), is the distance r at which the field amplitude is down by a factor $1/e$ compared to its value on the axis. We will consequently refer to it as the beam "spot size." The parameter ω_0 is the minimum spot size. It is the beam spot size at the plane $z = 0$. The parameter R in (3.2-14) is the radius of curvature of the very nearly spherical wavefronts[4] at z. We can verify this statement by deriving the radius of curvature of the constant phase surfaces (wavefronts) or, more simply, by considering the form of a spherical wave emitted by a point radiator placed at $z = 0$. It is given by

[4] Actually, it follows from (3.2-14) that, with the exception of the immediate vicinity of the plane $z = 0$, the wavefronts are parabolic since they are defined by $k[z + (r^2/2R)] =$ const. For $r^2 \ll z^2$, the distinction between parabolic and spherical surfaces is not important.

$$E \propto \frac{1}{R} e^{-ikR} = \frac{1}{R} \exp\left(-ik\sqrt{x^2 + y^2 + z^2}\right)$$
$$\simeq \frac{1}{R} \exp\left(-ikz - ik\frac{x^2 + y^2}{2R}\right) \qquad x^2 + y^2 \ll z^2 \tag{3-2-15}$$

since z is equal to R, the radius of curvature of the spherical wave. Comparing (3.2-15) with (3.2-14) we identify R as the radius of curvature of the Gaussian beam. The convention regarding the sign of $R(z)$ is that it is negative if the center of curvature occurs at $z' > z$ and vice versa.

The form of the fundamental Gaussian beam is, according to (3.2-14), uniquely determined once its minimum spot size ω_0 and its location—that is, the plane $z = 0$—are specified. The spot size ω and radius of curvature R at any plane z are then found from (3.2-11) and (3.2-12). Some of these characteristics are displayed in Figure 3-1. The hyperbolas shown in this figure correspond to the ray direction and are intersections of planes that include the z axis and the hyperboloids

$$x^2 + y^2 = \text{const. } \omega^2(z) \tag{3.2-16}$$

These hyperbolas correspond to the local direction of energy propagation. The spherical surfaces shown have radii of curvature given by (3.2-12). For large z the hyperboloids $x^2 + y^2 = \omega^2$ are asymptotic to the cone

$$r = \sqrt{x^2 + y^2} = \frac{\lambda}{\pi\omega_0 n} z \tag{3.2-17}$$

whose half-apex angle, which we take as a measure of the angular beam spread, is

$$\theta_{\text{beam}} = \tan^{-1}\left(\frac{\lambda}{\pi\omega_0 n}\right) \simeq \frac{\lambda}{\pi\omega_0 n} \qquad \text{for } \theta_{\text{beam}} \ll \pi \tag{3.2-18}$$

This last result is a rigorous manifestation of wave diffraction according to which a wave that is confined in the transverse direction to an aperture of radius ω_0 will spread (diffract) in the far field ($z \gg \pi\omega_0^2 n/\lambda$) according to (3.2-18).

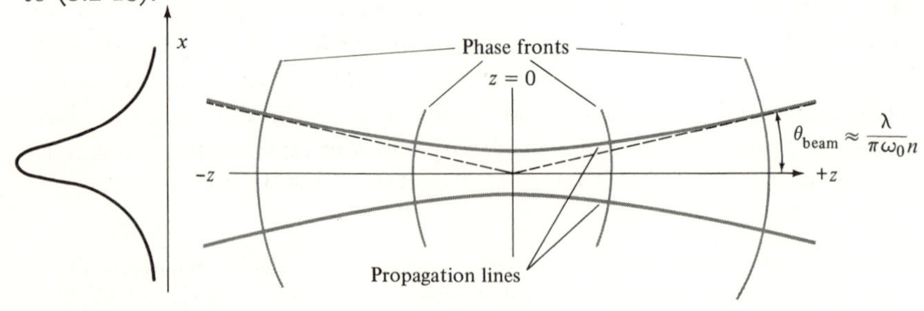

Figure 3-1 Propagating Gaussian beam.

3.3 Fundamental Gaussian Beam in a Lenslike Medium—The ABCD Law

We now return to the general case of a lenslike medium so that $k_2 \neq 0$. The P and q functions of (3.1-9) obey, according to (3.1-11)

$$\left(\frac{1}{q}\right)^2 + \left(\frac{1}{q}\right)' + \frac{k_2}{k} = 0$$
$$P' = -\frac{i}{q} \tag{3.3-1}$$

If we introduce the function s defined by

$$\frac{1}{q} = \frac{s'}{s} \tag{3.3-2}$$

we obtain from (3.3-1)

$$s'' + s\frac{k_2}{k} = 0$$

so that

$$s(z) = a \sin\sqrt{\frac{k_2}{k}}\,z + b\cos\sqrt{\frac{k_2}{k}}\,z$$
$$s'(z) = a\sqrt{\frac{k_2}{k}}\cos\sqrt{\frac{k_2}{k}}\,z - b\sqrt{\frac{k_2}{k}}\sin\sqrt{\frac{k_2}{k}}\,z \tag{3.3-3}$$

where a and b are arbitrary constants.

Using (3.3-3) in (3.3-2) and expressing the result in terms of an input value q_0 gives the following result for the complex beam radius $q(z)$

$$q(z) = \frac{\cos[(\sqrt{k_2/k})z]q_0 + \sqrt{k/k_2}\sin[(\sqrt{k_2/k})z]}{-\sin[(\sqrt{k_2/k})z]\sqrt{k_2/k}\,q_0 + \cos[(\sqrt{k_2/k})z]} \tag{3.3-4}$$

The physical significance of $q(z)$ in this case can be extracted from (3.1-9). We expand the part of $\psi(r, z)$ that involves r. The result is

$$\psi \propto e^{-ikr^2/2q(z)}$$

If we express the real and imaginary parts of $q(z)$ by means of

$$\frac{1}{q(z)} = \frac{1}{R(z)} - i\frac{\lambda}{\pi n\omega^2(z)} \tag{3.3-5}$$

we obtain

$$\psi \propto \exp\left[\frac{-r^2}{\omega^2(z)} - i\frac{kr^2}{2R(z)}\right]$$

so that $\omega(z)$ is the beam spot size and R its radius of curvature, as in the case of a homogeneous medium, which is described by (3.2-14). For the special case of a homogeneous medium ($k_2 = 0$), (3.3-4) reduces to (3.2-4).

Transformation of the Gaussian beam—the ABCD law. We have derived above the transformation law of a Gaussian beam (3.3-4) propagating through a lenslike medium that is characterized by k_2. We note first by comparing (3.3-4) to Table 2-1(6) and to (2.3-5) that the transformation can be described by

$$q_2 = \frac{Aq_1 + B}{Cq_1 + D} \tag{3.3-6}$$

where A, B, C, D are the elements of the ray matrix which relates the ray (r, r') at a plane 2 to the ray at plane 1. It follows immediately that the propagation through, or reflection from, any of the elements shown in Table 2-1 also obeys (3.3-6), since these elements can all be viewed as special cases of a lenslike medium. For future reference we note that by applying (3.3-6) to a thin lens of focal length f we obtain from (3.3-6) and Table 2-1(2)

$$\frac{1}{q_2} = \frac{1}{q_1} - \frac{1}{f} \tag{3.3-7}$$

so that using (3.3-5)

$$\begin{aligned} \omega_2 &= \omega_1 \\ \frac{1}{R_2} &= \frac{1}{R_1} - \frac{1}{f} \end{aligned} \tag{3.3-8}$$

These results apply, as well, to reflection from a mirror with a radius of curvature R if we replace f by $R/2$.

Consider next the propagation of a Gaussian beam through two lenslike media that are adjacent to each other. The ray matrix describing the first one is (A_1, B_1, C_1, D_1) while that of the second one is (A_2, B_2, C_2, D_2). Taking the input beam parameter as q_1 and the output beam parameter as q_3, we have from (3.3-6)

$$q_2 = \frac{A_1 q_1 + B_1}{C_1 q_1 + D_1}$$

for the beam parameter at the output of medium 1 and

$$q_3 = \frac{A_2 q_2 + B_2}{C_2 q_2 + D_2}$$

and after combining the last two equations,

$$q_3 = \frac{A_T q_1 + B_T}{C_T q_1 + D_T} \tag{3.3-9}$$

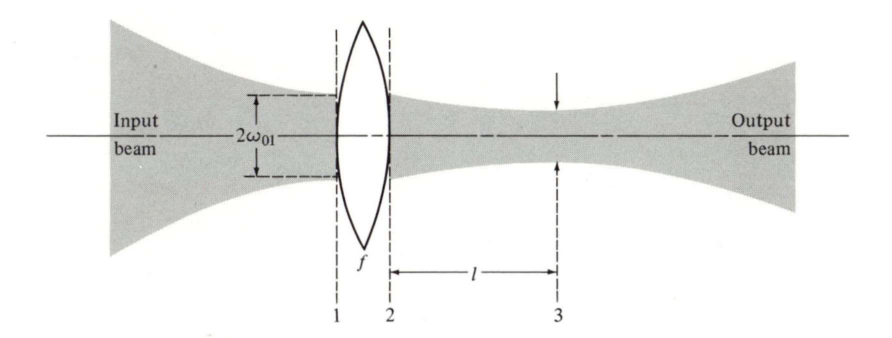

Figure 3-2 Focusing of a Gaussian beam.

where (A_T, B_T, C_T, D_T) are the elements of the ray matrix relating the output plane (3) to the input plane (1), that is,

$$\begin{vmatrix} A_T & B_T \\ C_T & D_T \end{vmatrix} = \begin{vmatrix} A_2 & B_2 \\ C_2 & D_2 \end{vmatrix} \begin{vmatrix} A_1 & B_1 \\ C_1 & D_1 \end{vmatrix} \tag{3.3-10}$$

It follows by induction that (3.3-9) applies to the propagation of a Gaussian beam through any arbitrary number of lenslike media and elements. The matrix (A_T, B_T, C_T, D_T) is the ordered product of the matrices characterizing the individual members of the chain.

The great power of the ABCD law is that it enables us to trace the Gaussian beam parameter $q(z)$ through a complicated sequence of lenslike elements. The beam radius $R(z)$ and spot size $\omega(z)$ at any plane z can be recovered through the use of (3.3-5). The application of this method will be made clear by the following example.

Example—Gaussian beam focusing. As an illustration of the application of the ABCD law, we consider the case of a Gaussian beam that is incident at its waist on a thin lens of focal length f, as shown in Figure 3-2. We will find the location of the waist of the output beam and the beam radius at that point.

At the input plane 1 $\omega = \omega_{01}$, $R_1 = \infty$ so that

$$\frac{1}{q_1} = \frac{1}{R_1} - i\frac{\lambda}{\pi\omega_{01}^2 n} = -i\frac{\lambda}{\pi\omega_{01}^2 n}$$

using (3.3-8) leads to

$$\frac{1}{q_2} = \frac{1}{q_1} - \frac{1}{f} = -\frac{1}{f} - i\frac{\lambda}{\pi\omega_{01}^2 n}$$

$$q_2 = \frac{1}{-1/f - i(\lambda/\pi\omega_{01}^2 n)} = \frac{-a + ib}{a^2 + b^2}$$

$$a \equiv \frac{1}{f} \qquad b \equiv \frac{\lambda}{\pi\omega_{01}^2 n}$$

At plane 3 we obtain, using (3.2-4),

$$q_3 = q_2 + l = \frac{-a}{a^2+b^2} + \frac{ib}{a^2+b^2} + l$$

$$\frac{1}{q_3} = \frac{1}{R_3} - i\frac{\lambda}{\pi\omega_3{}^2 n}$$

$$= \frac{[-a/(a^2+b^2)+l] - ib/(a^2+b^2)}{[-a/(a^2+b^2)+l]^2 + b^2/(a^2+b^2)^2}$$

Since plane 3 is, according to the statement of the problem, to correspond to the output beam waist, $R_3 = \infty$. Using this fact in the last equation leads to

$$l = \frac{a}{a^2+b^2} = \frac{f}{1+(f/\pi\omega_{01}^2 n/\lambda)^2} = \frac{f}{1+(f/z_{01})^2} \tag{3.3-11}$$

as the location of the new waist, and to

$$\frac{\omega_3}{\omega_{01}} = \frac{f\lambda/\pi\omega_{01}^2 n}{\sqrt{1+(f\lambda/\pi\omega_{01}^2 n)^2}} = \frac{f/z_{01}}{\sqrt{1+(f/z_{01})^2}} \tag{3.3-12}$$

for the output beam waist. The confocal beam parameter

$$z_{01} \equiv \frac{\pi\omega_{01}^2 n}{\lambda}$$

is, according to (3.2-11), the distance from the waist in which the input beam spot size increases by $\sqrt{2}$ and is a convenient measure of the convergence of the input beam. The smaller z_{01}, the "stronger" the convergence.

3.4 A Gaussian Beam in Lens Waveguide

As another example of the application of the ABCD law, we consider the propagation of a Gaussian beam through a sequence of thin lenses, as shown in Figure 2-2. The matrix, relating a ray in plane $s+1$ to the plane $s = 1$ is

$$\begin{vmatrix} A_T & B_T \\ C_T & D_T \end{vmatrix} = \begin{vmatrix} A & B \\ C & D \end{vmatrix}^s \tag{3.4-1}$$

where (A, B, C, D) is the matrix for propagation through a single two-lens, unit cell ($\Delta s = 1$) and is given by (2.1-6). We can use a well-known formula for the sth power of a matrix with a unity determinant (unimodular) to obtain

$$
\begin{aligned}
A_T &= \frac{A \sin (s\theta) - \sin [(s-1)\theta]}{\sin \theta} \\
B_T &= \frac{B \sin (s\theta)}{\sin \theta} \\
C_T &= \frac{C \sin (s\theta)}{\sin \theta} \\
D_T &= \frac{D \sin (s\theta) - \sin [(s-1)\theta]}{\sin \theta}
\end{aligned}
\tag{3.4-2}
$$

where

$$
\cos \theta = \tfrac{1}{2}(A + D) = \left(1 - \frac{d}{f_2} - \frac{d}{f_1} + \frac{d^2}{2f_1f_2}\right) \tag{3.4-3}
$$

and then use (3.4-2) in (3.3-9) with the result

$$
q_{s+1} = \frac{\{A \sin (s\theta) - \sin [(s-1)\theta]\}q_1 + B \sin (s\theta)}{C \sin (s\theta) q_1 + D \sin (s\theta) - \sin [(s-1)\theta]} \tag{3.4-4}
$$

The condition for the confinement of the Gaussian beam by the lens sequence is, from (3.4-4), that θ be real; otherwise, the sine functions will yield growing exponentials. From (3.4-3), this condition becomes $|\cos \theta| \leq 1$, or

$$
0 \leq \left(1 - \frac{d}{2f_1}\right)\left(1 - \frac{d}{2f_2}\right) \leq 1 \tag{3.4-5}
$$

that is, the same as condition (2.1-16) for stable-ray propagation.

3.5 High-Order Gaussian Beam Modes in a Homogeneous Medium

The Gaussian mode treated up to this point has a field variation that depends only on the axial distance z and the distance r from the axis. If we do not impose the condition $\partial/\partial\phi = 0$ (where ϕ is the azimuthal angle in a cylindrical coordinate system (r, ϕ, z)) and take $k_2 = 0$, the wave equation (3.1-3) has solutions in the form of [2]

$$
\begin{aligned}
E_{l,m}(x, y, z) &= E_0 \frac{\omega_0}{\omega(z)} H_l\left(\sqrt{2}\frac{x}{\omega(z)}\right) H_m\left(\sqrt{2}\frac{y}{\omega(z)}\right) \\
&\quad \times \exp\left[-ik\frac{x^2+y^2}{2q(z)} - ikz + i(l+m+1)\eta\right] \\
&= E_0 \frac{\omega_0}{\omega(z)} H_l\left(\sqrt{2}\frac{x}{\omega(z)}\right) H_m\left(\sqrt{2}\frac{y}{\omega(z)}\right) \\
&\quad \times \exp\left[-\frac{x^2+y^2}{\omega^2(z)} - \frac{ik(x^2+y^2)}{2R(z)} - ikz + i(l+m+1)\eta\right]
\end{aligned}
\tag{3.5-1}
$$

where H_l is the Hermite polynomial of order l, and $\omega(z)$, $R(z)$, $q(z)$, and η are defined as in (3.2-11) through (3.2-13).

We note for future reference that the phase shift on the axis is

$$
\begin{aligned}
\eta &= kz - (l+m+1)\tan^{-1}\left(\frac{z}{z_0}\right) \\
z_0 &= \frac{\pi\omega_0^2 n}{\lambda}
\end{aligned}
\tag{3.5-2}
$$

The transverse variation of the electric field along x (or y) is seen to be of the form $H_l(\xi)\exp(-\xi^2/2)$ where $\xi = \sqrt{2}x/\omega$. This function has been studied extensively, since it corresponds, also, to the quantum mechanical wavefunction $u_l(\xi)$ of the harmonic oscillator [3]. Some low-order functions normalized to represent the same amount of total beam power are shown in Figure 3-3. Photographs of actual field patterns are shown in Figure 3-4. Note that the first four correspond to the intensity $|u_l(\xi)|^2$ plots ($l = 0, 1, 2, 3$) of Figure 3-3.

3.6 High-Order Gaussian Beam Modes in Quadratic Index Media

In Section 3.3 we treated the propagation of a circularly symmetric Gaussian beam in lenslike media. Here we extend the treatment to higher order modes and limit our attention to steady-state (that is, $q(z)$ = const.) solutions in media whose index of refraction can be described by

$$
n^2(\mathbf{r}) = n^2\left(1 - \frac{n_2}{n}r^2\right) \qquad r^2 = x^2 + y^2 \tag{3.6-1(a)}
$$

that is consistent with (3.1-5) if we put $k_2 = 2\pi n_2/\lambda$.

The vector-wave equation (3.1-3) takes the form

$$
\nabla^2\mathbf{E} + k^2\left(1 - \frac{n_2}{n}r^2\right)\mathbf{E} = 0 \tag{3.6-1(b)}
$$

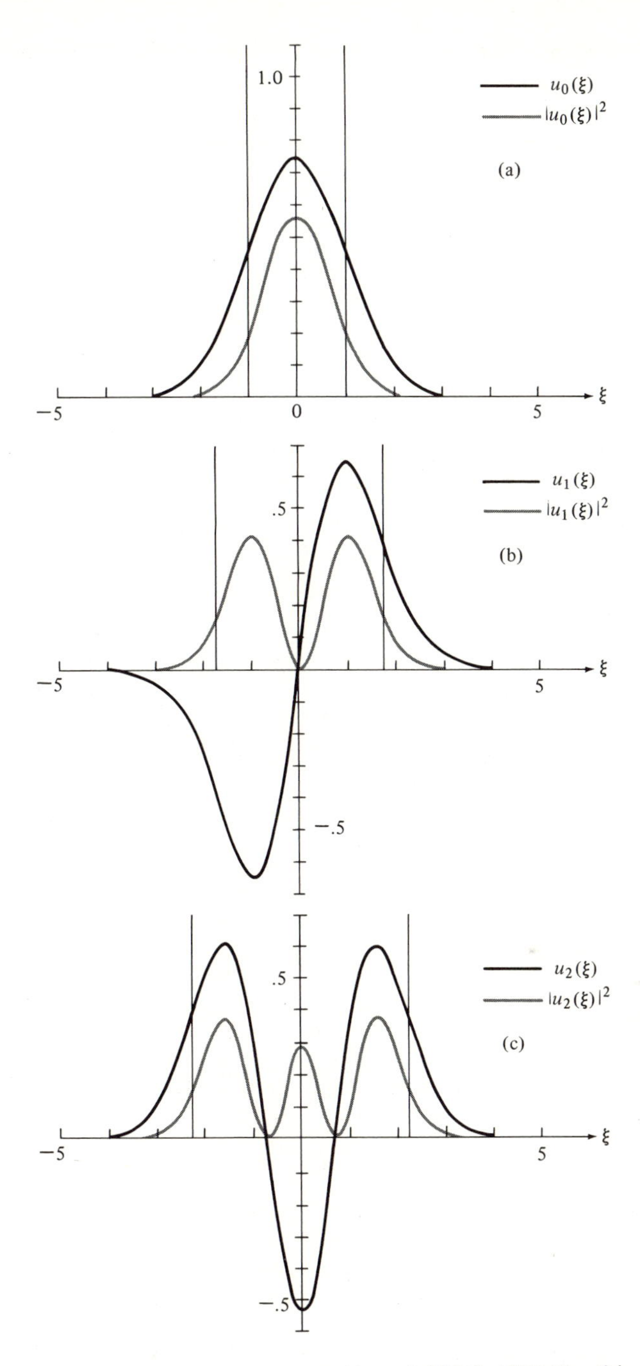

Figure 3-3 Hermite-Gaussian functions $u_l(\xi) = (\pi^{1/2}l!2^l)^{-1/2}H_l(\xi)e^{-\xi^2/2}$ corresponding to higher order beam solutions (Eq. 3.5-1). The curves are normalized so as to represent a fixed amount of total beam power in all the modes

$$\left(\int_{-\infty}^{\infty} u_l^2(\xi)\, d\xi = 1\right).$$

The solid curves are the functions $u_l(\xi)$ for $l = 0, 1, 2, 3,$ and 10. The dashed curves are $u_l^2(\xi)$.

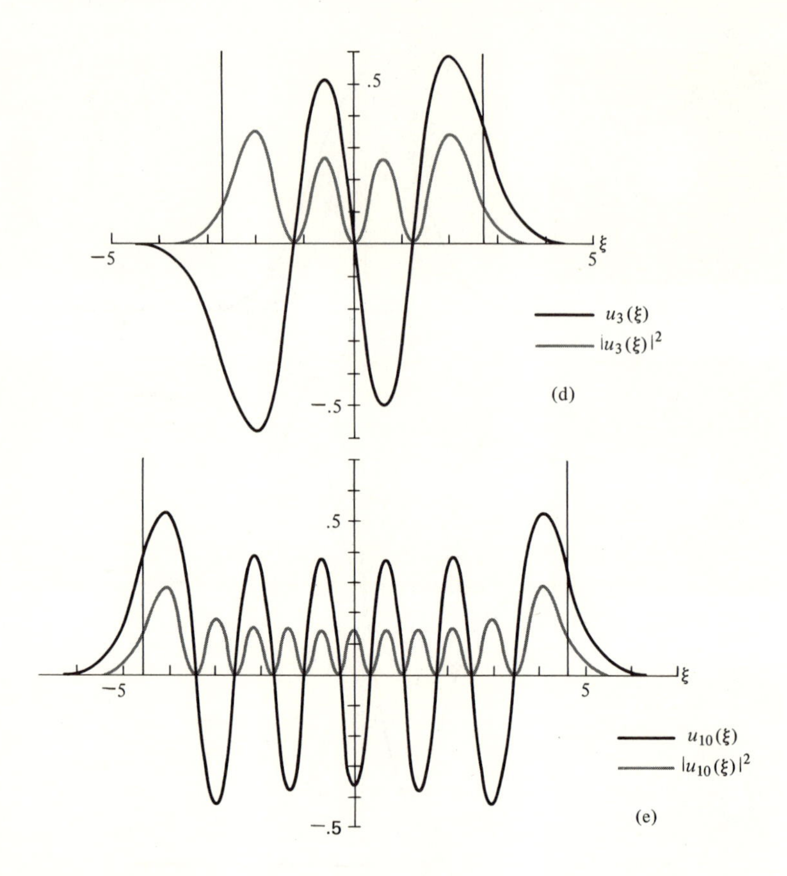

Figure 3-3 (*Continued*)

We consider some (scalar) component E of the last equation and assume a solution in the form

$$E(x, y) = \psi(x, y) \exp(-i\beta z)$$

Taking $\psi(x, y) = f(x)g(y)$ the wave equation [3.6-1(*b*)] becomes

$$\frac{1}{f}\frac{\partial^2 f}{\partial x^2} + \frac{1}{g}\frac{\partial^2 q}{\partial y^2} + k^2 - k^2\frac{n_2}{n}(x^2 + y^2) - \beta^2 = 0 \tag{3.6-2}$$

Since (3.6-2) is the sum of a y dependent part and an x dependent part, it follows that

$$\frac{1}{f}\frac{d^2 f}{dx^2} + \left(k^2 - \beta^2 - k^2\frac{n_2}{n}x^2\right) = C \tag{3.6-3}$$

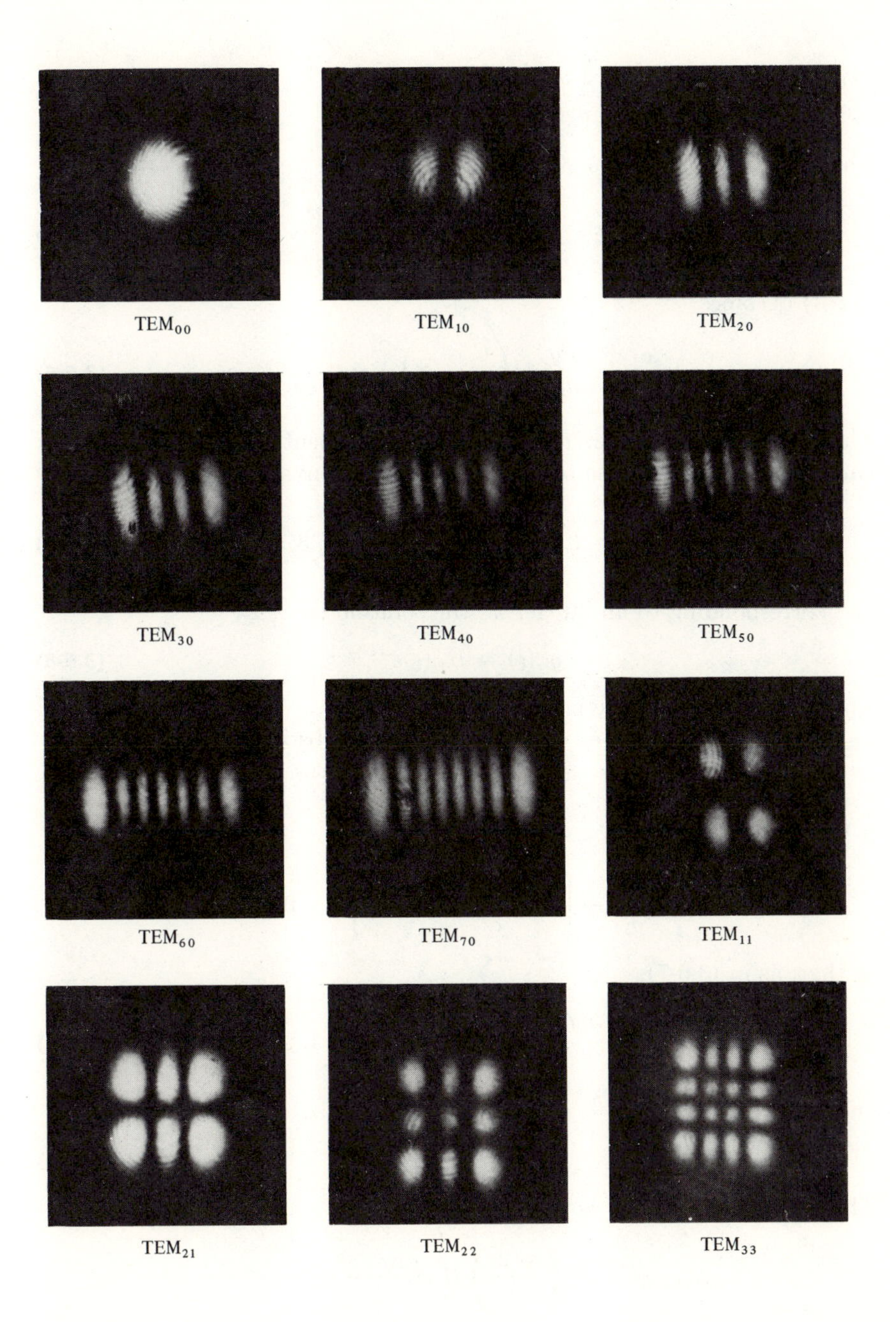

Figure 3-4 Intensity photographs of some low-order Gaussian beam modes. (After Reference [4].)

$$\frac{1}{g}\frac{d^2g}{dy^2} - k^2\frac{n_2}{n}y^2 = -C \tag{3.6-4}$$

where C is some constant. Consider first (3.6-4). Defining a variable ξ by

$$\xi = \alpha y \qquad \alpha \equiv k^{1/2}\left(\frac{n_2}{n}\right)^{1/4} \tag{3.6-5}$$

(3.6-4) becomes

$$\frac{d^2g}{d\xi^2} + \left(\frac{C}{\alpha^2} - \xi^2\right)g = 0 \tag{3.6-6}$$

This is a well-known differential equation and is identical to the Schrödinger equation of the harmonic oscillator [3]. The eigenvalue C/α^2 must satisfy

$$\frac{C}{\alpha^2} = (2m + 1) \qquad m = 1, 2, 3 \cdots \tag{3.6-7}$$

and corresponding to an integer m, the solution is

$$g_m(\xi) = H_m(\xi)e^{-\xi^2/2} \tag{3.6-8}$$

where H_m is the Hermite polynomial of order m.

We now repeat the procedure with (3.6-3). Substituting

$$\zeta = \alpha x$$

it becomes

$$\frac{\partial^2 f}{\partial \zeta^2} + \left[\frac{k^2 - \beta^2 - C}{\alpha^2} - \zeta^2\right]f = 0$$

so that, as in (3.6-7),

$$\frac{k^2 - \beta^2 - C}{\alpha^2} = (2l + 1) \qquad l = 1, 2, 3 \cdots \tag{3.6-9}$$

and

$$f_l(\zeta) = H_l(\zeta)e^{-\zeta^2/2} \tag{3.6-10}$$

The total solution for ψ is thus

$$\psi(x, y) = H_l\left(\frac{\sqrt{2}x}{\omega}\right)H_m\left(\frac{\sqrt{2}y}{\omega}\right)e^{-(x^2+y^2)/\omega^2}$$

where the "spot size" ω is, according to (3.6-5),

$$\omega = \frac{\sqrt{2}}{\alpha} = \sqrt{\frac{2}{k}}\left(\frac{n}{n_2}\right)^{1/4} = \sqrt{\frac{\lambda}{\pi}}\left(\frac{1}{nn_2}\right)^{1/4} \tag{3.6-11}$$

The total (complex) field is

$$E_{l,m}(x, y, z) = \psi_{l,m}(x, y)e^{-i\beta_{l,m}z}$$

$$= E_0 H_l\left(\sqrt{2}\,\frac{x}{\omega}\right) H_m\left(\sqrt{2}\,\frac{y}{\omega}\right) \exp\left(-\frac{x^2 + y^2}{\omega^2}\right) \exp\,(-i\beta_{l,m}z) \tag{3.6-12}$$

The propagation constant $\beta_{l,m}$ of the l, m mode is obtained from (3.6-7) and (3.6-9)

$$\beta_{l,m} = k\left[1 - \frac{2}{k}\sqrt{\frac{n_2}{n}}\,(l + m + 1)\right]^{1/2} \tag{3.6-13}$$

Two features of the mode solutions are noteworthy. (1) Unlike the homogeneous medium solution ($n_2 = 0$), the mode "spot size" ω is independent of z. This can be explained by the focusing action of the index variation ($n_2 > 0$), which counteracts the natural tendency of a confined beam to diffract (spread). In the case of an index of refraction which increases with r ($n_2 < 0$), it follows from (3.6-11) and (3.6-12) that $\omega^2 < 0$ and no confined solutions exist. The index profile in this case leads to defocusing, thus reinforcing the diffraction of the beam. (2) The dependence of β on the mode indices l, m causes the different modes to have phase velocities $v_{l,m} = \omega/\beta_{l,m}$ as well as group velocities $(v_g)_{l,m} = d\omega/d\beta_{l,m}$ which depend on l and m.

Let us consider the modal dispersion (that is, the dependence on l and m) of the group velocity of mode l, m

$$(v_g)_{l,m} = \frac{d\omega}{d\beta_{l,m}} \tag{3.6-14}$$

If the index variation is small so that

$$\frac{1}{k}\sqrt{\frac{n_2}{n}}\,(l + m + 1) \ll 1 \tag{3.6-15}$$

we can approximate (3.6-13) as

$$\beta_{l,m} \cong k - \sqrt{\frac{n_2}{n}}\,(l + m + 1) - \frac{n_2}{2kn}\,(l + m + 1)^2 \tag{3.6-16}$$

so that, according to (3.6-14),

$$(v_g)_{l,m} = \frac{c/n}{\left[1 + \dfrac{(n_2/n)}{2k^2}\,(l + m + 1)^2\right]} \tag{3.6-17}$$

The effect of the group velocity dispersion on pulse propagation is considered next.

Pulse spreading in quadratic index glass fibers. Glass fibers with quadratic index profiles [3.6-1(a)] are excellent channels for optical communication systems [5, 6]. The information is coded onto trains of optical pulses and the channel information capacity is thus fundamentally limited by the number of pulses which can be transmitted per unit time [7, 8].

There are two ways in which the group velocity dispersion limits the pulse repetition rate of the quadratic index channel.

1. *Modal dispersion.* If the optical pulses fed into the input end of the fiber excite a large number of modes (this will be the case if the input light is strongly focused so that the "rays" subtend a large angle), then each mode will travel with a group velocity $(v_g)_{l,m}$, as given by (3.6-17). If all the modes from (0, 0) to $(l_{\max}, m_{\max})$ are excited, the output pulse at $z = L$ will broaden to

$$\Delta\tau \cong L\left[\frac{1}{(v_g)_{l_{\max},m_{\max}}} - \frac{1}{(v_g)_{0,0}}\right] \tag{3.6-18}$$

We can use (3.6-17) and the condition $(n_2/n)(l + m + 1)^2/2k^2 \ll 1$ to obtain

$$\Delta\tau = \frac{nL}{c}\left[\frac{n_2}{2nk^2}(l_{\max} + m_{\max} + 1)^2\right] \tag{3.6-19}$$

The maximum number of pulses per second that can be transmitted without serious overlap of adjacent output pulse is thus $f_{\max} \sim 1/\Delta\tau$. High data rate transmission will thus require the use of single mode excitation, which can be achieved by the use of coherent single mode laser excitation [6, 7, 8].

Numerical example. Consider a 1 km long quadratic index fiber with $n = 1.5$, $n_2 = 5.1 \times 10^3$ cm^{-2}. Let the input optical pulses at $\lambda = 1\ \mu m$ excite the modes up to $l_{\max} = m_{\max} = 30$. Substitution in (3.6-19) gives

$$\Delta\tau = 8 \times 10^{-9} s$$

and $f_{\max} \sim (\Delta\tau)^{-1} = 1.25 \times 10^8$ pulses per second for the maximum pulse rate.

2. *Group velocity dispersion.* The pulse spreading (3.6-19) due to multimode excitation can be eliminated if one were to excite a single mode, say l, m only. In this case pulse spreading would still result from the dependence of $(v_g)_{l,m}$ on frequency. This spreading can be explained by the fact that a pulse with a spectral width $\Delta\omega$ will spread in a distance L by

$$\Delta\tau \approx 2L\left|\frac{d}{d\omega}\left(\frac{1}{v_g}\right)\right|\Delta\omega = \frac{2L}{v_g^2}\left|\frac{dv_g}{d\omega}\right|\Delta\omega \tag{3.6-20}$$

If the pulse derived from a coherent continuous source with a negligible spectral width, the pulse spectral width is related to the pulse duration τ by $\Delta\omega \sim 2/\tau$ and (3.6-20) becomes

$$\Delta\tau \approx \frac{4L}{v_g^2\tau}\left(\frac{dv_g}{d\omega}\right) \tag{3.6-21}$$

If the source bandwidth $\Delta\omega_s$ exceeds τ^{-1}, then we need to replace $\Delta\omega$ in (3.6-20) by $\Delta\omega_s$.

In order to check our semi-intuitive derivation of (3.6-21) and also as an instructive exercise in the mathematics of dispersive propagation, we will, in what follows, consider the problem of an optical pulse with a Gaussian envelope propagating in a dispersive channel.

The input pulse is taken as

$$\begin{aligned} E(z = 0, t) &= e^{-\alpha t^2} e^{i\omega_0 t} \\ &= e^{i\omega_0 t} \int_{-\infty}^{\infty} F(\Omega) e^{i\Omega t}\, d\Omega \end{aligned} \tag{3.6-22}$$

where $F(\Omega)$, the Fourier transform of the envelope $\exp(-\alpha t^2)$, is

$$F(\Omega) = \sqrt{\frac{1}{4\pi\alpha}}\, e^{-\Omega^2/4\alpha} \tag{3.6-23}$$

The field at a distance z is obtained by multiplying each frequency component $(\omega_0 + \Omega)$ in (3.6-22) by $\exp[-i\beta(\omega_0 + \Omega)z]$. If we expand $\beta(\omega_0 + \Omega)$ near ω_0 as

$$\beta(\omega_0 + \Omega) = \beta(\omega_0) + \left.\frac{d\beta}{d\omega}\right|_{\omega_0} \Omega + \frac{1}{2} \left.\frac{d^2\beta}{d\omega^2}\right|_{\omega_0} \Omega^2 + \cdots$$

we obtain

$$E(z, t) = e^{i(\omega_0 t - \beta_0 z)} \int_{-\infty}^{\infty} d\Omega F(\Omega) \exp\left\{i\left[\Omega t - \frac{\Omega z}{v_g} + \frac{1}{2}\frac{\partial}{\partial\omega}\left(\frac{1}{v_g}\right)\Omega^2 z\right]\right\} \tag{3.6-24}$$

where

$$\beta_0 \equiv \beta(\omega_0), \qquad \frac{d\beta}{d\omega} = \frac{1}{v_g} = \frac{1}{\text{group velocity}}$$

The field envelope is given by the integral in (3.6-24)

$$\begin{aligned} \mathcal{E}(z, t) &= \int_{-\infty}^{\infty} d\Omega F(\Omega) \exp\left\{i\Omega\left[\left(t - \frac{z}{v_g}\right) - \frac{1}{2}\frac{d}{d\omega}\left(\frac{1}{v_g}\right)\Omega z\right]\right\} \\ &= \int_{-\infty}^{\infty} d\Omega F(\Omega) \exp\left\{i\Omega\left[\left(t - \frac{z}{v_g}\right) - a\Omega z\right]\right\} \end{aligned} \tag{3.6-25}$$

where

$$a \equiv \frac{1}{2}\frac{d}{d\omega}\left(\frac{1}{v_g}\right) = -\frac{1}{2v_g^2}\frac{dv_g}{d\omega}$$

After substituting for $F(\Omega)$ from (3.6-23), the last equation becomes

$$\mathcal{E}(z, t) = \sqrt{\frac{1}{4\pi\alpha}} \int_{-\infty}^{\infty} \exp\left\{-\left[\Omega^2\left(\frac{1}{4\alpha} + iaz\right) - i\left(t - \frac{z}{v_g}\right)\Omega\right]\right\} d\Omega$$

Carrying out the integration yields

$$\mathcal{E}(z, t) = \frac{1}{\sqrt{1 + i4a\alpha z}} \exp\left[-\frac{(t - z/v_g)^2}{1/\alpha + 16a^2z^2\alpha}\right] \exp\left[i\frac{4az(t - z/v_g)^2}{1/\alpha^2 + 16a^2z^2}\right] \tag{3.6-26}$$

The pulse duration τ at z can be taken as the separation between the two times when the pulse envelope is smaller by a factor of e^{-1} from its peak value, that is,

$$\tau(z) = 2\sqrt{\frac{1}{\alpha} + 16a^2z^2\alpha} \tag{3.6-27}$$

At large distances such that $16a^2z^2\alpha \gg 1/\alpha$, we obtain

$$\tau(z) \cong 4\,|a|\,z\sqrt{\alpha} = \frac{8\,|a|\,z}{\tau(0)} \tag{3.6-28}$$

where $\tau(0) = 2\alpha^{-1/2}$ is the initial pulse duration. If we use the definition of the factor a (see line following 3.6-25), the last expression becomes

$$\tau(z) = \frac{4}{v_g^2}\left(\frac{dv_g}{d\omega}\right)\frac{z}{\tau(0)} \tag{3.6-29}$$

in agreement with (3.6-21).

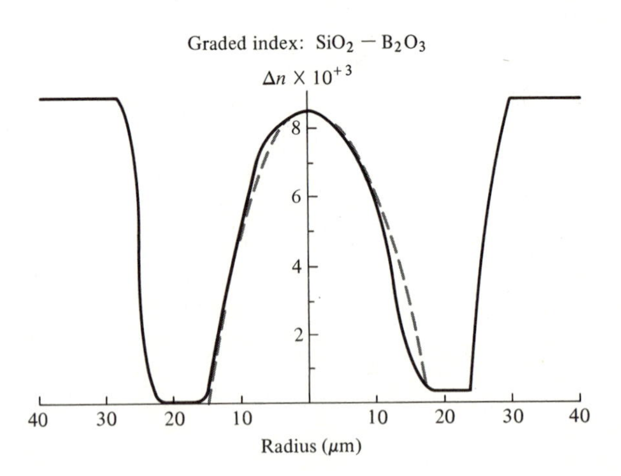

Figure 3-5 Graded refractive-index profile across the core of the SiO_2-B_2O_3 fiber. The dashed lines describe quadratic profiles. (After Reference [7].)

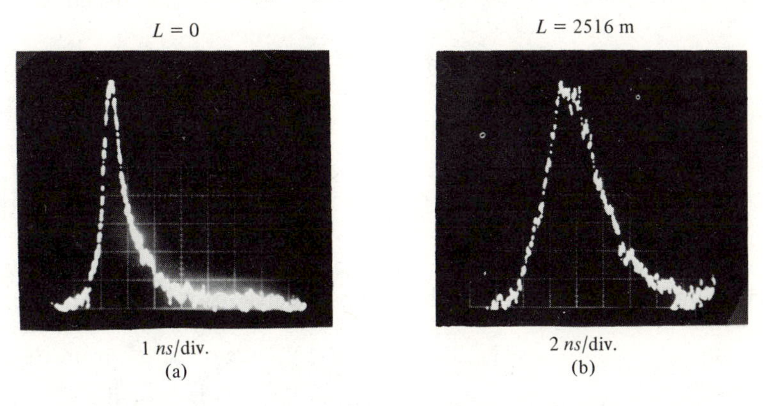

Figure 3-6 (**a**) Pulse shape at the input of the quadratic index fiber of Figure 3-5. (**b**) The output pulse after a propagation through 2516 m (note the change in the time scales). (After Reference [7].)

The dependence of v_g on ω, which according to (3.6-28) leads to pulse broadening, is due to two mechanisms:

1. v_g depends, according to (3.6-17), explicitly on ω since $k = \omega n/c$.
2. v_g depends implicitly on ω through the dependence of the material index of refraction n on ω. We thus write $dv_g/d\omega = \partial v_g/\partial\omega + (\partial v_g/\partial n)(dn/d\omega)$ and obtain from (3.6-20)

$$\Delta\tau = \frac{2L}{c}\left|\left[\frac{nn_2}{ck^3}(l + m + 1)^2 - \frac{dn}{d\omega}\right]\right|\Delta\omega \qquad \textbf{(3.6-30)}$$

where in the second term we assumed $(n_2/2k^2n)(l + m + 1)^2 \ll 1$. In most fibers the pulse spreading is dominated by the material dispersion term $dn/d\omega$.

The actual index variation in a parabolic index $SiO_2 - B_2O_3$ fiber with a value of $n_2 \simeq 5 \times 10^3\ \text{cm}^{-2}$ is shown in Figure 3-5. The broadening of an optical pulse after a propagation distance of $\sim$2.5 km is illustrated by Figure 3-6. The input optical pulse excites a large number of (l, m) modes and the broadening is of the type described by (3.6-19). The data are reproduced from Reference [7], which also describes the important consequence of intermode mixing on pulse broadening.

If we combine (3.6-26) with (3.6-24) we find that the total field at z is

$$E(z, t) = \mathcal{E}(z, t)e^{i(\omega_0 t - \beta_0 z)}$$

$$= \frac{e^{-i\beta_0 z}}{\sqrt{1 + i4a\alpha z}} \exp i\left[\omega_0 t + \frac{4az(t - z/v_g)^2}{\alpha^{-2} + 16a^2z^2}\right] - \frac{(t - z/v_g)^2}{1/\alpha + 16a^2z^2\alpha} \qquad \textbf{(3.6-31)}$$

The oscillation phase is thus

$$\Phi(z, t) = \omega_0 t + \frac{4az(t - z/v_g)^2}{\alpha^{-2} + 16a^2z^2} - \beta_0 z$$

The local "frequency" $\omega(z, t)$ is then

$$\omega(z, t) = \frac{\partial \Phi}{\partial t} = \omega_0 + \frac{8az(t - z/v_g)}{\alpha^{-2} + 16a^2z^2} \tag{3.6-32}$$

and consists of the original frequency ω_0 and a linear frequency sweep (chirp) which is proportional to the group velocity dispersion term a.

3.7 Propagation in Media with a Quadratic Gain Profile

In many laser media the gain is a strong function of position. This variation can be due to a variety of causes, among them: (1) the radial distribution of energetic electrons in the plasma region of gas lasers [9], (2) the variation of pumping intensity in solid state lasers, and (3) the dependence of the degree of gain saturation on the radial position in the beam.

We can account for an optical medium with quadratic gain (or loss) variation by taking the complex propagation constant $k(r)$ in (3.1-5) as

$$k(r) = k \pm i\left(\alpha_0 - \frac{1}{2}\alpha_2 r^2\right) \tag{3.7-1}$$

where the plus (minus) sign applies to the case of gain (loss). Assuming $k_2 r^2 \ll k$ in (3.1-5), we have $k_2 = i\alpha_2$. Using this value in (3.1-11) to obtain the steady state[5] ($(1/q)' = 0$) solution of the complex beam radius yields

$$\frac{1}{q} = -i\sqrt{\frac{k_2}{k}} = -i\sqrt{\frac{i\alpha_2}{k}} \tag{3.7-2}$$

The steady-state beam radius and spot size are obtained from (3.3-5) and (3.7-2)

$$\omega^2 = 2\sqrt{\frac{\lambda}{\pi n \alpha_2}}$$

$$R = 2\sqrt{\frac{\pi n}{\lambda \alpha_2}} \tag{3.7-3}$$

We thus find that the steady-state solution corresponds to beams with a constant spot size but with a finite radius of curvature.

[5] "Steady state" here refers not to the intensity, which according to (3.7-1) is growing or decaying with z, but to the beam radius of curvature and spot size.

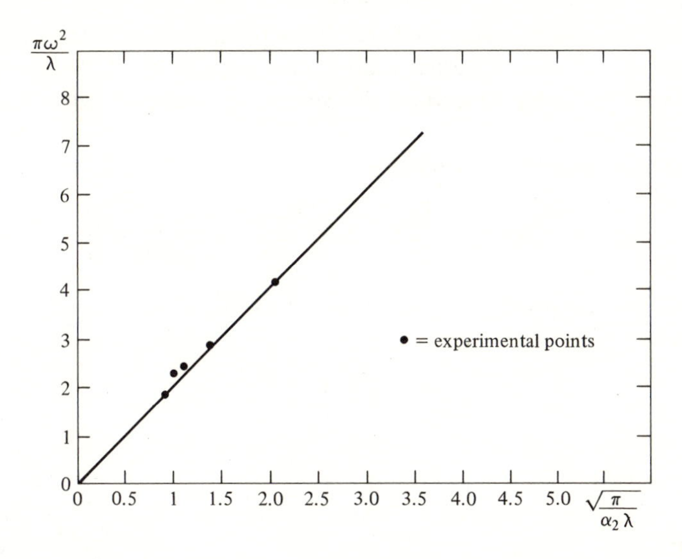

Figure 3-7 Theoretical curve showing the dependence of beam radius on quadratic gain constant α_2. Experimental points were obtained in a xenon 3.39 μm laser in which α_2 was varied by controlling the unsaturated laser gain. (After Reference [10].)

The general (nonsteady state) behavior of the Gaussian beam in a quadratic gain medium is described by (3.3-4), where $k_2 = i\alpha_2$.

Experimental data showing a decrease of the beam spot size with increasing gain parameter α_2 in agreement with (3.7-3) are shown in Figure 3-7.

3.8 Elliptic Gaussian Beams

All the beam solutions considered up to this point have one feature in common. The field drops off as in (3.5-1), according to

$$E_{m,n} \propto \exp\left[-\frac{x^2+y^2}{\omega^2(z)}\right] \tag{3.8-1}$$

so that the locus in the x-y plane of the points where the field is down by a factor of e^{-1} from its value on the axis is a circle of radius $\omega(z)$. We will refer to such beams as circular Gaussian beams.

The wave equation (3.1-8) also admits solutions in which the variation in the x and y directions is characterized by

$$E_{m,n} \propto \exp\left[-\frac{x^2}{\omega_x^2(z)} - \frac{y^2}{\omega_y^2(z)}\right] \tag{3.8-2}$$

with $\omega_x \neq \omega_y$. Such beams, which we name elliptic Gaussian, result, for example, when a circular Gaussian beam passes through a cylindrical lens or when a laser beam emerges from an astigmatic resonator—that is, one whose mirrors possess different radii of curvature in the z-y and z-x planes.

We will not repeat the whole derivation for this case, but will indicate the main steps.

Instead of (3.1-9) we assume a solution

$$\psi = \exp\left\{-i\left[P(z) + \frac{k}{2q_x(z)}x^2 + \frac{k}{2q_y(z)}y^2\right]\right\} \tag{3.8-3}$$

that results, in a manner similar to (3.1-11), in[6]

$$\begin{aligned} \left(\frac{1}{q_x}\right)^2 + \left(\frac{1}{q_x}\right)' + \frac{k_{2x}}{k} &= 0 \\ \left(\frac{1}{q_y}\right)^2 + \left(\frac{1}{q_y}\right)' + \frac{k_{2y}}{k} &= 0 \end{aligned} \tag{3.8-4}$$

and

$$\frac{dP}{dz} = -i\left(\frac{1}{q_x} + \frac{1}{q_y}\right) \tag{3.8-5}$$

In the case of a homogeneous ($k_{2x} = k_{2y} = 0$) beam we obtain as in (3.2-4),

$$q_x(z) = z + C_x \tag{3.8-6}$$

where C_x is an arbitrary constant of integration. We find it useful to write C_x as

$$C_x = -z_x + q_{0x} \tag{3.8-7}$$

where z_x is real and q_{0x} is imaginary. The physical significance of these two constants will become clear in what follows. A similar result with $x \to y$ is obtained for $q_y(z)$. Using the solutions of $q_x(z)$ and $q_y(z)$ in (3.8-5) gives

$$P = -\frac{i}{2}\left[\ln\left(1 + \frac{z - z_x}{q_{0x}}\right) + \ln\left(1 + \frac{z - z_y}{q_{0y}}\right)\right]$$

Proceeding straightforwardly, as in the derivation connecting (3.2-6,. . .14), results in

$$E(x, y, z) = E_0 \frac{\sqrt{\omega_{0x}\omega_{0y}}}{\sqrt{\omega_x(z)\omega_y(z)}} \exp\left\{-i[kz - \eta(z)] - \frac{ikx^2}{2q_x(z)} - \frac{iky^2}{2q_y(z)}\right\}$$

[6] The parameters k_{2x} and k_{2y} are defined by

$$k^2(x, y) = k^2 - kk_{2x}x^2 - kk_{2y}y^2$$

which is a generalization of (3.1-5).

$$= E_0 \frac{\sqrt{\omega_{0x}\omega_{0y}}}{\sqrt{\omega_x(z)\omega_y(z)}} \exp\left\{-i[kz - \eta(z)] - x^2\left(\frac{1}{\omega_x^2(z)} + \frac{ik}{2R_x(z)}\right)\right.$$

$$\left. - y^2\left(\frac{1}{\omega_y^2(z)} + \frac{ik}{2R_y(z)}\right)\right\} \quad \textbf{(3.8-8)}$$

where

$$q_{0x} = i\frac{\pi\omega_{0x}^2 n}{\lambda}$$

$$\omega_x^2(z) = \omega_{0x}^2\left[1 + \left(\frac{\lambda(z - z_x)}{\pi\omega_{0x}^2 n}\right)^2\right] \quad \textbf{(3.8-9)}$$

$$R_x(z) = z\left[1 + \left(\frac{\pi\omega_{0x}^2 n}{\lambda(z - z_x)}\right)^2\right]$$

with similar expression in which $x \to y$ for q_{0y}, ω_y, R_y.

The phase delay $\eta(z)$ in (3.8-8) is now given by

$$\eta(z) = \tfrac{1}{2}\tan^{-1}\left(\frac{\lambda(z - z_x)}{\pi\omega_{0x}^2 n}\right) + \tfrac{1}{2}\tan^{-1}\left(\frac{\lambda(z - z_y)}{\pi\omega_{0y}^2 n}\right) \quad \textbf{(3.8-10)}$$

It follows that *all* the results derived for the case of circular Gaussian beams apply, separately, to the x-z and to the y-z behavior of the elliptic Gaussian beam. For the purpose of analysis the elliptic beam can be considered as two independent "beams." The position of the waist is not necessarily the same for these two beams. It occurs at $z = z_x$ for the x-z beam and at $z = z_y$ for the y-z beam in the example of Figure 3-8, where z_x and z_y are arbitrary.

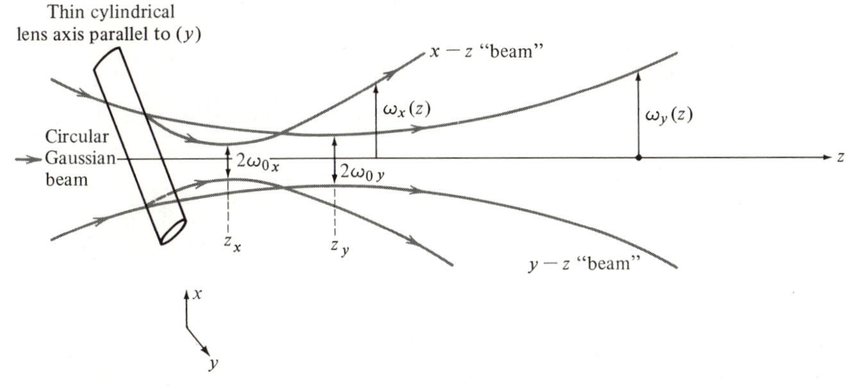

Figure 3-8 Illustration of an elliptic beam produced by cylindrical focusing of a circular Gaussian beam.

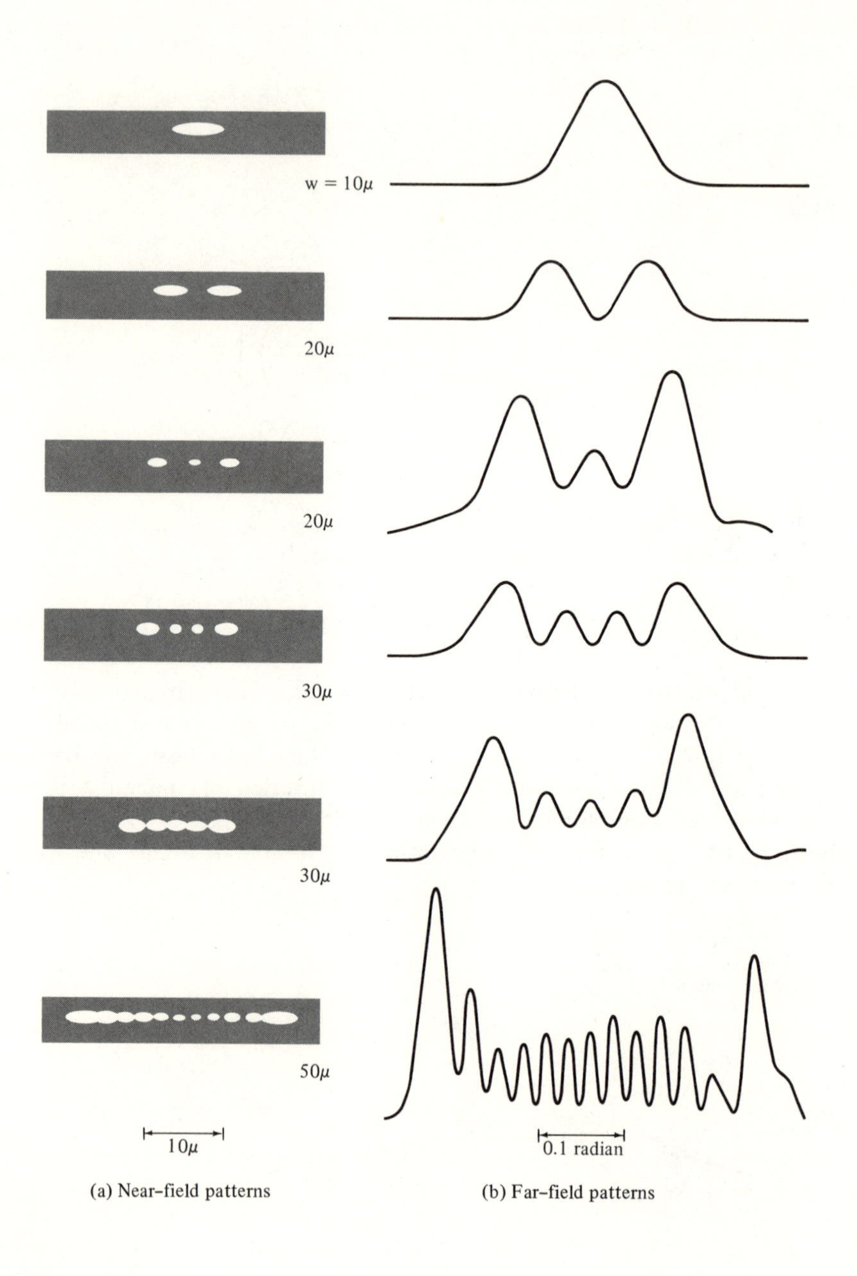

Figure 3-9 (**a**) Near-field and (**b**) far-field intensity distributions of the output of stripe contact GaAs-GaAlAs lasers. (After Reference [11].)

It also follows from the similarity between (3.8-4) and (3.1-11) that the ABCD transformation law (3.3-9) can be applied separately to $q_x(z)$ and $q_y(z)$ which, according to (3.8-8), are given by

$$\frac{1}{q_x(z)} = \frac{1}{R_x(z)} - i\frac{\lambda}{\pi n \omega_x^2(z)}$$
$$\frac{1}{q_y(z)} = \frac{1}{R_y(z)} - i\frac{\lambda}{\pi n \omega_y^2(z)} \tag{3.8-11}$$

Elliptic Gaussian beams in a quadratic lenslike medium. Here we consider the *steady-state* elliptic beam propagating in a medium whose index of refraction is given by

$$n^2(\mathbf{r}) = n^2\left(1 - \frac{n_{2x}}{n}x^2 - \frac{n_{2y}}{n}y^2\right) \tag{3.8-12}$$

The derivation is identical to that presented in Section 3.6, resulting in

$$E_{l,m}(\mathbf{r}) = E_0 e^{-i\beta_{l,m}z} H_l\left(\sqrt{2}\frac{x}{\omega_x}\right) H_m\left(\sqrt{2}\frac{y}{\omega_y}\right) \exp\left(-\frac{x^2}{\omega_x^2} - \frac{y^2}{\omega_y^2}\right) \tag{3.8-13}$$

where

$$\omega_x = \left(\frac{\lambda}{\pi}\right)^{1/2}\left(\frac{1}{nn_{2x}}\right)^{1/4}$$
$$\omega_y = \left(\frac{\lambda}{\pi}\right)^{1/2}\left(\frac{1}{nn_{2y}}\right)^{1/4} \tag{3.8-14}$$

$$\beta_{l,m} = k\left\{1 - \frac{2}{k}\left[\sqrt{\frac{n_{2x}}{n}}\left(l + \frac{1}{2}\right) + \sqrt{\frac{n_{2y}}{n}}\left(m + \frac{1}{2}\right)\right]\right\}^{1/2} \tag{3.8-15}$$

The beam, as in the solution of the homogeneous case (3.8-8), possesses different spot sizes and radii of curvature in the y-z and x-z planes. The beam parameters, however, are independent of z.

Elliptic Gaussian beams have been observed experimentally in the output of stripe geometry gallium arsenide junction lasers [11], [12], [13] . Near-field and far-field experimental intensity distributions corresponding to some $(0, m)$ modes are shown in Figure 3-9.

PROBLEMS

3-1 Solve the problem leading up to Equations (3.3-11) and (3.3-12) for the case where the lens is placed in an arbitrary position relative to the input beam (that is, not at its waist).

3-2 **a.** Assume a Gaussian beam incident normally on a solid prism with an index of refraction n as shown. What is the far-field diffraction angle of the output beam?

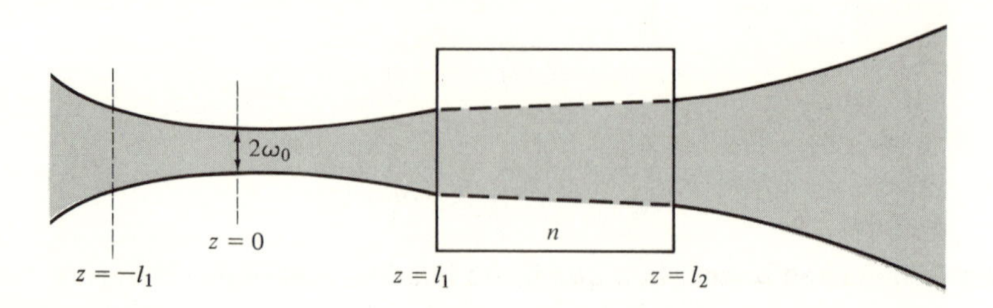

b. Assume that the prism is moved to the left until its input face is at $z = -l_1$. What is the new beam waist and what is its location? (Assume that the crystal is long enough that the beam waist is inside the crystal.)

3-3 A Gaussian beam with a wavelength λ is incident on a lens placed at $z = l$ as shown. Calculate the lens focal length, f, so that the output beam has a waist at the front surface of the sample crystal. Show that (given l and L) two solutions exist. Sketch the beam behavior for each of these solutions.

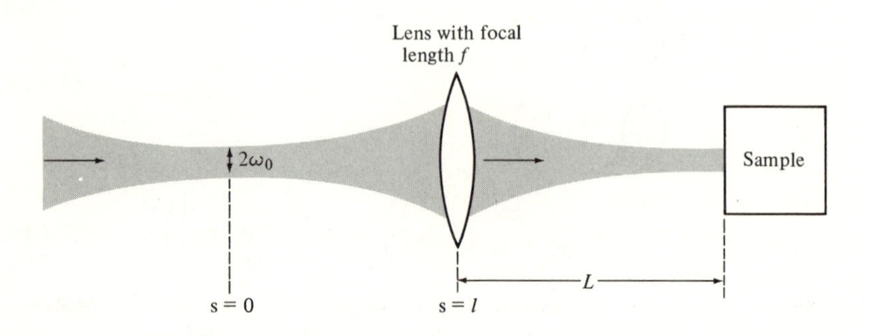

3-4 Complete all the missing steps in the derivation of Section 3.8.

3-5 Find the beam spot size and the maximum number of pulses per second that can be carried by an optical beam ($\lambda = 1\ \mu$m) propagating in a quadratic index glass fiber with $n = 1.5$, $n_2 = 5 \times 10^2\ \text{cm}^{-2}$. (a) in the case of a single mode excitation $l = m = 0$; (b) in the case where all the modes with $l, m \leq 5$ are excited.

Using dispersion data (n versus ω) of any typical commercial glass and taking $n_2 = 5 \times 10^3\ \text{cm}^{-2}$, $l_{\text{max}} = m_{\text{max}} = 30$, compare the relative contributions of modal and material dispersion to pulse broadening.

REFERENCES

[1] Kogelnik, H., "On the propagation of Gaussian beams of light through lenslike media including those with a loss and gain variation," *Appl. Opt.*, vol. 4, p. 1562, 1965.

[2] Marcuse, D., *Light Transmission Optics.* Princeton, N.J: Van Nostrand, 1972.

[3] Yariv, A., *Quantum Electronics,* 2d Ed. New York: Wiley, 1975, Section 2.2.

[4] Kogelnik, H., and W. Rigrod, *Proc. IRE,* vol. 50, p. 230, 1962.

[5] Kawakami, S., and J. Nishizawa, "An optical waveguide with the optimum distribution of the refractive index with reference to waveform distortion," *IEEE Trans. Microwave Theory and Technique,* MTT-16, vol. 10, p. 814, 1968.

[6] Miller, S. E., E. A. J. Marcatili, and T. Li, "Research toward optical fiber transmission systems," *Proc. IEEE,* vol. 61, p. 1703, 1973.

[7] Cohen, L. G., and H. M. Presby, "Shuttle pulse measurement of pulse spreading in a low loss graded index fiber," *Appl. Opt.*, vol. 14, p. 1361, 1975.

[8] Cohen, L. G., and S. D. Personick, "Length dependence of pulse dispersion in a long multimode optical fiber," *Appl. Opt.*, vol. 14, p. 1250, 1975.

[9] Bennett, W. R., "Inversion Mechanisms in Gas Lasers," *Appl. Opt.*, Suppl. 2 *Chemical Lasers*, p. 3, 1965.

[10] Casperson, L., and A. Yariv, "The Gaussian mode in optical resonators with a radial gain profile," *Appl. Phys. Lett.*, vol. 12, p. 355, 1968.

[11] Zachos, T. H., "Gaussian beams from GaAs junction lasers," *Appl. Phys. Letters*, vol. 12 (1969), p. 318.

[12] Zachos, T. H., and J. E. Ripper, "Resonant modes of GaAs junction lasers," *IEEE J. of Quantum Electron.*, *QE*-5 (1969), p. 29.

[13] H. Yonezu et al., "A GaAs–Al_xGa_{1-x} as double heterostructure planar stripe laser," *Jap. Journ. of Appl. Phys.*, vol. 12, p. 1585, 1973.

CHAPTER

4

Optical Resonators

4.0 Introduction

Optical resonators, like their low-frequency, radio-frequency, and microwave counterparts, are used primarily in order to build up large field intensities with moderate power inputs. A universal measure of this property is the quality factor Q of the resonator. Q is defined by the relation

$$Q = \omega \times \frac{\text{field energy stored by resonator}}{\text{power dissipated by resonator}} \tag{4.0-1}$$

As an example, consider the case of a simple resonator formed by bouncing a plane TEM wave between two perfectly conducting planes of separation l so that the field inside is

$$e(z, t) = E \sin \omega t \sin kz \tag{4.0-2}$$

According to (1.3-22), the average electric energy stored in the resonator is

$$\mathcal{E}_{\text{electric}} = \frac{A\epsilon}{2T} \int_0^l \int_0^T e^2(z, t)\, dz\, dt \tag{4.0-3}$$

where A is the cross-sectional area, ϵ is the dielectric constant, and $T = 2\pi/\omega$ is the period. Using (4.0-2) we obtain

$$\mathcal{E}_{\text{electric}} = \tfrac{1}{8}\epsilon E^2 V \tag{4.0-4}$$

where $V = lA$ is the resonator volume. Since the average magnetic energy stored in a resonator is equal to the electric energy [1], the total stored energy is

$$\mathcal{E} = \tfrac{1}{4}\epsilon E^2 V \tag{4.0-5}$$

Thus, designating the power input to the resonator by P, we obtain from (4.0-1)

$$Q = \frac{\omega \epsilon E^2 V}{4P}$$

The peak field is given by

$$E = \sqrt{\frac{4QP}{\omega \epsilon V}} \tag{4.0-6}$$

The main difference between an optical resonator and a microwave resonator—for example, one operating at $\lambda = 1$ cm ($\nu = 3 \times 10^{10}$ Hz)—is that in the latter case one can easily fabricate the resonator with typical dimensions comparable to λ. This leads to the presence of one, or just a few, resonances in the region of interest. In the optical regime, however, $\lambda \simeq 10^{-4}$ cm, so the resonator is likely to have typical dimensions that are very large in comparison to the wavelength. Under these conditions the number of resonator modes in a frequency interval $d\nu$ is given (see Problem 4-8 or, for example, Reference [2]) by

$$N \simeq \frac{8\pi n^3 \nu^2 V}{c^3} d\nu \tag{4.0-7}$$

where V is the volume of the resonator. For the case of $V = 1$ cm^3, $\nu = 3 \times 10^{14}$ Hz and $d\nu = 3 \times 10^{10}$, as an example, (4.0-7) yields $N \sim 2 \times 10^9$ modes. If the resonator were closed, all these modes would have similar values of Q. This situation is to be avoided in the case of lasers, since it will cause the atoms to emit power (thus causing oscillation) into a large number of modes, which may differ in their frequencies as well as in their spatial characteristics.

This objection is overcome to a large extent by the use of open resonators, which consist essentially of a pair of opposing flat or curved reflectors. In such resonators the energy of the vast majority of the modes does not travel at right angles to the mirrors and will thus be lost in essentially a single traversal. These modes will consequently possess a very low Q. If the mirrors are curved, the few surviving modes will, as shown below, have their energy localized near the axis; thus the diffraction losses caused by the open sides can be made small compared with other loss mechanisms such as mirror transmission.

4.1 Fabry–Perot Etalon

The Fabry–Perot etalon, or interferometer, named after its inventors [3] can be considered as the archetype of the optical resonator. It consists of a plane-parallel plate of thickness l and index n which is immersed in a medium of index n'.[1] Let a plane wave be incident on the etalon at an angle θ' to the normal, as shown in Figure 4-1. We can treat the problem of the transmission (and reflection) of the plane wave through the etalon by considering the infinite number of partial waves produced by reflections at the two end surfaces. The phase delay between two partial waves—which is attributable to one additional round trip—is given, according to Figure 4-2, by

$$\delta = \frac{4\pi nl \cos\theta}{\lambda} \tag{4.1-1}$$

where λ is the vacuum wavelength of the incident wave and θ is the internal angle of incidence. If the complex amplitude of the incident wave is taken as A_i, then the partial reflections B_1, B_2, and so forth, are given by

$$B_1 = rA_i \qquad B_2 = tt'r'A_ie^{i\delta} \qquad B_3 = tt'r'^3A_ie^{2i\delta} \qquad \ldots$$

where r is the reflection coefficient (ratio of reflected to incident amplitude), t is the transmission coefficient for waves incident from n' toward n, and r' and t' are the corresponding quantities for waves traveling from n toward n'. The complex amplitude of the (total) reflected wave is $A_r =$

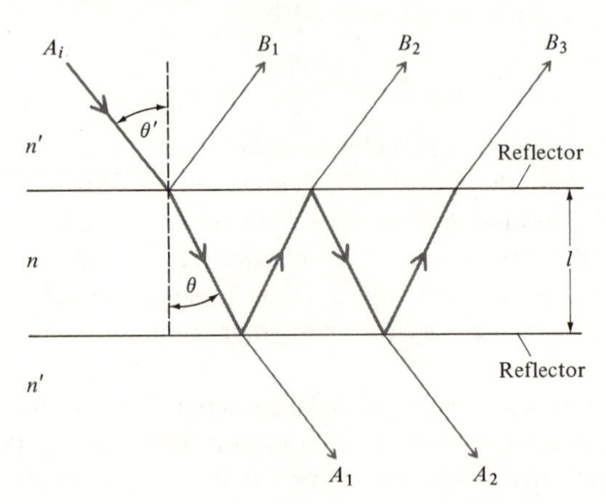

Figure 4-1

---

[1] In practice, one often uses etalons made by spacing two partially reflecting mirrors a distance l apart so that $n = n' = 1$. Another common form of etalon is produced by grinding two plane-parallel (or curved) faces on a transparent solid and then evaporating a metallic or dielectric layer (or layers) on the surfaces.

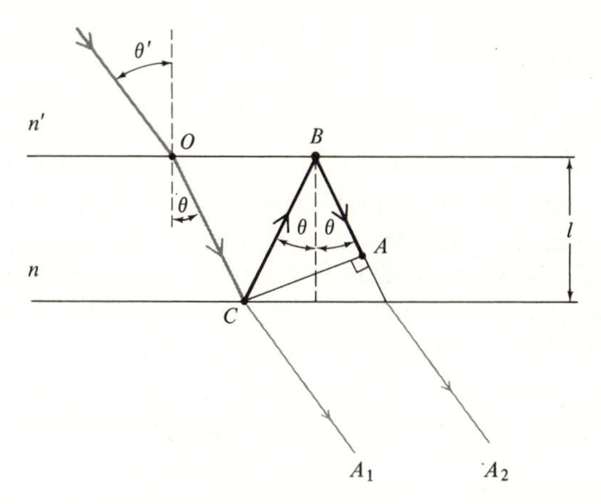

Figure 4-2 Two successive reflections, A_1 and A_2. Their path difference is given by

$$\delta L = AB + BC = l\frac{\cos 2\theta}{\cos\theta} + \frac{l}{\cos\theta} = 2l\cos\theta$$

$$\rightarrow \delta = \frac{2\pi(\delta L)n}{\lambda} = \frac{4\pi nl\cos\theta}{\lambda}$$

$B_1 + B_2 + B_3 + \cdots$, or

$$A_r = \{r + tt'r'e^{i\delta}(1 + r'^2e^{i\delta} + r'^4e^{2i\delta} + \cdots)\}A_i \tag{4.1-2}$$

For the transmitted wave,

$$A_1 = tt'A_i \qquad A_2 = tt'r'^2e^{i\delta}A_i \qquad A_3 = tt'r'^4e^{2i\delta}A_i$$

where a phase factor, exp $(i\delta)$, which corresponds to a single traversal of the plate and is common to all the terms, has been left out. Adding up the A terms, we obtain

$$A_t = tt'(1 + r'^2e^{i\delta} + r'^4e^{2i\delta} + \cdots) \tag{4.1-3}$$

for the complex amplitude of the total transmitted wave. We notice that the terms within the parentheses in (4.1-2) and (4.1-3) form an infinite geometric progression; adding them, we get

$$A_r = \frac{(1 - e^{i\delta})\sqrt{R}}{1 - Re^{i\delta}}A_i \tag{4.1-4}$$

and

$$A_t = \frac{T}{1 - Re^{i\delta}}A_i \tag{4.1-5}$$

where we used the fact that $r' = -r$, the conservation-of-energy relation that applies to lossless mirrors

$$r^2 + tt' = 1$$

as well as the definitions

$$R \equiv r^2 = r'^2 \qquad T \equiv tt'$$

R and T are, respectively, the fraction of the intensity reflected and transmitted at each interface and will be referred to in the following discussion as the mirrors' reflectance and transmittance.

If the incident intensity (watts per square meter) is taken as $A_i A_i^*$, we obtain from (4.1-4) the following expression for the fraction of the incident intensity that is reflected:

$$\frac{I_r}{I_i} = \frac{A_r A_r^*}{A_i A_i^*} = \frac{4R \sin^2 (\delta/2)}{(1 - R)^2 + 4R \sin^2 (\delta/2)} \tag{4.1-6}$$

Moreover, from (4.1-5),

$$\frac{I_t}{I_i} = \frac{A_t A_t^*}{A_i A_i^*} = \frac{(1 - R)^2}{(1 - R)^2 + 4R \sin^2 (\delta/2)} \tag{4.1-7}$$

for the transmitted fraction. Our basic model contains no loss mechanisms, so conservation of energy requires that $I_t + I_r$ be equal to I_i, as is indeed the case.

Let us consider the transmission characteristics of a Fabry–Perot etalon. According to (4.1-7) the transmission is unity whenever

$$\delta = \frac{4\pi n l \cos \theta}{\lambda} = 2m\pi \qquad m = \text{any integer} \tag{4.1-8}$$

Using (4.1-1), the condition (4.1-8) for maximum transmission can be written as

$$\nu_m = m \frac{c}{2nl \cos \theta} \qquad m = \text{any integer} \tag{4.1-9}$$

where $c = \nu\lambda$ is the velocity of light in vacuum and ν is the optical frequency. For a fixed l and θ, (4.1-9) defines the unity transmission (resonance) frequencies of the etalon. These are separated by the so-called free spectral range

$$\Delta\nu \equiv \nu_{m+1} - \nu_m = \frac{c}{2nl \cos \theta} \tag{4.1-10}$$

Theoretical transmission plots of a Fabry–Perot etalon are shown in Figure 4-3. The maximum transmission is unity, as stated previously. The minimum transmission, on the other hand, approaches zero as R approaches unity.

If we allow for the existence of losses in the etalon medium, we find that the peak transmission is less than unity. Taking the fractional intensity loss per pass as $(1 - A)$, we find that the maximum transmission drops from unity to

$$\left(\frac{I_t}{I_i}\right)_{\max} = \frac{(1 - R)^2 A}{(1 - RA)^2} \tag{4.1-11}$$

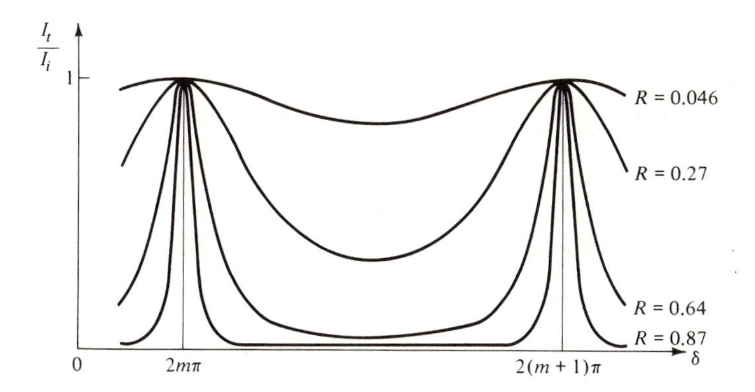

Figure 4-3 Transmission characteristics (theoretical) of a Fabry–Perot etalon. (After Reference[4].)

The proof of (4.1-11) is left as an exercise (Problem 4-2).

An experimental transmission plot of a Fabry–Perot etalon is shown in Figure 4-4.

4.2 Fabry–Perot Etalons as Optical Spectrum Analyzers

According to (4.1-8), the maximum transmission of a Fabry–Perot etalon occurs when

$$\frac{2nl\cos\theta}{\lambda} = m \tag{4.2-1}$$

Taking, for simplicity, the case of normal incidence ($\theta = 0°$), we obtain the following expression for the change $d\nu$ in the resonance frequency of a given transmission peak due to a length variation dl

$$\frac{d\nu}{\Delta\nu} = -\frac{dl}{(\lambda/2n)} \tag{4.2-2}$$

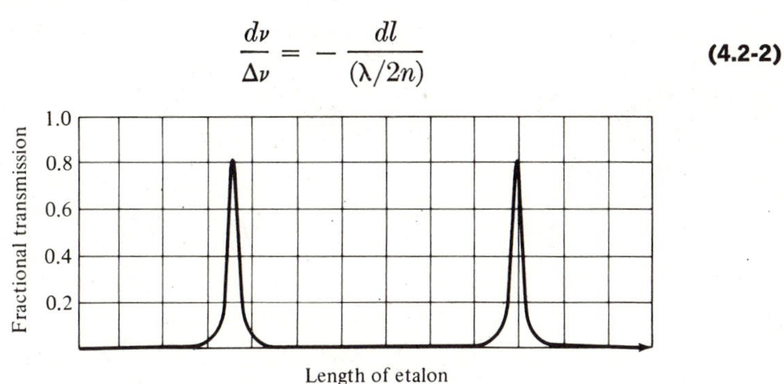

Figure 4-4 Experimental transmission characteristics of a Fabry–Perot etalon at 6328 Å as a function of the etalon optical length with $R = 0.9$ and $A = 0.98$. The two peaks shown correspond to a change in the optical length $\Delta(nl) = \lambda/2$. (After Reference [5].)

where $\Delta\nu$ is the intermode frequency separation as given by (4.1-10). According to (4.2-2), we can tune the peak transmission frequency of the etalon by $\Delta\nu$ by changing its length by half a wavelength. This property is utilized in operating the etalon as a scanning interferometer. The optical signal to be analyzed passes through the etalon as its length is being swept. If the width of the transmission peaks is small compared to that of the spectral detail in the incident optical beam signal, the output of the etalon will constitute a replica of the spectral profile of the signal. In this application it is important that the spectral width of the signal beam be smaller than the intermode spacing of the etalon ($c/2nl$) so that the ambiguity due to simultaneous transmission through more than one transmission peak can be avoided. For the same reason the total length scan is limited to $dl < \lambda/2n$. Figure 4-5 demonstrates the operation of a scanning Fabry–Perot etalon; Figure 4-6 shows intensity versus frequency data obtained by analyzing the output of a multimode He–Ne laser oscillating near 6328 Å. The peaks shown correspond to longitudinal laser modes, which will be discussed in Section 4.5.

It is clear from the foregoing that when operating as a spectrum analyzer the etalon resolution—that is, its ability to distinguish details in the spectrum—is limited by the finite width of its transmission peaks. If we take, somewhat arbitrarily,[2] the limiting resolution of the etalon as the separation $\Delta\nu_{1/2}$ between the two frequencies at which the transmission is down to half its peak value, from (4.1-7) we obtain

$$\sin^2\left(\frac{\delta_{1/2} - 2m\pi}{2}\right) = \frac{(1-R)^2}{4R} \tag{4.2-3}$$

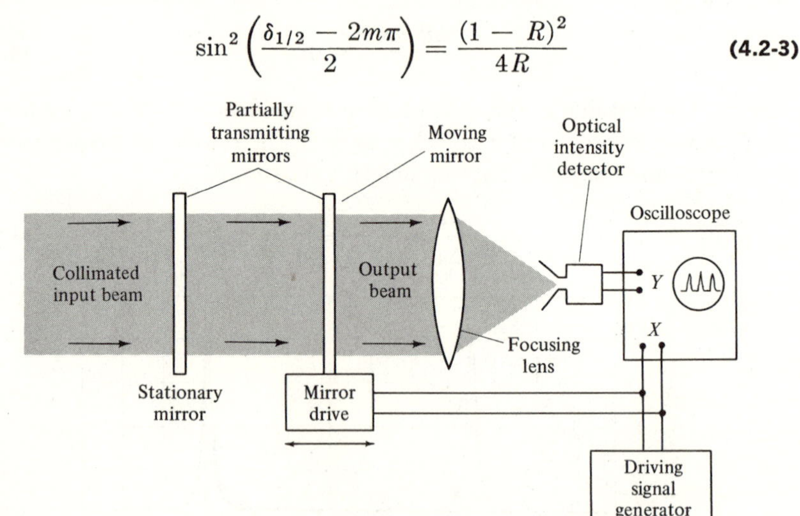

Figure 4-5 Typical scanning Fabry–Perot interferometer experimental arrangement.

---

[2] For a more complete discussion concerning the definition of resolution, see Reference [4].

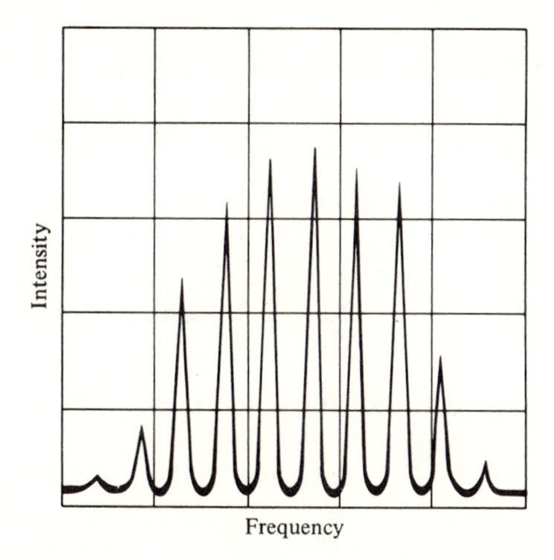

Figure 4-6 Intensity versus frequency analysis of the output of an He-Ne 6328 Å laser obtained with a scanning Fabry–Perot etalon. The horizontal scale is 250 MHz per division.

where $\delta_{1/2}$ is the value of δ corresponding to the two half-power points—that is, the value of δ at which the denominator of (4.1-7) is equal to $2(1 - R)^2$. If we assume $(\delta_{1/2} - 2m\pi) \ll \pi$, so that the width of the high-transmission regions in Figure 4-3 is small compared to the separation between the peaks, we obtain

$$\Delta\nu_{1/2} = \frac{c}{2\pi nl \cos\theta}(\delta_{1/2} - 2m\pi) \simeq \frac{c}{2\pi nl \cos\theta}\,\frac{1 - R}{\sqrt{R}} \tag{4.2-4}$$

Or using (4.1-10) and defining the etalon finesse as

$$F \equiv \frac{\pi\sqrt{R}}{1 - R} \tag{4.2-5}$$

we obtain

$$\Delta\nu_{1/2} = \frac{\Delta\nu}{F} = \frac{c}{2nl \cos\theta F} \tag{4.2-6}$$

for the limiting resolution. The finesse F (which is used as a measure of the resolution of Fabry–Perot etalon) is, according to (4.2-6), the ratio of the separation between peaks to the width of a transmission bandpass. This ratio can be read directly from the transmission characteristics such as those of Figure 4-4, for which we obtain $F \simeq 26$.

Numerical example—design of a Fabry–Perot etalon. Consider the problem of designing a scanning Fabry–Perot etalon to be used in studying the mode structure of a He–Ne laser with the following characteristics: $l_{\text{laser}} = 100$ cm and the region of oscillation $= \Delta\nu_{\text{gain}} \simeq 1.5 \times 10^9$ Hz.

The free spectral range of the etalon (that is, its intermode spacing) must exceed the spectral region of interest, so from (4.1-10) we obtain

$$\frac{c}{2nl_{\text{etal}}} \geqslant 1.5 \times 10^9 \qquad \text{or} \qquad 2nl_{\text{etal}} \leqslant 20 \text{ cm} \tag{4.2-7}$$

The separation between longitudinal modes of the laser oscillator is $c/2nl_{\text{laser}} = 1.5 \times 10^8$ Hz (here we assume $n = 1$). We choose the resolution of the etalon to be a tenth of this value, so spectral details as narrow as 1.5×10^7 Hz can be resolved. According to (4.2-6), this resolution can be achieved if

$$\Delta\nu_{1/2} = \frac{c}{2nl_{\text{etal}}F} \leqslant 1.5 \times 10^7 \qquad \text{or} \qquad 2nl_{\text{etal}}F \geqslant 2 \times 10^3 \tag{4.2-8}$$

To satisfy condition (4.2-7), we choose $2nl_{\text{etal}} = 20$ cm; thus (4.2-8) is satisfied when

$$F \geqslant 100 \tag{4.2-9}$$

A finesse of 100 requires, according to (4.2-5), a mirror reflectivity of approximately 97 percent.

As a practical note we may add that the finesse, as defined by the first equality in (4.2-6), depends not only on R but also on the mirror flatness and the beam angular spread. These points are taken up in Problems 4-2 and 4-4.

Another important mode of optical spectrum analysis performed with Fabry–Perot etalons involves the fact that a noncollimated monochromatic beam incident on the etalon will emerge simultaneously, according to (4.1-8), along many directions θ,[3] which correspond to the various orders m. If the output is then focused by a lens, each such direction θ will give rise to a circle in the focal plane of the lens, and therefore each frequency component present in the beam leads to a family of circles. This mode of spectrum analysis is especially useful under transient conditions where scanning etalons cannot be employed. Further discussion of this topic is included in Problem 4-6.

4.3 Optical Resonators with Spherical Mirrors

In this section we study the properties of optical resonators formed by two opposing spherical mirrors; see References [6] and [7]. We will show that the field solutions inside the resonators are those of the propagating Gaussian beams, which were considered in Chapter 3. It is, consequently, useful to start by reviewing the properties of these beams.

The field distribution corresponding to the (m, n) transverse mode is

---

[3] Each direction θ corresponds in three dimensions to the surface of a cone with a half-apex angle θ.

given, according to (3.5-1), by

$$E_{l,m}^{(x)}(\mathbf{r}) = E_0 \frac{\omega_0}{\omega(z)} H_l\left(\sqrt{2}\frac{x}{\omega(z)}\right) H_m\left(\sqrt{2}\frac{y}{\omega(z)}\right) \times \exp\left[-\frac{x^2+y^2}{\omega^2(z)} - ik\frac{x^2+y^2}{2R(z)} - ikz + i(l+m+1)\eta\right]$$

where the spot size $\omega(z)$ is **(4.3-1)**

$$\omega(z) = \omega_0\left[1+\left(\frac{z}{z_0}\right)^2\right]^{1/2} \qquad z_0 = \frac{\pi\omega_0{}^2 n}{\lambda} \tag{4.3-2}$$

and where ω_0, the minimum spot size, is a parameter characterizing the beam. The radius of curvature of the wavefronts is

$$R(z) = z\left[1+\left(\frac{\pi\omega_0{}^2 n}{\lambda z}\right)^2\right] = \frac{1}{z}[z^2+z_0{}^2] \tag{4.3-3}$$

and the phase factor η is as follows:

$$\eta = \tan^{-1}\left(\frac{\lambda z}{\pi\omega_0{}^2 n}\right) \tag{4.3-4}$$

The sign of $R(z)$ is taken as positive when the center of curvature is to the left of the wavefront, and vice versa. According to (4.3-1) and (4.3-2) the loci of the points at which the beam intensity (watts per square meter) is a given fraction of its intensity on the axis are the hyperboloids

$$x^2 + y^2 = \text{const.} \times \omega^2(z) \tag{4.3-5}$$

The hyperbolas generated by the intersection of these surfaces with planes that include the z axis are shown in Figure 4-7. These hyperbolas are normal to the phase fronts and thus correspond to the local direction of energy flow. The hyperboloid $x^2 + y^2 = \omega^2(z)$ is, according to (4.3-1), the locus of the points where the exponential factor in the field amplitude

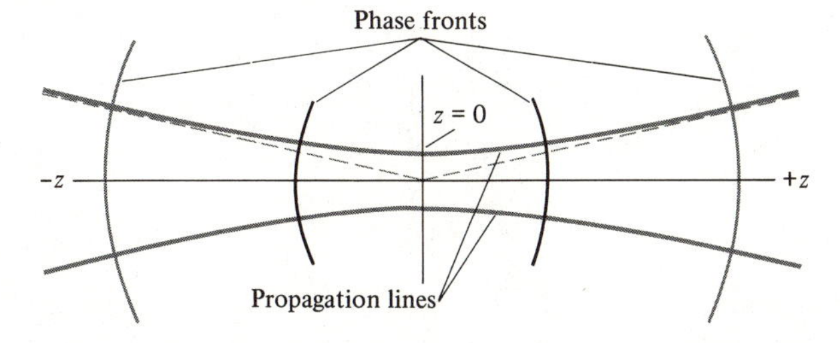

Figure 4-7 Hyperbolic curves corresponding to the local directions of propagation. The nearly spherical phase fronts represent possible positions for reflectors. Any two reflectors form a resonator with a transverse field distribution given by (4.3-1)

is down to e^{-1} from its value on the axis. The quantity $\omega(z)$ is thus defined as the *mode spot size* at the plane z.

Given a beam of the type described by (4.3-1) we can form an optical resonator merely by inserting at points z_1 and z_2 two reflectors with radii of curvature that match those of the propagating beam spherical phase fronts at these points. Since the surfaces are normal to the direction of energy propagation as shown in Figure 4-7, the reflected beam retraces itself; thus, if the phase shift between the mirrors is some multiple of 2π radians, a *self-reproducing stable field* configuration results.

Alternatively, given two mirrors with spherical radii of curvature R_1 and R_2 and some distance of separation l, we can, under certain conditions to be derived later, adjust the position $z = 0$ and the parameter ω_0 so that the mirrors coincide with two spherical wavefronts of the propagating beam defined by the position of the waist ($z = 0$) and ω_0. If, in addition, the mirrors can be made large enough to intercept the majority (99 percent, say) of the incident beam energy in the fundamental ($l = m = 0$) transverse mode, we may expect this mode to have a larger Q than higher order transverse modes, which, according to Figure 3-3, have fields extending farther from the axis and consequently lose a larger fraction of their energy by "spilling" over the mirror edges (diffraction losses).

Optical resonator algebra. As mentioned in the preceding paragraphs, we can form an optical resonator by using two reflectors, one at z_1 and the other at z_2, chosen so that their radii of curvature are the same as those of the beam wavefronts at the two locations. The propagating beam mode (4.3-1) is then reflected back and forth between the reflectors without a change in its transverse profile. The requisite radii of curvature, determined by (4.3-3), are

$$R_1 = +z_1 + \frac{z_0^2}{z_1}$$

$$R_2 = +z_2 + \frac{z_0^2}{z_2}$$

from which we get

$$\begin{aligned} z_1 &= +\frac{R_1}{2} \pm \frac{1}{2}\sqrt{R_1^2 - 4z_0^2} \\ z_2 &= +\frac{R_2}{2} \pm \frac{1}{2}\sqrt{R_2^2 - 4z_0^2} \end{aligned} \tag{4.3-6}$$

For a given minimum spot size $\omega_0 = (\lambda z_0/\pi n)^{1/2}$, we can use (4.3-6) to find the positions z_1 and z_2 at which to place mirrors with curvatures R_1 and R_2, respectively. In practice, we often start with given mirror curvatures R_1 and R_2 and a mirror separation l. The problem is then to find the minimum spot size ω_0, its location with respect to the reflectors, and the mirror spot sizes ω_1 and ω_2. Taking the mirror spacing as $l = z_2 - z_1$, we can solve (4.3-6) for z_0^2, obtaining

$$z_0^2 = \frac{l(-R_1 - l)(R_2 - l)(R_2 - R_1 - l)}{(R_2 - R_1 - 2l)^2} \tag{4.3-7}$$

where z_2 is to the right of z_1 (so that $l = z_2 - z_1 > 0$) and the mirror curvature is taken as positive when the center of curvature is to the left of the mirror.

The minimum spot size is $\omega_0 = (\lambda z_0/\pi n)^{1/2}$ and its position is next determined from (4.3-6). The mirror spot sizes are then calculated by the use of (4.3-2).

The symmetrical mirror resonator. The special case of a resonator with symmetrically (about $z = 0$) placed mirrors merits a few comments. The planar phase front at which the minimum spot size occurs is, by symmetry, at $z = 0$. Putting $R_2 = -R_1 = R$ in (4.3-7) gives

$$z_0^2 = \frac{(2R - l)l}{4} \tag{4.3-8}$$

and

$$\omega_0 = \left(\frac{\lambda z_0}{\pi n}\right)^{1/2} = \left(\frac{\lambda}{\pi n}\right)^{1/2} \left(\frac{l}{2}\right)^{1/4} \left(R - \frac{l}{2}\right)^{1/4} \tag{4.3-9}$$

which, when substituted in (4.3-2) with $z = l/2$, yields the following expression for the spot size at the mirrors:

$$\omega_{1,2} = \left(\frac{\lambda l}{2\pi n}\right)^{1/2} \left[\frac{2R^2}{l(R - l/2)}\right]^{1/4} \tag{4.3-10}$$

A comparison with (4.3-9) shows that, for $R \gg l$, $\omega \simeq \omega_0$ and the beam spread inside the resonator is small.

The value of R (for a given l) for which the mirror spot size is a minimum, is readily found from (4.3-10) to be $R = l$. When this condition is fulfilled we have what is called a symmetrical *confocal resonator*, since the two foci, occurring at a distance of $R/2$ from the mirrors, coincide. From (4.3-8) and the relation $\omega_0 = (\lambda z_0/\pi n)^{1/2}$ we obtain

$$(\omega_0)_{\text{conf}} = \left(\frac{\lambda l}{2\pi n}\right)^{1/2} \tag{4.3-11}$$

whereas from (4.3-10) we get

$$(\omega_{1,2})_{\text{conf}} = (\omega_0)_{\text{conf}} \sqrt{2} \tag{4.3-12}$$

so the beam spot size increases by $\sqrt{2}$ between the center and the mirrors.

Numerical example—design of a symmetrical resonator. Consider the problem of designing a symmetrical resonator for $\lambda = 10^{-4}$ cm with a mirror separation $l = 2m$. If we were to choose the confocal geometry with $R = l = 2m$, the minimum spot size (at the resonator center) would be, from (4.3-11) and for $n = 1$

$$(\omega_0)_{\text{conf}} = \left(\frac{\lambda l}{2\pi n}\right)^{1/2} = 0.0564 \text{ cm}$$

whereas, using (4.3-12), the spot size at the mirrors would have the value

$$(\omega_{1,2})_{\text{conf}} = \omega_0\sqrt{2} \simeq 0.0798 \text{ cm}$$

Assume next that a mirror spot size $\omega_{1,2} = 0.3$ cm is desired. Using this value in (4.3-10) and assuming $R \gg l$, we get

$$\frac{\omega_{1,2}}{(\lambda l n/2\pi)^{1/2}} = \frac{0.3}{0.056} = \left(\frac{2R}{l}\right)^{1/4}$$

whence

$$R \simeq 400l \simeq 799 \text{ meters}$$

so that the assumption $R \gg l$ is valid. The minimum beam spot size ω_0 is found, through (4.3-2) and (4.3-8), to be

$$\omega_0 = 0.994\omega_{1/2} \simeq 0.3 \text{ cm}$$

Thus, to increase the mirror spot size from its minimum (confocal) value of 0.0798 cm to 0.3 cm, we must use exceedingly plane mirrors ($R = 799$ meters). This also shows that even small mirror curvatures (that is, large R) give rise to "narrow" beams.

The numerical example we have worked out applies equally well to the case in which a plane mirror is placed at $z = 0$. The beam pattern is equal to that existing in the corresponding half of the symmetric resonator in the example, so the spot size on the planar reflector is ω_0.

4.4 Mode Stability Criteria

The ability of an optical resonator to support low (diffraction) loss[4] modes depends on the mirrors' separation l and their radii of curvature R_1 and R_2. To illustrate this point, consider first the symmetric resonator with $R_2 = R_1 = R$.

The ratio of the mirror spot size at a given l/R to its minimum confocal ($l/R = 1$) value, given by the ratio of (4.3-10) to (4.3-12), is

$$\frac{\omega_{1,2}}{\omega_{\text{conf}}} = \left[\frac{1}{(l/R)[2 - (l/R)]}\right]^{1/4} \quad \textbf{(4.4-1)}$$

This ratio is plotted in Figure 4-8. For $l/R = 0$ (plane-parallel mirrors) and for $l/R = 2$ (two concentric mirrors), the spot size becomes infinite.

[4] By diffraction loss we refer to the fact that due to the beam spread (see 3.2-18), a fraction of the Gaussian beam energy "misses" the mirror and is not reflected and is thus lost.

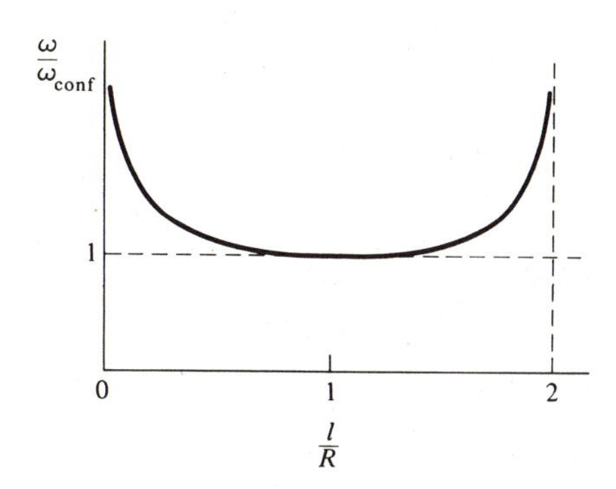

Figure 4-8 Ratio of beam spot size at the mirrors of a symmetrical resonator to its confocal ($l/R = 1$) value.

It is clear that the diffraction losses for these cases are very high, since most of the beam energy "spills over" the reflector edges. Since, according to Table 2.1, the reflection of a Gaussian beam from a mirror with a radius of curvature R is formally equivalent to its transmission through a lens with a focal length $f = R/2$, the problem of the existence of stable confined optical modes in a resonator is formally the same as that of the existence of stable solutions for the propagation of a Gaussian beam in a biperiodic lens sequence, as shown in Figure 4-9. This problem was considered in Section 3.3 and led to the stability condition (3.4-5).

If, in (3.4-5), we replace f_1 by $R_1/2$ and f_2 by $R_2/2$,[5] we obtain the

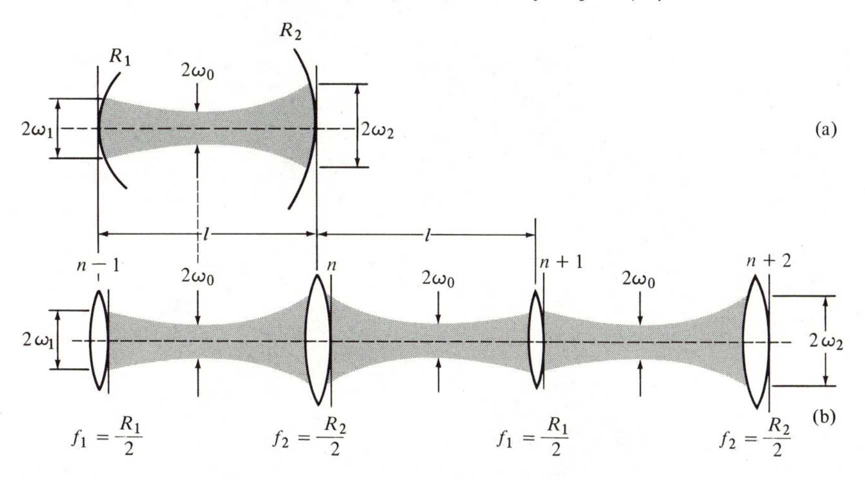

Figure 4-9 (**a**) Asymmetric resonator ($R_1 \neq R_2$) with mirror curvatures R_1 and R_2. (**b**) Biperiodic lens system (lens waveguide) equivalent to resonator shown in (**a**).

[5] This causes the sign convention of R_1 and R_2 to be different from that used in the preceding sections. The sign of R is the same as that of the focal length of the equivalent lens. This makes R_1 (or R_2) positive when the center of curvature of mirror 1 (or 2) is in the direction of mirror 2 (or 1), and negative otherwise.

stability condition for optical resonators

$$0 \leqslant \left(1 - \frac{l}{R_1}\right)\left(1 - \frac{l}{R_2}\right) \leqslant 1 \tag{4.4-2}$$

A convenient representation of the stability condition (4.4-2) is by means of the diagram [7] shown in Figure 4-10. From this diagram, for example, it can be seen that the symmetric concentric ($R_1 = R_2 = l/2$), confocal ($R_1 = R_2 = l$), and the plane-parallel ($R_1 = R_2 = \infty$) resonators are all on the verge of instability and thus may become extremely lossy by small deviations of the parameters in the direction of instability.

4.5 Modes in a Generalized Resonator—The Self-Consistent Method

Up to this point we have treated resonators consisting of two opposing spherical mirrors. We may, sometimes, wish to consider the properties of

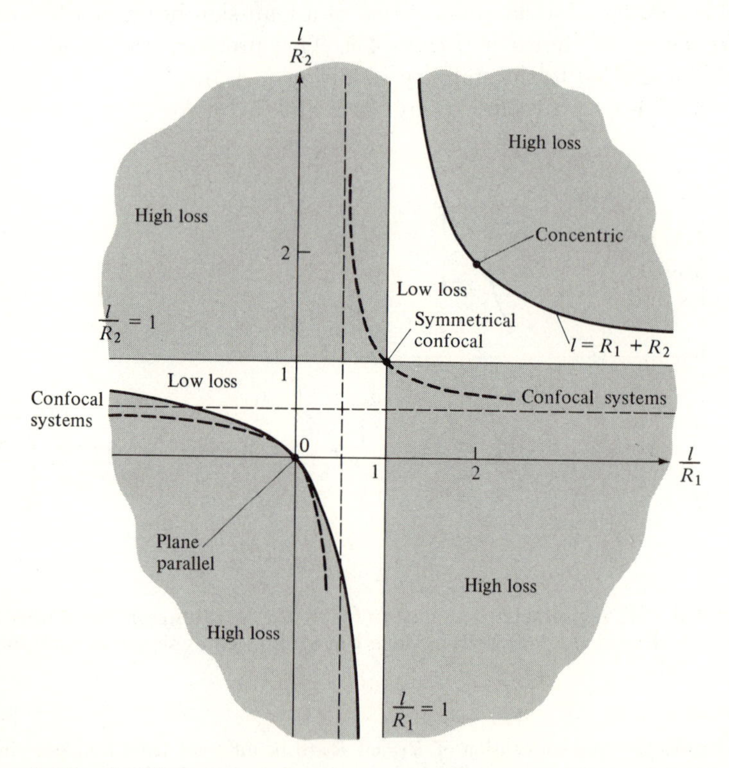

Figure 4-10 Stability diagram of optical resonator. Shaded (high-loss) areas are those in which the stability condition $0 \leq (1 - l/R_1)(1 - l/R_2) \leq 1$ is violated and the clear (low-loss) areas are those in which it is fulfilled. The sign convention for R_1 and R_2 is discussed in Footnote 5. (After Reference [7].)

more complex resonators made up of an arbitrary number of lenslike elements such as those shown in Table 2-1. A simple case of such a resonator may involve placing a lens between two spherical reflectors or constructing an off-axis three-reflector resonator. Yet another case is that of a traveling wave resonator in which the beam propagates in one sense only.

In either of these cases we need to find if low-loss (that is, "stable") modes exist in the complex resonator, and if so, to solve for the spot size $\omega(z)$ and the radius of curvature $R(z)$ everywhere.

We apply the self-consistency condition and require that a stable eigenmode of the resonator is one that *reproduces itself after one round trip.* We choose an *arbitrary* reference plane in the resonator, denote the steady-state complex beam parameter at this plane as q_s, and, using the ABCD law (3.3-9), require that

$$q_s = \frac{Aq_s + B}{Cq_s + D} \tag{4.5-1}$$

where A, B, C, D are the "ray" matrix elements for one complete round trip—starting and ending at the chosen reference plane.

Solving (4.5-1) for $1/q_s$ gives

$$\frac{1}{q_s} = \frac{(D - A) \pm \sqrt{(D - A)^2 + 4BC}}{2B} \tag{4.5-2}$$

since the individual elements in the resonator are described by unimodular matrices, that is, $A_iD_i - B_iC_i = 1$ (see Table 2-1), it follows that the matrix A, B, C, D, which is the product of individual matrices, satisfies

$$AD - BC = 1$$

and (4.5-2) can, consequently, be written as

$$\frac{1}{q_s} = \frac{D - A}{2B} \pm i\frac{\sqrt{1 - [(D + A)/2]^2}}{B} = \frac{D - A}{2B} + \frac{i \sin \theta}{B} \tag{4.5-3}$$

where

$$\cos \theta = \frac{D + A}{2}$$

$$\theta = \pm \left| \cos^{-1}\left(\frac{D + A}{2}\right) \right| \tag{4.5-4}$$

According to (3.2-14) the condition for a confined Gaussian beam is that the square of the beam spot size ω^2 be a finite positive number. Recalling that

$$\frac{1}{q} = \frac{1}{R} - i\frac{\lambda}{\pi\omega^2 n} \tag{4.5-4[a]}$$

we find by comparing the last expression to (4.5-3) that the condition for a confined beam is satisfied by choosing $\theta < 0$ in (4.5-4) provided

$$\left|\frac{D+A}{2}\right| < 1 \tag{4.5-5}$$

and the steady-state beam parameter is

$$\frac{1}{q_s} = \frac{D-A}{2B} - i\frac{\sqrt{1-[(D+A)/2]^2}}{B} = \frac{D-A}{2B} + \frac{i\sin\theta}{B} \qquad \theta < 0 \tag{4.5-6}$$

Equation (4.5-5) can thus be viewed as the generalization of the stability condition (4.4-2) to the case of an arbitrary resonator. When applied to a resonator composed of two spherical reflectors, it reduces to (4.4-2).

The radius of curvature R and the spot size ω at the reference plane are obtained from (4.5-6) by using (4.5-4a)

$$R = \frac{2B}{D-A}$$
$$\omega = \left(\frac{\lambda}{\pi n}\right)^{1/2} \frac{B^{1/2}}{[1-[(D+A)/2]^2]^{1/4}} \tag{4.5-7}$$

The complex beam parameter q, and hence ω and R, at any other plane can be obtained by applying the ABCD law (3.3-9) to q_s.

Stability of the resonator modes. The treatment just concluded dealt with the existence of steady-state (self-reproducing) resonator modes. Having found that such modes do exist, we need to inquire whether the modes are stable. This can be done by perturbing the steady-state solution $1/q_s$ as given by (4.5-6) and following the evolution of the perturbation with propagation [8].

We start with (3.3-9), which relates the beam parameter q_{out} to the beam parameter q_{in}, after one round trip

$$q_{\text{out}} = \frac{Aq_{\text{in}} + B}{Cq_{\text{in}} + D}$$

where A, B, C, D are the ray matrix elements for one complete round trip inside the optical resonator. Rewriting the last expression as

$$q_{\text{out}}^{-1} = \frac{C + Dq_{\text{in}}^{-1}}{A + Bq_{\text{in}}^{-1}} \tag{4.5-8}$$

we obtain by differentiation

$$\begin{aligned}\frac{dq_{\text{out}}^{-1}}{dq_{\text{in}}^{-1}} &= \frac{D - \left[(C + Dq_{\text{in}}^{-1})/(A + Bq_{\text{in}}^{-1})\right] B}{A + Bq_{\text{in}}^{-1}} \\ &= \frac{D - Bq_{\text{out}}^{-1}}{A + Bq_{\text{in}}^{-1}}\end{aligned} \tag{4.5-9}$$

At steady state $q_{\text{out}} = q_{\text{in}} \equiv q_s$

$$\left.\frac{dq_{\text{out}}^{-1}}{dq_{\text{in}}^{-1}}\right|_{q_{\text{in}}=q_s} = \frac{D - Bq_s^{-1}}{A + Bq_s^{-1}} \tag{4.5-10}$$

Using (4.5-6) we obtain

$$D - Bq_s^{-1} = \frac{D + A}{2} - i \sin\theta = e^{-i\theta}$$

$$A + Bq_s^{-1} = \frac{D + A}{2} + i \sin\theta = e^{i\theta}$$

so that

$$\left.\frac{dq_{\text{out}}^{-1}}{dq_{\text{in}}^{-1}}\right|_{q_{\text{in}}=q_s} = \bar{e}^{2i\theta} \tag{4.5-11}$$

Because confined modes require, according to (4.5-4) and (4.5-5), that θ be real, it follows from (4.5-11) that a small perturbation $\Delta q_{\text{in}}^{-1}$ of the beam parameter q^{-1} from the steady-state value q_s^{-1} does not decay, since the perturbation after one round trip $(\Delta q_{\text{out}}^{-1})$ satisfies

$$|\Delta q_{\text{out}}^{-1}| = |\Delta q_{\text{in}}^{-1}| \tag{4.5-12}$$

We thus find that the theory predicts that mode perturbations in Gaussian mode resonators do not decay. This does not agree with experience, which shows that the mode characteristics of laser oscillators are highly stable, thus implying a strong perturbational decay, that is, $|\Delta q_{\text{out}}^{-1}| < |\Delta q_{\text{in}}^{-1}|$. The discrepancy is resolved if we include in the analysis leading to (4.5-11) the fact that the resonator mirrors are of finite extent. This point is considered in Appendix A.

4.6 Resonance Frequencies of Optical Resonators

Up to this point we have considered only the dependence of the spatial mode characteristics on the resonator mirrors (their radii of curvature and separation). Another important consideration is that of determining the resonance frequencies of a given spatial mode.

The frequencies are determined by the condition that the complete round-trip phase delay of a resonant mode be some multiple of 2π. This requirement is equivalent to that in microwave waveguide resonators where the resonator length must be equal to an integral number of half-guide wavelengths [1]. This requirement makes it possible for a stable standing wave pattern to establish itself along the axis with a transverse field distribution equal to that of the propagating mode.

If we consider a spherical mirror resonator with mirrors at z_2 and z_1, the resonance condition for the l, m mode can be written as[6]

$$\eta_{l,m}(z_2) - \eta_{l,m}(z_1) = q\pi \tag{4.6-1}$$

where q is some integer and $\eta_{l,m}(z)$, the phase shift, is given according to (3.5-2) by

$$\eta_{l,m}(z) = kz - (l + m + 1)\tan^{-1}\frac{z}{z_0}$$
$$(z_0 = \pi\omega_0^2 n/\lambda) \tag{4.6-2}$$

The resonance condition (4.6-1) is thus

$$k_q d - (l + m + 1)\left(\tan^{-1}\frac{z_2}{z_1} - \tan^{-1}\frac{z_1}{z_0}\right) = q\pi \tag{4.6-3}$$

where $d = z_2 - z_1$ is the resonator length. It follows that

$$k_{q+1} - k_q = \frac{\pi}{d}$$

or, using $k = 2\pi\nu n/c$,

$$\nu_{q+1} - \nu_q = \frac{c}{2nd} \tag{4.6-4}$$

for the intermode frequency spacing.

Let us consider, next, the effect of varying the transverse mode indices l and m in a mode with a fixed q. We notice from (4.6-3) that the resonant frequencies depend on the sum $(l + m)$ and not on l and m separately, so for a given q all the modes with the same value of $l + m$ are degenerate (that is, they have the same resonance frequencies). Considering (4.6-3) at two different values of $l + m$ gives

$$k_1 d - (l + m - 1)_1\left(\tan^{-1}\frac{z_2}{z_0} - \tan^{-1}\frac{z_1}{z_0}\right) = q\pi$$

$$k_2 d - (l + m + 1)_2\left(\tan^{-1}\frac{z_2}{z_0} - \tan^{-1}\frac{z_1}{z_0}\right) = q\pi$$

[6] In obtaining (4.6-1) we did not allow for the phase shift upon reflection. This correction does not affect any of the results of this section.

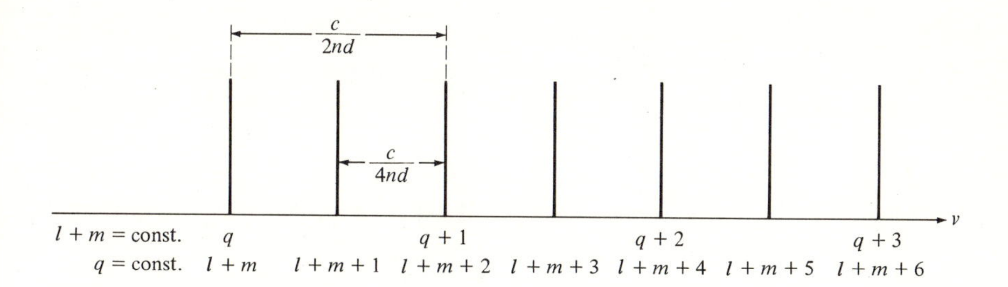

Figure 4-11 Position of resonance frequencies of a confocal ($d = R$) optical resonator as a function of the mode indices l, m, and q.

and, by subtraction,

$$(k_1 - k_2)d = [(l + m + 1)_1 - (l + m + 1)_2]\left(\tan^{-1}\frac{z_2}{z_0} - \tan^{-1}\frac{z_1}{z_0}\right) \quad \text{(4.6-5)}$$

and

$$\Delta\nu = \frac{c}{2\pi nd}\,\Delta(l + m)\left(\tan^{-1}\frac{z_2}{z_0} - \tan^{-1}\frac{z_1}{z_0}\right) \quad \text{(4.6-6)}$$

for the change $\Delta\nu$ in the resonance frequency caused by a change $\Delta(l + m)$ in the sum $(l + m)$. As an example, in the case of a confocal resonator ($R = d$) we have, according to (4.3-6), $z_2 = -z_1 = z_0$; therefore, $\tan^{-1}(z_2/z_0) = -\tan^{-1}(z_1/z_0) = \pi/4$, and (4.6-6) becomes

$$\Delta\nu_{\text{conf}} = \frac{1}{2}\,[\Delta(l + m)]\,\frac{c}{2nd} \quad \text{(4.6-7)}$$

Comparing (4.6-7) to (4.6-4) we find that in the confocal resonator the resonance frequencies of the transverse modes, resulting from changing l and m, either coincide or fall halfway between those which result from a change of the longitudinal mode index q. This situation is depicted in Figure 4-11.

To see what happens to the transverse resonance frequencies (that is, those due to a variation of l and m) in a confocal resonator, we may consider the nearly planar resonator in which $|z_1|$ and z_2 are small compared to z_0 (that is, $d \ll R_1$ and R_2). In this case, (4.6-6) becomes

$$\Delta\nu \simeq \frac{c}{2\pi n z_0}\,\Delta(l + m) \quad \text{(4.6-8)}$$

The mode grouping for this case is illustrated in Figure 4-12.

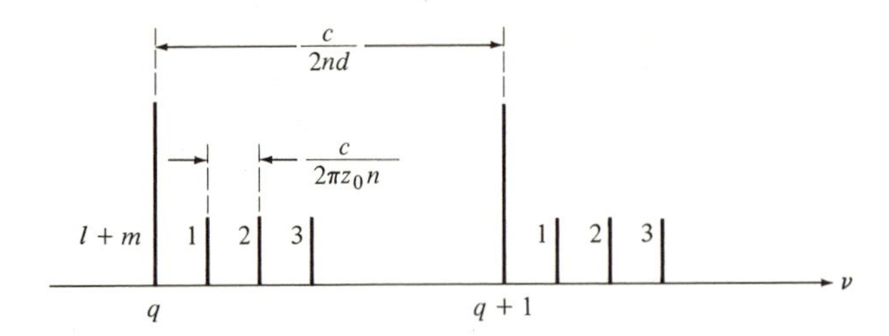

Figure 4-12 Resonant frequencies of a near-planar ($R \gg d$) optical resonator as a function of the mode indices l, m, and q.

The situation depicted in Figure 4-12 is highly objectionable if the resonator is to be used as a scanning interferometer. The reason is that in reconstructing the spectral profile of the unknown signal, an ambiguity is caused by the simultaneous transmission of more than one frequency. This ambiguity is resolved by using a confocal etalon whose mode spacing is as shown in Figure 4-11 and by choosing d to be small enough that the intermode spacing $c/4nd$ exceeds the width of the spectral region that is scanned.

4.7 Losses in Optical Resonators

An understanding of the mechanisms by which electromagnetic energy is dissipated in optical resonators and the ability to control them are of major importance in understanding and operating a variety of optical devices. For historical reasons as well as for reasons of convenience, these losses are often characterized by a number of different parameters. This book uses, in different places, the concepts of loss per pass, photon lifetime, and quality factor Q to describe losses in resonators. Let us see how these quantities are related to each other.

The decay lifetime (photon lifetime) t_c of a cavity mode is defined by means of the equation

$$\frac{d\mathcal{E}}{dt} = -\frac{\mathcal{E}}{t_c} \tag{4.7-1}$$

where $\mathcal{E}$ is the energy stored in the mode. If the fractional (intensity) loss per pass is L and the length of the resonator is l, then the fractional loss per unit time is cL/nl; therefore

$$\frac{d\mathcal{E}}{dt} = -\frac{cL}{nl}\mathcal{E}$$

and, from (4.7-1),

$$t_c = \frac{nl}{cL} \tag{4.7-2}$$

for the case of a resonator with mirrors' reflectivities R_1 and R_2 and an average distributed loss constant α, the average loss per pass is $L = \alpha l - ln\sqrt{R_1R_2}$ so that

$$t_c = \frac{n}{c[\alpha - (1/l)\, ln\, \sqrt{R_1R_2}]} \approx \frac{nl}{c[\alpha l + (1 - \sqrt{R_1R_2})]} \tag{4.7-3}$$

where the approximate equality applies when $R_1R_2 \approx 1$.

The quality factor of the resonator is defined universally as

$$Q = \frac{\omega \mathcal{E}}{P} = -\frac{\omega \mathcal{E}}{d\mathcal{E}/dt} \tag{4.7-4}$$

where $\mathcal{E}$ is the stored energy, ω is the resonant frequency, and $P = -d\mathcal{E}/dt$ is the power dissipated. By comparing (4.7-4) and (4.7-1) we obtain

$$Q = \omega t_c \tag{4.7-5}$$

The Q factor is related to the full width $\Delta\nu_{1/2}$ (at the half-power points) of the resonator's Lorentzian response curve as ([4] and Section 5.1).

$$\Delta\nu_{1/2} = \frac{\nu}{Q} = \frac{1}{2\pi t_c} \tag{4.7-6}$$

so that, according to (4.7-3)

$$\Delta\nu_{1/2} = \frac{c[\alpha - (1/l)\ \ln\ \sqrt{R_1R_2}]}{2\pi n} \tag{4.7-7}$$

The most common loss mechanisms in optical resonators are the following.

1. *Loss resulting from nonperfect reflection.* Reflection loss is unavoidable, since without some transmission no power output is possible. In addition, no mirror is ideal; and even when mirrors are made to yield the highest possible reflectivities, some residual absorption and scattering reduce the reflectivity to somewhat less than 100 percent.
2. *Absorption and scattering in the laser medium.* Transitions from some of the atomic levels, which are populated in the process of pumping, to higher lying levels constitute a loss mechanism in optical resonators when they are used as laser oscillators. Scattering from inhomogeneities and imperfections is especially serious in solid-state laser media.
3. *Diffraction losses.* From (4.3-1) or from Figure 3-3, we find that the energy of propagating-beam modes extends to considerable distances from the axis. When a resonator is formed by "trapping" a propagating beam between two reflectors, it is clear that for

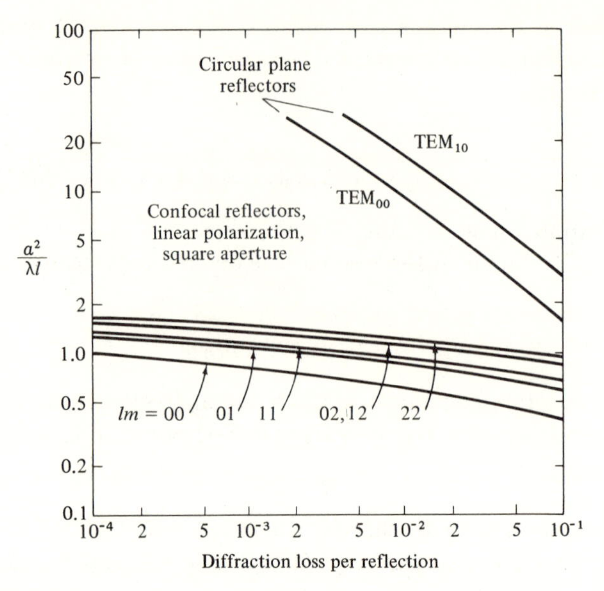

Figure 4-13 Diffraction losses for a plane-parallel and several low-order confocal resonators; a is the mirror radius and l is their spacing. The pairs of numbers under the arrows refer to the transverse-mode indices l, m. (After Reference [6].)

finite-dimension reflectors some of the beam energy will not be intercepted by the mirrors and will therefore be lost. For a given set of mirrors this loss will be greater, the higher the transverse mode indices l, m, since in this case the energy extends farther. This fact is used to prevent the oscillation of higher order modes by inserting apertures into the laser resonator whose opening is large enough to allow most of the fundamental $(0, 0, q)$ mode energy through but small enough to increase substantially the losses of the higher order modes. Figure 4-13 shows the diffraction losses of a number of low-order confocal resonators. Of special interest is the dramatic decrease of the diffraction losses that results from the use of spherical reflectors instead of the plane-parallel ones.

4.8 Unstable Optical Resonators

Optical resonators with parameters falling within the shaded regions of Figure 4-10—that is, those violating condition 4.4-2—were found to be unstable. This instability was demonstrated in Section 4.4 as manifesting itself in a sudden and steep increase of the beam spot size at the mirrors as the resonator parameters approach the unstable regime. This leads to large diffraction losses.

A number of important laser applications exist where the large diffraction losses attendant upon operation in the unstable region are acceptable or even desirable. Some of the reasons are:

1. Operation in the stable regime has been shown (see example of Section 4.3) to lead to narrow Gaussian beams. This situation is not compatible with the need for high-power output that requires large lasing volumes.
2. The losses in unstable resonators are dominated by diffraction (that is, beam power "missing" the reflectors) and are thus desirable in situations where the high gain prescribes large output coupling ratios (see Chapter 6).
3. The nature of the coupling results in an output beam with a large aperture that is consequently well collimated without the use of telescoping optics. This point will be made clear by the following discussion.

The more sophisticated theoretical analyses of this problem make use of Huygen's integral method to derive the diffraction losses and field distribution of the modes of the unstable resonator. In the following brief treatment we will use a geometrical output analysis advanced by Siegman [9] that emphasizes the essential physical characteristics of the resonator and that yields results in fair agreement with experiments.

Referring to Figure 4-14 we assume that the right-going wave leaving mirror M_1 is a spherical wave originating in a virtual center P_1 that is not, in general, the center of curvature of mirror M_1. This wave is incident on M_2, from which a fraction of the original intensity is reflected as a uniform spherical wave coming from a virtual center P_2. For self-consistency this wave, then, is reflected from M_1 as if it originated at P_1. The self-consistency condition is satisfied if the virtual image of P_1 upon reflection from M_2 is at P_2 and vice versa.

Applying the imaging formulas of geometrical optics to the configuration of Figure 4-14, the self-consistency condition becomes

$$
\begin{aligned}
\frac{1}{r_1} - \frac{1}{r_2 + 1} &= -\frac{2l}{R_1} = 2(g_1 - 1) \\
\frac{1}{r_2} - \frac{1}{r_1 + 1} &= -\frac{2l}{R_2} = 2(g_2 - 1)
\end{aligned}
\tag{4.8-1}
$$

where $g_i \equiv 1 - l/R_i$ $(i = 1, 2)$, and the sign of R is as discussed in footnote 5, so that in the example of Figure 4-14 R_1 and R_2 are negative.

Solving (4.8-1) for r_1 and r_2 gives

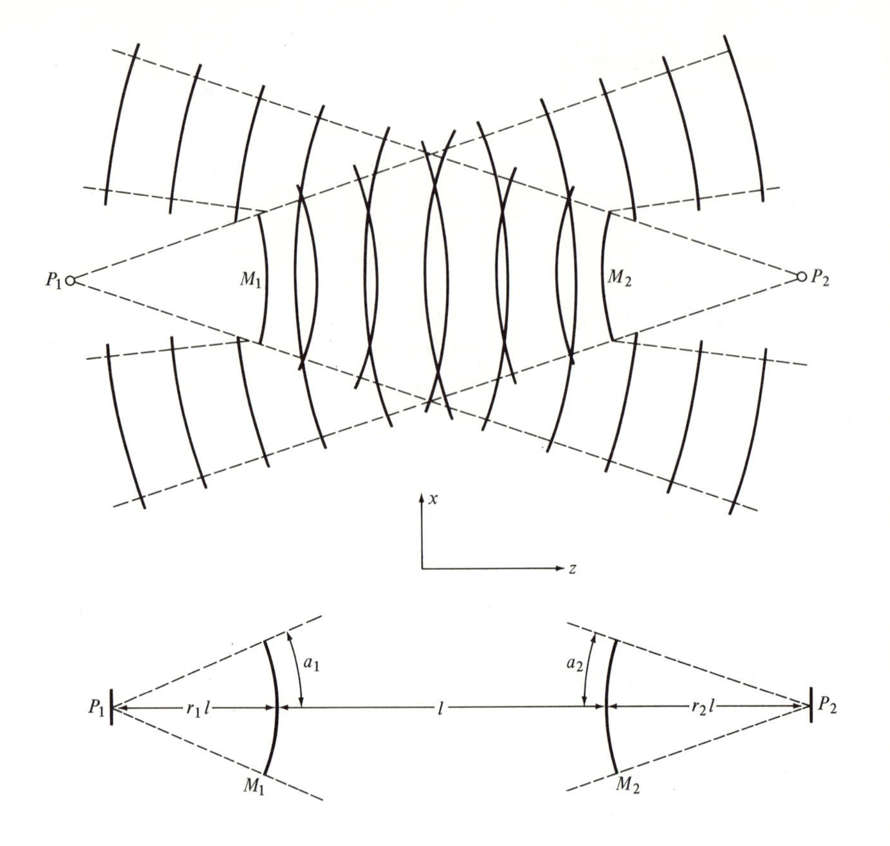

Figure 4-14 Spherical-wave picture of the mode in an unstable resonator. Points P_1 and P_2 are the virtual centers of the spherical waves. Each wave diverges so that a sizable fraction of its energy spills past the opposite mirror. (After Reference [9].)

$$r_1 = \frac{\pm\sqrt{g_1g_2(g_1g_2 - 1)} - g_1g_2 + g_2}{2g_1g_2 - g_1 - g_2}$$
$$r_2 = \frac{\pm\sqrt{g_1g_2(g_1g_2 - 1)} - g_1g_2 + g_1}{2g_1g_2 - g_1 - g_2} \tag{4.8-2}$$

The expressions (4.8-2) for r_1 and r_2 can be used to calculate, in the geometrical optics approximations, the loss per round trip of the unstable resonator. To demonstrate this we return to Figure 4-14 and consider first, for simplicity, the case of a strip geometry where the mirrors are infinitely long in the y direction and have curvature only along their width as shown.

The self-consistent wave, immediately following reflection from mirror 1, is taken to have a total energy of unity. Upon reflection from M_2 the total energy is reduced to

$$(\Gamma_1)_{\text{strip}} = \frac{r_1 a_2}{(r_1 + 1)a_1} \tag{4.8-3}$$

To arrive at the last result, we took into account the two dimensional beam spread due to virtual emanation from P_1 between M_1 and M_2. The total transmission factor per round trip due to both mirrors is thus

$$\Gamma_{\text{strip}} = (\Gamma_1\Gamma_2)_{\text{strip}} = \frac{r_1 r_2}{(r_1 + 1)(r_2 + 1)} \tag{4.8-4}$$

and for spherical mirrors (with curvature in both planes)

$$\Gamma_{1,2} = (\Gamma_{1,2})^2_{\text{strip}}$$

and

$$\Gamma = \Gamma_1\Gamma_2 = \frac{r_1^2 r_2^2}{(r_1 + 1)^2(r_2 + 1)^2} \tag{4.8-5}$$

The average fractional power loss *per pass* may be taken, following (4.7-3), as

$$\bar{\delta} = 1 - \Gamma^{1/2} = [1 - (\Gamma_1\Gamma_2)^{1/2}] \tag{4.8-6}$$

The unstable resonator loss is thus independent of the mirror dimensions, depending only on their radii of curvature and separation.

A plot of the losses of some symmetric unstable resonators that is obtained from (4.8-2) and (4.8-5) is shown in Figure 4-15.

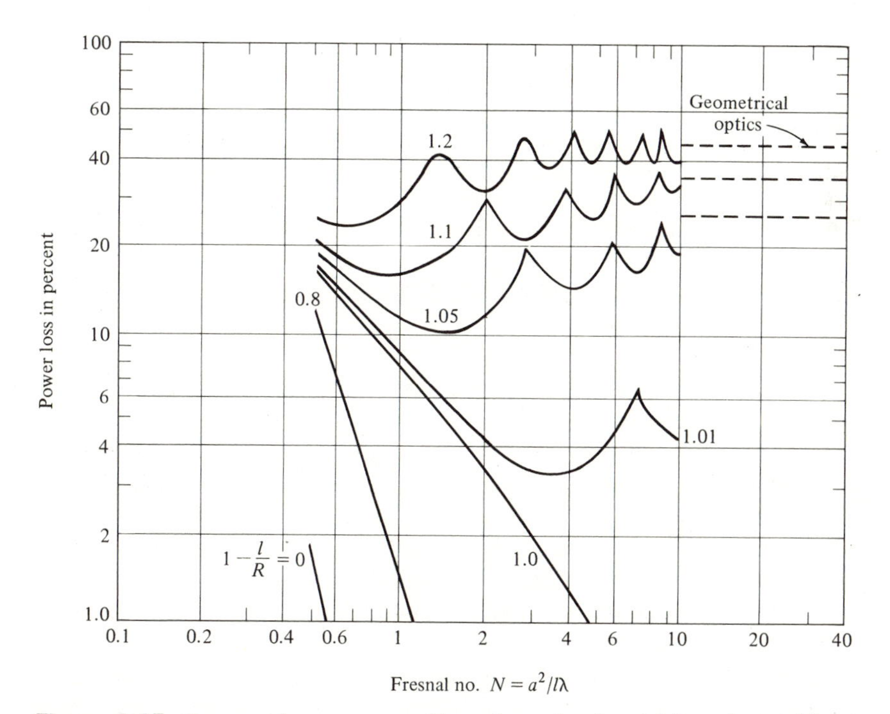

Figure 4-15 Loss per bounce versus Fresnel number for stable and unstable resonators. (After Reference [9].)

An electromagnetic analysis of unstable resonators is given in Appendix A. For more detailed theoretical treatments the reader is referred to References [10, 11, and 12].

PROBLEMS

4-1 Plot I_r/I_i vs. δ of a Fabry–Perot etalon with $R = 0.9$.

4-2 Show that if a Fabry–Perot etalon has a fractional intensity loss per pass of $(1 - A)$, its peak transmission is given as $(1 - R)^2A/(1 - RA)^2$.

4-3 Starting with the definition (4.2-6)

$$F \equiv \frac{\nu_{m+1} - \nu_m}{\Delta\nu_{1/2}}$$

for the finesse of a Fabry–Perot etalon and using semiquantitative arguments, show why in the case where the root-mean-square surface deviation from perfect flatness is approximately λ/N, the finesse cannot exceed $F \simeq N/2$. (*Hint:* Consider the spreading of the transmission peak due to a small number of etalons of nearly equal length transmitting in parallel.)

4-4 Show that the angular spread of a beam that is incident normally on a plane-parallel Fabry–Perot etalon must not exceed

$$\theta_{1/2} = \sqrt{\frac{2\lambda}{nlF}}$$

if its peak transmission is not to deviate substantially from unity.

4-5 Complete the derivation of Equations (4.1-4), (4.1-5), (4.1-6), and (4.1-7).

4-6 Consider a diverging monochromatic beam that is incident on a plane-parallel Fabry–Perot etalon.

a. Obtain an expression for the various angles along which the output energy is propagating. [*Hint:* These correspond to the different values of θ in (4.1-8) that result from changing m.]

b. Let the output beam in (a) be incident on a lens with a focal length f. Show that the energy distribution in the focal plane consists of a series of circles, each corresponding to a different value of m. Obtain an expression for the radii of the circles.

c. Consider the effect in (*b*) of having simultaneously two frequencies ν_1 and ν_2 present in the input beam. Derive an expression for the separation of the respective circles in the focal plane. Show that the smallest separation $\nu_1 - \nu_2$ that can be resolved by this technique is given by $(\Delta\nu)_{\min} \sim c/2nlF$.

4-7 a. Derive the phase shift between the incident and transmitted field amplitudes in a Fabry–Perot etalon as a function δ. Sketch it qualitatively for a number of different reflectivities.

b. Assume that the optical length of an etalon with $R = 0.9$ and $\theta = 0°$ is adjusted so that its transmission is a maximum and then modulated about this point according to

$$\Delta(nl) = \frac{\lambda}{100} \cos \omega_m t$$

Show that, to first-order, the output is phase-modulated. What is the modulation index? [For a definition of the modulation index δ see Equation (9.4-3).]

4-8 Show that the number of modes per unit frequency in a resonator whose dimensions are large compared to the wavelength is given by

$$N = \frac{8\pi n^3 \nu^2 V}{c^3} \tag{1}$$

where c is the velocity of light in vacuum and V is the volume of the resonator. (*Hint*: Assume a cube resonator with sides equal to L having perfectly conducting walls.) Taking the modes' fields as proportional to

$$\sin k_x x \sin k_y y \sin k_z z$$

show that in order for the fields to be zero at the boundaries, the conditions

$$k_x = \frac{2\pi l}{L} \qquad k_y = \frac{2\pi m}{L} \qquad k_z = \frac{2\pi n}{L}$$

where l, m, and n are any (positive) integers, must be satisfied. Each new combination of l, m, and n specifies a mode. Show that Equation (1) follows from the foregoing considerations and the fact that

$$k^2 = \frac{4\pi^2 n^2 \nu^2}{c^2} = k_x^2 + k_y^2 + k_z^2$$

so each frequency ν defines a sphere in the space k_x, k_y, k_z.

4-9 Calculate the fraction of the power of a fundamental ($l = m = 0$) Gaussian beam that passes through an aperture with a radius equal to the beam spot size.

4-10 Show that in the case of a conventional two-reflector resonator the stability condition (Equation 4.5-5) reduces to (Equation 4.4-2).

4-11 Consider a spherical mirror with a radius of curvature R whose reflectivity varies as

$$\rho(r) = \rho_0 \exp(-r^2/a^2)$$

where r is the radial distance from the center.

Show that the (A, B, C, D) matrix of this mirror is given by

$$\begin{vmatrix} A & B \\ C & D \end{vmatrix} = \begin{vmatrix} 1 & 0 \\ -\frac{2}{R} - i\frac{\lambda}{\pi a^2} & 1 \end{vmatrix}$$

REFERENCES

[1] Ramo, S., J. R. Whinnery, and T. Van Duzer, *Fields and Waves in Communication Electronics*. New York: Wiley, 1965.

[2] Yariv, A., *Quantum Electronics*, 2nd Ed. New York: Wiley, 1975, p. 96.

[3] Fabry, C., and A. Perot, "Théorie et applications d'une nouvelle methode de spectroscopie interférentielle," *Ann. Chim. Phys.*, vol. 16, p. 115, 1899.

[4] Born, M., and E. Wolf, *Principles of Optics*, 3d Ed. New York: Pergamon, 1965, Chap. 7.

[5] Peterson, D. G., and A. Yariv, "Interferometry and laser control with Fabry-Perot etalons," *Appl. Opt.*, vol. 5, p. 985, 1966.

[6] Boyd, G. D., and J. P. Gordon, "Confocal multimode resonator for millimeter through optical wavelength masers," *Bell System Tech. J.*, vol. 40, p. 489, 1961.

[7] Boyd, G. D., and H. Kogelnik, "Generalized confocal resonator theory," *Bell System Tech. J.*, vol. 41, p. 1347, 1962.

[8] Casperson, L., "Gaussian light beams in inhomogeneous media," *Appl. Opt.*, vol. 12, p. 2434, 1973.

[9] Siegman, A. E., "Unstable optical resonators for laser applications," *Proc. IEEE*, vol. 53, p. 277, 1965.

[10] Siegman, A. E., and H. Y. Miller, "Unstable optical resonator loss calculations using the prong method," *Appl. Optics*, vol. 9, p. 2729, 1970.

[11] Anan'ev, Y. A., "Unstable resonators and their applications," *Soviet J. Quant. Elec.*, vol. 1, p. 565, 1972.

[12] Yariv, A., and P. Yeh, "Confinement and stability in optical resonators employing mirrors with Gaussian reflectivity tapers," *Optics Comm.*, vol. 13, p. 30, 1975.

CHAPTER

5

Interaction of Radiation and Atomic Systems

5.0 Introduction

In this chapter we consider what happens to an electromagnetic wave propagating in an atomic medium. We are chiefly concerned with the possibility of growth (or attenuation) of the radiation resulting from its interaction with atoms. We also consider the changes in the velocity of propagation of light due to such interaction. The concepts derived in this chapter will be used in the next one in treating the laser oscillator.

5.1 Spontaneous Transitions between Atomic Levels—Homogeneous and Inhomogeneous Broadening

One of the basic results of the theory of quantum mechanics is that each physical system can be found, upon measurement, in only one of a predetermined set of energetic states—the so-called eigenstates of the system. With each of these states we associate an energy that corresponds to the total energy of the system when occupying the state. Some of the simpler systems, which are treated in any basic text on quantum mechanics, include the free electron, the hydrogen atom, and the harmonic oscillator. Examples of more complicated systems include the hydrogen molecule and the semiconducting crystal. With each state, the state i of the hydrogen atom say, we associate an eigenfunction [1]

$$\psi_i(\mathbf{r}, t) = u_i(\mathbf{r})e^{-iE_it/\hbar} \tag{5.1-1}$$

where $|u_i(\mathbf{r})|^2\, dx\, dy\, dz$ gives the probability of finding the electron, once it is known to be in the state i, within the volume element $dx\, dy\, dz$, which is centered on the point $\mathbf{r}$. E_i is the state energy described above and $\hbar = h/2\pi$ where $h = 6.626 \times 10^{-34}$ joule-second is Planck's constant.

One of the main tasks of quantum mechanics is the determination of the eigenfunctions $u_i(\mathbf{r})$ and the corresponding energies E_i of various physical systems. In this book, however, we will accept the existence of these states, their energy levels, as well as a number of other related results whose justification is provided by the experimentally proved formalism of quantum mechanics. Some of these results are discussed in the following.

The concept of spontaneous emission. In Figure 5-1 we show a system of energy levels that are associated with a given physical system—an atom, say. Let us concentrate on two of these levels—1 and 2, for example. If the atom is known to be in state 2 at $t = 0$ there is a finite probability per unit time that it will undergo a transition to state 1, emitting in the process a photon of energy $h\nu = E_2 - E_1$. This process, occurring as it does without the inducement of a radiation field, is referred to as *spontaneous emission.*

Another equivalent way of thinking about spontaneous transitions, and one corresponding more closely to experimental situations, is the following: Consider a large number N_2 of identical atoms that are known to be in state 2 at $t = 0$. The average number of these atoms undergoing spontaneous transition to state 1 per unit time is

$$-\frac{dN_2}{dt} = A_{21}N_2 \equiv \frac{N_2}{(t_{\text{spont}})_{21}} \qquad \textbf{(5.1-2)}$$

where A_{21} is the spontaneous transition rate and $(t_{\text{spont}})_{21} \equiv A_{21}{}^{-1}$ is called the spontaneous lifetime associated with the transition $2 \rightarrow 1$. It follows from quantum mechanical considerations that spontaneous transitions take place from a given state only to states lying lower in energy, so no spontaneous transitions take place from 1 to 2. The rate A_{21} can be calculated using the eigenfunctions of states 2 and 1. In this book we *accept* the existence of spontaneous emission A_{21} and regard A_{21} as a parameter characterizing the transition $2 \rightarrow 1$ of the given physical system.[1]

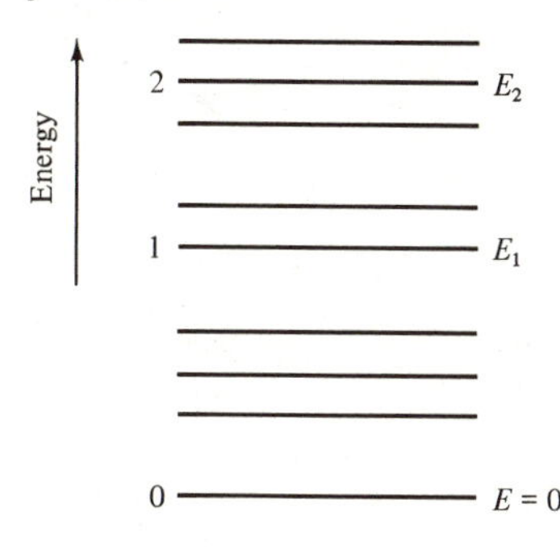

Figure 5-1 Some of the energy levels of an atomic system. Level 0, the ground state, is the lowest energy state. Levels 1 and 2 represent two excited states.

Lineshape function—homogeneous and inhomogeneous broadening. If one performs a spectral analysis of the radiation emitted by spontaneous $2 \rightarrow 1$ transitions, one finds that the radiation is not strictly monochromatic (that is, of one frequency) but occupies a finite frequency bandwidth. The function describing the distribution of emitted intensity versus the frequency ν is referred to as the lineshape function $g(\nu)$ (of the transition $2 \rightarrow 1$) and its arbitrary scale factor is usually chosen so that the function is normalized according to

$$\int_{-\infty}^{+\infty} g(\nu)\, d\nu = 1 \tag{5.1-3}$$

We can consequently view $g(\nu)\, d\nu$ as the *a priori* probability that a given spontaneous emission from level 2 to level 1 will result in a photon whose frequency is between ν and $\nu + d\nu$.

Another method of determining $g(\nu)$ is to apply an electromagnetic field to the sample containing the atoms and then plot the amount of energy absorbed by $1 \rightarrow 2$ transitions as a function of the frequency. This function, when normalized according to (5.1-3) is again $g(\nu)$.

The fact that both the emission and the absorption are described by the same lineshape function $g(\nu)$ can be verified experimentally, and follows from basic quantum mechanical considerations. The proof is beyond the scope of this book, but we can perhaps make a plausibility argument using the following example. Consider an RLC circuit that is excited into oscillation by connecting it to a signal source of frequency $\nu_0 = 1/2\pi\sqrt{LC}$. The excitation is then discontinued and the transient decay of the oscillation is observed. It is a straightforward problem to show that the intensity spectrum of the decaying oscillation, which is analogous to spontaneous emission since the total energy is decreasing, is the same as a plot of the absorption power vs. frequency of the same circuit; this last process being equivalent to induced absorption in the atomic system. It will be left as an exercise to show that in the case of the RLC circuit the spectrum characterizing the decay or absorption is proportional to

$$f(\nu) = \frac{1}{(\nu - \nu_0)^2 + (\nu_0/2Q)^2} \tag{5.1-4}$$

---

[1] The quantum mechanical derivation gives [1]

$$A_{21} = \frac{2e^2\omega^3(x_{12}{}^2 + y_{12}{}^2 + z_{12}{}^2)}{3hc^3\epsilon}$$

for a class of transitions known as electric dipole transitions. The parameter ϵ is the dielectric constant at ω and

$$x_{12} = \int_{\substack{\text{all}\\ \text{space}}} u_1^*(\mathbf{r})\, x u_2(\mathbf{r})\, d^3\mathbf{r}$$

where x, y, and z are the coordinates of the electron.

where $Q = 2\pi\nu_0 CR$ is the quality factor of the circuit.

The formal equivalence between an atomic transition and an oscillator goes even further than this *RLC* circuit example indicates. Later in this chapter we will use it extensively to describe the interaction between an atomic system and an electromagnetic field.

Homogeneous and inhomogeneous broadening [2]. One of the possible causes for the frequency spread of spontaneous emission is the finite lifetime τ of the emitting state. If we consider the emission from the excited state as that corresponding to a damped oscillator and choose the decay time of the oscillator as τ we can take the radiated field as

$$\begin{aligned} e(t) &= E_0 e^{-t/\tau} \cos \omega_0 t \\ &= \frac{E_0}{2} [e^{i(\omega_0 + i\sigma/2)t} + e^{-i(\omega_0 - i\sigma/2)t}] \end{aligned} \tag{5.1-5}$$

where $\sigma/2 = \tau^{-1}$ is the field decay rate (the intensity decay rate is σ). The Fourier transform of $e(t)$ is

$$\begin{aligned} E(\omega) &= \int_0^{+\infty} e(t) e^{-i\omega t}\, dt \\ &= \frac{E_0}{2} \left[\frac{i}{(\omega_0 - \omega + i\sigma/2)} - \frac{i}{(\omega_0 + \omega - i\sigma/2)} \right] \end{aligned} \tag{5.1-6}$$

where the lower limit of integration is taken as $t = 0$ (instead of $t = -\infty$) to correspond with the start of our observation period. The spectral density of the spontaneous emission is proportional to $|E(\omega)|^2$. If we limit our attention to the vicinity of the resonant frequency $\omega \simeq \omega_0$, we obtain

$$|E(\omega)|^2 \propto \frac{1}{(\omega - \omega_0)^2 + (\sigma/2)^2} \tag{5.1-7}$$

which is of the same form as (5.1-4).

Curves with the functional dependence of (5.1-7) are called Lorentzian. They occur often in physics and engineering, since, as shown, they characterize the response of damped resonant systems.

The separation $\Delta\nu$ between the two frequencies at which the Lorentzian is down to half its peak value is referred to as the linewidth and is given by

$$\Delta\nu = \frac{\sigma}{2\pi} = \frac{1}{\pi\tau} \tag{5.1-8}$$

In the case of atomic transitions between an upper level (u) and a lower level (l), the coherent interaction of an atom in either state (u or l) with the field can be interrupted by the finite lifetime of the state (τ_u, τ_l) or by an elastic collision which erases any phase memory (τ_{cu}, τ_{cl}). We thus generalize (5.1-8) to read

$$\Delta\nu = \frac{1}{\pi} (\tau_u^{-1} + \tau_l^{-1} + \tau_{cu}^{-1} + \tau_{cl}^{-1})$$

Rewriting (5.1-7) in terms of $\Delta\nu$ and, at the same time, normalizing it according to (5.1-3), we obtain the normalized Lorentzian lineshape function

$$g(\nu) = \frac{\Delta\nu}{2\pi[(\nu - \nu_0)^2 + (\Delta\nu/2)^2]} \tag{5.1-9}$$

The type of broadening (that is, the finite width of the emitted spectrum) described above is called *homogeneous broadening.* It is characterized by the fact that the spread of the response over a band $\sim\Delta\nu$ is characteristic of *each* atom in the sample. The function $g(\nu)$ thus describes the response of any of the atoms which are indistinguishable.

As mentioned above, homogeneous broadening is due most often to the finite interaction lifetime of the emitting or absorbing atoms. Some of the most common mechanisms are:

1. The spontaneous lifetime of the excited state.
2. Collision of an atom embedded in a crystal with a phonon. This may involve the emission or absorption of acoustic energy. Such a collision does not terminate the lifetime of the atom in its absorbing or emitting state. It does interrupt, however, the relative phase between the atomic oscillation (see Section 5.4) and that of the field, thus causing a broadening of the response according to (5.1-6) where τ now represents the mean uninterrupted interaction time.
3. Pressure broadening of atoms in a gas. At sufficiently high atomic densities the collisions between atoms become frequent enough that lifetime termination and phase interruption as in the preceding mechanism dominate the broadening mechanism.

There are, however, many physical situations in which the individual atoms are distinguishable, each having a slightly different transition frequency ν_0. If one observes, in this case, the spectrum of the spontaneous emission, its spectral distribution will reflect the spread in the individual transition frequencies and not the broadening due to the finite lifetime of the excited state. Two typical situations give rise to this type of broadening, referred to as *inhomogeneous.*

First of all, the energy levels, hence the transition frequencies, of ions present as impurities in a host crystal depend on the immediate crystalline surroundings. The ever-present random strain, as well as other types of crystal imperfections, cause the crystal surroundings to vary from one ion to the next, thus effecting a spread in the transition frequencies.

Second, the transition frequency ν of a gaseous atom (or molecule) is Doppler-shifted due to the finite velocity of the atom according to

$$\nu = \nu_0 + \frac{v_x}{c}\nu_0 \tag{5.1-10}$$

where v_x is the component of the velocity along the direction connecting the observer with the moving atom, c is the velocity of light in the medium, and ν_0 is the frequency corresponding to a stationary atom. The Maxwell velocity distribution function of a gas with atomic mass M which is at equilibrium at temperature T is [3]

$$f(v_x, v_y, v_z) = \left(\frac{M}{2\pi kT}\right)^{3/2} \exp\left[-\frac{M}{2kT}(v_x^2 + v_y^2 + v_z^2)\right] \quad \textbf{(5.1-11)}$$

$f(v_x, v_y, v_z)\, dv_x\, dv_y\, dv_z$ corresponds to the fraction of all the atoms whose x component of velocity is contained in the interval v_x to $v_x + dv_x$ while, simultaneously, their y and z components lie between v_y and $v_y + dv_y$, v_z and $v_z + dv_z$, respectively. Alternatively, we may view $f(v_x, v_y, v_z)$ $dv_x\, dv_y\, dv_z$ as the *a priori* probability that the velocity vector $\mathbf{v}$ of any given atom terminates within the differential volume $dv_x\, dv_y\, dv_z$ centered on $\mathbf{v}$ in velocity space so that

$$\iiint_{-\infty}^{\infty} f(v_x, v_y, v_z)\, dv_x\, dv_y\, dv_z = 1 \quad \textbf{(5.1-12)}$$

According to (5.1-10) the probability $g(\nu)\, d\nu$ that the transition frequency is between ν and $\nu + d\nu$ is equal to the probability that v_x will be found between $v_x = (\nu - \nu_0)(c/\nu_0)$ and $(\nu + d\nu - \nu_0)(c/\nu_0)$ irrespective of the values of v_y and v_z [since if $v_x = (\nu - \nu_0)(c/\nu_0)$, the Doppler-shifted frequency will be equal to ν regardless of v_y and v_z]. This probability is thus obtained by substituting $v_x = (\nu - \nu_0)c/\nu_0$ in $f(v_x, v_y, v_z)$ $dv_x\, dv_y\, dv_z$, and then integrating over all values of v_y and v_z. The result is

$$g(\nu)\, d\nu = \left(\frac{M}{2\pi kT}\right)^{3/2} \int_{-\infty}^{\infty}\int_{-\infty}^{\infty} e^{-(M/2kT)(v_y^2+v_z^2)}\, dv_y\, dv_z \times e^{-(M/2kT)(c^2/\nu_0^2)(\nu-\nu_0)^2}\left(\frac{c}{\nu_0}\right) d\nu \quad \textbf{(5.1-13)}$$

Using the definite integral

$$\int_{-\infty}^{\infty} e^{-(M/2kT)v_z^2}\, dv_z = \left(\frac{2\pi kT}{M}\right)^{1/2}$$

we obtain, from (5.1-13),

$$g(\nu) = \frac{c}{\nu_0}\left(\frac{M}{2\pi kT}\right)^{1/2} e^{-(M/2kT)(c^2/\nu_0^2)(\nu-\nu_0)^2} \quad \textbf{(5.1-14)}$$

for the *normalized Doppler-broadened lineshape*. The functional dependence of $g(\nu)$ in (5.1-14) is referred to as Gaussian. The width of $g(\nu)$ in this case is taken as the frequency separation between the points where $g(\nu)$ is down to half its peak value. It is obtained from (5.1-14) as

$$\Delta\nu_D = 2\nu_0\sqrt{\frac{2kT}{Mc^2}\ln 2} \quad \textbf{(5.1-15)}$$

where the subscript D stands for "Doppler." We can reexpress $g(\nu)$ in terms of $\Delta\nu_D$, obtaining

$$g(\nu) = \frac{2(\ln 2)^{1/2}}{\pi^{1/2}\,\Delta\nu_D} e^{-[4(\ln 2)(\nu-\nu_0)^2/\Delta\nu_D{}^2]} \tag{5.1-16}$$

In Figure 5-2 we show, as an example of a lineshape function, the spontaneous emission spectrum of Nd^{3+} when present as an impurity ion in a $CaWO_4$ lattice. The spectrum consists of a number of transitions, which are partially overlapping.

Numerical example—the Doppler linewidth of Ne. Consider the 6328 Å transition in Ne, which is used in the popular He–Ne laser. Using the atomic mass 20 for neon in (5.1-15) and taking $T = 300°K$, we obtain

$$\Delta\nu_D \simeq 1.5 \times 10^9 \text{ Hz}$$

for the Doppler linewidth. The 10.6 μm transition in the CO_2 laser has, according to (5.1-15), a linewidth $\Delta\nu_D \approx 6 \times 10^7$ Hz.

5.2 Induced Transitions

In the presence of an electromagnetic field of frequency $\nu \sim (E_2 - E_1)/h$ an atom whose energy levels are shown in Figure 5-1 can undergo a transition from state 1 to 2, *absorbing* in the process a quantum of excitation (photon) with energy $h\nu$ from the field. If the atom happens to occupy state 2 at the moment when it is first subjected to the electromagnetic

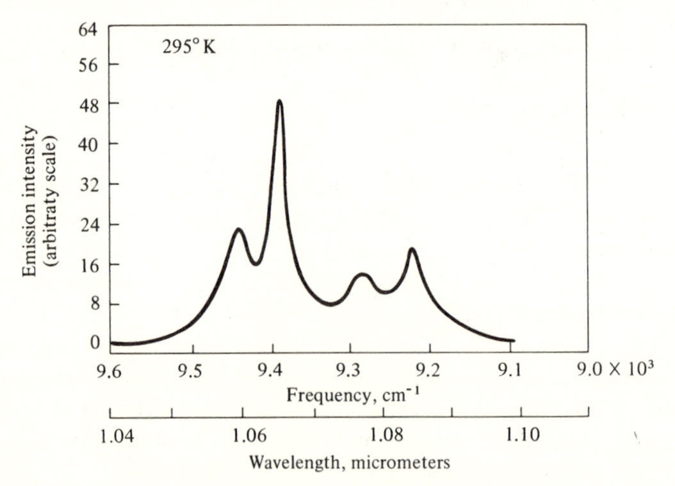

Figure 5-2 Emission spectrum of Nd^{3+}:$CaWO_4$ in the vicinity of the 1.06-μm laser transition. The main peak responds to the laser transition. (After Reference [4].)

field, it will make a downward transition to state 1, *emitting* a photon of energy $h\nu$.

What distinguishes the process of induced transition from the spontaneous one described in the last section is the fact that the induced rate for $2 \rightarrow 1$ and $1 \rightarrow 2$ transitions is *equal*, whereas the spontaneous $1 \rightarrow 2$ (that is, the one in which the atomic energy increases) transition rate is zero. Another fundamental difference—one that, again, follows from quantum mechanical considerations—is that the induced rate is *proportional* to the *intensity* of the electromagnetic field, whereas the spontaneous rate is independent of it. The relationship between the induced transition rate and the (inducing) field intensity is of fundamental importance in treating the interaction of atomic systems with electromagnetic fields. Its derivation follows.

Consider first the interaction of an assembly of identical atoms with a radiation field whose energy density is distributed uniformly in frequency in the vicinity of the transition frequency. Let the energy density per unit frequency be $\rho(\nu)$. We assume that the induced transition rates per atom from $2 \rightarrow 1$ and $1 \rightarrow 2$ are both proportional to $\rho(\nu)$ and take them as

$$\begin{aligned}(W'_{21})_{\text{induced}} &= B_{21}\rho(\nu)\\(W'_{12})_{\text{induced}} &= B_{12}\rho(\nu)\end{aligned} \tag{5.2-1}$$

where B_{21} and B_{12} are constants to be determined. The total downward $(2 \rightarrow 1)$ transition rate is the sum of the induced and spontaneous contributions

$$W'_{21} = B_{21}\rho(\nu) + A_{21} \tag{5.2-2}$$

The spontaneous rate A_{21} was discussed in Section 5.1. The total upward $(1 \rightarrow 2)$ transition rate is

$$W'_{12} = (W'_{12})_{\text{induced}} = B_{12}\rho(\nu) \tag{5.2-3}$$

Our first task is to obtain an expression for B_{12} and B_{21}. Since the magnitude of the coefficients B_{21} and B_{12} depends on the atoms and not on the radiation field, we consider, without loss of generality, the case where the atoms are in thermal equilibrium with a blackbody (thermal) radiation field at temperature T. In this case the radiation density is given by [5]

$$\rho(\nu) = \frac{8\pi n^3 h\nu^3}{c^3} \frac{1}{e^{h\nu/kT} - 1} \tag{5.2-4}$$

Since at thermal equilibrium the average populations of levels 2 and 1 are constant with time, it follows that the number of $2 \rightarrow 1$ transitions in a given time interval is equal to the number of $1 \rightarrow 2$ transitions; that is,

$$N_2 W'_{21} = N_1 W'_{12} \tag{5.2-5}$$

where N_1 and N_2 are the population densities of level 1 and 2, respectively. Using (5.2-2) and (5.2-3) in (5.2-5), we obtain

$$N_2[B_{21}\rho(\nu) + A_{21}] = N_1 B_{12}\rho(\nu)$$

and, substituting for $\rho(\nu)$ from (5.2-4),

$$N_2\left[B_{21}\frac{8\pi n^3 h\nu^3}{c^3(e^{h\nu/kT}-1)} + A_{21}\right] = N_1\left[B_{12}\frac{8\pi n^3 h\nu^3}{c^3(e^{h\nu/kT}-1)}\right] \tag{5.2-6}$$

Since the atoms are in thermal equilibrium, the ratio N_2/N_1 is given by the Boltzmann factor [5] as

$$\frac{N_2}{N_1} = e^{-h\nu/kT} \tag{5.2-7}$$

Equating (N_2/N_1) as given by (5.2-6) to (5.2-7) gives

$$\frac{8\pi n^3 h\nu^3}{c^3(e^{h\nu/kT}-1)} = \frac{A_{21}}{B_{12}e^{h\nu/kT} - B_{21}} \tag{5.2-8}$$

The last equality can be satisfied only when

$$B_{12} = B_{21} \tag{5.2-9}$$

and simultaneously

$$\frac{A_{21}}{B_{21}} = \frac{8\pi n^3 h\nu^3}{c^3} \tag{5.2-10}$$

The last two equations were first given by Einstein [6]. We can, using (5.2-10) rewrite the induced transition rate (5.2-1) as

$$W_i' = \frac{A_{21}c^3}{8\pi n^3 h\nu^3}\rho(\nu) = \frac{c^3}{8\pi n^3 h\nu^3 t_{\text{spont}}}\rho(\nu) \tag{5.2-11}$$

where, because of (5.2-9) the distinction between $2 \rightarrow 1$ and $1 \rightarrow 2$ induced transition rates is superfluous.

Equation (5.2-11) gives the transition rate per atom due to a field with a uniform (white) spectrum with energy density per unit frequency $\rho(\nu)$. In quantum electronics our main concern is in the transition rates that are induced by a monochromatic (that is, single-frequency) field of frequency ν. Let us denote this transition rate as $W_i(\nu)$. We have established in Section 5.1 that the strength of interaction of a monochromatic field of frequency ν with an atomic transition is proportional to the lineshape function $g(\nu)$, so $W_i(\nu) \propto g(\nu)$. Furthermore, we would expect $W_i(\nu)$ to go over into W_i' as given by (5.2-11) if the spectral width of the radiation field is gradually increased from zero to a point at which it becomes large compared to the transition linewidth. These two requirements are satisfied if we take $W_i(\nu)$ as

$$W_i(\nu) = \frac{c^3\rho_\nu}{8\pi n^3 h\nu^3 t_{\text{spont}}} g(\nu) \tag{5.2-12}$$

where ρ_ν is the energy density (joules per cubic meter) of the electromagnetic field inducing the transitions. To show that $W_i(\nu)$ as given by (5.2-12) indeed goes over smoothly into (5.2-11) as the spectrum of the field broadens, we may consider the broad spectrum field as made up of a large number of closely spaced monochromatic components at ν_k with random phases and then by adding the individual transition rates obtained from (5.2-12)

$$W_i' = \sum_{\nu_k} W_i(\nu_k) = \frac{c^3}{8\pi n^3 h t_{\text{spont}}} \sum_k \frac{\rho_{\nu_k}}{\nu_k{}^3} g(\nu_k) \tag{5.2-13}$$

where ρ_{ν_k} is the energy density of the field component oscillating at ν_k. We can replace the summation of (5.2-13) by an integral if we replace ρ_{ν_k} by $\rho(\nu)\, d\nu$ where $\rho(\nu)$ is the energy density per unit frequency; thus, (5.2-13) becomes

$$W_i' = \frac{c^3}{8\pi n^3 h t_{\text{spont}}} \int_{-\infty}^{+\infty} \frac{\rho(\nu) g(\nu)\, d\nu}{\nu^3} \tag{5.2-14}$$

In situations where $\rho(\nu)$ is sufficiently broad compared with $g(\nu)$, and thus the variation of $\rho(\nu)/\nu^3$ over the region of interest [where $g(\nu)$ is appreciable] can be neglected, we can pull $\rho(\nu)/\nu^3$ outside the integral sign, obtaining

$$W_i' = \frac{c^3}{8\pi n^3 h \nu^3 t_{\text{spont}}} \rho(\nu)$$

where we used the normalization condition

$$\int_{-\infty}^{+\infty} g(\nu)\, d\nu = 1$$

This agrees with (5.2-11).

Returning to our central result, Equation (5.2-12), we can rewrite it in terms of the intensity $I_\nu = c\rho_\nu/n$ (watts per square meter) of the optical wave as

$$W_i(\nu) = \frac{A_{21} c^2 I_\nu}{8\pi n^2 h \nu^3} g(\nu) = \frac{\lambda^2 I_\nu}{8\pi n^2 h \nu t_{\text{spont}}} g(\nu) \tag{5.2-15}$$

where c is the velocity of propagation of light in vacuum and $t_{\text{spont}} \equiv 1/A_{21}$.

5.3 Absorption and Amplification

Consider the case of a monochromatic plane wave of frequency ν and intensity I_ν propagating through an atomic medium with N_2 atoms per unit volume in level 2 and N_1 in level 1. According to (5.2-15) there will occur $N_2 W_i$ induced transitions per unit time per unit volume from level 2 to level 1 and $N_1 W_i$ transitions from 1 to 2. The net power generated within a unit volume is thus

$$\frac{P}{\text{volume}} = (N_2 - N_1)W_i h\nu$$

This radiation is added coherently (that is, with a definite phase relationship) to that of the traveling wave so that it is equal, in the absence of any dissipation mechanisms, to the increase in the intensity per unit length, or, using (5.2-15),

$$\frac{dI_\nu}{dz} = (N_2 - N_1)\frac{c^2 g(\nu)}{8\pi n^2 \nu^2 t_{\text{spont}}} I_\nu \tag{5.3-1}$$

The solution of (5.3-1) is

$$I_\nu(z) = I_\nu(0)e^{\gamma(\nu)z} \tag{5.3-2}$$

where

$$\gamma(\nu) = (N_2 - N_1)\frac{c^2}{8\pi n^2 \nu^2 t_{\text{spont}}} g(\nu) \tag{5.3-3}$$

that is, the intensity grows exponentially when the population is inverted ($N_2 > N_1$) or is attenuated when $N_2 < N_1$. The first case corresponds to laser-type amplification, whereas the second case is the one encountered in atomic systems at thermal equilibrium. The two situations are depicted in Figure 5-3. We recall that at thermal equilibrium

$$\frac{N_2}{N_1} = e^{-h\nu/kT} \tag{5.3-4}$$

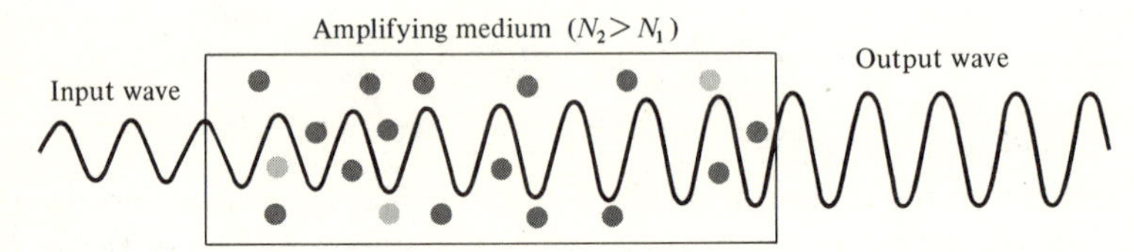

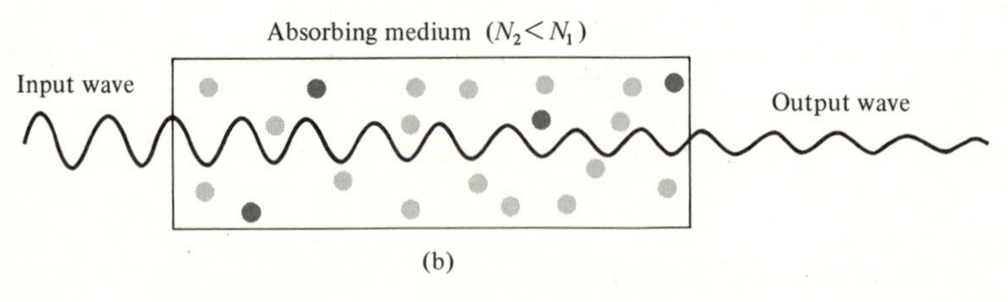

(b)

Legend:

● Atom in upper State 2

● Atom in lower State 1

Figure 5-3 Amplification of a traveling electromagnetic wave in (**a**) an inverted population ($N_2 > N_1$), and (**b**) its attenuation in an absorbing ($N_2 < N_1$) medium.

so that systems at thermal equilibrium are always absorbing. The inversion condition $N_2 > N_1$ can still be represented by (5.3-4), provided we take T as negative. As a matter of fact, the condition $N_2 > N_1$ is often referred to as one of "negative temperature"—the "temperature" in this case serving as an indicator of the population ratio, in accordance with (5.3-4).

The absorption, or amplification, of electromagnetic radiation by an atomic transition can be described not only by means of the exponential gain constant $\gamma(\nu)$ but also, alternatively, in terms of the imaginary part of the electric susceptibility $\chi''(\nu)$ of the propagation medium. According to (1.2-19) the density of absorbed power is

$$\frac{\overline{\text{Power}}}{\text{Volume}} = \frac{\omega\epsilon_0\chi''(\nu)}{2}|E|^2 \tag{5.3-5}$$

where, since we are concerned here only with electric susceptibilities, we replace $\chi_e(\nu)$ by the symbol $\chi(\nu)$. This last result must agree with a derivation using the concept of the induced transition rate $W_i(\nu)$ according to which

$$\frac{\overline{\text{Power}}}{\text{Volume}} = (N_1 - N_2)W_i(\nu)h\nu \tag{5.3-6}$$

Equating (5.3-5) to (5.3-6), substituting (5.2-15) for $W_i(\nu)$, and using the relation $I_\nu = (c/n)\epsilon|E|^2/2$ [see (1.3-26)], we obtain

$$\chi''(\nu) = \frac{(N_1 - N_2)\lambda^3}{16\pi^2 n t_{\text{spont}}} g(\nu) \tag{5.3-7}$$

where $n^2 \equiv \epsilon/\epsilon_0$ and λ is the wavelength in vacuum. In the case of a Lorentzian lineshape function $g(\nu)$, we use (5.1-9) to rewrite the last result as

$$\chi''(v) = \frac{(N_1 - N_2)\lambda^3}{8\pi^3 n t_{\text{spont}}\,\Delta\nu} \frac{1}{1 + [4(\nu - \nu_0)^2]/(\Delta\nu)^2} \tag{5.3-8}$$

Numerical example. Let us estimate the exponential gain constant at line center of a ruby (Al_2O_3 doped with Cr^{3+} ions) crystal having the following characteristics:

$$N_2 - N_1 = 5 \times 10^{17}/\text{cm}^3$$

$$\Delta\nu \simeq \frac{1}{g(\nu_0)} = 2 \times 10^{11} \text{ Hz at } 300°\text{K}$$

$$t_{\text{spont}} = 3 \times 10^{-3} \text{ second}$$

$$\nu = 4.326 \times 10^{14} \text{ Hz}$$

$$\frac{c}{n} \text{ (in ruby)} \simeq 1.69 \times 10^{10} \text{ cm/sec}$$

Using these values in (5.3-3) gives

$$\gamma(\nu) \simeq 5 \times 10^{-2} \text{ cm}^{-1}$$

Thus, the intensity of a wave with a frequency corresponding to the center of the transition is amplified by approximately 5 percent per cm in its passage through a ruby rod with the foregoing characteristics.

5.4 The Electron Oscillator Model of an Atomic Transition

The interaction of an electromagnetic field with an atomic transition is accompanied not only by absorption (or emission) of energy, but also by a dispersive effect in which the phase velocity of the incident wave depends on the frequency. The reason is that when the frequency ω of the wave is near that of the atomic transition, the atoms acquire large dipole moments that oscillate at ω, and the total field is now the sum of the incident and the field radiated by the dipoles. Since the radiated field is not necessarily in phase with the incident one, the effect is to change the phase velocity $(\mu\epsilon)^{-1/2}$ of the incident wave or, equivalently, to change the real part of the dielectric constant ϵ.

In treating this problem analytically we need to solve first for the dipole moment of an atom that is induced by an incident field. This problem is usually handled by the sophisticated quantum mechanical formalism of the density matrix [7]. We will employ, instead, the electron oscillator model for the atomic transition. According to this model which has been used by atomic physicists before the advent of quantum mechanics, we account for the dipole moment induced in a single atom by replacing the atom with an electron oscillating in a harmonic potential well [8]. The resonance frequency and absorption width of the electronic oscillator are chosen to agree with those of the real transition. We will also find it necessary to introduce an additional parameter, the so-called "oscillator strength," which characterizes the strength of the interaction between the oscillator and the field so that the calculated absorption (or emission) strength agrees with the experimental value.

The equation of motion of the one-dimensional electronic oscillator is

$$\frac{d^2x(t)}{dt^2} + \sigma \frac{dx(t)}{dt} + \frac{k}{m} x(t) = -\frac{e}{m} e(t) \tag{5.4-1}$$

where $x(t)$ is the deviation of the electron from its equilibrium position, σ is the damping coefficient, kx is the restoring force, $e(t)$ is the instantaneous electric field, and the electronic charge is $-e$. Taking the electric field $e(t)$ and the deviation $x(t)$ as

$$\begin{aligned} e(t) &= \operatorname{Re}[Ee^{i\omega t}] \\ x(t) &= \operatorname{Re}[X(\omega)e^{i\omega t}] \end{aligned} \tag{5.4-2}$$

respectively, and defining the resonant frequency by

$$\omega_0 \equiv \sqrt{\frac{k}{m}} \tag{5.4-3}$$

Equation (5.4-1) becomes

$$(\omega_0^2 - \omega^2)X + i\omega\sigma X = -\frac{e}{m}E \tag{5.4-4}$$

so that the deviation amplitude is given by

$$X(\omega) = \frac{-(e/m)E}{\omega_0^2 - \omega^2 + i\omega\sigma} \tag{5.4-5}$$

We are interested primarily in the response at frequencies near resonance. Thus, if we put $\omega \simeq \omega_0$, (5.4-5) becomes

$$X(\omega \simeq \omega_0) = \frac{-(e/m)E}{2\omega_0(\omega_0 - \omega) + i\omega_0\sigma}$$

The dipole moment of a single electron is

$$\mu(t) = -ex(t)$$

Therefore, in the case of N oscillators per unit volume there results a polarization (dipole moment per unit volume)

$$p(t) = \operatorname{Re}\,[P(\omega)e^{i\omega t}] \tag{5.4-6}$$

where the complex polarization $P(\omega)$ is given by

$$\begin{aligned} P(\omega) = -NeX(\omega) &= \frac{Ne^2/m}{2\omega_0(\omega_0 - \omega) + i\omega_0\sigma}E \\ &= \frac{-i(Ne^2/m\omega_0\sigma)}{1 + i[2(\omega - \omega_0)/\sigma]}E \end{aligned} \tag{5.4-7}$$

The electronic susceptibility $\chi(\omega)$ is defined as the ratio of the complex amplitude of the induced polarization to that of the inducing field (multiplied by ϵ_0)

$$P(\omega) = \epsilon_0\chi(\omega)E \tag{5.4-8}$$

and is consequently a complex number. If we separate $\chi(\omega)$ into its real and imaginary components according to

$$\chi(\omega) = \chi'(\omega) - i\chi''(\omega) \tag{5.4-9}$$

we obtain, from (5.4-6), (5.4-8), and (5.4-9),

$$\begin{aligned} p(t) &= \operatorname{Re}\,[\epsilon_0\chi(\omega)Ee^{i\omega t}] \\ &= \epsilon_0 \mathrm{E}\chi'(\omega)\cos\omega t + \epsilon_0 E\chi''(\omega)\sin\omega t \end{aligned} \tag{5.4-10}$$

Therefore, $\chi'(\omega)$ and $\chi''(\omega)$ are associated, respectively, with the in-phase and quadrature components of the polarization.

From (5.4-7) and (5.4-8) it follows that

$$\chi(\omega) = -i\left(\frac{Ne^2}{m\omega_0\sigma\epsilon_0}\right)\frac{1}{1 + 2i(\omega - \omega_0)/\sigma} \tag{5.4-11}$$

and thus

$$\chi'(\omega) = \left(\frac{Ne^2}{m\omega_0\sigma\epsilon_0}\right)\frac{2(\omega_0 - \omega)/\sigma}{1 + 4(\omega - \omega_0)^2/\sigma^2} \tag{5.4-12}$$

$$\chi''(\omega) = \left(\frac{Ne^2}{m\omega_0\sigma\epsilon_0}\right)\frac{1}{1 + 4(\omega - \omega_0)^2/\sigma^2} \tag{5.4-13}$$

Expressing $\chi(\omega)$ in terms of $\nu = \omega/2\pi$ and introducing $\Delta\nu = \sigma/2\pi$, the frequency separation between the two points where $\chi''(\nu)$ is down to half its peak intensity, we obtain

$$\chi''(\nu) = \left(\frac{Ne^2}{16\pi^2 m\nu_0\epsilon_0}\right)\frac{\Delta\nu}{(\Delta\nu/2)^2 + (\nu - \nu_0)^2} \tag{5.4-14}$$

$$\chi'(\nu) = \frac{2(\nu_0 - \nu)}{\Delta\nu}\chi''(\nu) = \left(\frac{Ne^2}{8\pi^2 m\nu_0\epsilon_0}\right)\frac{(\nu_0 - \nu)}{(\Delta\nu/2)^2 + (\nu - \nu_0)^2} \tag{5.4-15}$$

By comparing (5.4-14) with (5.1-9), we find that $\chi''(\nu)$ has a Lorentzian shape. The Lorentzian was found in Section 5.1 to represent the absorption as a function of frequency of an *RLC* oscillator, so it should not come as a surprise to find, in the next section, that the power absorption of the electronic oscillator is also proportional to $\chi''(\nu)$.

A normalized plot of $\chi'(\nu)$ and $\chi''(\nu)$ is shown in Figure 5-4. The most noteworthy features of the plot are that the extrema of $\chi'(\nu)$ occur at the half-power frequencies and the effects of dispersion $\chi'(\nu)$ are also "felt" at frequencies at which the absorption $\chi''(\nu)$ is negligible; and that the magnitude of $\chi'(\nu)$ at its extrema is half the peak value of $\chi''(\nu)$.

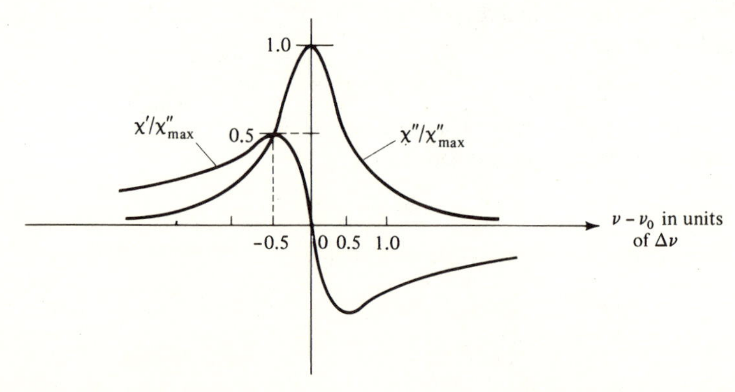

Figure 5-4 A plot of the real (χ') and (negative) imaginary (χ'') parts of the electronic susceptibility.

The significance of $\chi(\nu)$. According to (1.2-3), the electric displacement vector is defined by

$$\mathbf{D} = \epsilon_0\mathbf{E} + \mathbf{P} + \mathbf{P}_{\text{transition}} = \epsilon\mathbf{E} + \epsilon_0\chi\mathbf{E}$$

where the complex notation is used and the polarization is separated into a resonant component $\mathbf{P}_{\text{transition}}$ due to the specific atomic transition and a nonresonant component $\mathbf{P}$ that accounts for all the other contributions to the polarization. We can rewrite the last equation as

$$\mathbf{D} = \epsilon\left[1 + \frac{\epsilon_0}{\epsilon}\chi(\omega)\right]\mathbf{E} = \epsilon'(\omega)\mathbf{E} \tag{5.4-16}$$

so that the complex dielectric constant becomes

$$\epsilon'(\omega) = \epsilon\left[1 + \frac{\epsilon_0}{\epsilon}\chi(\omega)\right] \tag{5.4-17}$$

We have thus accounted for the effect of the atomic transition by modifying ϵ according to (5.4-17). Having derived $\chi(\omega)$, using detailed atomic information, we can ignore its physical origin and proceed to treat the wave propagation in the medium with ϵ' given by (5.4-17), using Maxwell's equations.

As an example of this point of view we consider the propagation of a plane electromagnetic wave in a medium with a dielectric constant $\epsilon'(\omega)$. According to (1.3-17), the wave has the form of

$$e(z, t) = \text{Re}\,[Ee^{i(\omega t - k'z)}] \tag{5.4-18}$$

where, using (1.3-13) and (5.4-17) and assuming $(\epsilon_0/\epsilon)|\chi| \ll 1$, we obtain

$$k' = \omega\sqrt{\mu\epsilon'} \simeq k\left[1 + \frac{\epsilon_0}{2\epsilon}\chi\right]$$

where $k = \omega\sqrt{\mu\epsilon}$.

Expressing $\chi(\omega)$ in terms of its real and imaginary components as in (5.4-9) leads to

$$k' = k\left[1 + \frac{\chi'(\omega)}{2n^2}\right] - i\frac{k\chi''(\omega)}{2n^2} \tag{5.4-19}$$

where $n = (\epsilon/\epsilon_0)^{1/2}$ is the index of refraction in the medium[2] far away from resonance. Substituting (5.4-19) back into (5.4-18), we find that the atomic transition results in a wave propagating according to

$$e(z, t) = \text{Re}\,[Ee^{i\omega t - i(k+\Delta k)z}e^{(\gamma/2)z}] \tag{5.4-20}$$

---

[2] Since the velocity of light is $c = (\mu\epsilon)^{-1/2}$, n is the ratio of the velocity of light in vacuum to that in the medium at frequencies sufficiently removed from resonance that the effect of the specific atomic transition can be ignored.

The result of the atomic polarization is thus to change the phase delay per unit length from k to $k + \Delta k$, where

$$\Delta k = \frac{k\chi'(\omega)}{2n^2} \tag{5.4-21}$$

as well as to cause the amplitude to vary exponentially with distance according to $e^{(\gamma/2)z}$, where

$$\gamma(\omega) = -\frac{k\chi''(\omega)}{n^2} \tag{5.4-22}$$

It is quite instructive to rederive (5.4-22) using a different approach. According to (1.2-13), the average power absorbed per unit volume from an electromagnetic field with an x component only is

$$\overline{\frac{\text{Power}}{\text{Volume}}} = \overline{e_x(t)\frac{dp_x(t)}{dt}} = \tfrac{1}{2}\,\text{Re}\,[E(i\omega P)^*] \tag{5.4-23}$$

where E and P are the complex electric field and polarization in the x direction, respectively, and horizontal bars denote time-averaging. Using (5.4-8) and (5.4-9) in (5.4-23), we obtain

$$\overline{\frac{\text{Power}}{\text{Volume}}} = \frac{\omega\epsilon_0}{2}\chi''|E|^2 \tag{5.4-24}$$

The absorption of energy at a rate given by (5.4-24) must lead to an attenuation of the wave intensity I, according to

$$I(z) = I_0 e^{\omega(\nu)z} \tag{5.4-25}$$

where

$$\gamma(\omega) = I^{-1}\frac{dI}{dz} \tag{5.4-26}$$

Conservation of energy thus requires that

$$\frac{dI}{dz} = -(\text{power absorbed per unit volume}) = -\frac{\omega\epsilon_0}{2}\chi''|E|^2$$

Using the last result in (5.4-26), as well as relation (1.3-26),

$$I = \frac{c\epsilon}{2n}|E|^2$$

where $c/n = \omega/k$ is the velocity of light in the medium, gives

$$\gamma(\omega) = -\frac{k\chi''(\omega)}{n^2}$$

in agreement with (5.4-22).

5.5 Atomic Susceptibility

In (5.4-14) and (5.4-15) we derived an expression for the electric susceptibility $\chi(\nu)$ of a medium made up of idealized electron oscillators. This expression cannot be used to represent the susceptibility of an actual atomic transition. This can be seen by comparing the classical expression for $\chi''(\nu)$ as given by (5.4-14) to the quantum mechanical expression

$$\chi''(\nu) = \frac{(N_1 - N_2)\lambda^3}{8\pi^3 t_{\text{spont}} \Delta\nu n} \frac{1}{1 + [4(\nu - \nu_0)^2/(\Delta\nu)^2]} \tag{5.5-1}$$

which was derived in Section 5.3. We find that in the case of the electron oscillator, $\chi''(\nu) > 0$, so power can only be absorbed from the field. In the quantum mechanical model, on the other hand, the sign of $\chi''(\nu)$ is the same as that of $(N_1 - N_2)$ so, for $N_2 > N_1$, power is actually added to the field. The quantum mechanical derivation also shows that the absorption (or emission) strength is inversely proportional to t_{spont}, which is characteristic of a given transition.

We accept the quantum mechanical expression for $\chi''(\nu)$ given by (5.5-1) as a correct representation of the absorption due to an atomic transition. We justify this choice by its agreement with experiment. We use the classical analysis, however, to obtain a relationship between $\chi'(\nu)$ and $\chi''(\nu)$, assuming that such a relationship is correct in spite of the objections raised previously.[3] From (5.4-15) and (5.5-1) we obtain

$$\begin{aligned} \chi'(\nu) &= \frac{2(\nu_0 - \nu)}{\Delta\nu} \chi''(\nu) \\ &= \frac{(N_1 - N_2)\lambda^3}{4\pi^3 n(\Delta\nu)^2 t_{\text{spont}}} \frac{(\nu_0 - \nu)}{1 + [4(\nu - \nu_0)^2/(\Delta\nu)^2]} \end{aligned} \tag{5.5-2}$$

This expression will be used to represent the dispersion of a homogeneously broadened transition with a Lorentzian lineshape.

5.6 Gain Saturation in Homogeneous Laser Media

In Section 5.3 we derived an expression (5.3-3) for the exponential gain constant due to a population inversion. It is given by

$$\gamma(\nu) = (N_2 - N_1) \frac{c^2}{8\pi n^2 \nu^2 t_{\text{spont}}} g(\nu) \tag{5.6-1}$$

[3] That this assumption is correct can be shown using the Kramers–Kronig relation (see Problem 5-3 and Reference [1], p. 155). Physically, this is attributable to the fact that both $\chi'(\nu)$ and $\chi''(\nu)$, as deduced from the electron oscillator model, are off by the same factor, so the ratio $\chi''(\nu)/\chi'(\nu)$ is correct.

where N_2 and N_1 are the population densities of the two atomic levels involved in the induced transition. There is nothing in (5.6-1) to indicate what causes the inversion $(N_2 - N_1)$, and this quantity can be considered as a parameter of the system. In practice the inversion is caused by a "pumping" agent, hereafter referred to as the pump, which can take various forms such as the electric current in injection lasers, the flashlamp light in pulsed ruby lasers, or the energetic electrons in plasma-discharge gas lasers.

Consider next the situation prevailing at some point *inside* a laser medium in the presence of an optical wave. The pump establishes a population inversion, which in the absence of any optical field has a value ΔN^0. The presence of the optical field induces $2 \rightarrow 1$ and $1 \rightarrow 2$ transitions. Since $N_2 > N_1$ and the induced rates for $2 \rightarrow 1$ and $1 \rightarrow 2$ transitions are equal, it follows that more atoms are induced to undergo a transition from level 2 to level 1 than in the opposite direction and that, consequently, the new equilibrium population inversion is smaller than ΔN^0.

The reduction in the population inversion and hence of the gain constant brought about by the presence of an electromagnetic field is called gain saturation. Its understanding is of fundamental importance in quantum electronics. As an example, which will be treated in the next chapter, we may point out that gain saturation is the mechanism which reduces the gain inside laser oscillators to a point where it just balances the losses so that steady oscillation can result.

In Figure 5-5 we show the ground state 0 as well as the two laser levels 2 and 1 of a four-level laser system. The density of atoms pumped per unit time into level 2 is taken as R_2, and that pumped into 1 is R_1. Pumping into 1 is, of course, undesirable since it leads to a reduction of the inversion. In many practical situations it cannot be avoided. The actual decay life-

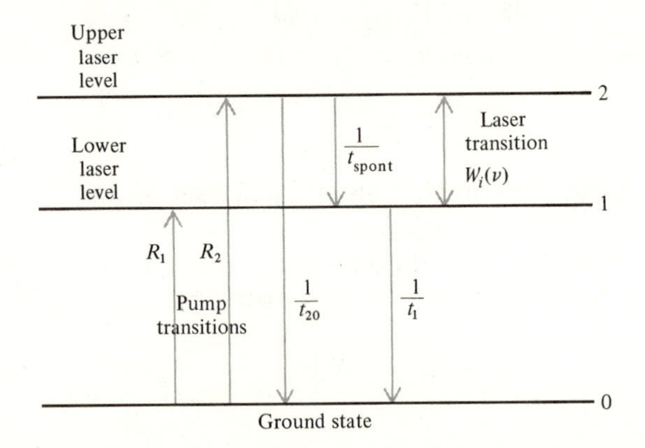

Figure 5-5 Energy levels and transition rates of a four-level laser system (the fourth level, which is involved in the original excitation by the pump, is not shown and the pumping is shown as proceeding directly into levels 1 and 2). The total lifetime of level 2 is t_2, where $1/t_2 = 1/t_{spont} + 1/t_{20}$.

time of atoms in level 2 at the absence of any radiation field is taken as t_2. This decay rate has a contribution t_{spont}^{-1} which is due to spontaneous (photon emitting) $2 \rightarrow 1$ transitions as well as to additional nonradiative relaxation from 2 to 1. The lifetime of atoms in level 1 is t_1. The induced rate for $2 \rightarrow 1$ and $1 \rightarrow 2$ transitions due to a radiation field at frequency ν is denoted by $W_i(\nu)$ and, according to (5.2-15), is given by

$$W_i(\nu) = \frac{\lambda^2 g(\nu)}{8\pi n^2 h\nu t_{\text{spont}}} I_\nu \tag{5.6-2}$$

where $g(\nu)$ is the normalized lineshape of the transition and I_ν is the intensity (watts per square meter) of the optical field.

The equations describing the populations of level 2 and 1 in the combined presence of a radiation field at ν and a pump are:

$$\frac{dN_2}{dt} = R_2 - \frac{N_2}{t_2} - (N_2 - N_1)W_i(\nu) \tag{5.6-3}$$

$$\frac{dN_1}{dt} = R_1 - \frac{N_1}{t_1} + \frac{N_2}{t_{\text{spont}}} + (N_2 - N_1)W_i(\nu) \tag{5.6-4}$$

N_2 and N_1 are the population densities (m^{-3}) of levels 2 and 1 respectively. R_2 and R_1 are the pumping rates ($m^{-3} - \text{s}^{-1}$) into these levels. N_2/t_2 is the change per unit time in the population of 2 due to decay out of level 2 to all levels. This includes spontaneous transitions to 1 but *not* induced transitions. The rate for the latter is $N_2 W_i(\nu)$ so that the net change in N_2 due to induced transitions is given by the last term of (5.6-3). At steady state the populations are constant with time, so putting $d/dt = 0$ in the two preceding equations, we can solve for N_1, N_2, and obtain[4]

$$N_2 - N_1 = \frac{[R_2 t_2 - (R_1 + \delta R_2)t_1]}{1 + [t_2 + (1 - \delta)t_1]W_i(\nu)} \tag{5.6-5}$$

where $\delta = t_2/t_{\text{spont}}$. If the optical field is absent, $W_i(\nu) = 0$, and the inversion density is given by

$$\Delta N^0 = R_2 t_2 - (R_1 + \delta R_2)t_1 \tag{5.6-6}$$

we can use (5.6-6) to rewrite (5.6-5) as

$$N_2 - N_1 = \frac{\Delta N^0}{1 + \phi t_{\text{spont}} W_i(\nu)} \tag{5.6-7}$$

where the parameter ϕ is defined by

$$\phi = \delta \left[1 + (1 - \delta) \frac{t_1}{t_2} \right]$$

[4] Levels 1 and 2 are assumed to be high enough (in energy) that the role of thermal processes in populating them can be neglected.

We note that in efficient laser systems $t_2 \simeq t_{\text{spont}}$, so $\delta \simeq 1$, and that $t_1 \ll t_2$, so $\phi \simeq 1$. Substituting (5.6-2) for $W_i(\nu)$ the last equation becomes

$$N_2 - N_1 = \frac{\Delta N^0}{1 + [\phi\lambda^2 g(\nu)/8\pi n^2 h\nu] I_\nu} = \frac{\Delta N^0}{1 + [I_\nu/I_s(\nu)]} \tag{5.6-8}$$

where $I_s(\nu)$, the saturation intensity, is given by

$$I_s(\nu) = \frac{8\pi n^2 h\nu}{\phi\lambda^2 g(\nu)} = \frac{8\pi n^2 h\nu}{(t_2/t_{\text{spont}})\lambda^2 g(\nu)} = \frac{8\pi n^2 h\nu \Delta\nu}{(t_2/t_{\text{spont}})\lambda^2} \tag{5.6-9}$$

and corresponds to the intensity level (watts per square meter) that causes the inversion to drop to one half of its nonsaturated value (ΔN^0). By using (5.6-8) in the gain expression (5.6-1), we obtain our final result

$$\begin{aligned} \gamma(\nu) &= \frac{1}{1 + [I_\nu/I_s(\nu)]} \frac{\Delta N^0 \lambda^2}{8\pi n^2 t_{\text{spont}}} g(\nu) \\ &= \frac{\gamma_0(\nu)}{1 + [I_\nu/I_s(\nu)]} \end{aligned} \tag{5.6-10}$$

which shows the dependence of the gain constant on the optical intensity.

In closing we recall that (5.6-10) applies to a homogeneous laser system. This is due to the fact that in the rate equations (5.6-3) and (5.6-4) we considered all the atoms as equivalent and, consequently, exercising the same transition rates. This assumption is no longer valid in inhomogeneous laser systems. This case is treated in the next section.

5.7 Gain Saturation in Inhomogeneous Laser Media

In Section 5.6 we considered the reduction in optical gain—that is, saturation—due to the optical field in a homogeneous laser medium. In this section we treat the problem of gain saturation in inhomogeneous systems.

According to the discussion of Section 5.1, in an inhomogeneous atomic system the individual atoms are distinguishable, with each atom having a unique transition frequency $(E_2 - E_1)/h$. We can thus imagine the inhomogeneous medium as made up of classes of atoms each designated by a continuous variable ξ.[5] Furthermore, we define a function $p(\xi)$ so that the *a priori* probability that an atom has its ξ parameter between ξ and $\xi + d\xi$ is $p(\xi)\, d\xi$. It follows that

$$\int_{-\infty}^{\infty} p(\xi)\, d\xi = 1 \tag{5.7-1}$$

since any atom has a unit probability of having its ξ value between $-\infty$ and ∞.

---

[5] The variable ξ can, as an example, correspond to the center frequency of the lineshape function $g^\xi(\nu)$ of atoms in group ξ.

The atoms within a given class ξ are considered as homogeneously broadened, having a lineshape function $g^\xi(\nu)$ that is normalized so that

$$\int_{-\infty}^{\infty} g^\xi(\nu)\, d\nu = 1 \tag{5.7-2}$$

In Section 5.1 we defined the transition lineshape $g(\nu)$ by taking $g(\nu)\, d\nu$ to represent the *a priori* probability that a spontaneous emission will result in a photon whose frequency is between ν and $\nu + d\nu$. Using this definition we obtain

$$g(\nu)\, d\nu = \left[\int_{-\infty}^{\infty} p(\xi) g^\xi(\nu)\, d\xi\right] d\nu \tag{5.7-3}$$

which is a statement of the fact that the probability of emitting a photon of frequency between ν and $\nu + d\nu$ is equal to the probability $g^\xi(\nu)\, d\nu$ of this occurrence, given that the atom belongs to class ξ, summed up over all the classes.

Next we proceed to find the contribution to the inversion which is due to a single class ξ. The equations of motion [9] are

$$\begin{aligned} \frac{dN_2^\xi}{dt} &= R_2 p(\xi) - \frac{N_2^\xi}{t_2} - [N_2^\xi - N_1^\xi] W_i^\xi(\nu) \\ \frac{dN_1^\xi}{dt} &= R_1 p(\xi) - \frac{N_1^\xi}{t_1} + \frac{N_2^\xi}{t_{\text{spont}}} + [N_2^\xi - N_1^\xi] W_i^\xi(\nu) \end{aligned} \tag{5.7-4}$$

and are similar to (5.6-3) and (5.6-4), except that N_2^ξ and N_1^ξ refer to the upper and lower level densities of atoms in class ξ only. The pumping rate (atoms/m^3-sec) into levels 2 and 1 is taken to be proportional to the probability of finding an atom in class ξ and is given by $R_2 p(\xi)$ and $R_1 p(\xi)$, respectively. The total pumping rate into level 2 is, as in Section 5.6, R_2 since

$$\int_{-\infty}^{\infty} R_2 p(\xi)\, d\xi = R_2 \int_{-\infty}^{\infty} p(\xi)\, d\xi = R_2$$

where we made use of (5.7-1). The induced transition rate $W_i^\xi(\nu)$ is given, according to (5.2-15), by

$$W_i^\xi(\nu) = \frac{\lambda^2}{8\pi n^2 h\nu t_{\text{spont}}}\, g^\xi(\nu) I_\nu \tag{5.7-5}$$

which is of a form identical to (5.6-2) except that $g^\xi(\nu)$ refers to the lineshape function of atoms in class ξ. The steady-state $d/dt = 0$ solution of (5.7-4) yields

$$N_2^\xi - N_1^\xi = \frac{\Delta N^0 p(\xi)}{1 + \phi t_{\text{spont}} W_i^\xi(\nu)} \tag{5.7-6}$$

where ΔN^0 and ϕ have the same significance as in Section 5.6. The total power emitted by induced transitions per unit volume by atoms in class ξ

is thus

$$\frac{P^{\xi}(\nu)}{V} = (N_2^{\xi} - N_1^{\xi})h\nu W_i^{\xi}(\nu) = \frac{\Delta N^0 p(\xi)h\nu}{[1/W_i^{\xi}(\nu)] + \phi t_{\text{spont}}} \tag{5.7-7}$$

where the spontaneous lifetime is assumed the same for all the groups ξ.

Summing (5.7-7) over all the classes, we obtain an expression for the total power at ν per unit volume emitted by the atoms

$$\frac{P(\nu)}{V} = \frac{\Delta N^0 h\nu}{t_{\text{spont}}} \int_{-\infty}^{\infty} \frac{p(\xi)\,d\xi}{[1/W_i(\nu)t_{\text{spont}}] + \phi} \tag{5.7-8}$$

which, by the use of (5.7-5), can be rewritten as

$$\frac{P(\nu)}{V} = \frac{\Delta N^0 h\nu}{t_{\text{spont}}} \int_{-\infty}^{\infty} \frac{p(\xi)\,d\xi}{[8\pi n^2 h\nu/\lambda^2 I_\nu g^{\xi}(\nu)] + \phi} \tag{5.7-9}$$

The stimulated emission of power causes the intensity of the traveling optical wave to increase with distance z according to $I_\nu = I_\nu(0) \exp[\gamma(\nu)z]$, where

$$\gamma(\nu) = \frac{dI_\nu}{dz}\Big/ I_\nu = \frac{P(\nu)}{V}\Big/ I_\nu$$

$$= \frac{\Delta N^0 \lambda^2}{8\pi t_{\text{spont}}} \int_{-\infty}^{\infty} \frac{p(\nu_\xi)\,d\nu_\xi}{[1/g^{\xi}(\nu)] + (\phi\lambda^2 I_\nu/8\pi n^2 h\nu)} \tag{5.7-10}$$

where we replaced $p(\xi)\,d\xi$ by $p(\nu_\xi)\,d\nu_\xi$.

This is our basic result.

As a first check on (5.7-10), we shall consider the case in which $I_\nu \ll 8\pi n^2 h\nu/\phi\lambda^2 g^{\xi}(\nu)$ and therefore the effects of saturation can be ignored. Using (5.7-3) in (5.7-10), we obtain

$$\gamma(\nu) = \frac{\Delta N^0 \lambda^2}{8\pi n^2 t_{\text{spont}}} g(\nu)$$

which is the same as (5.3-3). This shows that in the absence of saturation the expressions for the gain of a homogeneous and an inhomogeneous atomic system are identical.

Our main interest in this treatment is in deriving the saturated gain constant for an inhomogeneously broadened atomic transition. If we assume that in each class ξ all the atoms are identical (homogeneous broadening), we can use (5.1-9) for the lineshape function $g^{\xi}(\nu)$, and therefore,

$$g^{\xi}(\nu) = \frac{\Delta\nu}{2\pi[(\Delta\nu/2)^2 + (\nu - \nu_\xi)^2]} \tag{5.7-11}$$

where $\Delta\nu$ is called the homogeneous linewidth of the inhomogeneous line. Atoms with transition frequencies that are clustered within $\Delta\nu$ from each other can be considered as indistinguishable. The term "homogeneous packet" is often used to describe them. Using (5.7-11) in (5.7-10) leads to

$$\gamma(\nu) = \frac{\Delta N^0 \lambda^2 \Delta\nu}{16\pi^2 n^2 t_{\text{spont}}} \int_{-\infty}^{\infty} \frac{p(\nu_\xi)\, d\nu_\xi}{(\nu - \nu_\xi)^2 + (\Delta\nu/2)^2 + (\phi\lambda^2 I_\nu \Delta\nu/16\pi^2 n^2 h\nu)} \tag{5.7-12}$$

In the extreme inhomogeneous case, the width of $p(\nu_\xi)$ is by definition very much larger than the remainder of the integrand in (5.7-12) and thus it is essentially a constant over the region in which the integrand is a maximum. In this case we can pull $p(\nu_\xi)_{\nu_\xi=\nu} = p(\nu)$ outside the integral sign in (5.7-12), obtaining

$$\gamma(\nu) = \frac{\Delta N^0 \lambda^2 \Delta\nu}{16\pi^2 n^2 t_{\text{spont}}} p(\nu) \times \int_{-\infty}^{\infty} \frac{d\nu_\xi}{(\nu - \nu_\xi)^2 + (\Delta\nu/2)^2 + (\phi\lambda^2 \Delta\nu I_\nu/16\pi^2 n^2 h\nu)} \tag{5.7-13}$$

Using the definite integral

$$\int_{-\infty}^{\infty} \frac{dx}{x^2 + a^2} = \frac{\pi}{a}$$

to evaluate (5.7-13), we obtain

$$\gamma(\nu) = \frac{\Delta N^0 \lambda^2 p(\nu)}{8\pi n^2 t_{\text{spont}}} \frac{1}{\sqrt{1 + (\phi\lambda^2 I_\nu/4\pi^2 n^2 h\nu\Delta\nu)}} \tag{5.7-14}$$

$$= \gamma_0(\nu) \frac{1}{\sqrt{1 + (I_\nu/I_s)}} \tag{5.7-15}$$

where $I_s = 4\pi^2 n^2 h\nu\Delta\nu/\phi\lambda^2$ is the saturation intensity. A comparison of (5.7-15) with (5.7-10) shows that, because of the square root, the saturation—that is, decrease in gain—sets in more slowly as the intensity I_ν is increased.

PROBLEMS

5-1 Consider a parallel *RLC* circuit, which is connected to a signal generator so that the voltage across it is

$$v(t) = V_0 \cos 2\pi\nu t$$

At $t = 0$ the circuit is disconnected from the signal generator.

a. What is the voltage $v(t)$ for $t > 0$?

b. Find the Fourier transform $V(\omega)$ of $v(t)$. Show that in the high-Q case (where $Q = 2\pi\nu_0 RC$) and for frequencies $\nu \simeq \nu_0 \equiv 1/2\pi\sqrt{LC}$,

$$|V(\nu)|^2 \propto \frac{1}{(\nu - \nu_0)^2 + (\nu_0/2Q)^2}$$

c. Obtain the expression for the amount of average power $P(\nu)$ absorbed by the *RLC* circuit from a signal generator with an output current

$$i(t) = I_0 \cos 2\pi\nu t$$

Show that the expression for $P(\nu)$ is proportional to that of $|V(\nu)|^2$ obtained in (b).

5-2 Calculate the maximum absorption coefficient for the R_1 transition in pink ruby with a Cr^{3+} concentration of 2×10^{19} cm^{-3}. Assume that $t_{spont} = 3 \times 10^{-3}$ second and $\Delta\nu = 11$ cm^{-1}. Compare the result to the absorption data of Figure 7-4.

5-3 Show that Equations (5.5-1) and (5.5-2) are derivable from each other using the Kramers–Kronig relationship. For a discussion of this relation see, for example, Reference [1], page 155.

REFERENCES

[1] See, for example, A. Yariv, *Quantum Electronics*, 2d Ed. New York: Wiley, 1975.

[2] Portis, A. M., "Electronic structure of F centers: Saturation of the electron skin resonance," *Phys. Rev.*, vol. 91, p. 1071, 1953.

[3] See, for example, R. Kubo, *Statistical Mechanics*. Amsterdam: North Holland, 1964, p. 31.

[4] Johnson, L. F., "Optically pumped pulsed crystal lasers other than ruby," in *Lasers*, vol. 1, A. K. Levine, ed. New York: Marcel Dekker, Inc., 1966, p. 137.

[5] Kittel, C., *Elementary Statistical Physics*. New York: Wiley, 1958, p. 197.

[6] Einstein, A., "Zur Quantentheorie der Strahlung," *Phys. Z.*, vol. 18, pp. 121–128, March 1917.

[7] See, for example, R. H. Pantell and H. E. Puthoff, *Fundamentals of Quantum Electronics*. New York: Wiley, 1969, p. 31.

[8] Ditchburn, R. W., *Light*. New York: Interscience, 1963, Chap. 15.

[9] Gordon, J. P., unpublished memorandum, Bell Telephone Laboratories.

[10] *Lasers and Light—Readings from Scientific American*. San Francisco: Freeman, 1969.

[11] Mitchell, A. C. G., and M. W. Zemansky, *Resonance Radiation and Excited Atoms*. New York: Cambridge, 1961.

CHAPTER

6

Theory of Laser Oscillation

6.0 Introduction

In Chapter 5 we found that an atomic medium with an inverted population ($N_2 > N_1$) is capable of amplifying an electromagnetic wave if the latter's frequency falls within the transition lineshape. Consider next the case in which the laser medium is placed inside an optical resonator. As the electromagnetic wave bounces back and forth between the two reflectors it passes through the laser medium and is amplified. If the amplification exceeds the losses caused by imperfect reflection in the mirrors and scattering in the laser medium, the field energy stored in the resonator will increase with time. This causes the amplification constant to decrease as a result of gain saturation [see (5.6-10) and the discussion surrounding it.] The oscillation level will keep increasing until the saturated gain per pass just equals the losses. At this point the net gain per pass is unity and no further increase in the radiation intensity is possible—that is, steady-state oscillation obtains.

In this chapter we will derive the start-oscillation inversion needed to sustain laser oscillation, beginning with the theory of the Fabry–Perot etalon. We will also obtain an expression for the oscillation frequency of the laser oscillator and show how it is affected by the dispersion of the atomic medium. We will conclude by considering the problem of optimum output coupling and laser pulses.

6.1 Fabry–Perot Laser

A laser oscillator is basically a Fabry–Perot etalon, as studied in detail in Chapter 4, in which the space between the two mirrors contains an amplifying medium with an inverted atomic population. We can account for the inverted population by using (5.4-19). Taking the propagation constant of the medium as

$$k'(\omega) = k + k\frac{\chi'(\omega)}{2n^2} - ik\frac{\chi''(\omega)}{2n^2} - i\frac{\alpha}{2} \tag{6.1-1}$$

where $k - i\alpha/2$ is the propagation constant of the medium at frequencies well removed from that of the laser transition, $\chi(\omega) = \chi'(\omega) - i\chi''(\omega)$ is the complex dielectric susceptibility due to the laser transition and is given by (5.5-1) and (5.5-2). Since α accounts for the distributed passive losses of the medium,[1] the intensity loss factor per pass is exp $(-\alpha l)$.

Figure 6-1 shows a plane wave of (complex) amplitude E_i which is incident on the left mirror of a Fabry–Perot etalon containing a laser medium. The ratio of transmitted to incident fields at the left mirror is taken as t_1 and that at the right mirror as t_2. The ratios of reflected to incident fields inside the laser medium at the left and right boundaries are r_1 and r_2, respectively.

The propagation factor corresponding to a single transit is exp $(-ik'l)$ where k' is given by (6.1-1) and l is the length of the etalon.

Adding the partial waves at the output to get the total outgoing wave E_t we obtain

$$E_t = t_1 t_2 E_i e^{-ik'l}[1 + r_1 r_2 e^{-i2k'l} + r_1{}^2 r_2{}^2 e^{-i4k'l} + \cdots]$$

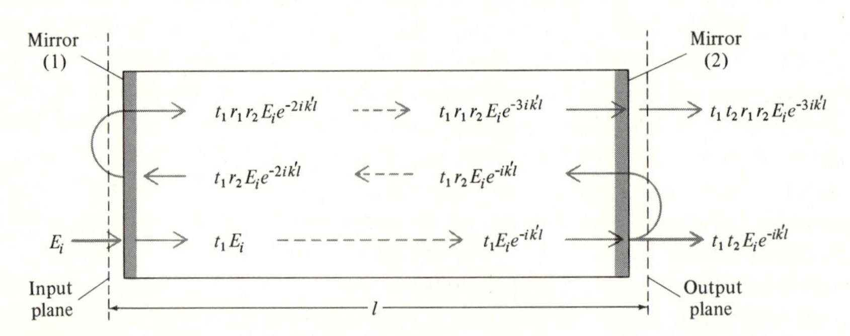

Figure 6-1 Model used to analyze a laser oscillator. A laser medium (that is, one with an inverted atomic population) with a complex propagation constant $k'(\omega)$ is placed between two reflecting mirrors.

---

[1] In addition to and in the presence of the gain attributable to the inverted laser transition, the medium may possess a residual attenuation due to a variety of mechanisms, such as scattering at imperfections, absorption by excited atomic levels, and others. The attenuation resulting from all of these mechanisms is lumped into the distributed loss constant α.

which is a geometric progression with a sum

$$E_t = E_i \left[\frac{t_1 t_2 e^{-ik'l}}{1 - r_1 r_2 e^{-i2k'l}} \right]$$

$$= E_i \left[\frac{t_1 t_2 e^{-i(k+\Delta k)l} e^{(\gamma-\alpha)l/2}}{1 - r_1 r_2 e^{-2i(k+\Delta k)l} e^{(\gamma-\alpha)l}} \right] \quad \textbf{(6.1-2)}$$

where we used (5.3-3), (6.1-1), and the relation $k' = k + \Delta k + i(\gamma - \alpha)/2$ with

$$\Delta k = k \frac{\chi'(\omega)}{2n^2}$$

$$\gamma = -k \frac{\chi''(\omega)}{n^2} \quad \textbf{(6.1-3)}$$

$$= (N_2 - N_1) \frac{\lambda^2}{8\pi n^2 t_{\text{spont}}} g(\nu) \quad \textbf{(6.1-4)}$$

If the atomic transition is inverted ($N_2 > N_1$), then $\gamma > 0$ and the denominator of (6.1-2) can become very small. The transmitted wave E_t can thus become larger than the incident wave E_i. The Fabry–Perot etalon (with the laser medium) in this case acts as an amplifier with a power gain $|E_t/E_i|^2$. We recall that in the case of the passive Fabry–Perot etalon (that is, one containing no laser medium), whose transmission is given by (4.1-7), $|E_t| \leqslant |E_i|$ and thus no power gain is possible. In the case considered here, however, the inverted population constitutes an energy source, so the transmitted wave can exceed the incident one.

If the denominator of (6.1-2) becomes zero, which happens when

$$r_1 r_2 e^{-2i[k+\Delta k(\omega)]l} e^{[\gamma(\omega)-\alpha]l} = 1 \quad \textbf{(6.1-5)}$$

then the ratio E_t/E_i becomes infinite. This corresponds to a finite transmitted wave E_t with a *zero* incident wave ($E_i = 0$)—that is, to *oscillation.* Physically, condition (6.1-5) represents the case in which a wave making a complete round trip inside the resonator returns to the starting plane with the *same amplitude* and, except for some integral multiple of 2π, with the *same phase.* Separating the oscillation condition (6.1-5) into the amplitude and phase requirements gives

$$r_1 r_2 e^{[\gamma_t(\omega)-\alpha]l} = 1 \quad \textbf{(6.1-6)}$$

for the threshold gain constant $\gamma_t(\omega)$ and

$$2[k + \Delta k(\omega)]l = 2\pi m \qquad m = 1, 2, 3, \cdots \quad \textbf{(6.1-7)}$$

for the phase condition. The amplitude condition (6.1-6) can be written as

$$\gamma_t(\omega) = \alpha - \frac{1}{l} \ln r_1 r_2 \quad \textbf{(6.1-8)}$$

which, using (6.1-4), becomes

$$N_t \equiv (N_2 - N_1)_t = \frac{8\pi n^2 t_{spont}}{g(\nu)\lambda^2}\left(\alpha - \frac{1}{l}\ln r_1 r_2\right) \tag{6.1-9}$$

This is the population inversion density at threshold.[2] It is often stated in a different form.[3]

Numerical example—population inversion. To get an order of magnitude estimate of the critical population inversion $(N_2 - N_1)_t$ we use data typical of a 6328 Å He–Ne laser (which is discussed in Section 7.5). The appropriate constants are

$$\begin{aligned} \lambda &= 6.328 \times 10^{-5} \text{ cm} \\ t_{spont} &= 10^{-7} \text{ sec} \\ l &= 10 \text{ cm} \\ 1/g(\nu_0) &\simeq \Delta\nu \simeq 10^9 \text{ Hz} \end{aligned}$$

(The last figure is the Doppler broadened width of the laser transition.)

The cavity decay time t_c is calculated from (6.1-10) assuming $\alpha = 0$ and $R_1 = R_2 = 0.98$. Since $R_1 = R_2 \simeq 1$, we can use the approximation $-\ln x = 1 - x$, $x \simeq 1$, to write

$$t_c \simeq \frac{nl}{c(1-R)} = 2 \times 10^{-8} \text{ second}$$

Using the foregoing data in (6.1-11), we obtain

$$N_t \simeq 10^9 \text{ cm}^{-3}$$

---

[2] It was derived originally by Schawlow and Townes in their classic paper showing the feasibility of lasers; see Reference [1].

[3] Consider the case in which the mirror losses and the distributed losses are all small, and therefore $r_1^2 \simeq 1$, $r_2^2 \simeq 1$ and $\exp(-\alpha l) \simeq 1$. A wave starting with a unit intensity will return after one round trip with an intensity $R_1R_2 \exp(-2\alpha l)$, where $R_1 \equiv r_1^2$ and $R_2 \equiv r_2^2$ are the mirrors' reflectivities. The fractional intensity loss per round trip is thus $1 - R_1R_2 \exp(-2\alpha l)$. Since this loss occurs in a time $2ln/c$, it corresponds to an exponential decay time constant t_c (of the intensity) given by

$$\frac{1}{t_c} = \frac{(1 - R_1R_2e^{-2\alpha l})c}{2ln}$$

Therefore, the energy $\mathcal{E}$ stored in the passive resonator decays as $d\mathcal{E}/dt = -\mathcal{E}/t_c$. Since $R_1R_2e^{-2\alpha l} \simeq 1$, we can use the relation $1 - x \simeq -\ln x$, $x \simeq 1$, to write $1/t_c$ as

$$\frac{1}{t_c} \simeq \frac{c}{n}\left[\alpha - \frac{1}{l}\ln r_1 r_2\right] \tag{6.1-10}$$

and the threshold condition (6.1-9) becomes

$$N_t \equiv (N_2 - N_1)_t = \frac{8\pi n^3 \nu^2 t_{spont}}{c^3 t_c g(\nu)} \tag{6.1-11}$$

where $N \equiv N_2 - N_1$ and the subscript t signifies threshold.

6.2 Oscillation Frequency

The phase part of the start oscillation condition as given by (6.1-7) is satisfied at an infinite set of frequencies, which correspond to the different value of the integer m. If, in addition, the gain condition (6.1-6) is satisfied at one or more of these frequencies, the laser will oscillate at this frequency.

To solve for the oscillation frequency we use (6.1-3) to rewrite (6.1-7) as

$$kl\left[1+\frac{\chi'(\omega)}{2n^2}\right]=m\pi \qquad \textbf{(6.2-1)}$$

Introducing

$$\nu_m=\frac{mc}{2ln} \qquad \textbf{(6.2-2)}$$

so that it corresponds to the mth resonance frequency of the passive $[N_2-N_1=0]$ resonator and, using relations (5.4-15) and (5.4-22)

$$\chi'(\omega)=\frac{2(\nu_0-\nu)}{\Delta\nu}\chi''(\omega)$$

$$\gamma(\omega)=-\frac{k\chi''(\omega)}{n^2}$$

we obtain from (6.2-1)

$$\nu\left[1-\left(\frac{\nu_0-\nu}{\Delta\nu}\right)\frac{\gamma(\nu)}{k}\right]=\nu_m \qquad \textbf{(6.2-3)}$$

where ν_0 is the center frequency of the atomic lineshape function. Let us assume that the laser length is adjusted so that one of its resonance frequencies ν_m is very near ν_0. We anticipate that the oscillation frequency ν will also be close to ν_0 and take advantage of the fact that when $\nu\simeq\nu_0$ the gain constant $\gamma(\nu)$ is a slowly varying function of ν; see Figure 5-4 for $\chi''(\nu)$, which is proportional to $\gamma(\nu)$. We can consequently replace $\gamma(\nu)$ in (6.2-3) by $\gamma(\nu_m)$, obtaining

$$\nu=\nu_m-(\nu_m-\nu_0)\frac{\gamma(\nu_m)c}{2\pi n\Delta\nu} \qquad \textbf{(6.2-4)}$$

as the solution for the oscillation frequency ν.

We can recast (6.2-4) in a slightly different, and easier to use, form by starting with the gain threshold condition (6.1-6). Taking, for simplicity $r_1=r_2=\sqrt{R}$ and assuming that $R\simeq 1$ and $\alpha=0$, we can write (6.1-6) as[4]

$$\gamma_t(\nu)\simeq\frac{1-R}{l}$$

We also take advantage of the relation

$$\Delta\nu_{1/2}\simeq\frac{c(1-R)}{2\pi nl}$$

which relates the passive resonator linewidth $\Delta\nu_{1/2}$ to R [this relation

[4] This result can be obtained by putting $R=1-\Delta$, where $\Delta\ll 1$. Equation (6.1-6) becomes $1+\gamma_t l\simeq 1+\Delta\Rightarrow\gamma_t\simeq\Delta/l=(1-R)/l$.

follows from (4.7-7) for $\alpha = 0$ and $R \approx 1$] and write (6.2-4) as

$$\nu = \nu_m - (\nu_m - \nu_0)\frac{\Delta\nu_{1/2}}{\Delta\nu} \tag{6.2-5}$$

A study of (6.2-5) shows that if the passive cavity resonance ν_m coincides with the atomic line center—that is, $\nu_m = \nu_0$—oscillation takes place at $\nu = \nu_0$. If $\nu_m \neq \nu_0$, oscillation takes place near ν_m but is shifted slightly toward ν_0. This phenomenon is referred to as *frequency pulling* and is demonstrated by Figure 6-2.

6.3 Three- and Four-Level Lasers

Lasers are commonly classified into the so-called "three-level" or "four-level" lasers. An idealized model of a four-level laser is shown in Figure 6-3. The feature characterizing this laser is that the separation E_1 of the terminal laser level from the ground state is large enough that at the temperature T at which the laser is operated, $E_1 \gg kT$. This guarantees that the thermal equilibrium population of level 1 can be neglected. If, in addition, the lifetime t_1 of atoms in level 1 is short compared to t_2, we can neglect N_1 compared to N_2 and the threshold condition (6.1-11) is satisfied when

$$N_2 \simeq N_t \tag{6.3-1}$$

Therefore, laser oscillation begins when the upper laser level acquires, by pumping, a population density equal to the threshold value N_t.

A three-level laser is one in which the lower laser level is either the ground state or a level whose separation E_1 from the ground state is small

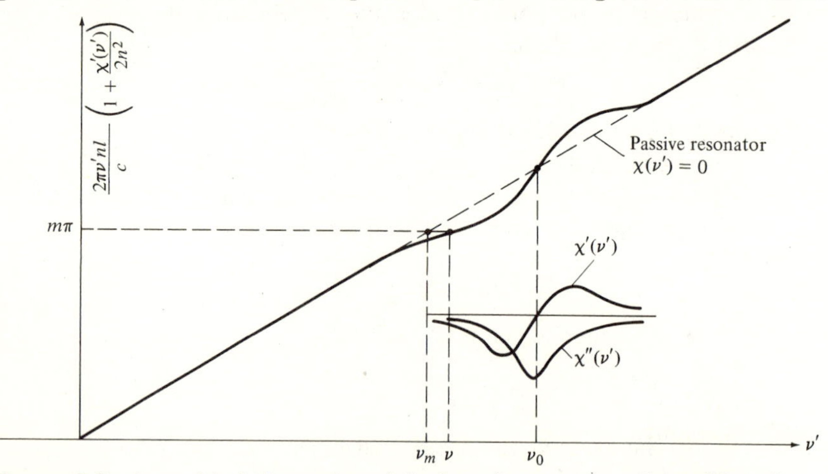

Figure 6-2 A graphical illustration of the laser frequency condition (Equation 6.2-1) showing how the atomic dispersion $\chi'(\nu)$ "pulls" the laser oscillation frequency, ν, from the passive resonator value, ν_m, toward that of the atomic resonance at ν_0.

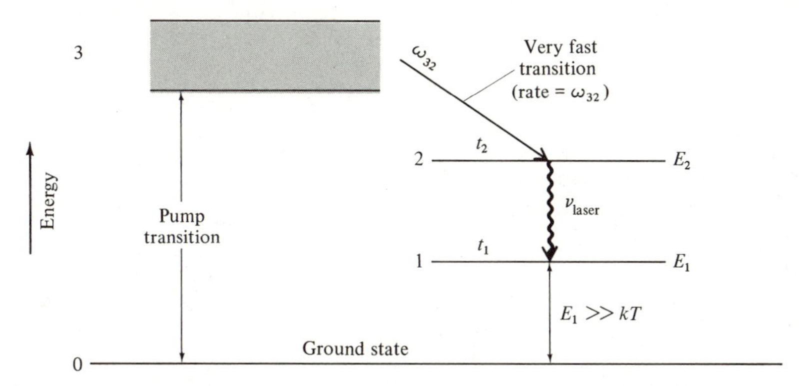

Figure 6-3 Energy-level diagram of an idealized four-level laser.

compared to kT, so that at thermal equilibrium a substantial fraction of the total population occupies this level. An idealized three-level laser system is shown in Figure 6-4.

At a pumping level that is strong enough to create a population $N_2 = N_1 = N_0/2$ in the upper laser level,[5] the optical gain γ is zero, since $\gamma \propto N_2 - N_1 = 0$. To satisfy the oscillation condition the pumping rate has to be further increased until

$$N_2 = \frac{N_0}{2} + \frac{N_t}{2}$$

and

$$N_1 = \frac{N_0}{2} - \frac{N_t}{2} \tag{6.3-2}$$

so $N_2 - N_1 = N_t$. Since in most laser systems $N_0 \gg N_t$, we find by comparing (6.3-1) to (6.3-2) that the pump rate at threshold in a three-level

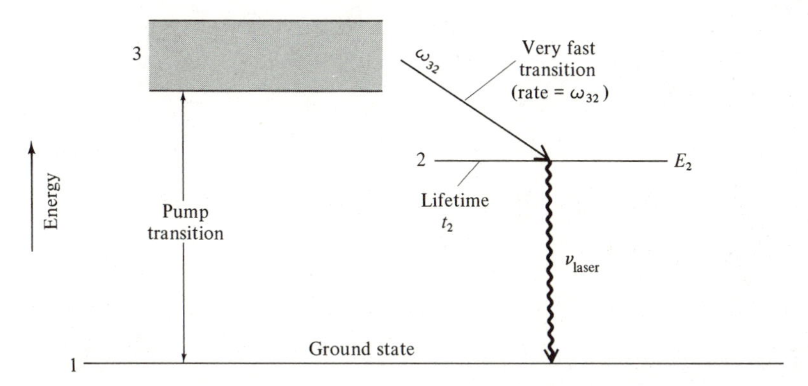

Figure 6-4 Energy-level diagram of an idealized three-level laser.

---

[5] Here we assume that because of the very fast transition rate ω_{32} out of level 3, the population of this level is negligible and $N_1 + N_2 = N_0$, where N_0 is the density of the active atoms.

laser must exceed that of a four-level laser—all other factors being equal—by

$$\frac{(N_2)_{3\text{-level}}}{(N_2)_{4\text{-level}}} \sim \frac{N_0}{2N_t}$$

In the numerical example given in the next chapter we will find that in the case of the ruby laser this factor is ~ 100.

The need to maintain about $N_0/2$ atoms in the upper level of a three-level laser calls for a *minimum* expenditure of power of

$$(P_s)_{3\text{-level}} = \frac{N_0 h\nu V}{2t_2} \tag{6.3-3}$$

and of

$$(P_s)_{4\text{-level}} = \frac{N_t h\nu V}{t_2} = \frac{8\pi n^3 \Delta\nu \nu^2 h\nu}{c^3 t_c (t_2/t_{\text{spont}})} \tag{6.3-4}$$

in a four-level laser. V is the volume. The last two expressions are derived by multiplying the decay rate (atoms per second) from the upper level at threshold, which is $N_0V/2t_2$ and N_tV/t_2 in the two cases, by the energy $h\nu$ per transition. If the decay rate per atom t_2^{-1} (seconds^{-1}) from the upper level is due to spontaneous emission only, we can replace t_2 by t_{spont}. P_s is then equal to the power emitted through fluorescence by atoms within the (mode) volume V at threshold. We will refer to it as the *critical fluorescence* power. In the case of the four-level laser we use (6.1-11) for N_t and, putting $t_2 = t_{\text{spont}}$, obtain

$$(P_s)_{4\text{-level}} = \frac{N_t h\nu V}{t_2} = \frac{8\pi n^3 h \Delta\nu V}{\lambda^3 t_c} \frac{t_{\text{spont}}}{t_2} \tag{6.3-5}$$

where $\Delta\nu = 1/g(\nu_0)$ is the width of the laser transition lineshape.

Numerical example—critical fluorescence power of an Nd^{3+}:glass laser. The critical fluorescence power of an Nd^{3+}:glass laser is calculated using the following data:

$l = 10$ cm

$V = 10$ cm^3

$\lambda = 1.06 \times 10^{-6}$ meter

$R = $ (mirror reflectivity) $= 0.95$

$n \simeq 1.5$

$$t_c \simeq \frac{nl}{(1-R)c} = 10^{-8} \text{ second}$$

$\Delta\nu = 3 \times 10^{12}$ Hz

The Nd^{3+}:glass is a four-level laser system (see Figure 7-11), since level 1 is about 2,000 cm^{-1} above the ground state so that at room temperature

$E_1 \approx 10kT$. We can thus use (6.3-5), obtaining $N_t = 8.5 \times 10^{15}$ cm^{-3} and

$$P_s \simeq 150 \text{ watts}$$

6.4 Power in Laser Oscillators

In Section 6.1 we derived an expression for the threshold population inversion N_t at which the laser gain becomes equal to the losses. We would expect that as the pumping intensity is increased beyond the point at which $N_2 - N_1 = N_t$ the laser will break into oscillation and emit power. In this section we obtain the expression relating the laser power output to the pumping intensity. We also treat the problem of optimum coupling—that is, of the mirror transmission that results in the maximum power output.

Rate equations. Consider an ideal four-level laser such as the one shown in Figure 6-3. We take $E_1 \gg kT$ so that the thermal population of the lower laser level 1 can be neglected. We assume that the critical inversion density N_t is very small compared to the ground-state population, so during oscillation the latter is hardly affected. We can consequently characterize the pumping intensity by R_2 and R_1, the density of atoms pumped per second into levels 2 and 1, respectively. Process R_1, which populates the lower level 1, causes a reduction of the gain and is thus detrimental to the laser operation. In many laser systems, such as discharge gas lasers, considerable pumping into the lower laser level is unavoidable, and therefore a realistic analysis of such systems must take R_1 into consideration.

The rate equations that describe the populations of levels 1 and 2 become

$$\frac{dN_2}{dt} = -N_2\omega_{21} - W_i(N_2 - N_1) + R_2 \qquad \textbf{(6.4-1)}$$

$$\frac{dN_1}{dt} = -N_1\omega_{10} + N_2\omega_{21} + W_i(N_2 - N_1) + R_1 \qquad \textbf{(6.4-2)}$$

ω_{ij} is the decay rate per atom from level i to j; thus the density of atoms per second undergoing decay from i to j is $N_i\omega_{ij}$. If the decay rate is due entirely to spontaneous transitions, then ω_{ij} is equal to the Einstein A_{ij} coefficient introduced in Section 5.1. W_i is the probability per unit time that an atom in level 2 will undergo an *induced* (stimulated) transition to level 1 (or vice versa). W_i, given by (5.2-15), is proportional to the energy density of the radiation field inside the cavity.

Implied in the foregoing rate equations is the fact that we are dealing with a homogeneously broadened system. In an inhomogeneously broadened atomic transition, atoms with different transition frequencies $(E_2 - E_1)/h$ experience different induced transition rates and a single parameter W_i is not sufficient to characterize them.

In a steady-state situation we have $\dot{N}_1 = \dot{N}_2 = 0$. In this case we can solve (6.4-1) and (6.4-2) for N_1 and N_2, obtaining

$$N_2 - N_1 = \frac{R_2\{1 - (\omega_{21}/\omega_{10})[1 + (R_1/R_2)]\}}{W_i + \omega_{21}} \tag{6.4-3}$$

A necessary condition for population inversion in our model is thus $\omega_{21} < \omega_{10}$, which is equivalent to requiring that the lifetime of the upper laser level $\omega_{21}{}^{-1}$ exceed that of the lower one. The effectiveness of the pumping is, according to (6.4-3), reduced by the finite pumping rate R_1 and lifetime $\omega_{10}{}^{-1}$ of level 1 to an effective value

$$R = R_2\left[1 - \frac{\omega_{21}}{\omega_{10}}\left(1 + \frac{R_1}{R_2}\right)\right] \tag{6.4-4}$$

so (6.4-3) can be written as

$$N_2 - N_1 = \frac{R}{W_i + \omega_{21}} \tag{6.4-5}$$

Below the oscillation threshold the induced transition rate W_i is zero (since the oscillation energy density is zero) and $N_2 - N_1$ is, according to (6.4-5), proportional to the pumping rate R. This state of affairs continues until $R = N_t\omega_{21}$, at which point $N_2 - N_1$ reaches the threshold value [see (6.1-11)]

$$N_t = \frac{8\pi n^3\nu^2 t_{\text{spont}}}{c^3 t_c g(\nu_0)} = \frac{8\pi n^3\nu^2 t_{\text{spont}}\Delta\nu}{c^3 t_c} \tag{6.4-6}$$

This is the point at which the gain at ν_0 due to the inversion is large enough to make up *exactly* for the cavity losses (the criterion that was used to derive N_t). Further increase of $N_2 - N_1$ with pumping is impossible in a *steady-state situation*, since it would result in a rate of induced (energy) emission that exceeds the losses so that the field energy stored in the resonator will increase with time in violation of the steady-state assumption.

This argument suggests that, under steady-state conditions, $N_2 - N_1$ must remain equal to N_t regardless of the amount by which the threshold pumping rate is exceeded. An examination of (6.4-5) shows that this is possible, provided W_i is allowed to increase once R exceeds its threshold value $\omega_{21}N_t$, so that the equality

$$N_t = \frac{R}{W_i + \omega_{21}} \tag{6.4-7}$$

is satisfied. Since, according to (5.2-15), W_i is proportional to the energy density in the resonator, (6.4-7) relates the electromagnetic energy stored in the resonator to the pumping rate R. To derive this relationship we first solve (6.4-7) for W_i, obtaining

$$W_i = \frac{R}{N_t} - \omega_{21} \qquad R \geqslant N_t\omega_{21} \tag{6.4-8}$$

The total power generated by stimulated emission is

$$P_e = (N_tV)W_ih\nu \tag{6.4-9}$$

where V is the volume of the oscillating mode. Using (6.4-8) in (6.4-9) gives

$$\frac{P_e}{Vh\nu} = N_t\omega_{21}\left[\frac{R}{N_t\omega_{21}} - 1\right] \qquad R \geqslant N_t\omega_{21} \tag{6.4-10}$$

This expression may be recast in a slightly different form, which we will find useful later on. We use expression (6.4-6) for N_t and, recalling that in our idealized model $\omega_{21}{}^{-1} = t_{\text{spont}}$, obtain

$$\frac{P_e}{Vh\nu} = N_t\omega_{21}\left[\frac{R}{(p/t_c)} - 1\right] \qquad R \geqslant \frac{p}{t_c} \tag{6.4-11}$$

where

$$p = \frac{8\pi n^3\nu^2}{c^3 g(\nu_0)} = \frac{8\pi n^3\nu^2\Delta\nu}{c^3} \tag{6.4-12}$$

According to (4.0-7), p corresponds to the density (meters^{-3}) of radiation modes whose resonance frequencies fall within the atomic transition linewidth $\Delta\nu$—that is, the density of radiation modes that are capable of interacting with the transition.

Returning to the expression for the power output of a laser oscillator (6.4-11), we find that the term $R/(p/t_c)$ is the factor by which the pumping rate R exceeds its threshold value p/t_c. In addition, in an ideal laser system, $\omega_{21} = t_{\text{spont}}{}^{-1}$, so we can identify $N_t\omega_{21}h\nu V$ with the power P_s emitted by the spontaneous emission at threshold, which is defined by (6.3-5). We can consequently rewrite (6.4-11) as

$$P_e = P_s\left(\frac{R}{R_t} - 1\right) \tag{6.4-13}$$

The main attraction of (6.4-13) is in the fact that, in addition to providing an extremely simple expression for the power emitted by the laser atoms, it shows that for each increment of pumping, measured relative to the threshold value, the power increases by P_s. An experimental plot showing the linear relation predicted by (6.4-13) is shown in Figure 6-5.

In the numerical example of Section 6.3, which was based on an Nd^{3+}:glass laser, we obtained $P_s = 150$ watts. We may expect on this basis that the power from this laser for, say $(R/R_t) \simeq 2$ (that is, twice above threshold) will be of the order of 150 watts.

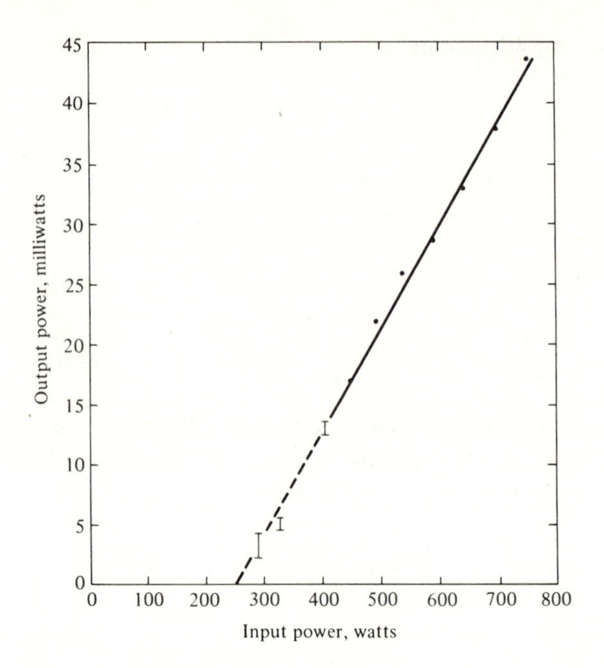

Figure 6-5 Plot of output power versus electric power input to a xenon lamp in a CW $CaF_2:U^{3+}$ laser. Mirror transmittance at 2.61 μm is 0.2 percent, $T = 77°K$. (After Reference [2].)

6.5 Optimum Output Coupling in Laser Oscillators

The total loss encountered by the oscillating laser mode can conveniently be attributed to two different sources: (a) the inevitable residual loss due to absorption and scattering in the laser material, in the mirrors, as well as diffraction losses in the finite diameter reflectors; (b) the (useful) loss due to coupling of output power through the partially transmissive reflector. It is obvious that loss (a) should be made as small as possible since it raises the oscillation threshold without contributing to the output power. The problem of the coupling loss (b), however, is more subtle. At zero coupling (that is, both mirrors have zero transmission) the threshold will be at its minimum value and the power P_e emitted by the atoms will be maximum. But since none of this power is available as output, this is not a useful state of affairs. If, on the other hand, we keep increasing the coupling loss, the increasing threshold pumping will at some point exceed the actual pumping level. When this happens, oscillation will cease and the power output will again be zero. Between these two extremes there exists an optimum value of coupling (that is, mirror transmission) at which the power output is a maximum.

The expression for the population inversion was shown in (6.4-5) to have the form

$$N_2 - N_1 = \frac{R/\omega_{21}}{1 + (W_i/\omega_{21})} \tag{6.5-1}$$

Since the exponential gain constant $\gamma(\nu)$ is, according to (5.3-3), proportional to $N_2 - N_1$, we can use (6.5-1) to write it as

$$\gamma = \frac{\gamma_0}{1 + (W_i/\omega_{21})} \tag{6.5-2}$$

where γ_0 is the unsaturated ($W_i = 0$) gain constant (that is, the gain exercised by a very weak field, so that $W_i \ll \omega_{21}$). We can use (6.4-9) to express W_i in (6.5-2) in terms of the total emitted power P_e and then, in the resulting expression, replace $N_t V h\nu\omega_{21}$ by P_s. The result is

$$\gamma = \frac{\gamma_0}{1 + (P_e/P_s)} \tag{6.5-3}$$

where P_s, the saturation power, is given by (6.3-4). The oscillation condition (6.1-6) can be written as

$$e^{\gamma_t l}(1 - L) = 1 \tag{6.5-4}$$

where $L = 1 - r_1 r_2 \exp(-\alpha l)$ is the fraction of the intensity lost per pass. In the case of small losses ($L \ll 1$), (6.5-4) can be written as

$$\gamma_t l = L \tag{6.5-5}$$

According to the discussion in the introduction to this chapter, once the oscillation threshold is exceeded, the actual gain γ exercised by the laser oscillation is clamped at the threshold value γ_t regardless of the pumping. We can thus replace γ by γ_t in (6.5-3) and, solving for P_e, obtain

$$P_e = P_s\left(\frac{g_0}{L} - 1\right) \tag{6.5-6}$$

where $g_0 = \gamma_0 l$ (that is, the unsaturated gain per pass in nepers). P_e, we recall, is the *total* power given off by the atoms due to stimulated emission. The total loss per pass L can be expressed as the sum of the residual (unavoidable) loss L_i and the useful mirror transmission[6] T, so

$$L = L_i + T \tag{6.5-7}$$

The fraction of the total power P_e that is coupled out of the laser as useful output is thus $T/(T + L_i)$. Therefore, using (6.5-6) we can write the

---

[6] For the sake of simplicity we can imagine one mirror as being perfectly reflecting, whereas the second (output) mirror has a transmittance T.

(useful) power output as

$$P_o = P_s\left(\frac{g_0}{L_i + T} - 1\right)\frac{T}{T + L_i} \tag{6.5-8}$$

Replacing P_s in (6.5-8) by the right side of (6.3-5), and recalling from (4.7-2) that for small losses

$$t_c = \frac{nl}{(L_i + T)c} = \frac{nl}{Lc} \tag{6.5-9}$$

Equation (6.5-8) becomes

$$P_o = I_s A T\left(\frac{g_0}{L_i + T} - 1\right) = \frac{8\pi n^2 h\nu \Delta\nu A}{\lambda^2(t_2/t_{\text{spont}})} T\left(\frac{g_0}{L_i + T} - 1\right) \tag{6.5-10}$$

where $A = V/l$ is the cross-sectional area of the mode (assumed constant) and I_s is the saturation intensity as given in (5.6-9). Maximizing P_0 with respect to T by setting $\partial P_0/\partial T = 0$ yields

$$T_{\text{opt}} = -L_i + \sqrt{g_0 L_i} \tag{6.5-11}$$

as the condition for the mirror transmission that yields the maximum power output.

The expression for the power output at optimum coupling is obtained by substituting (6.5-11) for T in (6.5-10). The result, using (5.6-9), is

$$(P_o)_{\text{opt}} = \frac{8\pi n^2 h\nu \Delta\nu A}{(t_2/t_{\text{spont}})\lambda^2}(\sqrt{g_0} - \sqrt{L_i})^2 = I_s A(\sqrt{g_0} - \sqrt{L_i})^2$$

$$\equiv S(\sqrt{g_0} - \sqrt{L_i})^2 \tag{6.5-12}$$

where the parameter S is defined by (6.5-12) and is independent of the excitation level (pumping) or losses.

Theoretical plots of (6.5-10) with L_i as a parameter are shown in Figure 6-6. Also shown are experimental data points obtained in a He–Ne

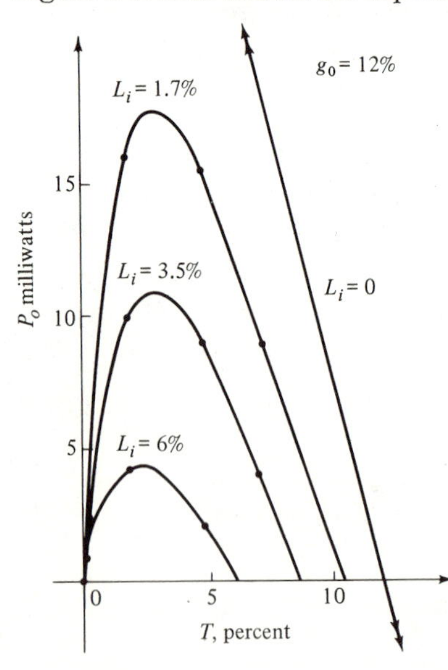

Figure 6-6 Useful power output (P_o) versus mirror transmission T for various values of internal loss L_i in an He-Ne 6328 Å laser. (After Laures, *Phys. Letters*, vol. 10, p. 61, 1964.)

6328 Å laser. Note that the value of g_0 is given by the intercept of the $L_i = 0$ curve and is equal to 12 percent. The existence of an optimum coupling resulting in a maximum power output for each L_i is evident.

It is instructive to consider what happens to the energy $\mathcal{E}$ stored in the laser resonator as the coupling T is varied. A little thinking will convince us that $\mathcal{E}$ is proportional to P_o/T.[7] A plot of P_o (taken from Figure 6-6) and $\mathcal{E} \propto P_o/T$ as a function of the coupling T is shown in Figure 6-7. As we may expect, $\mathcal{E}$ is a monotonically decreasing function of T.

6.6 Multimode Laser Oscillation and Mode Locking

In this section we contemplate the effect of homogeneous or inhomogeneous broadening (in the sense described in Section 5.1) on the laser oscillation.

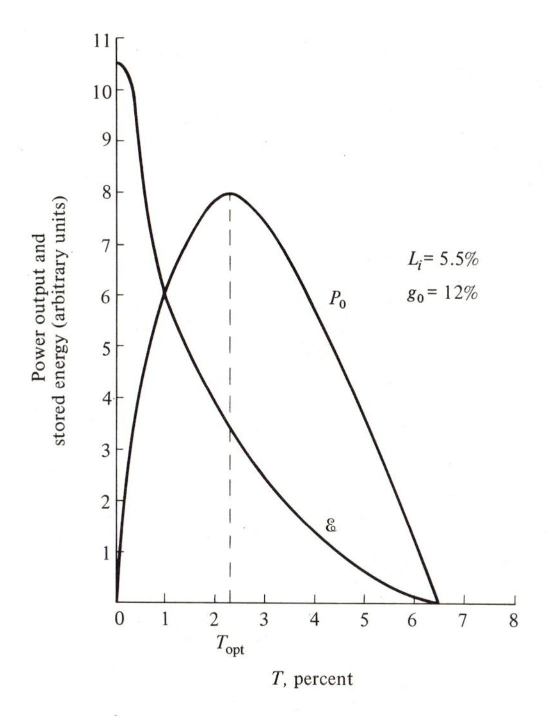

Figure 6-7 Power output P_o and stored energy $\mathcal{E}$ plotted against mirror transmission T.

[7] The internal one-way power P_i incident on the mirrors is related, by definition, to P_o by $P_o = P_iT$. The total energy $\mathcal{E}$ is proportional to P_i.

We start by reminding ourselves of some basic results pertinent to this discussion:

1. The actual gain constant prevailing inside a laser oscillator *at the oscillation frequency* ν is clamped, at steady state, at a value

$$\gamma_t(\nu) = \alpha - \frac{1}{l} \ln r_1 r_2 \tag{6.1-8}$$

2. The gain constant of a distributed laser medium is

$$\gamma(\nu) = (N_2 - N_1) \frac{c^2}{8\pi n^2 \nu^2 t_{\text{spont}}} g(\nu) \tag{5.3-3}$$

3. The optical resonator can support oscillations, provided sufficient gain is present to overcome losses, at frequencies[8] ν_q separated by

$$\nu_{q+1} - \nu_q = \frac{c}{2nl} \tag{4.6-4}$$

Now consider what happens to the gain constant $\gamma(\nu)$ inside a laser oscillator as the pumping is increased from some value below threshold. Operationally, we can imagine an extremely weak wave of frequency ν launched into the laser medium and then measuring the gain constant $\gamma(\nu)$ as "seen" by this signal as ν is varied.

We treat first the case of a homogeneous laser. Below threshold the inversion $N_2 - N_1$ is proportional to the pumping rate and $\gamma(\nu)$, which is given by (5.3-3), is proportional to $g(\nu)$. This situation is illustrated by curve A in Figure 6-8(a). The spectrum (4.6-4) of the passive resonances is shown in Figure 6-8(b). As the pumping rate is increased, the point is reached at which the gain per pass at the center resonance frequency ν_0 is equal to the average loss per pass. This is shown in curve B. At this point, oscillation at ν_0 starts. An increase in the pumping cannot increase the inversion since this will cause $\gamma(\nu_0)$ to increase beyond its clamped value as given by Equation (6.1-8). Since the spectral lineshape function $g(\nu)$ describes the response of each individual atom, all the atoms being identical, it follows that the gain profile $\gamma(\nu)$ above threshold as in curve C is identical to that at threshold curve B.[9] The gain at other frequencies—such as ν_{-1}, ν_1, ν_{-2}, ν_2, and so forth—remains below the threshold value so that the ideal homogeneously broadened laser can oscillate only at a single frequency.

---

[8] The high-order transverse modes discussed in Section 4.5 are ignored here.

[9] Further increase in pumping, and the resulting increase in optical intensity, will eventually cause a broadening of $\gamma(\nu)$ due to the shortening of the lifetime by induced emission.

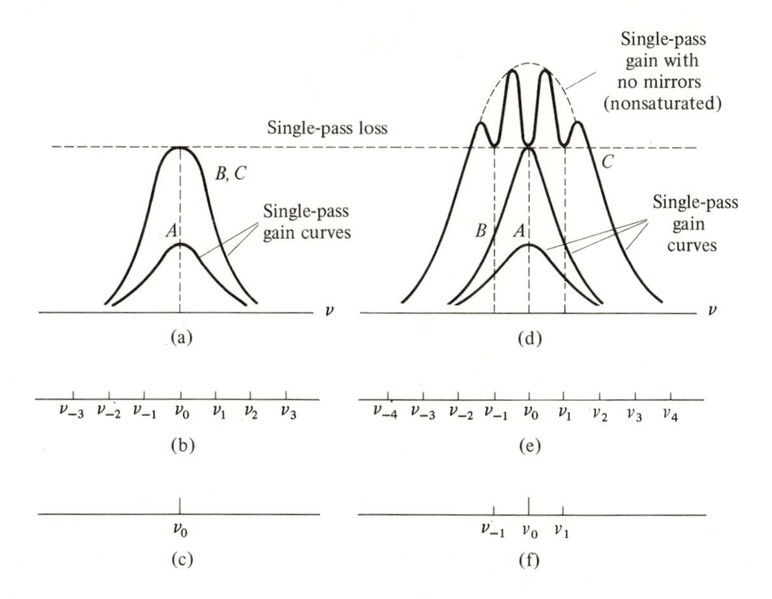

Figure 6-8 (**a**) Single-pass gain curves for a homogeneous atomic system (*A*—below threshold; *B*—at threshold; *C*—well above threshold). (**b**) Mode spectrum of optical resonator. (**c**) Oscillation spectrum (only one mode oscillates). (**d**) Single-pass gain curves for an inhomogeneous atomic system (*A*—below threshold; *B*—at threshold; *C*—well above threshold). (**e**) Mode spectrum of optical resonator. (**f**) Oscillation spectrum for pumping level *C*, showing three oscillating modes.

In the extreme inhomogeneous case, the individual atoms can be considered as being all different from one another and as acting independently. The lineshape function $g(\nu)$ reflects the distribution of the transition frequencies of the individual atoms. The gain profile $\gamma(\nu)$ below threshold is proportional to $g(\nu)$ and its behavior is similar to that of the homogeneous case. Once threshold is reached as in curve B, the gain at ν_0 remains clamped at the threshold value. There is no reason, however, why the gain at other frequencies should not increase with further pumping. This gain is due to atoms that do not communicate with those contributing to the gain at ν_0. Further pumping will thus lead to oscillation at additional longitudinal-mode frequencies as shown in curve C. Since the gain at each oscillating frequency is clamped, the gain profile curve acquires depressions at the oscillation frequencies. This phenomenon is referred to as "hole burning" [7].

A plot of the output frequency spectrum showing the multimode oscillation of a He–Ne 0.6328-μm laser is shown in Figure 6-9.

Mode locking. We have argued above that in an inhomogeneously broadened laser, oscillation can take place at a number of frequencies, which are

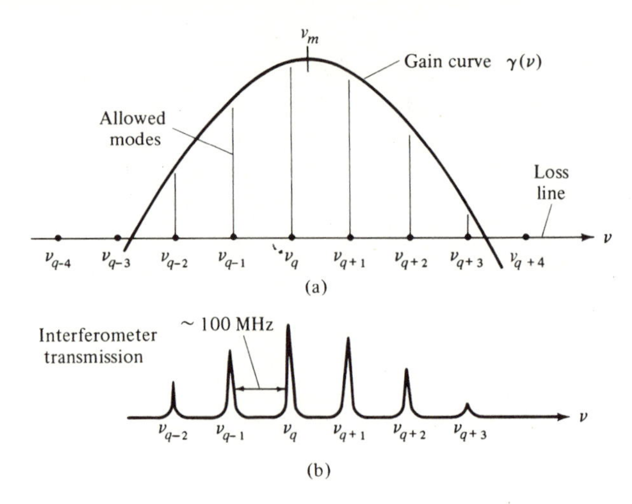

Figure 6-9 (**a**) Inhomogeneously broadened Doppler gain curve of the 6328 Å Ne transition and position of allowed longitudinal-mode frequencies.
(**b**) Intensity versus frequency profile of an oscillating He-Ne laser. Six modes have sufficient gain to oscillate (After Reference [8].)

separated by (assuming $n = 1$)

$$\omega_q - \omega_{q-1} = \frac{\pi c}{l} \equiv \omega$$

Now consider the total optical electric field resulting from such multimode oscillation at some arbitrary point, say next to one of the mirrors, in the optical resonator. It can be taken, using complex notation, as

$$e(t) = \sum_n E_n e^{i[(\omega_o + n\omega)t + \phi_n]} \tag{6.6-1}$$

where the summation is extended over the oscillating modes and ω_0 is chosen, arbitrarily, as a reference frequency. ϕ_n is the phase of the nth mode. One property of (6.6-1) is that $e(t)$ is periodic in $T \equiv 2\pi/\omega = 2l/c$, which is the round-trip transit time inside the resonator.

$$\begin{aligned} e(t + T) &= \sum_n E_n \exp\left\{i\left[(\omega_0 + n\omega)\left(t + \frac{2\pi}{\omega}\right) + \phi_n\right]\right\} \\ &= \sum_n E_n \exp\{i[(\omega_0 + n\omega)t + \phi_n]\} \exp\left\{i\left[2\pi\left(\frac{\omega_0}{\omega} + n\right)\right]\right\} \\ &= e(t) \end{aligned} \tag{6.6-2}$$

Since ω_0/ω is an integer ($\omega_0 = m\pi c/l$)

$$\exp\left[2\pi i\left(\frac{\omega_0}{\omega} + n\right)\right] = 1$$

Note that the periodic property of $e(t)$ depends on the fact that the phases ϕ_n are fixed. In typical lasers the phases ϕ_n are likely to vary randomly with time. This causes the intensity of the laser output to fluctuate randomly[10] and greatly reduces its usefulness for many applications where temporal coherence is important.

Two ways in which the laser can be made coherent are: First, make it possible for the laser to oscillate at a single frequency only so that mode interference is eliminated. This can be achieved in a variety of ways, including shortening the resonator length l, thus increasing the mode spacing ($\omega = \pi c/l$) to a point where only one mode has sufficient gain to oscillate. The second approach is to force the modes phases ϕ_n to maintain their relative values. This is the so-called "mode locking" technique, which (as shown previously) causes the oscillation intensity to consist of a periodic train with a period of $T = 2l/c = 2\pi/\omega$.

One of the most useful forms of mode locking results when the phases ϕ_n are made equal to zero. To simplify the analysis of this case assume that there are N oscillating modes with equal amplitudes. Taking $E_n = 1$ and $\phi_n = 0$ in (6.6-1) gives

$$e(t) = \sum_{-(N-1)/2}^{(N-1)/2} e^{i(\omega_0 + n\omega)t} \tag{6.6-3}$$

$$= e^{i\omega_0 t} \frac{\sin(N\omega t/2)}{\sin(\omega t/2)} \tag{6.6-4}$$

The average laser power output is proportional to $e(t)e^*(t)$ and is given by[11]

$$P(t) \propto \frac{\sin^2(N\omega t/2)}{\sin^2(\omega t/2)} \tag{6.6-5}$$

Some of the analytic properties of $P(t)$ are immediately apparent:

1. The power is emitted in a form of a train of pulses with a period $T = 2\pi/\omega = 2l/c$.
2. The peak power, $P(sT)$ (for $s = 1, 2, 3, \cdots$), is equal to N times the average power, where N is the number of modes locked together.
3. The peak field amplitude is equal to N times the amplitude of a single mode.
4. The individual pulse width, defined as the time from the peak to the first zero is $\tau = T/N$. The number of oscillating modes can be estimated by $N \simeq \Delta\omega/\omega$—that is, the ratio of the transition

---

[10] It should be noted that this fluctuation takes place because of random interference between modes and not because of intensity fluctuations of individual modes.

[11] The averaging is performed over a time that is long compared with the optical period $2\pi/\omega_0$ but short compared with the modulation period $2\pi/\omega$.

lineshape width $\Delta\omega$ to the frequency spacing ω between modes. Using this relation, as well as $T = 2\pi/\omega$ in $\tau = T/N$, we obtain

$$\tau \sim \frac{2\pi}{\Delta\omega} = \frac{1}{\Delta\nu} \qquad \textbf{(6.6-6)}$$

Thus the length of the mode-locked pulses is approximately the inverse of the gain linewidth.

A theoretical plot of $\sqrt{P(t)}$ as given by (6.6-5) for the case of five modes ($N = 5$) is shown in Figure 6-10. The ordinate may also be considered as being proportional to the instantaneous field amplitude.

The foregoing discussion was limited to the consideration of mode locking as a function of time. It is clear, however, that since the solution of Maxwell's equation in the cavity involves traveling waves (a standing wave can be considered as the sum of two waves traveling in opposite directions), mode locking causes the oscillation energy of the laser to be condensed into a packet that travels back and forth between the mirrors with the velocity of light c. The pulsation period $T = 2l/c$ corresponds simply to the time interval between two successive arrivals of the pulse at the mirror. The spatial length of the pulse L_p must correspond to its time duration multiplied by its velocity c. Using $\tau = T/N$ we obtain

$$L_p \sim c\tau = \frac{cT}{N} = \frac{2\pi c}{\omega N} = \frac{2l}{N} \qquad \textbf{(6.6-7)}$$

We can verify the last result by taking the basic resonator mode as being proportional to $\sin k_n x \sin \omega_n t$; the total optical field is then

$$e(z, t) = \sum_{n=-(N-1)/2}^{(N-1)/2} \sin\left[\frac{(m+n)\pi}{l} z\right] \sin\left[(m+n)\frac{\pi c}{l} t\right] \qquad \textbf{(6.6-8)}$$

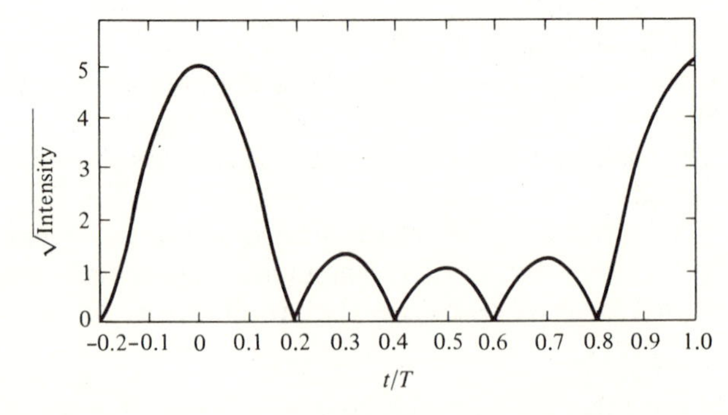

Figure 6-10 Theoretical plot of optical field amplitude [$\sqrt{P(t)} \propto \sin(N\omega t/2)/\sin(\omega t/2)$] resulting from phase locking of five ($N = 5$) equal-amplitude modes separated from each other by a frequency interval $\omega = 2\pi/T$.

where, using (4.6-4), $\omega_n = (m + n)(\pi c/l)$, $k_n = \omega_n/c$, and m is the integer corresponding to the central mode. We can rewrite (6.6-8) as

$$e(z, t) = \frac{1}{2} \sum_{n=-(N-1)/2}^{(N-1)/2} \left\{ \cos\left[(m + n) \frac{\pi}{l} (z - ct) \right] - \cos\left[(m + n) \frac{\pi}{l} (z + ct) \right] \right\} \quad \textbf{(6.6-9)}$$

which can be shown to have the spatial and temporal properties described previously. Figure 6-11 shows a spatial plot of (6.6-9) at time t.

Methods of mode locking. In the preceding discussion we considered the consequences of fixing the phases of the longitudinal modes of a laser-mode locking. Mode locking can be achieved by modulating the losses (or gain) of the laser at a radian frequency $\omega = \pi c/l$, which is equal to the intermode frequency spacing. The theoretical proof of mode locking by loss modulation (References [2], [9], and [10]) is rather formal, but a good plausibility argument can be made as follows: As a form of loss modulation consider a thin shutter inserted inside the laser resonator. Let the shutter be closed (high optical loss) most of the time except for brief periodic openings for a duration of τ_{open} every $T = 2l/c$ seconds. This situation is illustrated by Figure 6-12. A single laser mode will not oscillate in this case because of the high losses (we assume that τ_{open} is too short to allow the oscillation to build up during each opening). The same applies to multimode oscillation with arbitrary phases. There is one exception, however. If the phases were "locked" as in (6.6-3), the energy distribution inside the resonator would correspond to that shown in Figure 6-11 and would consist of a narrow ($L_p \simeq 2l/N$) traveling pulse. If this pulse should arrive at the shutter's position when it is open and if the pulse (temporal) length τ is short compared to the opening time τ_{open}, the mode-locked pulse will be "unaware" of the shutter's existence and, consequently, *will not be attenuated by it.* We

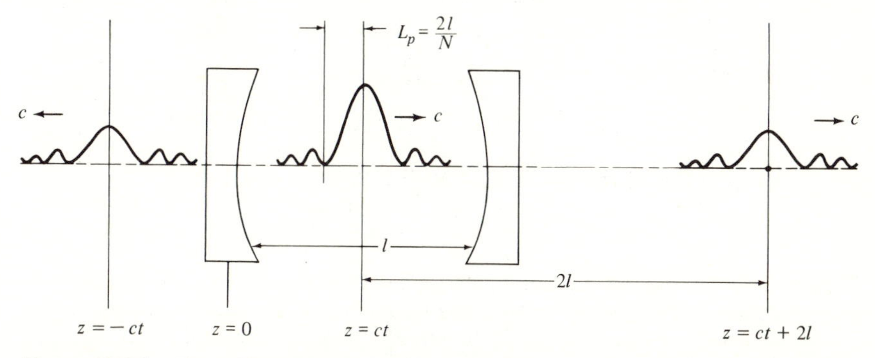

Figure 6-11 Traveling pulse of energy resulting from the mode locking of N laser modes; based on Equation (6.6-9).

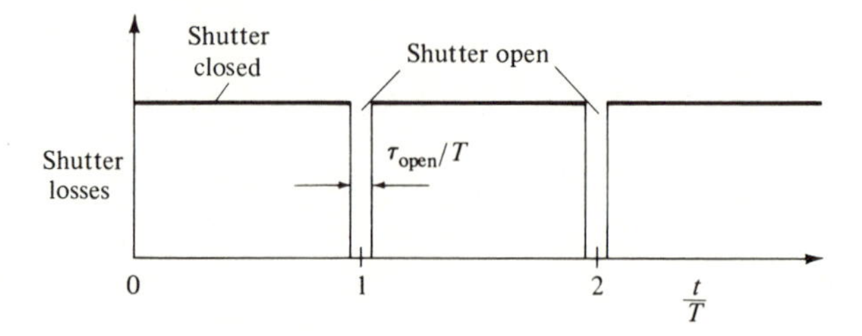

Figure 6-12 Periodic losses introduced by a shutter to induce mode locking. The presence of these losses favors the choice of mode phases that results in a pulse passing through the shutter during open intervals—that is, mode locking.

may thus reach the conclusion that loss modulation causes mode locking through some kind of "survival of the fittest" mechanism. In reality the periodic shutter chops off any intensity tails acquired by the mode-locked pulses due to a "wandering" of the phases from their ideal ($\phi_n = 0$) values. This has the effect of continuously restoring the phases.

An experimental setup used to mode-lock a He–Ne laser is shown in Figure 6-13; the periodic loss [11] is introduced by Bragg diffraction (see

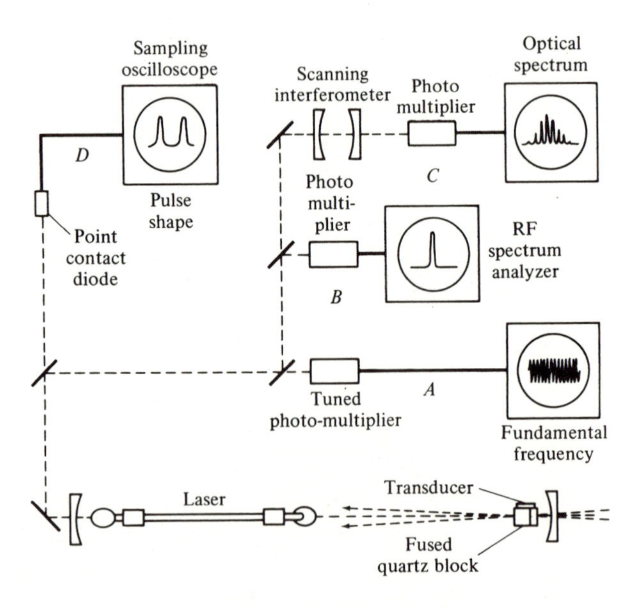

Figure 6-13 Experimental setup for laser mode locking by acoustic (Bragg) loss modulation. The loss is due to Bragg diffraction of the main laser beam by a standing acoustic wave. Parts A, B, C, and D of the experimental setup are designed to display the fundamental component of the intensity modulation, the power spectrum of the intensity modulation, the power spectrum of the optical field $e(t)$, and the optical intensity, respectively. (After Reference [12].)

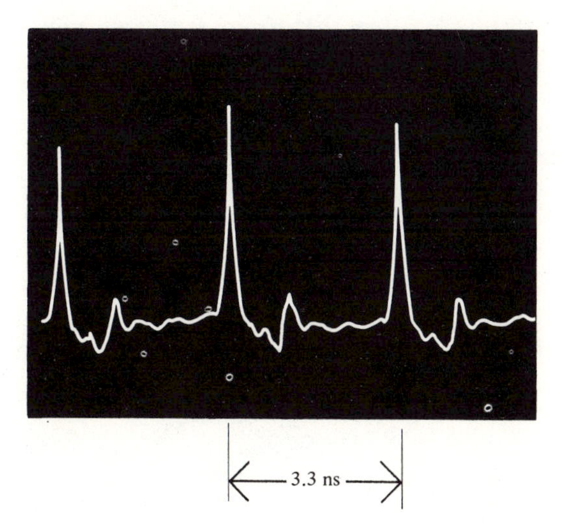

Figure 6-14 Power output as a function of time of a mode-locked Nd^{3+}:YAG laser. Width of pulse in display is limited by the detector. (After Reference [12].)

Sections 12.2 and 12.3) of a portion of the laser intensity from a standing acoustic wave. The standing-wave nature of the acoustic oscillation causes the strain to have a form

$$S(z, t) = S_0 \cos \omega_a t \cos k_a z \tag{6.6-10}$$

where the acoustic velocity is $v_a = \omega_a/k_a$. Since the change in the index of refraction is to first order, proportional to the strain $S(z, t)$, we can interpret (6.6-10) as a phase diffraction grating (see Sections 12.2, 3) with a spatial period $2\pi/k_a$, which is equal to the acoustic wavelength. The diffraction loss of the incident laser beam due to the grating reaches its peak twice in each acoustic period when $S(z, t)$ has its maximum and minimum values. The loss modulation frequency is thus $2\omega_a$ and mode locking occurs when $2\omega_a = \omega$, where ω is the (radian) frequency separation between two longitudinal laser modes.

Figure 6-14 shows the pulses resulting from mode locking a Nd^{3+}:YAG laser.

Mode locking occurs spontaneously in some lasers if the optical path contains a saturable absorber (an absorber whose opacity decreases with increasing optical intensity). This method is used to induce mode locking in the high-power pulsed solid state lasers [13, 15] and in continuous dye lasers. This is due to the fact that such a dye will absorb less power from a mode-locked train of pulses than from a random phase oscillation of many modes [2], since the first form of oscillation leads to the highest possible peak intensities, for a given average power from the laser, and is consequently attenuated less severely. For arguments identical with those advanced in connection with the periodic shutter (see discussion following [6.6-9]), it follows that the presence of a saturable absorber in the laser cavity will "force" the laser, by a "survival of the fittest" mechanism, to lock its modes' phases as in (6.6-9).

The use of passive mode locking in the continuous Rhodamine 6G dye laser yields pulses with a duration of $\sim 3 \times 10^{-13}$ seconds [28].

Table 6-1 lists some of the lasers commonly used in mode locking and the observed pulse durations. An analysis of mode locking in homogeneously broadened lasers is given in Appendix B.

Table 6-1 SOME LASER SYSTEMS, THEIR GAIN LINEWIDTH $\Delta\nu$, AND THE LENGTH OF THEIR PULSES IN THE MODE-LOCKED OPERATION

LASER MEDIUM	$\Delta\nu$, Hz	$(\Delta\nu)^{-1}$, SECONDS	OBSERVED PULSE DURATION, SECONDS
He–Ne (0.6328 μm) CW	$\sim 1.5 \times 10^{9}$	6.66×10^{-10}	$\sim 6 \times 10^{-10}$
Nd:YAG (1.06 μm) CW	$\sim 1.2 \times 10^{10}$	8.34×10^{-11}	$\sim 7.6 \times 10^{-11}$
Ruby (0.6934 μm) pulsed	6×10^{10}	1.66×10^{-11}	$\sim 1.2 \times 10^{-11}$
Nd^{3+}:glass pulsed	3×10^{12}	3.33×10^{-13}	$\sim 3 \times 10^{-13}$
Rhodamine 6G (dye laser) (0.6 μm)	5×10^{12}	2×10^{-13}	4×10^{-13}

6.7 Giant Pulse (*Q*-switched) Lasers

The technique of "*Q*-switching" is used to obtain intense and short bursts of oscillation from lasers; see References [16]–[18]. The quality factor Q of the optical resonator is degraded (lowered) by some means during the pumping so that the gain (that is, inversion $N_2 - N_1$) can build up to a very high value without oscillation. (The spoiling of the Q raises the threshold inversion to a value higher than that obtained by pumping.) When the inversion reaches its peak, the Q is restored abruptly to its (ordinary) high value. The gain (per pass) in the laser medium is now well above threshold. This causes an extremely rapid buildup of the oscillation and a simultaneous exhaustion of the inversion by stimulated $2 \rightarrow 1$ transitions. This process converts most of the energy that was stored by atoms pumped into the upper laser level into photons, which are now inside the optical resonator. These proceed to bounce back and forth between the reflectors with a fraction $(1 - R)$ "escaping" from the resonator each time. This causes a decay of the pulse with a characteristic time constant (the "photon lifetime") given in (4.7-3) as

$$t_c \simeq \frac{nl}{c(1 - R)}$$

Both experiment and theory indicate that the total evolution of giant laser pulse as described above is typically completed in $\sim 2 \times 10^{-8}$ second. We will consequently neglect the effect of population relaxation and pumping that take place during the pulse. We will also assume that the switching of the Q from the low to the high value is accomplished instantaneously.

The laser is characterized by the following variables: ϕ; the total number of photons in the optical resonator, $n \equiv (N_2 - N_1)V$; the total inversion; and t_c, the decay time constant for photons in the *passive* resonator. The exponential gain constant γ is proportional to n. The radiation intensity I thus grows with distance as $I(z) = I_0 \exp(\gamma z)$ and $dI/dz = \gamma I$. An observer traveling with the wave velocity will see it grow at a rate

$$\frac{dI}{dt} = \frac{dI}{dz}\frac{dz}{dt} = \gamma\left(\frac{c}{n}\right)I$$

and thus the temporal exponential growth constant is $\gamma(c/n)$. If the laser rod is of length L while the resonator length is l, then only a fraction L/l of the photons is undergoing amplification at any one time and the average growth constant is $\gamma c(L/nl)$. We can thus write

$$\frac{d\phi}{dt} = \phi\left(\frac{\gamma cL}{nl} - \frac{1}{t_c}\right) \tag{6.7-1}$$

where $-\phi/t_c$ is the decrease in the number of resonator photons per unit time due to incidental resonator losses and to the output coupling. Defining a dimensionless time by $\tau = t/t_c$ we obtain, upon multiplying (6.7-1) by t_c,

$$\frac{d\phi}{d\tau} = \phi\left[\left(\frac{\gamma}{nl/cLt_c}\right) - 1\right] = \phi\left[\frac{\gamma}{\gamma_t} - 1\right]$$

where $\gamma_t = (nl/cLt_c)$ is the minimum value of the gain constant at which oscillation (that is, $d\phi/d\tau = 0$) can be sustained. Since, according to (5.3-3) γ is proportional to the inversion n, the last equation can also be written as

$$\frac{d\phi}{d\tau} = \phi\left[\frac{n}{n_t} - 1\right] \tag{6.7-2}$$

where $n_t = N_t V$ is the total inversion at threshold as given by (6.1-9).

The term $\phi(n/n_t)$ in (6.7-2) gives the number of photons generated by induced emission per unit of normalized time. Since each generated photon results from a single transition, it corresponds to a decrease of $\Delta n = -2$ in the total inversion. We can thus write directly

$$\frac{dn}{d\tau} = -2\phi\frac{n}{n_t} \tag{6.7-3}$$

The coupled pair of equations, (6.7-2) and (6.7-3), describes the evolution of ϕ and n. It can be solved easily by numerical techniques. Before we proceed to give the results of such calculation we will consider some of the consequences that can be deduced analytically.

Dividing (6.7-2) by (6.7-3) results in

$$\frac{d\phi}{dn} = \frac{n_t}{2n} - \frac{1}{2}$$

and, by integration,

$$\phi - \phi_i = \frac{1}{2}\left[n_t \ln\frac{n}{n_i} - (n - n_i)\right]$$

Assuming that ϕ_i, the initial number of photons in the cavity, is negligible, we obtain

$$\phi = \frac{1}{2}\left[n_t \ln\frac{n}{n_i} - (n - n_i)\right] \tag{6.7-4}$$

for the relation between the number of photons ϕ and the inversion n at any moment. At $t \gg t_c$ the photon density ϕ will be zero so that setting $\phi = 0$ in (6.7-4) results in the following expression for the final inversion n_f:

$$\frac{n_f}{n_i} = \exp\left[\frac{n_f - n_i}{n_t}\right] \tag{6.7-5}$$

This equation is of the form $(x/a) = \exp(x - a)$, where $x = n_f/n_t$ and $a = n_i/n_t$, so that it can be solved graphically (or numerically) for n_f/n_i as a function of n_i/n_t.[12] The result is shown in Figure 6-15. We notice that the fraction of the energy originally stored in the inversion which is converted into laser oscillation energy is $(n_i - n_f)/n_i$ and that it tends to unity as n_i/n_t increases.

The instantaneous power output of the laser is given by $P = \phi h\nu/t_c$, or, using (6.7-4), by

$$P = \frac{h\nu}{2t_c}\left[n_t \ln\frac{n}{n_i} - (n - n_i)\right] \tag{6.7-6}$$

Of special interest to us is the peak power output. Setting $\partial P/\partial n = 0$ we find that maximum power occurs when $n = n_t$. Putting $n = n_t$ in (6.7-6)

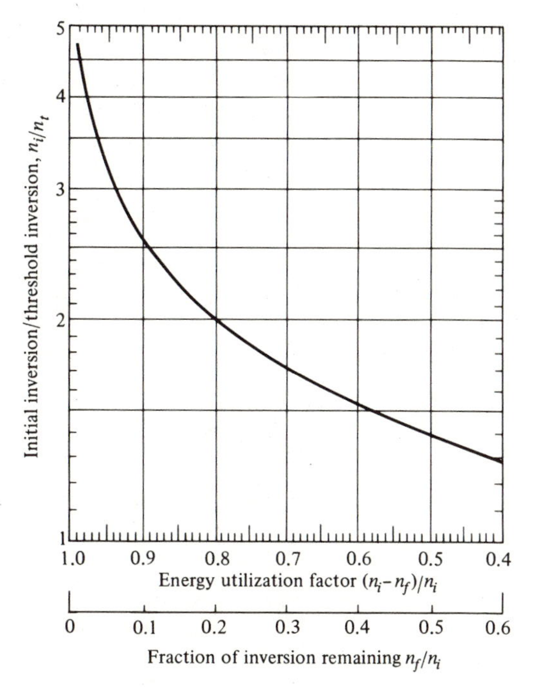

Figure 6-15 Energy utilization factor $(n_i - n_f)/n_i$ and inversion remaining after the giant pulse. (After Reference [19].)

[12] This can be done by assuming a value of a and finding the corresponding x at which the plots of x/a and $\exp(x - a)$ intersect.

gives

$$P_p = \frac{h\nu}{2t_c}\left[n_t \ln \frac{n_t}{n_i} - (n_t - n_i)\right] \tag{6.7-7}$$

for the peak power. If the initial inversion is well in excess of the (high-Q) threshold value (that is, $n_i \gg n_t$), we obtain from (6.7-7)

$$(P_p)_{n_i \gg n_t} \simeq \frac{n_i h\nu}{2t_c} \tag{6.7-8}$$

Since the power P at any moment is related to the number of photons ϕ by $P = \phi h\nu/t_c$ it follows from (6.7-8) that the maximum number of stored photons inside the resonator is $n_i/2$. This can be explained by the fact that if $n_i \gg n_t$, the buildup of the pulse to its peak value occurs in a time short compared to t_c so that at the peak of the pulse, when $n = n_t$, most of the photons that were generated by stimulated emission are still present in the resonator. Moreover, since $n_i \gg n_t$, the number of these photons $(n_i - n_t)/2$ is very nearly $n_i/2$.

A typical numerical solution of (6.7-2) and (6.7-3) is given in Figure 6-16.

To initiate the pulse we need, according to (6.7-2) and (6.7-3), to have $\phi_i \neq 0$. Otherwise the solution is trivial ($\phi = 0$, $n = n_i$). The appropriate value of ϕ_i is usually estimated on the basis of the number of spontaneously emitted photons within the acceptance solid angle of the laser mode at $t = 0$. We also notice, as discussed above, that the photon density, hence

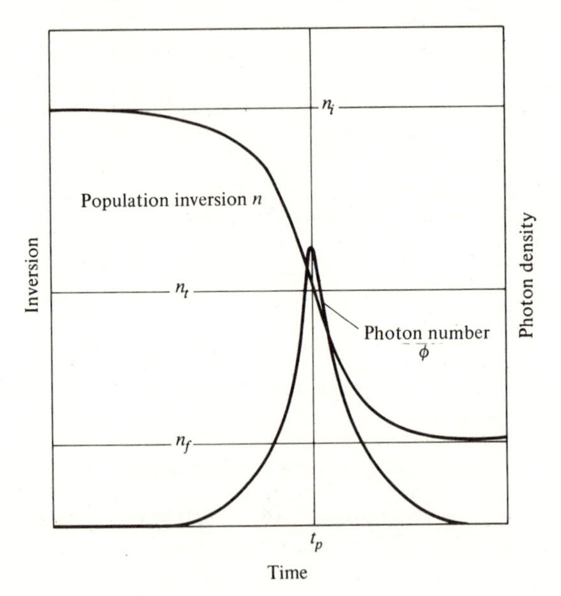

Figure 6-16 Inversion and photon density during a giant pulse. (After Reference [19].)

the power, reaches a peak when $n = n_t$. The energy stored in the cavity ($\propto \phi$) at this point is maximum, so stimulated transitions from the upper to the lower laser levels continue to reduce the inversion to a final value $n_f < n_t$.

Numerical solutions of (6.7-2) and (6.7-3) corresponding to different initial inversions n_i/n_t are shown in Figure 6-17. We notice that for $n_i \gg n_t$ the rise time becomes short compared to t_c but the fall time approaches a value nearly equal to t_c. The reason is that the process of stimulated emission is essentially over at the peak of the pulse ($\tau = 0$)

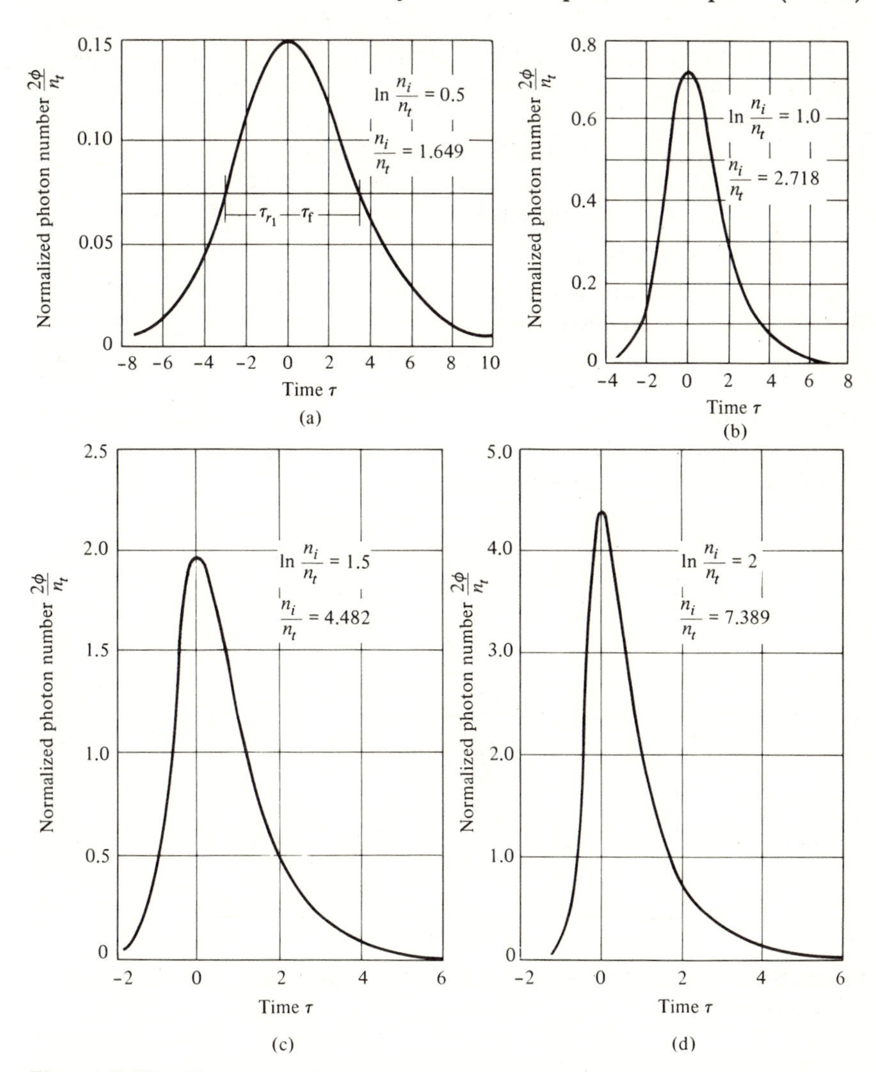

Figure 6-17 Photon number vs. time in central region of giant pulse. Time is measured in units of photon lifetime. (After Reference [19].)

and the observed output is due to the free decay of the photons in the resonator.

In Figure 6-18 we show an actual oscilloscope trace of a giant pulse. Giant laser pulses are used extensively in applications which require high peak powers and short duration. These applications include experiments in nonlinear optics, ranging, material machining and drilling, initiation of chemical reactions, and plasma diagnostics.

Numerical example—giant pulse ruby laser. Consider the case of pink ruby with a chromium ion density of $N = 1.58 \times 10^{19}$ cm^{-3}. Its absorption coefficient is taken from Figure 7-4, where it corresponds to that of the R_1 line at 6943 Å, and is $\alpha \simeq 0.2$ cm^{-1} (at 300°K). Other assumed characteristics are:

$$
\begin{aligned}
l &= \text{length of ruby rod} = 10 \text{ cm} \\
A &= \text{cross-sectional area of mode} = 1 \text{ cm}^2 \\
(1 - R) &= \text{fractional intensity loss per pass}^{13} = 20 \text{ percent} \\
n &= 1.78
\end{aligned}
$$

Since, according to (5.3-3), the exponential loss coefficient is proportional to $N_1 - N_2$, we have

$$\alpha\ (\text{cm}^{-1}) = 0.2 \frac{N_1 - N_2}{1.58 \times 10^{19}} \tag{6.7-9}$$

Thus, at room temperature, where $N_2 \ll N_1$ when $N_1 - N_2 \cong 1.58 \times 10^{19}$ cm^{-3}, and (6.7-9) yields $\alpha = 0.2$ cm^{-1} as observed. The expression for the gain coefficient follows directly from (6.7-9):

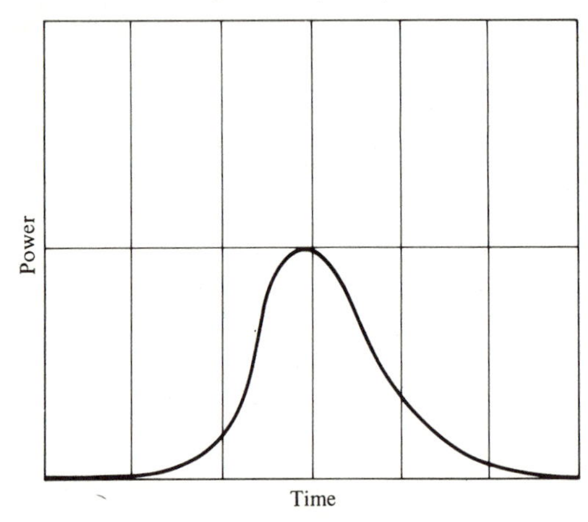

Figure 6-18 An oscilloscope trace of the intensity of a giant pulse. Time scale is 20 ns per division.

[13] We express the loss in terms of an effective reflectivity even though it is due to a number of factors, as discussed in Section 4.7.

$$\gamma\ (\text{cm}^{-1}) = 0.2\frac{N_2 - N_1}{1.58 \times 10^{19}} = 0.2\frac{n}{1.58 \times 10^{19}V} \qquad \textbf{(6.7-10)}$$

where n is the total inversion and $V = AL$ is the crystal volume in cm³.

Threshold is achieved when the net gain per pass is unity. This happens when

$$e^{\gamma_t l}R = 1 \qquad \text{or} \qquad \gamma_t = -\frac{1}{l}\ln R \qquad \textbf{(6.7-11)}$$

where the subscript t indicates the threshold condition.

Using (6.7-10) in the threshold condition (6.7-11) plus the appropriate data from above gives

$$n_t = 1.8 \times 10^{19} \qquad \textbf{(6.7-12)}$$

Assuming that the initial inversion is $n_i = 5n_t = 9 \times 10^{19}$ we find from (6.7-8) that the peak power is approximately

$$P_p = \frac{n_i h\nu}{2t_c} = 5.1 \times 10^9 \text{ watts} \qquad \textbf{(6.7-13)}$$

where $t_c = nl/c(1 - R) \simeq 2.5 \times 10^{-9}$ second.

The total pulse energy is

$$\mathcal{E} \sim \frac{n_i h\nu}{2} \sim 13 \text{ joule}$$

while the pulse duration (see Figure 6-17) $\simeq 3t_c \simeq 7.5 \times 10^{-9}$ second.

Methods of *Q*-switching. Some of the schemes used in Q-switching are:

1. Mounting one of the two end reflectors on a rotating shaft so that the optical losses are extremely high except for the brief interval in each rotation cycle in which the mirrors are nearly parallel.
2. The inclusion of a saturable absorber (bleachable dye) in the optical resonator; see References [13]–[15]. The absorber whose opacity decreases (saturates) with increasing optical intensity prevents rapid inversion depletion due to buildup of oscillation by presenting a high loss to the early stages of oscillation during which the slowly increasing intensity is not high enough to saturate the absorption. As the intensity increases the loss decreases, and the effect is similar, but not as abrupt, as that of a sudden increase of Q.
3. The use of an electrooptic crystal (or liquid Kerr cell) as a voltage-controlled gate inside the optical resonator. It provides a

more precise control over the losses (Q) than schemes 1 and 2. Its operation is illustrated by Figure 6-19 and is discussed in some detail in the following. The control of the phase delay in the electrooptic crystal by the applied voltage is discussed in detail in Chapter 9.

During the pumping of the laser by the light from a flashlamp, a voltage is applied to the electrooptic crystal of such magnitude as to introduce a $\pi/2$ relative phase shift (retardation) between the two mutually orthogonal components (x' and y') that make up the linearly polarized (x) laser field. On exiting from the electrooptic crystal at point f the light traveling to the right is circularly polarized. After reflection from the right mirror the light passes once more through the crystal. The additional retardation of $\pi/2$ adds to the earlier one to give a total retardation of π thus causing the emerging beam at d to be linearly polarized along y and consequently to be blocked by the polarizer.

It follows that with the voltage on, the losses are high, so oscillation is prevented. The Q-switching is timed to coincide with the point at which the inversion reaches its peak and is achieved by a removal of the voltage applied to the electrooptic crystal. This reduces the retardation to zero so that state of polarization of the wave passing through the crystal is unaffected and the Q regains its high value associated with the ordinary losses of the system.

6.8 Hole-Burning and the Lamb Dip in Doppler Broadened Gas Lasers

In this section we concern ourselves with some of the consequences of Doppler broadening in low pressure gas lasers.

Consider an atom with a transition frequency $\nu_0 = (E_2 - E_1)/h$ where 2 and 1 refer to the upper and lower laser levels, respectively. Let the component of the velocity of the atom parallel to the wave propagation direction be v. This component, thus, has the value

$$v = \frac{\mathbf{v}_{\text{atom}} \cdot \mathbf{k}}{k} \tag{6.8-1}$$

where the electromagnetic wave is described by

$$\mathbf{E} = \mathbf{E}e^{i(2\pi\nu t - \mathbf{k}\cdot\mathbf{r})} \tag{6.8-2}$$

An atom moving with a constant velocity $\mathbf{v}$, so that $\mathbf{r} = \mathbf{v}t + \mathbf{r}_0$, will exercise a field

$$\begin{aligned} \mathbf{E}_{\text{atom}} &= \mathbf{E}e^{i[2\pi\nu t - \mathbf{k}\cdot(\mathbf{r}_0 + \mathbf{v}t)]} \\ &= \mathbf{E}e^{i[(2\pi\nu - \mathbf{v}\cdot\mathbf{k})t - \mathbf{k}\cdot\mathbf{r}_0]} \end{aligned} \tag{6.8-3}$$

and will thus "see" a Doppler shifted frequency

$$\nu_D = \nu - \frac{\mathbf{v} \cdot \mathbf{k}}{2\pi} = \nu - \frac{v}{c}\nu \tag{6.8-4}$$

where in the second equality we took $n = 1$ so that $k = 2\pi\nu/c$ and used (6.8-1).

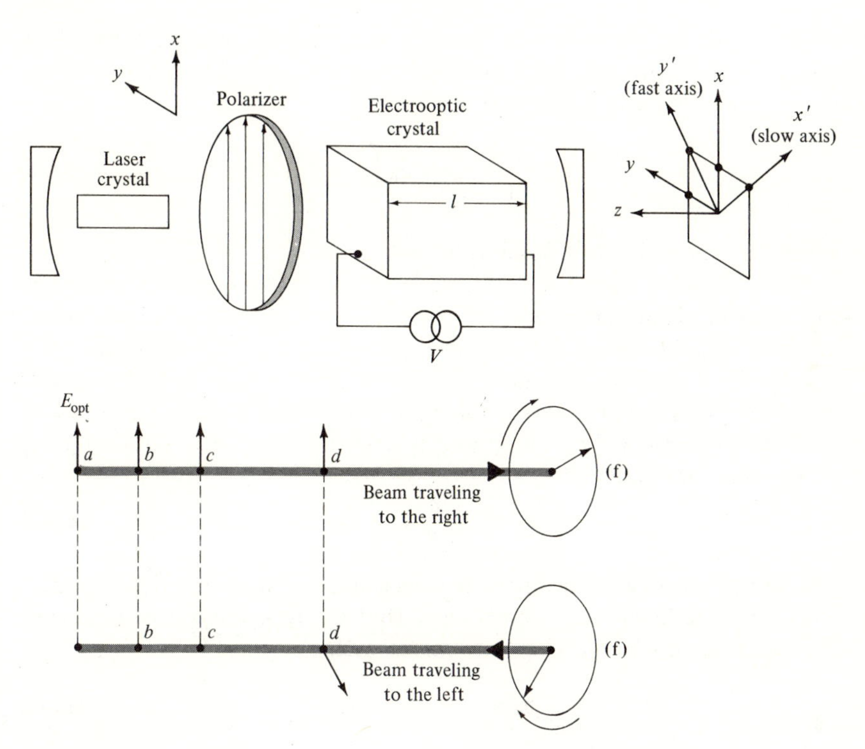

For beam traveling to right:

At point d,

$$E_x' = \frac{E}{\sqrt{2}} \cos \omega t, \qquad E_y' = \frac{E}{\sqrt{2}} \cos \omega t$$

The optical field is linearly polarized with its electric field vector parallel to x

At point f,

$$E_x' = \frac{E}{\sqrt{2}} \cos \left(\omega t + kl + \frac{\pi}{2}\right), \qquad E_y' = \frac{E}{\sqrt{2}} \cos (\omega t + kl)$$

Circularly polarized

For beam traveling to left:

At point f,

$$E_x' = -\frac{E}{\sqrt{2}} \cos \left(\omega t + kl + \frac{\pi}{2}\right), \qquad E_y' = -\frac{E}{\sqrt{2}} \cos (\omega t + kl)$$

Circularly polarized

At point d,

$$E_x' = -\frac{E}{\sqrt{2}} \cos (\omega t + 2kl + \pi), \qquad E_y' = -\frac{E}{\sqrt{2}} \cos (\omega t + 2kl)$$

Linearly polarized along y

Figure 6-19 Electrooptic crystal used as voltage-controlled gate in Q-switching a laser.

The condition for the maximum strength of interaction (that is, emission or absorption) between the moving atom and the wave is that the apparent (Doppler) frequency ν_D "seen" by the atom be equal to the atomic resonant frequency ν_0

$$\nu_0 = \nu - \frac{v}{c}\nu \tag{6.8-5}$$

or reversing the argument, a wave of frequency ν moving through an ensemble of atoms will "seek out" and interact most strongly with those atoms whose velocity component v satisfies

$$\nu = \frac{\nu_0}{1 - \frac{v}{c}} \approx \nu_0\left(1 + \frac{v}{c}\right) \tag{6.8-6}$$

where the approximation is valid for $v \ll c$.

Now consider a gas laser oscillating at a single frequency ν where, for the sake of definiteness, we take $\nu > \nu_0$. The standing wave electromagnetic field at ν inside the laser resonator consists of two waves traveling in opposite directions. Consider, first, the wave traveling in the positive x direction (the resonator axis is taken parallel to the x axis). Since $\nu > \nu_0$ the wave interacts, according to Equation (6.8-6) with atoms having $v > 0$, that is, atoms with

$$v_x = +\frac{c}{\nu}(\nu - \nu_0) \tag{6.8-7}$$

The wave traveling in the opposite direction ($-x$) must also interact with atoms moving in the same direction so that the Doppler shifted frequency is reduced from ν to ν_0. These are atoms with

$$v_x = -\frac{c}{\nu}(\nu - \nu_0) \tag{6.8-8}$$

We conclude that due to the standing wave nature of the field inside a conventional two-mirror laser oscillator, a given frequency of oscillation interacts with two velocity classes of atoms.

Consider, next, a four-level gas laser oscillating at a frequency $\nu > \nu_0$. At negligibly low levels of oscillation and at low gas pressure, the velocity distribution function of atoms in the upper laser level is given, according to (5.1-11), by

$$f(v_x) \propto e^{-Mv_x^2/2kT} \tag{6.8-9}$$

where $f(v_x)\,dv_x$ is proportional to the number of atoms (in the upper laser level) with x component of velocity between v_x and $v_x + dv_x$. As the oscillation level is increased, say by reducing the laser losses, we expect the number of atoms in the upper laser level, with x velocities near $v_x = \pm(c/\nu)(\nu - \nu_0)$, to decrease from their equilibrium value as given by (6.8-9). This is due to the fact that these atoms undergo stimulated downward transitions from level 2 to 1, thus reducing the number of atoms in level 2. The velocity distribution function under conditions of oscillation has consequently two depressions as shown schematically in Figure 6-20.

If the oscillation frequency ν is equal to ν_0, only a single "hole" exists in the velocity distribution function of the inverted atoms. This "hole" is centered on $v_x = 0$. We may, thus, expect the power output of a laser oscillating at $\nu = \nu_0$ to be less than that of a laser in which ν is tuned slightly to one side or the other of ν_0 (this tuning can be achieved by moving one of the laser mirrors). This power dip first predicted by Lamb [20] is indeed observed in gas lasers [21]. An experimental plot of the power versus frequency in a He–Ne 1.15-μm laser is shown in Figure 6-21. The phenomenon is referred to as the "Lamb dip" and is used in frequency stabilization schemes of gas lasers [22].

6.9 Relaxation Oscillation in Lasers

Relaxation oscillation of the intensity has been observed in most types of lasers [23] [24]. This oscillation takes place characteristically with a period which is considerably longer than the cavity decay time t_c (see Section 4.7) or the resonator round trip time $2l/c$. Typical values range between 0.1 μs to 10 μs.

The basic physical mechanism is an interplay between the oscillation field in the resonator and the atomic inversion. An increase in the field

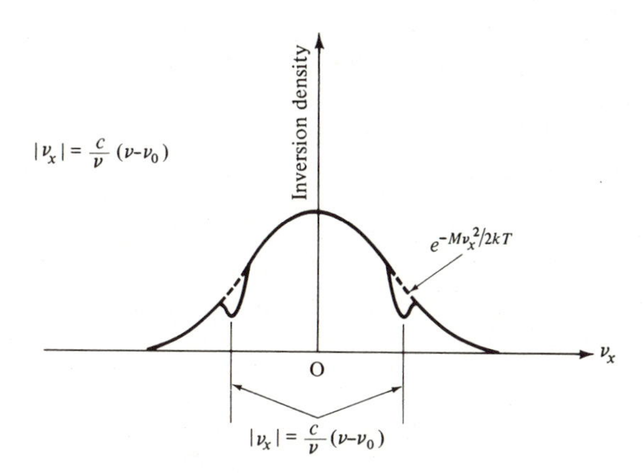

Figure 6-20 The distribution of inverted atoms as a function of v_x. The dashed curve which is proportional to exp $(-Mv_x^2/2kT)$ corresponds to the case of zero field intensity. The solid curve corresponds to a standing wave field at $\nu = \nu_0/(1 - v_x/c)$ or one at $\nu = \nu_0/(1 + v_x/c)$.

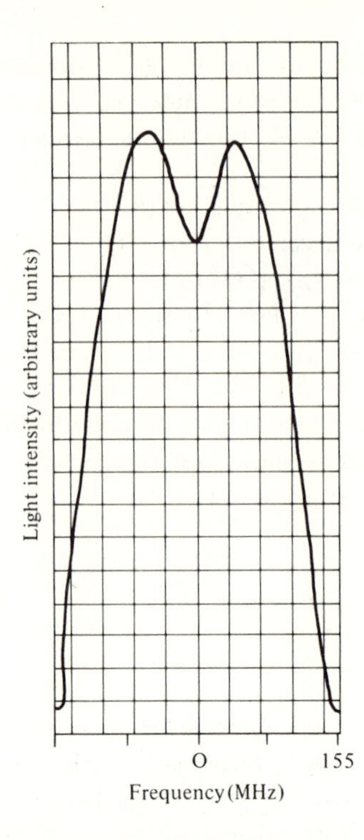

Figure 6-21 The power output as a function of the frequency of a single mode 1.15 μm He-Ne laser using the 20 Ne isotope. (After Reference [21].)

intensity causes a reduction in the inversion due to the increased rate of stimulated transitions. This causes a reduction in the gain which in turn tends to decrease the field intensity.

In the mathematical modeling of this phenomenon, we assume an ideal homogeneously broadened four-level laser such as described in Section 6.4. We also assume that the lower level population is negligible (that is, $W_i \ll \omega_{10} \gg \omega_{21}$) and take the inversion density $N \equiv N_2 - N_1 = N_2$. The pumping rate into level 2 (atoms/s $- m^3$) is R and the lifetime, due to all causes except stimulated emission, of atoms in level 2 is τ. Taking the induced transition rate per atom as W_i we have

$$\frac{dN}{dt} = R - W_i N - \frac{N}{\tau} \tag{6.9-1}$$

The transition rate W_i is, according to (5.2-15), proportional to the field intensity I and hence to the photon density q in the optical resonator. We can, consequently, rewrite (6.9-1) as

$$\frac{dN}{dt} = R - qBN - \frac{N}{\tau} \tag{6.9-2}$$

where B is a proportionality constant defined by $W_i \equiv Bq$. Since qBN is also the rate ($\text{s}^{-1} - m^{-3}$) at which photons are generated, we have

$$\frac{dq}{dt} = qBN - \frac{q}{t_c} \tag{6.9-3}$$

where t_c is the decay time constant for photons in the optical resonator as discussed in Sections 4.7 and 6.1. Equations (6.9-2) and (6.9-3) describe the interplay between the photon density q and the inversion N [25].

First we notice that in equilibrium, $dq/dt = dN/dt = 0$, the following relations are satisfied

$$N_0 = \frac{1}{Bt_c}$$
$$q_0 = \frac{RBt_c - \dfrac{1}{\tau}}{B} \tag{6.9-4}$$

From (6.9-4) it follows that when $R = (Bt_c\tau)^{-1}$, $q_0 = 0$. We denote this threshold pumping rate by R_t and define the pumping factor $r \equiv R/R_t$[14] so that the second of (6.9-4) can also be written as

$$q_0 = \frac{(r-1)}{B\tau} \tag{6.9-5}$$

Next, we consider the behavior of small perturbations from equilibrium. We take

$$N(t) = N_0 + N_1(t), \qquad N_1 \ll N_0$$

and

$$q(t) = q_0 + q_1(t), \qquad q_1 \ll q_0$$

Substituting these relations in Equations (6.9-2) and (6.9-3), and making use of Equation (6.9-4) we obtain

$$\frac{dN_1}{dt} = -RBt_cN_1 - \frac{q_1}{t_c} \tag{6.9-6}$$

$$\frac{dq_1}{dt} = \left(RBt_c - \frac{1}{\tau}\right)N_1 \tag{6.9-7}$$

Taking the derivative of (6.9-7), substituting (6.9-6) for dN_1/dt, and using (6.9-4) leads to

$$\frac{d^2q_1}{dt^2} + RBt_c\frac{dq_1}{dt} + \left(RB - \frac{1}{\tau t_c}\right)q_1 = 0 \tag{6.9-8}$$

or in terms of the pumping factor $r = RBt_c\tau$ introduced above,

$$\frac{d^2q_1}{dt^2} + \frac{r}{\tau}\frac{dq_1}{dt} + \frac{1}{\tau t_c}(r-1)q_1 = 0 \tag{6.9-9}$$

This is the differential equation describing a damped harmonic oscillator so that assuming a solution $q \propto e^{pt}$ we obtain

$$p^2 + \frac{r}{\tau}p + \frac{1}{\tau t_c}(r-1) = 0$$

---

[14] r is equal to the ratio of the unsaturated ($q = 0$) gain to the saturated gain (the saturated gain is the actual gain "seen" by the laser field and is equal to the loss).

with the solutions

$$p(\pm) = -\alpha \pm i\omega_m$$

$$\alpha = \frac{r}{2\tau}, \qquad \omega_m = \sqrt{\frac{1}{t_c\tau}(r-1) - \left(\frac{r}{2\tau}\right)^2}$$
$$\approx \sqrt{\frac{1}{t_c\tau}(r-1)} \qquad \frac{1}{t_c\tau}(r-1) \gg \left(\frac{r}{2\tau}\right)^2 \tag{6.9-10}$$

so that $q_1(t) \propto e^{-\alpha t}\cos\omega_m t$. The predicted perturbation in the power output (which is proportional to the number of photons q) is, thus, a damped sinusoid with the damping rate α and the oscillation frequency ω_m increasing with excess pumping.

While some lasers display the damped sinusoidal perturbation of intensity described above, in many other laser systems the perturbation is undamped. An example of the first is illustrated in Figure 6-22 which shows the output of a $CaWO_4$:Nd^{3+} laser.

Numerical example—relaxation oscillation. Considering the case shown in Figure 6-22 with the following parameters

$$\tau = 1.6 \times 10^{-4} \text{ second}$$
$$t_c \simeq 10^{-8} \text{ second}$$
$$r \simeq 2$$

which using (6.9-10) gives $T_m \equiv 2\pi/\omega_m \approx 8 \times 10^{-6}$ second.

The undamped relaxation oscillation observed in many cases can be understood, at least qualitatively, by considering Equation (6.9-9). As it stands, the equation is identical in form to that describing a damped nondriven harmonic oscillator or equivalently, a resonant *RLC* circuit.[15] Persistent, that is, nondamped, oscillation is possible when the "oscillator"

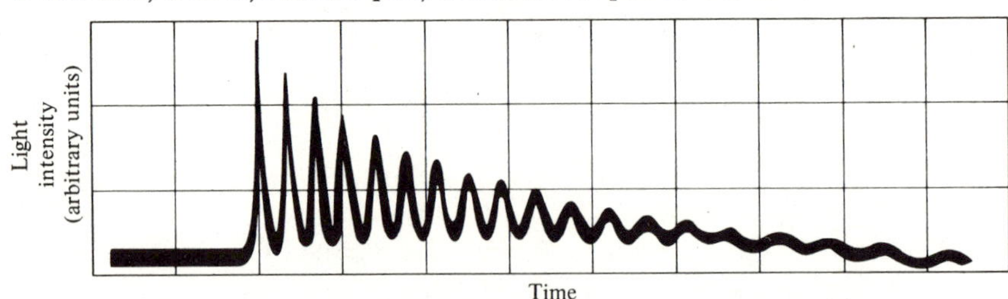

Figure 6-22 Intensity Relaxation Oscillation in a $CaWO_4$:Nd^{3+} laser at 1.06 μm. Horizontal scale = 20 μsec/division. (After Reference [26].)

---

[15] The differential equation describing an oscillator is given in (5.4–1).

is driven. In this case, the driving function will replace the zero on the right side of (6.9-9). One such driving mechanism may be due to time variation in the pumping rate R. In this case, we may take the pumping in the form

$$R = R_0 + R_1(t) \tag{6.9-11}$$

where R_0 is the average pumping and $R_1(t)$ is the deviation.

Retracing the steps leading to (6.9-6) but using (6.9-11), we find that the inversion equation is now

$$\frac{dN_1}{dt} = R_1 - R_0 B t_c N_1 - \frac{q_1}{t_c}$$

and that Equation (6.9-9) takes the form

$$\frac{d^2 q_1}{dt^2} + \frac{r}{\tau}\frac{dq_1}{dt} + \frac{1}{\tau t_c}(r-1)q_1 = \frac{1}{\tau}(r-1)R_1 \tag{6.9-12}$$

Taking the Fourier transform of both sides of Equation (6.9-12), defining $Q(\omega)$ and $R(\omega)$ as the transforms of $q_1(t)$ and $R_1(t)$, respectively, and then solving for $Q(\omega)$, gives

$$Q(\omega) = \frac{-\frac{1}{\tau}(r-1)R(\omega)}{\omega^2 - i\frac{r}{\tau}\omega - \frac{1}{\tau t_c}(r-1)} \tag{6.9-13}$$

$$= \frac{-\frac{1}{\tau}(r-1)R(\omega)}{(\omega - \omega_m - i\alpha)(\omega + \omega_m - i\alpha)}$$

$$\omega_m = \sqrt{\frac{1}{t_c\tau}(r-1) - \left(\frac{r}{2\tau}\right)^2} \tag{6.9-14}$$

$$\approx \sqrt{\frac{1}{t_c\tau}(r-1)}, \qquad \frac{1}{t_c\tau}(r-1) \gg \left(\frac{r}{2\tau}\right)^2$$

$$\alpha = \frac{r}{2\tau} \tag{6.9-15}$$

where we notice that ω_m and α correspond to the oscillation frequency and damping rate, respectively, of the transient case as given by (6.9-10). If we assume that the spectrum $R(\omega)$ of the driving function $R(t)$ is uniform (that is, like "white" noise) near $\omega \approx \omega_m$, we may expect the intensity spectrum $Q(\omega)$ to have a peak near $\omega = \omega_m$ with a width $\Delta\omega \approx 2\alpha \equiv r/\tau$. In addition, if $\Delta\omega \ll \omega_m$, we may expect the intensity fluctuation $q(t)$ as observed in the time domain to be modulated at a frequency ω_m[16] since for frequencies $\omega \approx \omega_m Q(\omega)$ is a maximum.

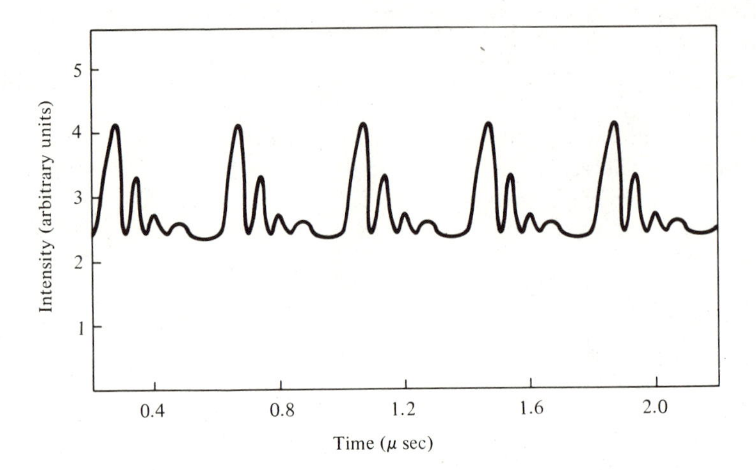

Figure 6-23 Intensity relaxation oscillation in a xenon 3.51 μm laser.

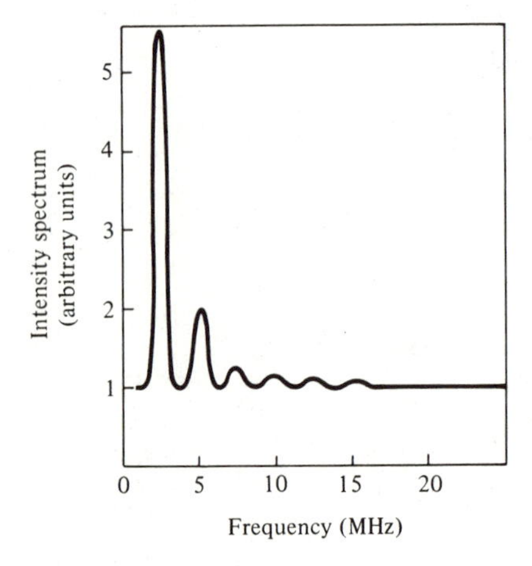

Figure 6-24 The intensity fluctuation spectrum of the laser output shown in Figure 6-23.

---

[16] To verify this statement, assume that $R(t)$ is approximated by a superposition of uncorrelated sinusoids $R(t) \propto \sum_n a_n e^{i\omega_n t}$ and using $R(\omega) \propto \int_{-\infty}^{\infty} R(t)e^{-i\omega t}dt$, we get $R(\omega) \propto \sum_n a_n \delta(\omega - \omega_n)$. From the inverse transform relation $q(t) \propto \int_{-\infty}^{\infty} Q(\omega)e^{i\omega t}d\omega$ and Equation (6.9-13), we get

$$q(t) \propto \sum_n \frac{a_n e^{i\omega_n t}}{[\omega_n - \omega_m - i\alpha][\omega_n + \omega_m - i\alpha]}$$

so that in the limit $\omega_m \gg \alpha$, $q(t)$ is a quasi-sinusoidal oscillation with a frequency ω_m.

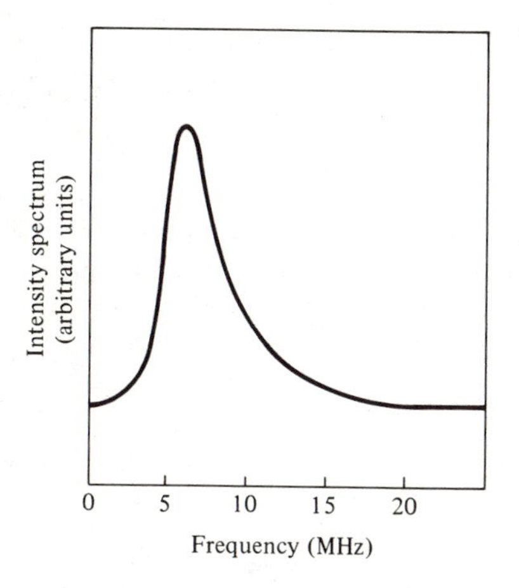

Figure 6-25 Same as Figure 6-24 except at increased pumping.

These conclusions are verified in experiments on different laser systems. In Figure 6-23 we show the intensity fluctuations of a xenon 3.51-μm laser. The repetition frequency is 2.5×10^6 Hz. A spectral analysis of the intensity yielding $Q(\omega)$ is shown in Figure 6-24. It consists of a narrow peak centered on $f_m = 2.5 \times 10^6$ Hz. An increase in the pumping strength is seen (Figure 6-25) to cause a broadening of the spectrum as well as a shift to higher frequencies consistent with the discussion following (6.9-15).

PROBLEMS

6-1 Show that the effect of frequency pulling by the atomic medium is to reduce the intermode frequency separation from $c/2l$ to

$$\frac{c}{2l}\left(1 - \frac{\gamma c}{2\pi\Delta\nu}\right)$$

where the symbols are defined in Section 6.2. Calculate the reduction for the case of a laser with $\Delta\nu = 10^9$ Hz, $\gamma = 4 \times 10^{-2}$ meter^{-1}, and $l =$ 100 cm.

6-2 Derive Equation (6.4-3).

6-3 Derive the optimum coupling condition (Equation 6.5-11).

6-4 Calculate the saturation power P_s of the He–Ne laser operating at 6328 Å. Assume $V = 2$ cm^3, $L = 1$ percent per pass, and $\Delta\nu = 1.5 \times 10^9$ Hz.

6-5 Calculate the critical inversion density N_t of the He–Ne laser described in Problem 6-4.

6-6 Derive an expression for the finesse of a Fabry–Perot etalon containing an inverted population medium. Assume that $r_1^2 = r_2^2 \simeq 1$ and that the inversion is insufficient to result in oscillation. Compare the finesse to that of a passive Fabry–Perot etalon.

6-7 Derive an expression for the maximum gain–bandwidth product of a Fabry–Perot regenerative amplifier. Define the bandwidth as the frequency region in which the intensity gain $(E_t E_t^*)/(E_i E_i^*)$ exceeds half its peak value. Assume that $\nu_0 = \nu_m$.

6-8 **a.** Derive Equation (6.6-4).
b. Show that if in (6.6-3) the phases are taken as $\phi_n = n\phi$, where ϕ is some constant, instead of $\phi_n = 0$, the result is merely one of delaying the pulses by $-\phi/\omega$.

6-9 **a.** Describe qualitatively what one may expect to see in parts A, B, C, and D of the mode-locking experiment sketched in Figure 6-13. (The reader may find it useful to read first the section on photomultipliers in Chapter 11.)
b. What is the effect of mode locking on the intensity of the beat signal (at $\omega = \pi c/l$) displayed by the RF spectrum analyzer in B. Assume N equal amplitude modes spaced by ω whose phases before mode locking are random. (*Answer:* Mode locking increases the beat signal power by N.)
c. Show that a standing wave at $\nu_0 + \delta$ (ν_0 is the center frequency of the Doppler broadened lineshape function) in a gas laser will burn the same two holes in the velocity distribution function (see Figure 6-20) as a field at $\nu_0 - \delta$.
d. Can two traveling waves, one at $\nu_0 + \delta$ the other at $\nu_0 - \delta$, interact with the same class of atoms? If the answer is yes, under what conditions?

6-10 Design a frequency stabilization scheme for gas lasers based on the Lamb dip (see Figure 6-21). [*Hint:* You may invent a new scheme, but, failing that, consider what happens to the phase of the modulation in the power output when the cavity length is modulated sinusoidally near the bottom of the Lamb dip. Can you derive an error correction signal from this phase which will control the cavity length?]

REFERENCES

[1] Schawlow, A. L., and C. H. Townes, "Infrared and optical masers," *Phys. Rev.*, vol. 112, p. 1940, 1958.

[2] Yariv, A., *Quantum Electronics*, 2d Ed. New York: Wiley, 1975.

[3] Smith, W. V., and P. P. Sorokin, *The Laser*. New York: McGraw-Hill, 1966.

[4] Lengyel, B. A., *Introduction to Laser Physics*. New York: Wiley, 1966.

[5] Birnbaum, G., *Optical Masers*. New York: Academic Press, 1964.

[6] *Lasers and Light—Readings from Scientific American*. San Francisco: Freeman, 1969.

[7] Bennett, W. R., Jr., "Gaseous optical masers," in *Appl. Opt. Suppl. 1, Optical Masers*, 1962, p. 24.

[8] Fork, R. L., D. R. Herriott, and H. Kogelnik, "A scanning spherical mirror interferometer for spectral analysis of laser radiation," *Appl. Optics*, vol. 3, p. 1471, 1964.

[9] DiDomenico, M., Jr., "Small signal analysis of internal modulation of lasers," *J. Appl. Phys.*, vol. 35, p. 2870, 1964.

[10] Yariv, A., "Internal modulation in multimode laser oscillators," *J. Appl. Phys.*, vol. 36, p. 388, 1965.

[11] Hargrove, L. E., R. L. Fork, and M. A. Pollack, "Locking of He–Ne laser modes induced by synchronous intracavity modulation," *Appl. Phys. Letters*, vol. 5, p. 4, 1964.

[12] DiDomenico, M., Jr., J. E. Geusic, H. M. Marcos, and R. G. Smith, "Generation of ultrashort optical pulses by mode locking the Nd^{3+}:YAG laser," *Appl. Phys. Letters*, vol. 8, p. 180, 1966.

[13] Mocker, H., and R. J. Collins, "Mode competition and self-locking effects in a Q-switched ruby laser," *Appl. Phys. Letters*, vol. 7, p. 270, 1965.

[14] DeMaria, A. J., "Picosecond laser pulses," Proc. IEEE, vol. 57, p. 3, 1969.

[15] DeMaria, A. J., "Mode locking," *Electronics*, Sept. 16, 1968, p. 112.

[16] Hellwarth, R. W., "Control of fluorescent pulsations," in *Advances in Quantum Electronics*, J. R. Singer, ed: New York: Columbia University Press, 1961, p. 334.

[17] McClung, F. J., and R. W. Hellwarth, *J. Appl. Phys.*, vol. 33, p. 828, 1962.

[18] Hellwarth, R. W., "Q modulation of lasers," in *Lasers*, vol. 1, A. K. Levine, ed. New York: Marcel Dekker, Inc., 1966, p. 253.

[19] Wagner, W. G., and B. A. Lengyel, "Evolution of the giant pulse in a laser," *J. Appl. Phys.*, vol. 34, p. 2042, 1963.

[20] Lamb, W. E., Jr., "Theory of an optical maser," *Phys. Rev.*, vol. 134, *A1429* (1964).

[21] Szöke, A., and A. Javan, "Isotope shift and saturation behavior of the 1.15μ transition of neon," *Phys. Rev. Letters*, vol. 10, *512* (1963).

[22] Bloom, A., *Gas Lasers*. New York: Wiley, 1963, p. 93.

[23] Collins, R. J., D. F. Nelson, A. L. Schawlow, W. Bond, C. G. B. Garrett, and W. Kaiser, *Phys. Rev. Letters*, *303* (1960).

[24] For additional references on relaxation oscillation the reader should consult
(a) Birnbaum, G., *Optical Masers.* New York: Academic Press, 1964, p. 191.
(b) Evtuhov, V., "Pulsed ruby lasers," in *Lasers,* edited by A. K. Levine. New York: M. Dekker, Inc., 1966, p. 76.

[25] The explanation of "spiking" in terms of these equations is due to:
(a) Dunsmuir, R., in *J. Elec. Control,* vol. 10, *453* (1961).
(b) Statz, H., and G. de Mars, *Quantum Electronics.* New York: Columbia University Press, 1960, p. 530.

[26] Johnson, L. F., in *Lasers,* edited by A. K. Levine. New York: M. Dekker, Inc., 1966, p. 174.

[27] Casperson, L., and A. Yariv, "Relaxation oscillation in a xenon 3.51μm laser," *J. Quantum Electronics,* 1971.

[28] Dienes, A., "Mode-locked CW dye lasers," *Opto Electronics,* vol. 6, 1974, p. 99.

CHAPTER

7

Some Specific Laser Systems

7.0 Introduction

The pumping of the atoms into the upper laser level is accomplished in a variety of ways, depending on the type of laser. In this chapter we will review some of the more common laser systems and in the process describe their pumping mechanisms. The laser systems described include: ruby, Nd^{3+}:YAG, Nd^{3+}:glass, He–Ne, CO_2, Ar^+, and the GaAs semiconductor junction laser.

7.1 Pumping and Laser Efficiency

Figure 7-1 shows the pumping–oscillation cycle of some (hypothetical) representative laser. The pumping agency elevates the atoms into some excited state 3 from which they relax into the upper laser level 2. The stimulated laser transition takes place between levels 2 and 1 and results in the emission of a photon of frequency ν_{21}.

It is evident from this figure that the minimum energy input per output photon is $h\nu_{30}$, so the power efficiency of the laser cannot exceed

$$\eta_{\text{atomic}} = \frac{\nu_{21}}{\nu_{30}} \tag{7.1-1}$$

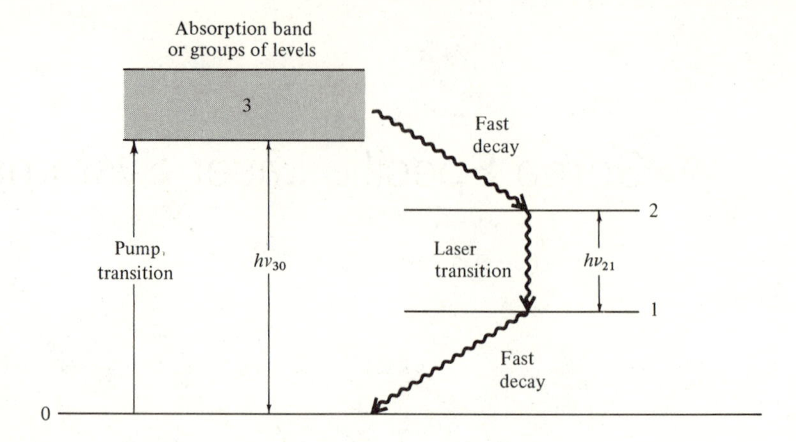

Figure 7-1 Pumping–oscillation cycle of a typical laser.

to which quantity we will refer as the "atomic quantum efficiency." The overall laser efficiency depends on the fraction of the total pump power which is effective in transferring atoms into level 3 and on the pumping quantum efficiency defined as the fraction of the atoms which, once in 3, make a transition to 2. The product of the last two factors, which constitutes an upper limit on the efficiency of optically pumped lasers, ranges from about 1 percent for solid-state lasers such as Nd^{3+}:YAG to about 30 percent in the CO_2 laser and to near unity in the GaAs junction laser. We shall discuss these factors when we get down to some specific laser systems. We may note, however, that according to (7.1-1), in an efficient laser system ν_{21} and ν_{30} must be of the same order of magnitude, so the laser transition should involve low-lying levels.

7.2 Ruby Laser

The first material in which laser action was demonstrated [1] and still one of the most useful laser materials is ruby, whose output is at $\lambda_0 = 0.6943\ \mu m$. The active laser particles are Cr^{3+} ions present as impurities in Al_2O_3 crystal. Typical Cr^{3+} concentrations are $\sim$0.05 percent by weight. The pertinent energy level diagram is shown in Figure 7-2.

The pumping of ruby is performed usually by subjecting it to the light of intense flashlamps (quite similar to the types used in flash photography). A portion of this light which corresponds in frequency to the two absorption bands $^4F_2$ and $^4F_1$ is absorbed, thereby causing Cr^{3+} ions to be transferred into these levels. The ions proceed to decay, within an average time of $\omega_{32}{}^{-1} \simeq 5 \times 10^{-8}$ seconds [2], into the upper laser level ^2E. The level ^2E is composed of two separate levels $2\bar{A}$ and $\bar{E}$ separated by

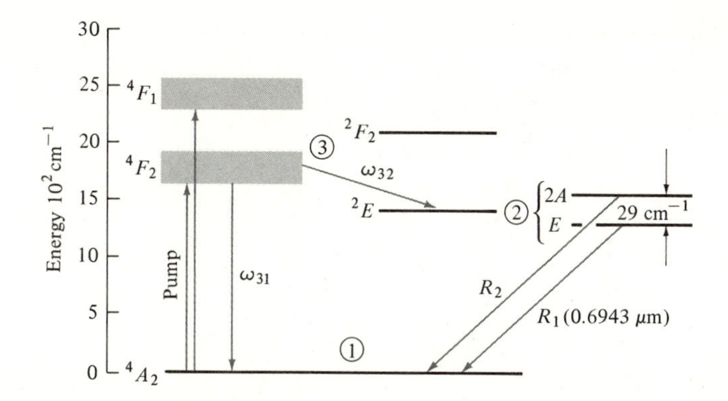

Figure 7-2 Energy levels pertinent to the operation of a ruby laser. (After Reference [2].)

29 cm^{-1}.[1] The lower of these two, $\bar{E}$, is the upper laser level. The lower laser level is the ground state, and thus, according to the discussion of Section 6.3, ruby is a three-level laser. The lifetime of atoms in the upper laser level $\bar{E}$ is $t_2 \simeq 3 \times 10^{-3}$ second. Each decay results in the (spontaneous) emission of a photon, so $t_2 \simeq t_{\text{spont}}$.

An absorption spectrum of a typical ruby with two orientations of the optical field relative to the c (optic) axis is shown in Figure 7-3. The two main peaks correspond to absorption into the useful $^4F_1$ and $^4F_2$ bands, which are responsible for the characteristic (ruby) color.

The ordinate is labeled in terms of the absorption coefficient and in terms of the transition cross section σ which may be defined as the absorption coefficient per unit inversion per unit volume and has consequently the dimension of area. According to this definition, $\alpha(\nu)$ is given by

$$\alpha(\nu) = (N_1 - N_2)\sigma(\nu) \tag{7.2-1}$$

A more detailed plot of the absorption near the laser emission wavelength is shown in Figure 7-4. The width $\Delta\nu$ of the laser transition as a function of temperature is shown in Figure 7-5. At room temperature, $\Delta\nu = 11$ cm^{-1}.

We can use ruby to illustrate some of the considerations involved in optical pumping of solid-state lasers. Figure 7-6 shows a typical setup of an optically pumped laser, such as ruby. The helical flashlamp surrounds the ruby rod. The flash excitation is provided by the discharge of the charge stored in a capacitor bank across the lamp.

The typical flash output consists of a pulse of light of duration $t_{\text{flash}} \simeq 5 \times 10^{-4}$ second. Let us, for the sake of simplicity, assume that the flash pulse is rectangular in time and of duration t_{flash}, and that it results in an

---

[1] The unit 1 cm^{-1} (one wavenumber) is the frequency corresponding to $\lambda_0 = 1$ cm, so 1 cm^{-1} is equivalent to $\nu = 3 \times 10^{10}$ Hz. It is also used as a measure of energy where 1 cm^{-1} corresponds to the energy $h\nu$ of a photon with $\nu = 3 \times 10^{10}$ Hz.

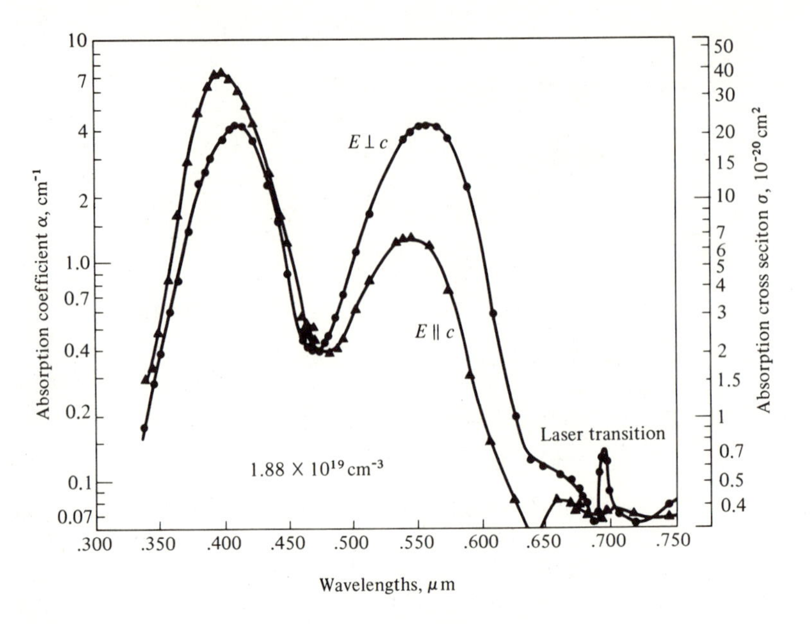

Figure 7-3 Absorption coefficient and absorption cross section as functions of wavelength for $E \parallel c$ and $E \perp c$. The 300°K data were derived from transmittance measurements on pink ruby with an average Cr ion concentration of 1.88×10^{19} cm^{-3}. (After Reference [3].)

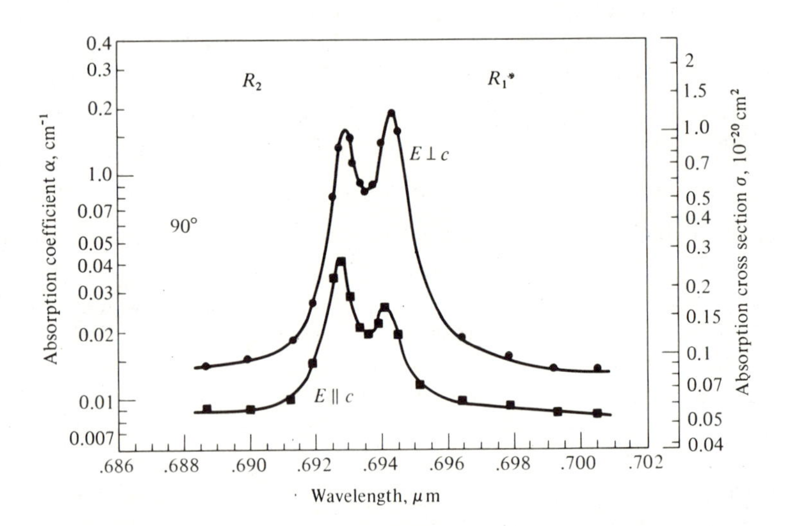

Figure 7-4 Absorption coefficient and absorption cross section as functions of wavelength for $E \parallel c$ and $E \perp c$. Sample was a pink ruby laser rod having a 90° c-axis orientation with respect to the rod axis and a Cr concentration of 1.58×10^{19} cm^{-3}. (After Reference [3].)

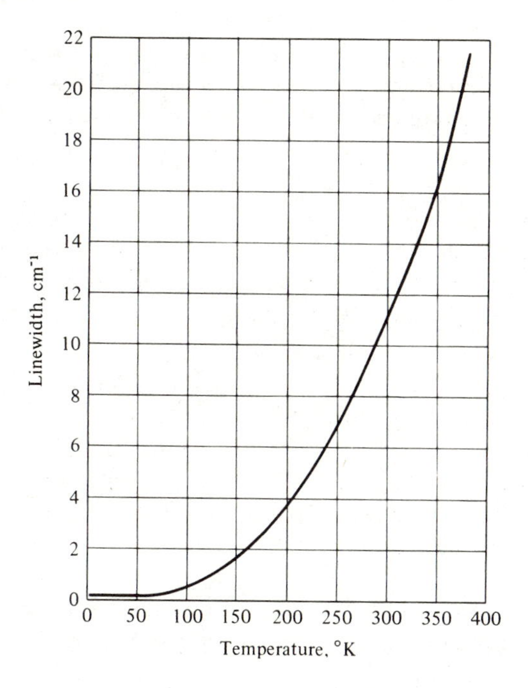

Figure 7-5 Line width of the R_1 line of ruby as a function of temperature. (After Reference [4].)

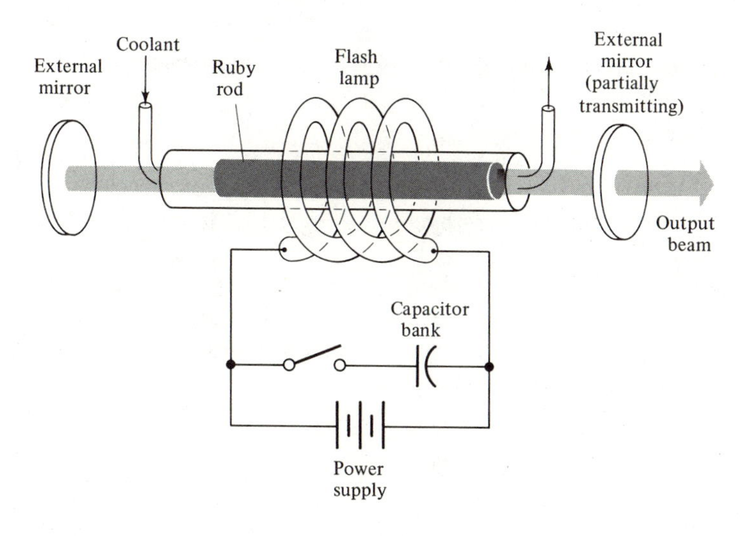

Figure 7-6 Typical setup of a pulsed ruby laser using flashlamp pumping and external mirrors.

optical flux at the crystal surface having $s(\nu)$ watts per unit area per unit frequency at the frequency ν. If the absorption coefficient of the crystal is $\alpha(\nu)$, then the amount of energy absorbed by the crystal per unit volume is[2]

$$t_{\text{flash}} \int_0^\infty s(\nu)\alpha(\nu)\, d\nu$$

If the absorption quantum efficiency (the probability that the absorption of a pump photon at ν results in transferring one atom into the upper laser level) is $\eta(\nu)$, the number of atoms pumped into level 2 per unit volume is

$$N_2 = t_{\text{flash}} \int_0^\infty \frac{s(\nu)\alpha(\nu)\eta(\nu)}{h\nu}\, d\nu \tag{7.2-2}$$

Since the lifetime $t_2 = 3 \times 10^{-3}$ second of atoms in level 2 is considerably longer than the flash duration ($\sim 5 \times 10^{-4}$ second) we may neglect the spontaneous decay out of level 2 during the time of the flash pulse, so N_2 represents the population of level 2 after the flash.

Numerical example—flash pumping of a pulsed ruby laser. Consider the case of a ruby laser with the following parameters:

$$N_0 = 2 \times 10^{19}\ \text{atoms/cm}^3$$

$$t_2 = t_{\text{spont}} = 3 \times 10^{-3}\ \text{second}$$

$$t_{\text{flash}} = 5 \times 10^{-4}\ \text{second}$$

If the useful absorption is limited to relatively narrow spectral regions, we may approximate (7.2-2) by

$$N_2 = \frac{t_{\text{flash}}\overline{s(\nu)}\,\overline{\alpha(\nu)}\,\overline{\eta(\nu)}\,\overline{\Delta\nu}}{h\bar{\nu}} \tag{7.2-3}$$

where the bars represent average values over the useful absorption region whose width is $\overline{\Delta\nu}$.

From Figure 7-3 we deduce an average absorption coefficient of $\overline{\alpha(\nu)} \simeq 1\ \text{cm}^{-1}$ over the two central peaks. Since ruby is a three-level laser the upper level population is, according to (6.3-2), $N_2 \simeq N_0/2 = 10^{19}\ \text{cm}^{-3}$. Using $\bar{\nu} \simeq 5 \times 10^{14}$ Hz, (7.2-3) yields

$$\bar{s}\,\overline{\Delta\nu}\, t_{\text{flash}} \simeq 3\ \text{J/cm}^2$$

for the pump energy in the useful absorption region that must fall on each square centimeter of crystal surface in order to obtain threshold inversion. To calculate the total lamp energy that is incident on the crystal we need

---

[2] We assume that the total absorption in passing the crystal is small, so $s(\nu)$ is taken to be independent of the distance through the crystal.

to know the spectral characteristics of the lamp output. Typical data of this sort are shown in Figure 7-7. The mercury-discharge lamp is seen to contain considerable output in the useful absorption regions (near 4000 Å and 5500 Å) of ruby. If we estimate the useful fraction of the lamp output at 10 percent, the fraction of the lamp light actually incident on the crystal as 20 percent, and the conversion of electrical-to-optical energy as 50 percent, we find the threshold electric energy input to the flashlamp per square centimeter of laser surface is

$$\frac{3}{0.1 \times 0.2 \times 0.5} = 300 \text{ J/cm}^2$$

These are, admittedly, extremely crude calculations. They are included not only to illustrate the order of magnitude numbers involved in laser pumping, but also as an example of the quick and rough estimates needed to discriminate between feasible ideas and "pie in the sky" schemes.

7.3 Nd^{3+}:YAG Laser

One of the most important laser systems is that using trivalent neodymium ions (Nd^{3+}) which are present as impurities in yttrium aluminum garnet (YAG = $Y_3Al_5O_{12}$); see References [6] and [7]. The laser emission occurs at $\lambda_0 = 1.0641\ \mu m$ at room temperature. The relevant energy levels are

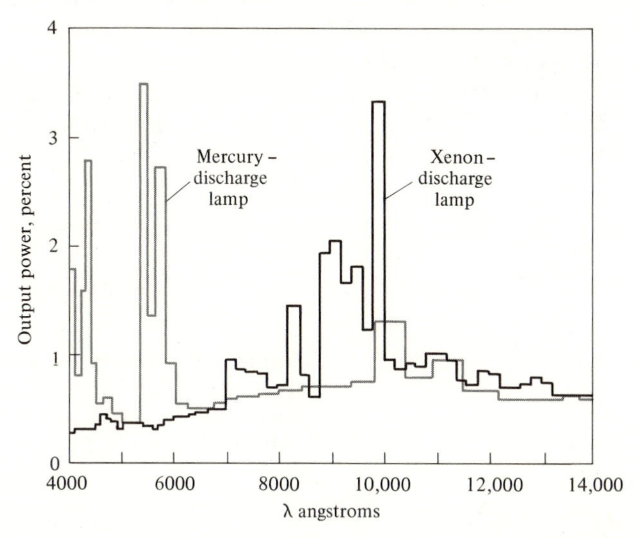

Figure 7-7 Spectral output characteristics of two commercial high-pressure lamps. Output is plotted as a fraction of electrical input to lamp over certain wavelength intervals (mostly 200 Å) between 0.4 and 1.4 μm. (After Reference [5].)

shown in Figure 7-8. The lower laser level is at $E_2 \simeq 2111 \text{ cm}^{-1}$ from the ground state so that at room temperature its population is down by a factor of $\exp(-E_2/kT) \simeq e^{-10}$ from that of the ground state and can be neglected. The Nd^{3+}:YAG thus fits our definition (see Section 6.3) of a four-level laser.

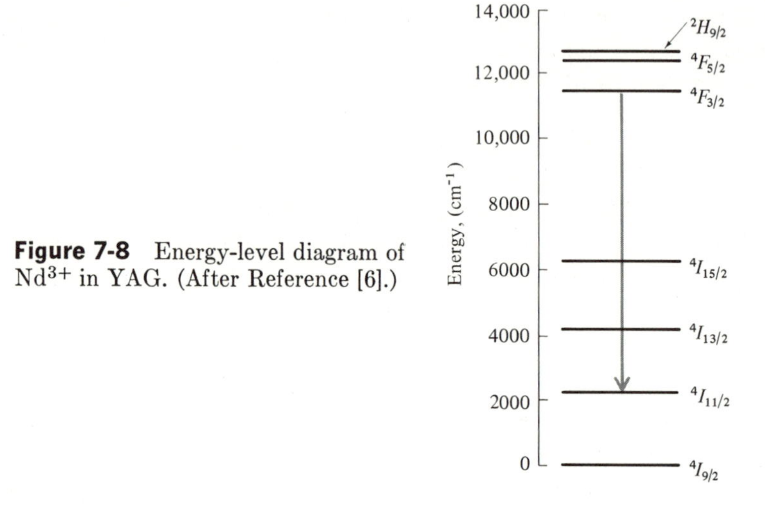

Figure 7-8 Energy-level diagram of Nd^{3+} in YAG. (After Reference [6].)

The spontaneous emission spectrum of the laser transition is shown in Figure 7-9. The width of the gain linewidth at room temperature is $\Delta\nu \simeq 6 \text{ cm}^{-1}$. The spontaneous lifetime for the laser transition has been measured [7] as $t_{\text{spont}} = 5.5 \times 10^{-4}$ second. The room-temperature cross section at the center of the laser transition is $\sigma = 9 \times 10^{-19} \text{ cm}^2$. If we compare this number to $\sigma = 1.22 \times 10^{-20} \text{ cm}^2$ in ruby (see Figure 7-4), we expect that at a given inversion the optical gain constant γ in

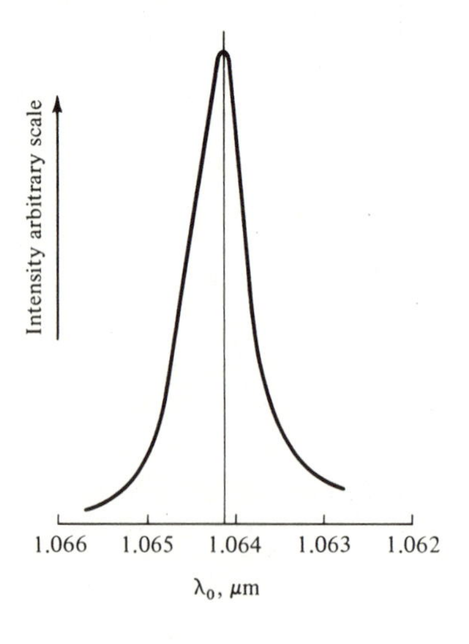

Figure 7-9 Spontaneous-emission spectrum of Nd^{3+} in YAG near the laser transition at $\lambda_0 = 1.064\ \mu$m. (After Reference [7].)

Nd^{3+}:YAG is approximately 75 times that of ruby. This causes the oscillation threshold to be very low and explains the easy continuous (CW) operation of this laser compared to ruby.

The absorption responsible for populating the upper laser level takes place in a number of bands between 13,000 and 25,000 cm^{-1}.

Numerical example—threshold of an Nd^{3+}:YAG laser

(*a*) *Pulsed threshold.* First we estimate the energy needed to excite a typical Nd^{3+}:YAG laser on a pulse basis so that we can compare it with that of ruby. We use the following data:

$$\left.\begin{aligned} l &= 20 \text{ cm (length optical resonator)} \\ L &= 4 \text{ percent } (= \text{loss per pass}) \\ n &= 1.5 \end{aligned}\right\} t_c = \frac{nl}{Lc} = 2.48 \times 10^{-8} \text{ second}$$

$$\Delta\nu = 6 \text{ cm}^{-1} \ (= 6 \times 3 \times 10^{10} \text{ Hz})$$

$$t_{\text{spont}} = 5.5 \times 10^{-4} \text{ second}$$

$$\lambda = 1.06\ \mu\text{m}$$

Using the foregoing data in (6.1-11) gives

$$N_t = \frac{8\pi n^3 t_{\text{spont}} \Delta\nu}{c t_c \lambda^2} \simeq 1.0 \times 10^{15} \text{ cm}^{-3}$$

Assuming that 5 percent of the exciting light energy falls within the useful absorption bands, that 5 percent of this light is actually absorbed by the crystal, that the average ratio of laser frequency to the pump frequency is 0.5, and that the lamp efficiency (optical output/electrical input) is 0.5, we obtain

$$\mathcal{E}_{\text{lamp}} = \frac{N_t h\nu_{\text{laser}}}{5 \times 10^{-2} \times 5 \times 10^{-2} \times 0.5 \times 0.5} \simeq 0.3 \text{ J/cm}^3$$

for the energy input to the lamp at threshold.

It is interesting to compare this last number to the figure of 300 joules per square centimeter of surface area obtained in the ruby example of Section 7.2. For reasonable dimension crystals (say, length = 5 cm, r = 2 mm) we obtain $\mathcal{E}_{\text{lamp}} = 0.25$ J. We expect the ruby threshold to exceed that of Nd^{3+}:YAG by three orders of magnitude, which is indeed the case.

(*b*) *Continuous operation.* The critical fluorescence power—that is, the actual power given off by spontaneous emission just below threshold

—is given by (6.3-4) as

$$\left(\frac{P_s}{V}\right) = \frac{N_t h\nu}{t_{\text{spont}}} \simeq 0.34 \text{ W/cm}^3$$

Taking the crystal diameter as 0.25 cm and its length as 3 cm and using the same efficiency factors assumed in the first part of this example, we can estimate the power input to the lamp at threshold as

$$P_{(\text{to lamp})} = \frac{0.4 \times (\pi/4) \times (0.25)^2 \times 3}{5 \times 10^{-2} \times 5 \times 10^{-2} \times 0.5 \times 0.5} \simeq 83 \text{ watts}$$

which is in reasonable agreement with experimental values [6].

A typical arrangement used in continuous solid state lasers is shown in Figure 7-10. The highly polished elliptic cylinder is used to concentrate the light from the lamp, which is placed along one focal axis, onto the laser rod, which occupies the other axis. This configuration guarantees that most of the light emitted by the lamp passes through the laser rod. The reflecting mirrors are placed outside the cylinder.

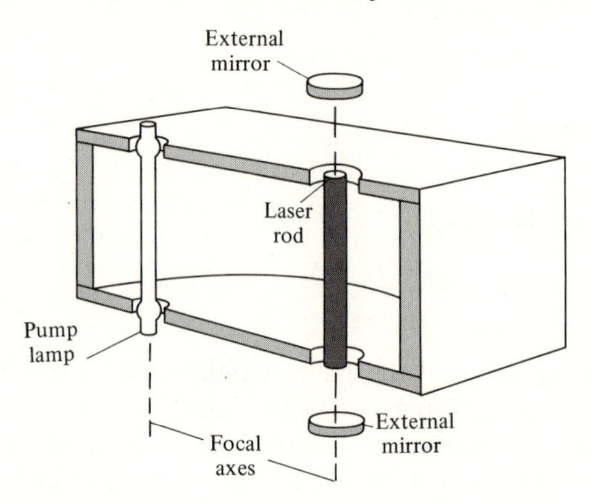

Figure 7-10 Typical continuous solid-state laser arrangement employing an elliptic cylinder housing for concentrating lamp light onto laser.

7.4 Neodymium-Glass Laser

One of the most useful laser systems is that which results when the Nd^{3+} ion is present as an impurity atom in glass [8].

The energy levels involved in the laser transition in a typical glass are shown in Figure 7-11. The laser emission wavelength is at $\lambda = 1.059\ \mu m$ and the lower level is approximately 1950 cm^{-1} above the ground state. As in the case of Nd^{3+}:YAG described in Section 7.3, we have here a four-level laser, since the thermal population of the lower laser level is

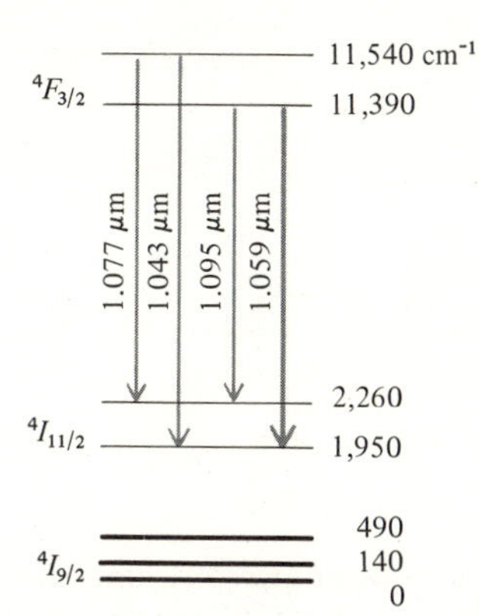

Figure 7-11 Energy-level diagram for the ground state and the states involved in laser emission at 1.059 μm for Nd^{3+} in a rubidium potassium barium silicate glass. (After Reference [8].)

negligible. The fluorescent emission near $\lambda_0 = 1.06\,\mu$m is shown in Figure 7-12. The fluorescent linewidth can be measured off directly and ranges, for the glasses shown, around 300 cm^{-1}. This width is approximately a factor of 50 larger than that of Nd^{3+} in YAG. This is due to the amorphous structure of glass, which causes different Nd^{3+} ions to "see" slightly different surroundings. This causes their energy splittings to vary slightly. Different ions consequently radiate at slightly different frequencies, causing a broadening of the spontaneous emission spectrum. The absorption bands responsible for pumping the laser level are shown in Figure 7-13. The probability that the absorption of a photon in any of these bands will result in pumping an atom to the upper laser level (that is, the absorption quantum efficiency) has been estimated [8] at about 0.4.

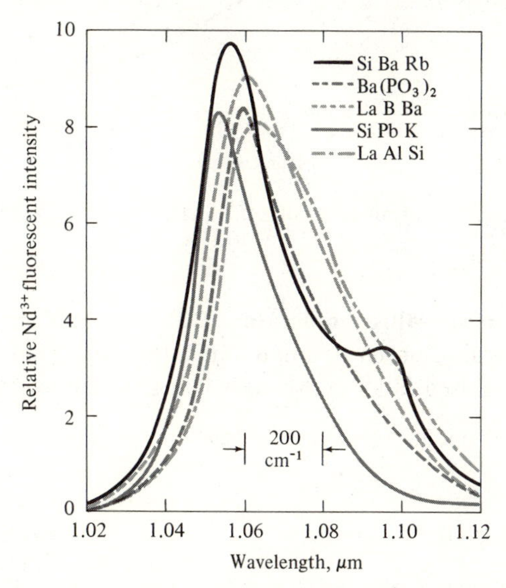

Figure 7-12 Fluorescent emission of the 1.06-μm line of Nd^{3+} at 300°K in various glass bases. (After Reference [8].)

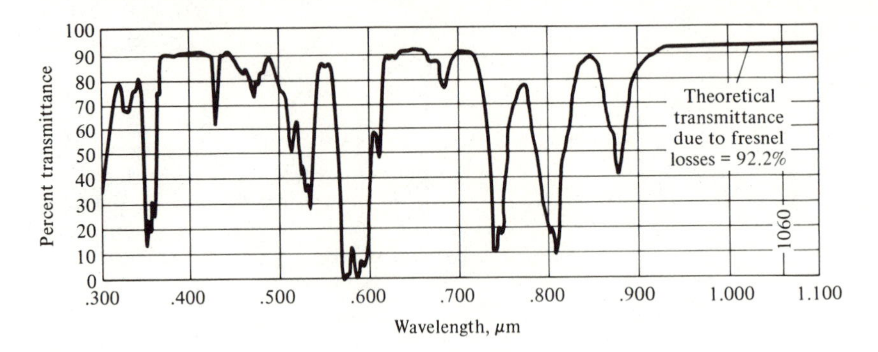

Figure 7-13 Nd^{3+} absorption spectrum for a sample of glass 6.4 mm thick with the composition 66 wt % SiO_2, 5 wt. % Nd_2O_3, 16 wt. % Na_2O, 5 wt. % BaO, 2 wt. % Al_2O_3, and 1 wt. % Sb_2O_3. (After Reference [8].)

The lifetime t_2 of the upper laser level depends on the host glass and on the Nd^{3+} concentration. This variation in two glass series is shown in Figure 7-14.

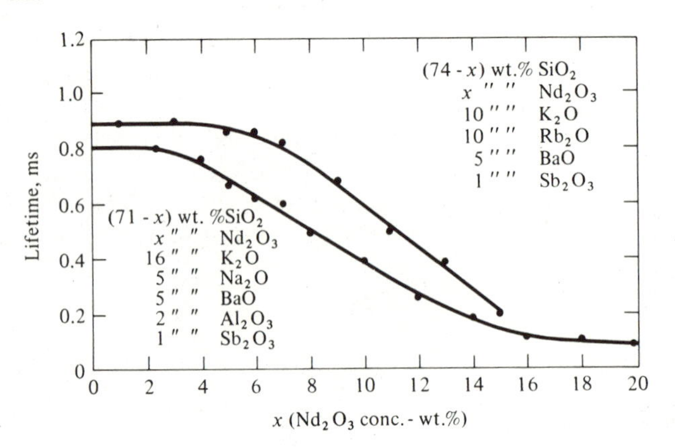

Figure 7-14 Lifetime as function of concentration for two glass series. (After Reference [8].)

Numerical example—thresholds for CW and pulsed operation of Nd^{3+}:glass lasers. Let us estimate first the threshold for continuous (CW) laser action in a Nd^{3+} glass laser using the following data:

$$\Delta\nu = 200 \text{ cm}^{-1} \text{ (see Figure 7-12)}$$

$$n = 1.5$$

$$t_{\text{spont}} \simeq t_2 = 3 \times 10^{-4} \text{ second}$$

$$\left.\begin{aligned} l &= \text{length of resonator} = 20 \text{ cm} \\ L &= \text{loss per pass} = 2 \text{ percent} \end{aligned}\right\} t_c \simeq \frac{nl}{Lc} = 5 \times 10^{-8} \text{ second}$$

Using (6.1-11) we obtain

$$N_t = \frac{8\pi t_{\text{spont}} n^3 \Delta\nu}{c t_c \lambda^2} = 9.05 \times 10^{15} \text{ atoms/cm}^3$$

for the critical inversion. The fluorescence power at threshold P_s is thus [see (6.3-5)]

$$P_s = \frac{N_t h\nu V}{t_{\text{spont}}} = 5.65 \text{ watts}$$

in a crystal volume $V = 1 \text{ cm}^3$.

We assume (a) that only 10 percent of the pump light lies within the useful absorption bands; (b) that because of the optical coupling inefficiency and the relative transparency of the crystal only 10 percent of the energy leaving the lamp within the absorption bands is actually absorbed; (c) that the absorption quantum efficiency is 40 percent; and (d) that the average pumping frequency is twice that of the emitted radiation. The lamp output at threshold is thus

$$\frac{2 \times 5.65}{0.1 \times 0.1 \times 0.4} = 2825 \text{ watts}$$

If the efficiency of the lamp in converting electrical to optical energy is about 50 percent, we find that continuous operation of the laser requires about 5 kW of power. This number is to be contrasted with a threshold of approximately 100 watts for the Nd:YAG laser, which helps explain why Nd:glass lasers are not operated continuously.

If we consider the pulsed operation of a Nd:glass laser by flash excitation we have to estimate the minimum energy needed to pump the laser at threshold. Let us assume here that the losses (attributable mostly to the output mirror transmittance) are $L = 20$ percent.[3] A recalculation of N_t gives

$$N_t = 9.05 \times 10^{16} \text{ atoms/cm}^3$$

The minimum energy needed to pump N_t atoms into level 2 is then

$$\frac{\mathcal{E}_{\min}}{V} = N_t(h\nu) = 1.7 \times 10^{-2} \text{ J/cm}^3$$

Assuming a crystal volume $V = 10 \text{ cm}^3$ and the same efficiency factors used in the CW example above, we find that the input energy to the flashlamp at threshold $\simeq 2 \times 1.7 \times 10^{-2} \times 10/0.1 \times 0.1 \times 0.4 = 85$ J. Typical Nd^{3+}:glass lasers with characteristics similar to those used in this example are found to require an input of about 150–300 joules at threshold.

[3] Because of the higher pumping rate available with flash pumping, optimum coupling (see Section 6.5) calls for larger mirror transmittances compared to the CW case.

7.5 He-Ne Laser

The first CW laser, as well as the first gas laser, was one in which a transition between the $2S$ and the $2p$ levels in atomic Ne resulted in the emission of 1.15 μm radiation [9]. Since then transitions in Ne were used to obtain laser oscillation at $\lambda_0 = 0.6328\ \mu$m [10] and at $\lambda_0 = 3.39\ \mu$m. The operation of this laser can be explained with the aid of Figure 7-15. A dc (or rf) discharge is established in the gas mixture containing typically, 1.0 mm Hg of He and 0.1 mm of Ne. The energetic electrons in the discharge excite helium atoms into a variety of excited states. In the normal cascade of these excited atoms down to the ground state, many collect in the long-lived metastable states 2^3S and 2^1S whose lifetimes are 10^{-4} second and 5×10^{-6} second, respectively. Since these long-lived (metastable) levels nearly coincide in energy with the $2S$ and $3S$ levels of Ne they can excite Ne atoms into these two excited states. This excitation takes place when an excited He atom collides with a Ne atom in the ground state and ex-

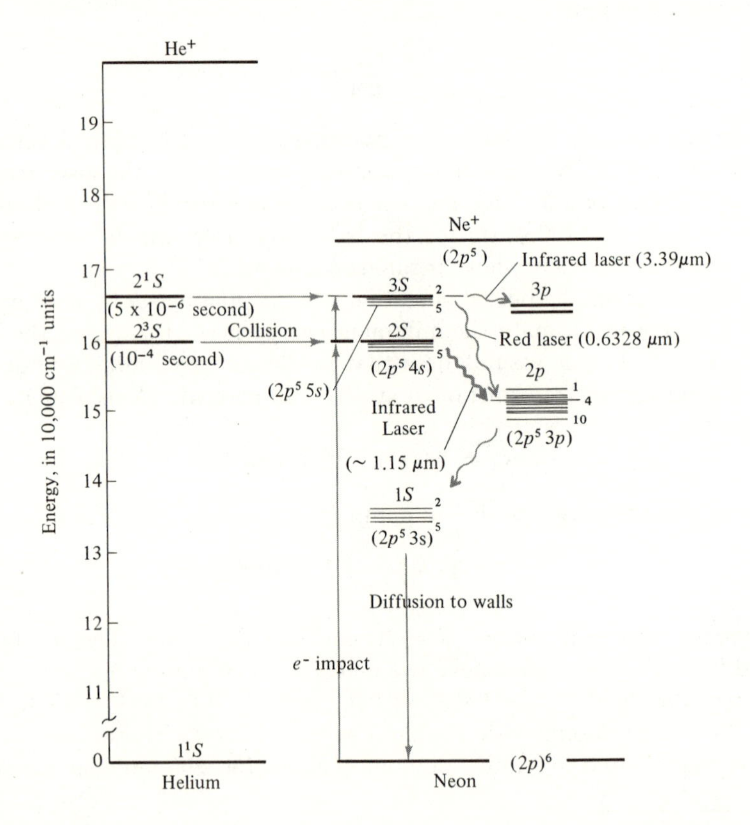

Figure 7-15 He-Ne energy levels. The dominant excitation paths for the red and infrared laser–maser transitions are shown. (After Reference [11].)

changes energy with it. The small difference in energy ($\sim 400\ \text{cm}^{-1}$ in the case of the $2S$ level) is taken up by the kinetic energy of the atoms after the collision. This is the main pumping mechanism in the He–Ne system.

1. *The 0.6328 μm oscillation.* The upper level is one of the Ne $3S$ levels, whereas the terminal level belongs to the $2p$ group. The terminal ($2p$) level decays radiatively with a time constant of about 10^{-8} second into the long-lived $1S$ state. This time is much shorter than the 10^{-7} second lifetime of the upper laser level $3S$. The condition $t_1 < t_2$ for population inversion in the $3S$–$2p$ transition (see Section 6.4) is thus fulfilled.

Another important point involves the level $1S$. Because of its long life it tends to collect atoms reaching it by radiative decay from the lower laser level $2p$. Atoms in $1S$ collide with discharge electrons and are excited back into the lower laser level $2p$. This reduces the inversion. Atoms in the $1S$ states relax back to the ground state mostly in collisions with the wall of the discharge tube. For this reason the gain in the 0.6328 μm transition is found to increase with decreasing tube diameter.

2. *The 1.15 μm oscillation.* The upper laser level $2S$ is pumped by resonant (that is, energy-conserving) collisions with the metastable 2^3S He level. It uses the same lower level as the 0.6328 μm transition and, consequently, also depends on wall collisions to depopulate the $1S$ Ne level.

3. *The 3.39 μm oscillation.* This involves a $3S$–$3p$ transition and thus uses the same upper level as the 0.6328 μm oscillation. It is remarkable for the fact that it provides a small-signal[4] optical gain of about 50 dB/m. This large gain reflects partly the inverse dependence of γ on ν^2 [see Equation (5.3-3)] as well as the short lifetime of the $3p$ level, which allows the buildup of a large inversion.

Because of the high gain in this transition, oscillation would normally occur at 3.39 μm rather than at 0.6328 μm. The reason is that the threshold condition will be reached first at 3.39 μm and, once that happens, the gain "clamping" will prevent any further buildup of the population of $3S$. The 0.6328 μm lasers overcome this problem by introducing into the optical path elements, such as glass or quartz Brewster windows, which absorb strongly at 3.39 μm but not at 0.6328 μm. This raises the threshold pumping level for the 3.39 μm oscillation above that of the 0.6328 μm oscillation.

[4] This is not the actual gain that exists inside the laser resonator, but the one-pass gain exercised by a very small input wave propagating through the discharge. In the laser the gain per pass is reduced by saturation until it equals the loss per pass.

A typical gas laser setup is illustrated by Figure 7-16. The gas envelope windows are tilted at Brewster's angle θ_B, so radiation with the electric field vector in the plane of the paper suffers no reflection losses at the windows. This causes the output radiation to be polarized in the sense shown, since the orthogonal polarization (the E vector out of the plane of the paper) undergoes reflection losses at the windows and, consequently, has a higher threshold.

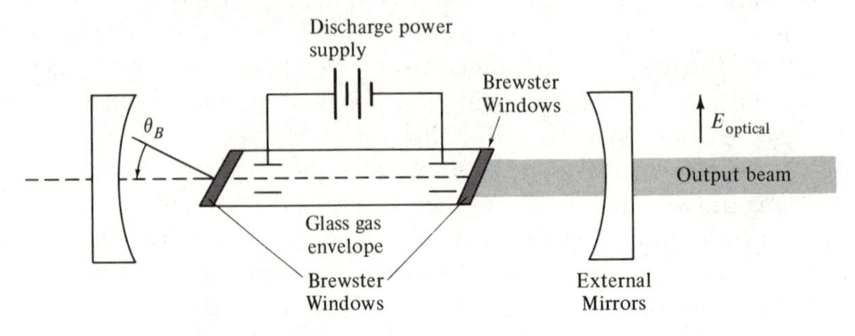

Figure 7-16 Typical gas laser.

7.6 Carbon Dioxide Laser

The lasers described so far in this chapter depend on electronic transitions between states in which the electronic orbitals (that is, charge distributions around the atomic nucleus) are different. As an example, consider the red (0.6328 μm) transition in Ne shown in Figure 7-15. It involves levels $2p^5 5s$ and $2p^5 3p$ so that in making a transition from the upper to the lower laser level one of the six outer electrons changes from a hydrogen-like state $5s$ (that is, $n = 5$, $l = 0$) to one in which $n = 3$ and $l = 1$.

The CO_2 laser [12] is representative of the so-called molecular lasers in which the energy levels of concern involve the internal vibration of the molecules—that is, the relative motion of the constituent atoms. The atomic electrons remain in their lowest energetic states and their degree of excitation is not affected.

As an illustration, consider the simple case of the nitrogen molecule. The molecular vibration involves the relative motion of the two atoms with respect to each other. This vibration takes place at a characteristic frequency of $\nu_0 = 2326\ \text{cm}^{-1}$ which depends on the molecular mass as well as the elastic restoring force between the atoms [13]. According to basic quantum mechanics the degrees of vibrational excitation are discrete (that is, quantized) and the energy of the molecule can take on the values $h\nu_0\ (v + \frac{1}{2})$, where $v = 0, 1, 2, 3, \cdots$. The energy-level diagram of N_2 (in its lowest electronic state) would then ideally consist of an equally spaced set of levels with a spacing of $h\nu_0$. The ground state ($v = 0$) and the first excited state ($v = 1$) are shown on the right side of Figure 7-17.

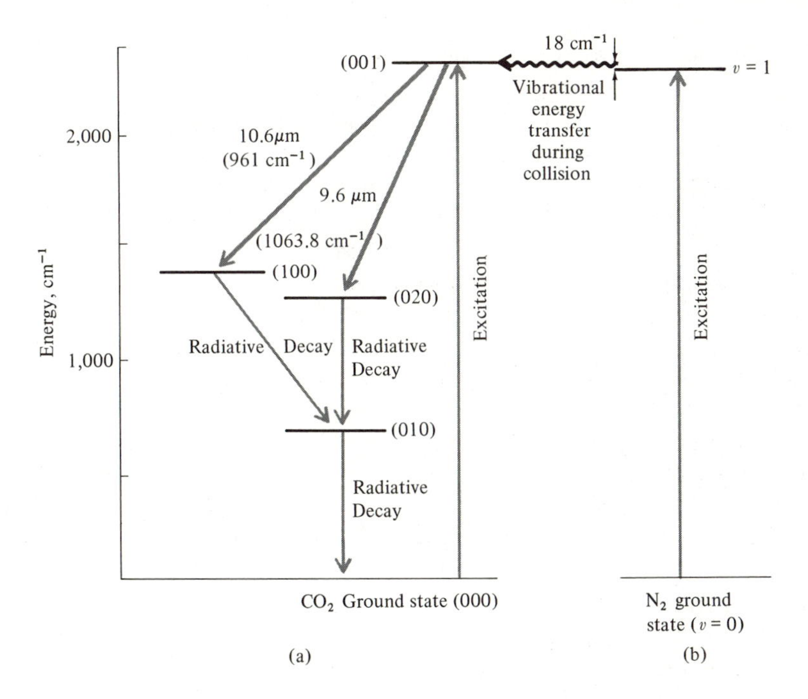

Figure 7-17 (**a**) Some of the low-lying vibrational levels of the carbon dioxide (CO_2) molecule, including the upper and lower levels for the 10.6-μm and 9.6-μm laser transitions. (**b**) Ground state ($v = 0$) and first excited state ($v = 1$) of the nitrogen molecule, which plays an important role in the selective excitation of the (001) CO_2 level.

The CO_2 molecule presents a more complicated case. Since it consists of three atoms, it can execute three basic internal vibrations, the so-called normal modes of vibration. These are shown in Figure 7-18. In (a) the molecule is at rest. In (b) the atoms vibrate along the internuclear axis in a symmetric manner. In (c) the molecules vibrate symmetrically along an axis perpendicular to the internuclear axis—the bending mode. In (d) the atoms vibrate asymmetrically along the internuclear axis. This mode is referred to as the asymmetric stretching mode. In the first approximation one can assume that the three normal modes are independent of each other, so the state of the CO_2 molecule can be described by a set of three integers (v_1, v_2, v_3), which correspond respectively to the degree of excitation of the three modes described. The total energy of the molecule is thus

$$E(v_1, v_2, v_3) = h\nu_1(v_1 + \tfrac{1}{2}) + h\nu_2(v_2 + \tfrac{1}{2}) + h\nu_3(v_3 + \tfrac{1}{2}) \qquad \textbf{(7.6-1)}$$

where ν_1, ν_2, and ν_3 are the frequencies of the symmetric stretch, bending, and asymmetric stretch modes, respectively.

Some of the low vibrational levels of CO_2 are shown in Figure 7-17. The upper laser level (001) is thus one in which only the asymmetric stretch mode, Figure 7-18(d), is excited and contains a single quantum $h\nu_3$ of energy.

The laser transition at 10.6 μm takes place between the (001) and (100) levels of CO_2. The excitation is provided usually in a plasma discharge which, in addition to CO_2, typically contains N_2 and He. The CO_2 laser possesses a high overall working efficiency of about 30 percent. This efficiency results primarily from three factors: (a) The laser levels are all near the ground state, and the atomic quantum efficiency ν_{21}/ν_{30}, which was discussed in Section 7.1, is about 45 percent; (b) a large fraction of the CO_2 molecules excited by electron impact cascade down the energy ladder from their original level of excitation and tend to collect in the long-lived (001) level; (c) a very large fraction of the N_2 molecules that are excited by the discharge tend to collect in the $v = 1$ level. Collisions with ground-state CO_2 molecules result in transferring their excitation to the latter, thereby exciting them to the (001) state as shown in Figure 7-17. The slight deficiency in energy (about 18 cm^{-1}) is made up by a decrease of the total kinetic energy of the molecules following the collision. This collision can be represented by

$$(v = 1) + (000) + \text{K.E.} = (v = 0) + (001) \qquad \textbf{(7.6-2)}$$

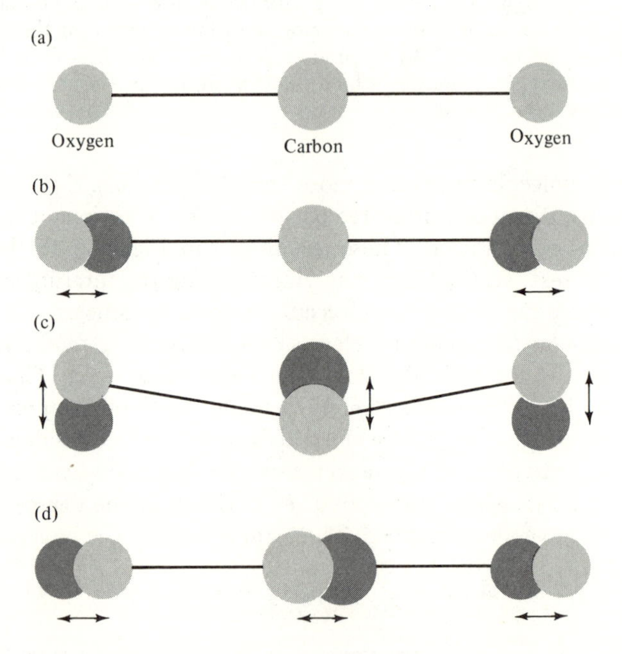

Figure 7-18 (**a**) Unexcited CO_2 molecule. (**b**), (**c**), and (**d**) The three normal modes of vibration of the CO_2 molecule. (After Reference [14].)

and has a sufficiently high cross section[5] that at the pressures and temperatures involved in the operation of a CO_2 laser most of the N_2 molecules in the $v = 1$ lose their excitation energy by this process.

Carbon dioxide lasers are not only efficient but can emit large amounts of power. Laboratory size lasers with discharge envelopes of a few feet in length can yield an output of a few kilowatts. This is due not only to the very *selective* excitation of the low-lying upper laser level, but also to the fact that once a molecule is stimulated to emit a photon it returns quickly to the ground state, where it can be used again. This is accomplished mostly through collisions with other molecules—such as that of He, which is added to the gas mixture.

7.7 Ar^+ Laser

Transitions between highly excited states of the singly ionized argon atom can be used to obtain oscillation at a number of visible (or near visible) wavelengths between 0.35 μm and 0.52 μm; see References [15] and [16]. The Ar^+ laser is consequently one of the most important lasers in use today. The pertinent energy level scheme is shown in Figure 7-19. The most prominent transition is the one at 4880 Å.

The Ar^+ laser can be operated in a pure Ar discharge that contains no other gases. The excitation mechanism involves collisions with energetic

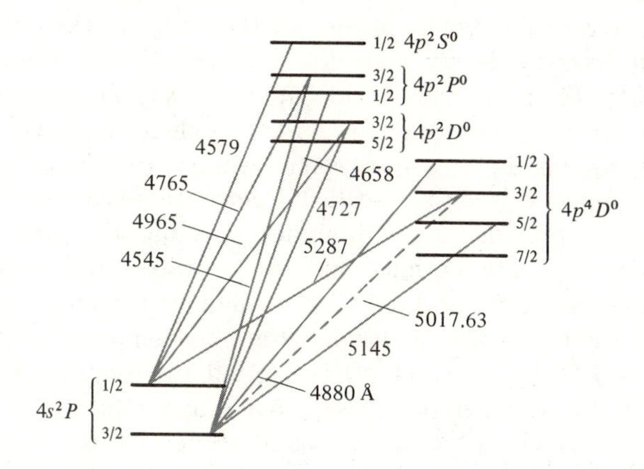

Figure 7-19 Energy level of Ar^+ and nine laser transitions. (Reference [15].)

[5] The cross section σ was defined in Section 7.2. In the present context it follows directly from that definition that the number of collisions of the type described by (7.6-2) per unit volume per unit time is equal to $N(v = 1)N(000)\sigma\bar{v}$ where $N(v = 1)$ and $N(000)$ are the densities of molecules in the states $v = 1$ of N_2 and (000) of CO_2, respectively. $\bar{v}$ is the (mean) relative velocity of the colliding molecules.

(~4–5 eV) electrons. Since the mean electron energy is small compared to the energy of the upper laser level (~20 eV above the ground state of the ion), it is clear that pumping is achieved by multiple collisions of Ar^+ ground-state ions with electrons followed by a number of cascading paths. The details of the collision and cascading processes are not clearly understood.

7.8 Semiconductor Junction Lasers[6]

In a semiconductor laser (see References [17]–[19]) the induced transitions take place in a *p-n* junction between occupied electron states in the conduction band and empty states in the valence band. One main difference between the semiconductor laser and the other lasers (atomic, molecular) considered in this chapter is that the transition takes place between states that are distributed in energy rather than between two well-defined energy levels.

The distribution of the electron states in a semiconductor is shown in Figure 7-20. The small dots in (a) indicate the position in energy of allowed electron states as a function of the propagation constant k of the electron for both the conduction and valence bands. The properties of a semiconductor depend critically on the manner in which the single electron states of 7-20(a) are filled. In Figure 7-20(b) is shown an intrinsic (undoped) semiconductor at a very low temperature. All the states in the valence band are filled with electrons (heavy black dots), whereas those in the conduction band are empty. In (c) we show what happens when the semiconductor is doped heavily with a donor impurity. All the valence band states as well as the conduction band states up to some level E_F, the Fermi energy,[7] are filled with electrons. This is the so-called degenerate *n*-type semiconductor. A heavy *p*-type doping density with an acceptor impurity causes a lowering of the boundary between filled and empty states—that is, the Fermi energy—into the valence band, so all the conduction band states as well as states near the top of the valence band are now empty. A feature common to Figure 7-20(b), (c), and (d) is that all the electron states up to the Fermi level E_F are filled, but those above it are empty. In (e) we show a new situation, in which the valence-band occupation is like that of the *p*-type degenerate semiconductor shown in (d), whereas the conduction band resembles the *n*-type degenerate case shown in (c). We have here a *doubly degenerate* semiconductor, which must be characterized with two Fermi energies E_{Fv} and E_{Fc} as shown. This situation can exist only under

[6] A familiarity with the basic concepts of semiconductor energy bands and *p-n* junction theory is assumed. A good treatment can be found in Reference [20].

[7] At zero temperature the Fermi level E_F marks the transition from occupied electron states below it to the empty states above it.

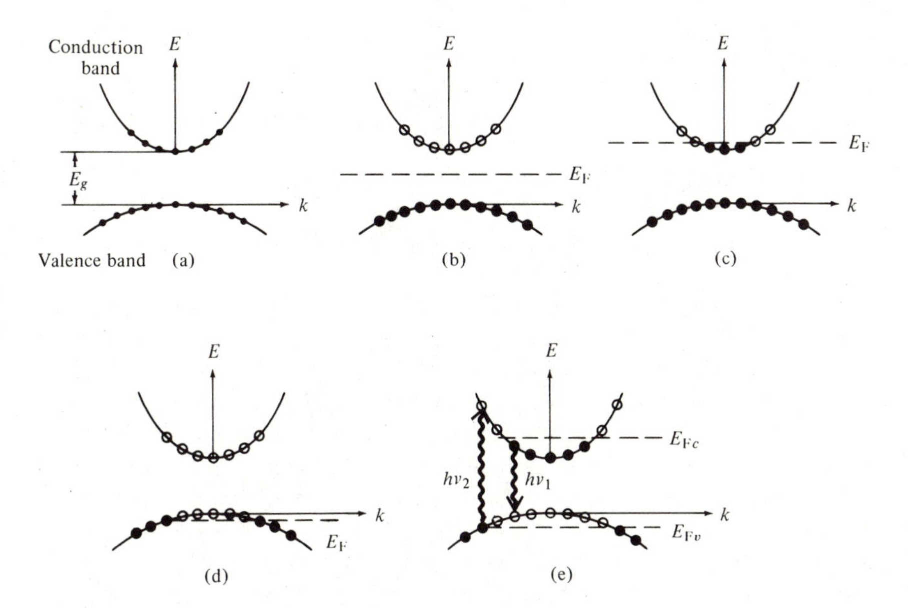

Figure 7-20 (**a**) Position of energy levels (small dots) near the energy gap of electrons in a direct semiconductor (such as GaAs) as a function of the electron propagation constant k. (**b**) An intrinsic semiconductor at 0°K. (**c**) An n-type degenerate semiconductor. (**d**) A p-type degenerate semiconductor. (**e**) A doubly degenerate semiconductor.

nonthermal equilibrium conditions. One example of double degeneracy is the case of an intrinsic semiconductor that is illuminated with light of frequency $\nu > E_g/h$. Absorption of this light proceeds by lifting electrons from the valence to the conduction band, leaving unoccupied states behind them. Since relaxation processes within each band are very fast compared with those between bands, the electrons within each band proceed to seek the lowest possible energies, which results in a distribution such as shown in Figure 7-20(e). Of course, if the light is turned off, all the electrons will relax back to the valence band and the electron distribution will return to that depicted by (b). Another example of a doubly degenerate distribution will be described below in connection with the p-n junction laser.

Now consider what happens to a wave propagating through a semiconductor at some frequency ν. The electrons can be induced by the radiation field to make transitions into *empty* states only. In Figure 7-20(b), (c), and (d), the empty states are always above the filled ones, so energy can only be absorbed from the wave. This no longer applies in the case of the doubly degenerate semiconductor (e). Frequencies ν such that $E_g < h\nu < E_{Fc} - E_{Fv}$ can induce only *downward* transitions from occupied conduction-band states to empty valence-band states. One such frequency ν_1 is shown in the figure. These frequencies cause the electrons to give up their energy and are thus amplified. Frequencies $\nu > (E_{Fc} - E_{Fv})/h$, such as ν_2, are absorbed. The condition for amplification is thus

$$E_{Fc} - E_{Fv} > h\nu > E_g \qquad \textbf{(7.8-1)}$$

A more formal derivation shows condition (7.8-1) to be valid at any temperature [21].

The p and n regions in a p-n junction laser are both degenerate. The position of the energy bands and electron occupation across the junction with no applied bias are shown in Figure 7-21(a). The Fermi energy has the same value all the way across the sample as appropriate to thermal equilibrium. Figure 7-21(b) shows what happens when a forward bias voltage V_{appl} that is nearly equal to the energy-gap voltage (E_g/e) is applied. The Fermi level in the n region is raised by eV_{appl} with respect to that in the p region. There now exists a narrow zone, called the active region, that contains both electrons and holes and is doubly degenerate. Electromagnetic radiation of frequency $E_g/h < \nu < (E_{Fc} - E_{Fv})/h$ propagating through this region is amplified. The thickness t of the active region can be approximated by the diffusion distance of electrons that are

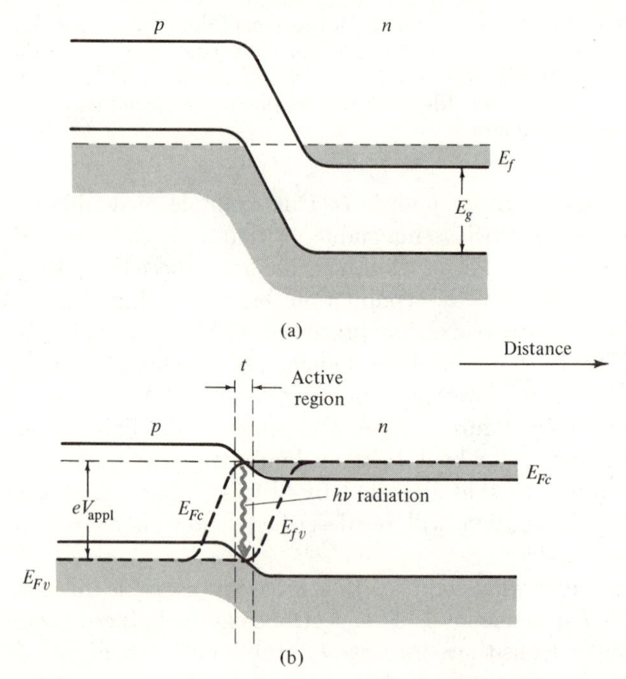

Figure 7-21 Degenerate p-n junction at (**a**) zero applied bias and (**b**) at a forward bias voltage $V_{\text{appl}} \simeq E_g/e$. The region containing both electrons and holes is called the active region. The oscillatory arrow indicates a recombination of an electron with a hole in the active region, leading to an emission of a photon with energy $h\nu$.

injected into a degenerate p region (that is, the mean distance traversed by the electrons before making a transition to an empty state in the valence band—the last event being referred to as an electron–hole recombination). This distance is given by $\sqrt{D\tau}$, where D is the diffusion coefficient and τ is the recombination time [20]. In GaAs, $D = 10\ \text{cm}^2/\text{s}$ and $\tau \simeq 10^{-9}$ second, so $t \sim 10^{-4}$ cm.

A typical GaAs laser whose emission is near 0.84 μm is shown in Figure 7-22. It is fabricated starting with a degenerate n-type semiconductor containing about 10^{18} tellurium (donor) atoms per cubic centimeter. The p region of the junction is obtained by diffusing an acceptor impurity such as Zn starting with a surface concentration of $\sim 10^{20}\ \text{cm}^{-3}$. Two end surfaces, normal to the junction plane, are polished (or cleaved) and serve as the laser reflectors.

Spontaneous recombination. The effect of applying a large ($\sim E_g/e$) forward bias voltage has been shown in Figure 7-21 to create a situation in which a large density of conduction-band electrons are present in a region—the socalled "active region"—that also contains a large number of unfilled valence-band states (holes). As in the case of the excited atoms discussed in Section 5.1, there is a finite probability per unit time that an electron in the conduction band in the active region will undergo a spontaneous transition to the valence band, giving off one photon in the process. This process results in the simultaneous annihilation of a conduction-band electron and an unfilled state (hole) in the valence band. It is referred to, consequently, as electron–hole recombination. The mean lifetime of an electron in the active region is the inverse of the spontaneous transition probability per unit time and is called the recombination lifetime. In GaAs this time is approximately 10^{-9} sec.

The recombination radiation spectrum of an InSb junction is shown in Figure 7-23.

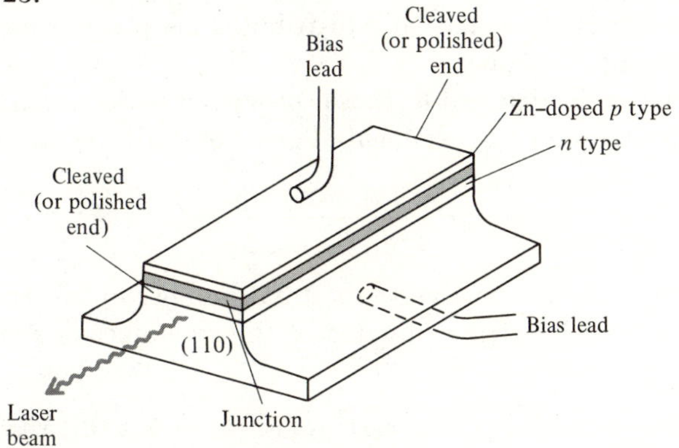

Figure 7-22 Typical p-n junction laser made of GaAs. Two parallel (110) faces are cleaved and serve as reflectors.

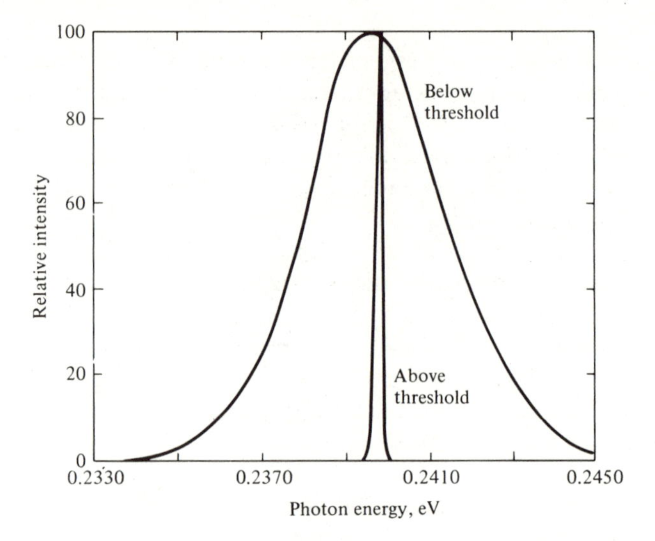

Figure 7-23 Infrared-emission spectra of a forward-biased InSb diode at 1.7°K below and above threshold. The broad spectrum corresponds to 300 mA and the narrow line to 400 mA. Width of line above threshold is limited by resolution of spectrometer. (After Reference [26].)

Amplification and oscillation in *p-n* junctions. In considering the problem of the amplification of electromagnetic radiation in *p-n* junctions we encounter a new situation. The inverted population of electrons is localized to within a distance $t \sim 1\ \mu$m of the junction center (the active region). The electromagnetic mode, on the other hand, is confined to a larger distance d.[8] The situation is depicted in Figure 7-24. Experiments reveal that, in GaAs, $d \simeq 2$–$5\ \mu$m and is thus larger than the active region thickness t.

Consider a crystal such as that shown in Figure 7-24, of a length l and a width in the y direction of w. Let us assume, for a moment, that $d = t$, and thus the inverted population is distributed more or less uniformly over the mode volume $V = dlw$.

This situation is identical to those considered in Section 5.3, so we can take the expression for the exponential gain constant, following (5.3-3), as

$$\gamma(\nu) = \frac{c^2(n_2 - n_1)/dlw}{8\pi n^2 \nu^2 t_{\text{recombination}}}\, g(\nu) \tag{7.8-2}$$

where n_2 and n_1 are, respectively, the total number of inverted electrons, $g(\nu)$ is the natural lineshape function of the spontaneous emission of the

[8] The transverse confinement of the mode is due to the fact that near the junction mid-plane the index of refraction decreases as a function of the distance from the center. This type of index variation is known to give rise to the so-called dielectric waveguide modes, which can be confined to small distances; see References [22]–[24].

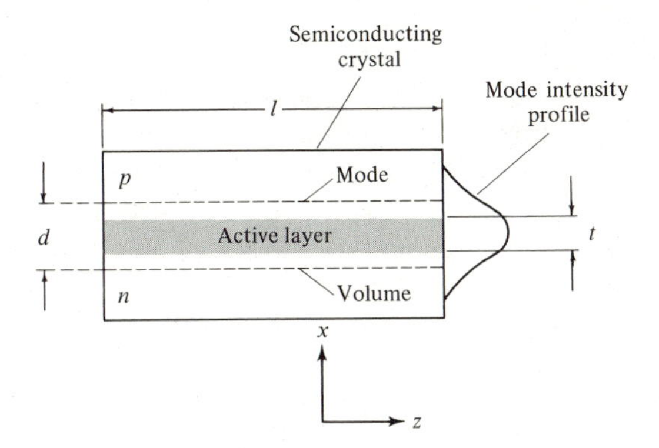

Figure 7-24 Schematic diagram showing the active layer and the transverse (x) intensity distribution of the fundamental laser mode.

junction, and $t_{\text{recombination}}$ is the lifetime of a conduction electron in the p region before making a spontaneous transition to an empty state in the valence band.

Next consider what happens to $\gamma(\nu)$ if the mode height d is made larger than t, as is the case in junction lasers. If the total mode power (watts) is kept a constant, the radiation intensity I (watts per square meter) as "seen" by the inverted electrons in the active region decreases. This causes, according to (5.2-15), a decrease in the induced transition rate W_i per electron, so the total power $W_i(n_2 - n_1)h\nu$ emitted by the electrons decreases. It follows, then, that when $d > t$ the gain constant $\gamma(\nu)$ is inversely proportional to d and is given by (7.8-2). Similar reasoning shows that if $d < t$, we need to replace d in (7.8-2) by t. This is the case in most other types of lasers in which the active region—that is, the region containing the inverted atomic population—is larger than that occupied by the electromagnetic mode.

The magnitude of the inversion $(n_2 - n_1)$ is not easily determined in an injection laser. It is possible, however, to relate it to the current through the diode. Assuming a low enough temperature that $n_1 = 0$, the total number of electrons injected into the diode in a given time interval must be equal in equilibrium to the number of spontaneous recombinations occurring during the same time

$$\frac{n_2}{t_{\text{recombination}}} = \frac{I\eta_i}{e} \tag{7.8-3}$$

where η_i, the internal quantum efficiency, is the fraction of the injected carriers (electrons or holes) that recombine radiatively and e is the electron charge. Using (7.8-3) in (7.8-2), putting $n_1 = 0$, and recalling that $lw = A$ is the junction area, we get

$$\gamma(\nu) = \frac{c^2 g(\nu)\eta_i}{8\pi n^2 \nu^2 ed}\left(\frac{I}{A}\right) \tag{7.8-4}$$

so that I/A (amperes per square meter) is the current density.

Threshold condition. Before deriving the start-oscillation condition of the injection laser we need to understand the origin of the optical losses. According to Figure 7-24, only a portion of the mode energy travels within the active region where it is amplified. Most of the energy propagates through the p and n regions and undergoes attenuation, which is characteristic of these regions.[9] We denote the distributed loss constant of the laser mode as α.[10] The other source of mode loss is the transmission through the end reflectors. Taking the reflectivity of the mirrors as R, the threshold condition (6.1-6) becomes

$$e^{(\gamma_t-\alpha)l}R = 1 \tag{7.8-5}$$

where the subscript t denotes threshold. Taking the logarithm of the last equation and using (7.8-4) we obtain

$$\frac{I_t}{A} = \frac{8\pi n^2\nu^2 ed\Delta\nu}{c^2\eta_i}\left(\alpha - \frac{1}{l}\ln R\right) \tag{7.8-6}$$

for the current density at threshold, where $\Delta\nu$, the transition linewidth, is defined by $\Delta\nu = g(\nu_0)^{-1}$. We note that the threshold current is proportional to the mode confinement distance d.

Numerical example—threshold current of a GaAs junction laser. Let us estimate the threshold injection current of a GaAs junction laser with the following characteristics:

$$\Delta\nu = 200 \text{ cm}^{-1}$$

$$\eta_i \simeq 1$$

$$\left(\alpha - \frac{1}{l}\ln R\right) = 20 \text{ cm}^{-1}$$

$$\lambda = 0.84 \ \mu\text{m}$$

$$n = 3.35$$

$$d = 2\mu\text{m}$$

Using these data in (7.8-6) gives

$$\frac{I_t}{A} \simeq 150 \text{ A/cm}^2$$

a value quite near that measured at low temperatures in GaAs injection lasers.

---

[9] This attenuation is due mostly to the presence of free carriers (electrons in the n, and holes in the p regions), which are accelerated by the optical field and dissipate energy through collisions [25].

[10] See Problem 7.1.

Table 7-1 contains a list of semiconductor junction lasers and their operating wavelengths.

Table 7-1 OSCILLATION WAVELENGTH AND OPERATING TEMPERATURE OF A NUMBER OF SEMICONDUCTOR *p-n* JUNCTION LASERS

MATERIAL	OSCILLATION WAVELENGTH, MICROMETERS		REFERENCES
GaAs	0.837 (4.2°K)	0.843 (77°K)	[17]–[19]
InP		0.907 (77°K)	[27]
InAs		3.1 (77°K)	[28], [29]
InSb	5.26 (10°K)		[30]
PbSe	8.5 (4.2°K)		[31]
PbTe	6.5 (12°K)		[32]
$Ga(As_xP_{1-x})$	0.65–0.84		[33]
$(Ga_xIn_{1-x})As$	0.84–3.5		[34]
$In(As_xP_{1-x})$	0.91–3.5		[35]
GaSb		1.6 (77°K)	[36]
$Pb_{1-x}Sn_xTe$	9.5–28 (~12°K) ↓ ↓ $x = 0.15$ $x = 0.27$		[37]
InGaP	0.5–0.7		

Heterojunction lasers. According to (7.8-6) and the discussion following (7.8-2) the threshold current density in a semiconductor junction laser is proportional to the electromagnetic mode confinement distance d or the thickness t of the active layer, whichever is larger.

One method of reducing both d and t utilizes the double heterostructure junction technique [38, 39, 40, 41]. The active region of the laser is usually a thin (0.1 μm–0.4 μm) layer of GaAs. This layer is sandwiched between a p type $Ga_{1-x}Al_xAs$ layer and an n type $Ga_{1-y}Al_yAs$ layer ($Ga_{1-x}Al_xAs$ is a GaAs crystal in which a fraction x of the Ga atoms has been replaced by Al) as shown in Figure 7-25. Since the lattice constant of $Ga_{1-x}Al_xAs$ is nearly the same (within a fraction of a percent) as that of GaAs, the composite crystal can be grown epitaxially with a minimum of lattice disruption.

Now there are two important properties of $Ga_{1-x}Al_xAs$ which bring about the reduction of t and d. The energy gap of the crystal increases monotonically with x. This results in a potential barrier for injected electrons at the GaAs-p $Ga_{1-x}Al_xAs$ interface and a similar barrier for injected holes at the GaAs-n $Ga_{1-y}Al_yAs$ interface. This situation, shown in Figure 7-25(b), results in a tight confinement, that is, small t, of the injected electrons and holes and prevents their diffusion as minority carriers away from the active region.

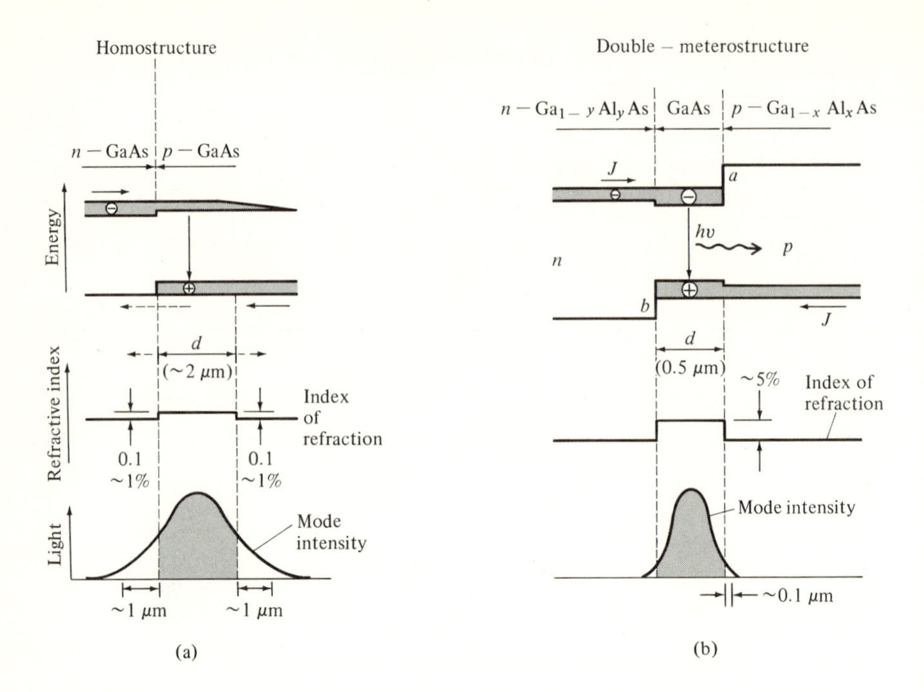

Figure 7-25 Schematic representation of the band edges with forward bias, refractive index changes, and optical field distribution in a homostructure and a double heterostructure diode. (After Reference [40].)

The second physical property utilized is the dependence of the index of refraction of $Ga_{1-x}Al_xAs$ on x. This dependence is approximately

$$\Delta n \approx -0.4x \tag{7.8-12}$$

where Δn is the change in the index relative to that of GaAs. The resulting index profile across the three layers is illustrated in Figure 7-25(b). The large index discontinuity gives rise to strong dielectric waveguiding (see Chapter 13) and a mode whose intensity is highly localized, that is, a small value of d. The practical exploitation of these ideas has led to dramatic reduction of the threshold of injection lasers [39, 40] to a point where continuous room temperature operation is now easily achieved.

A schematic representation of a GaAs double heterostructure laser is shown in Figure 7-26.

Power output of injection lasers. The considerations of saturation and power output in an injection laser are basically the same as that of conventional lasers, which were described in Section 6.4. As the injection current is increased above the threshold value (7.8-6), the laser oscillation

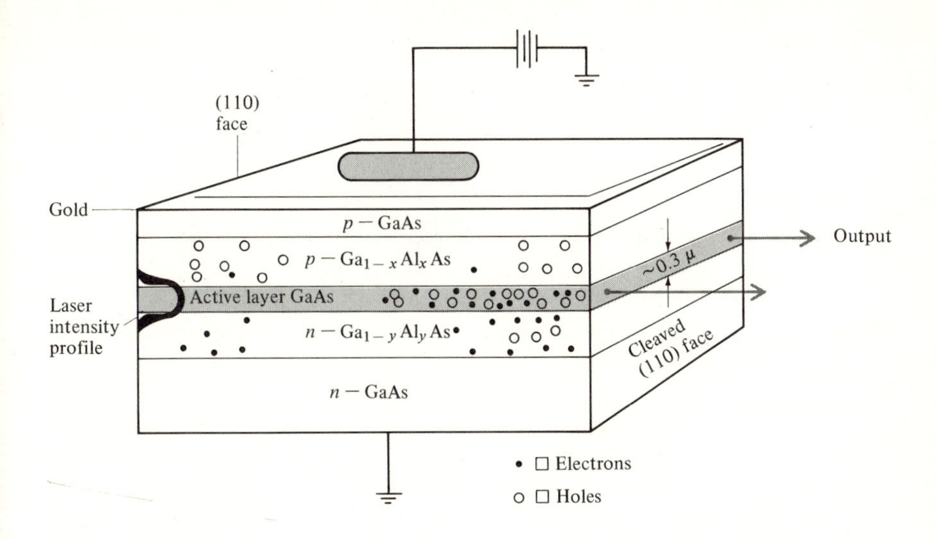

Figure 7-26 A typical double heterostructure GaAs-GaAlAs laser. Electrons and holes are injected into the active GaAs layer from the n and p $Ga_{1-x}Al_xAs$ layers, respectively. Frequencies near $v = E_g/h$ are amplified by stimulating electron-hole recombination.

intensity builds up. The resulting stimulated emission shortens the lifetime of the inverted carriers to the point where the magnitude of the inversion is clamped at its threshold value. Taking the probability that an injected carrier recombines radiatively within the active region as η_i (this is the internal quantum efficiency defined in Section 7.8), we can write the following expression for the power emitted by stimulated emission:

$$P_e = \frac{(I - I_t)\eta_i}{e} h\nu \tag{7.8-7}$$

Part of this power is dissipated inside the laser resonator, and the rest is coupled out through the end reflectors. These two powers are, according to (7.8-6), proportional to α and to $-l^{-1} \ln R$, respectively. We can thus write the output power as

$$P_o = \frac{(I - I_t)\eta_i h\nu}{e} \left(\frac{(1/l) \ln (1/R)}{\alpha + (1/l) \ln (1/R)} \right) \tag{7.8-8}$$

The external differential quantum efficiency η_{ex} is defined as the ratio of the photon output rate that results from an increase in the injection rate (carriers per second) to the increase in the injection rate:

$$\eta_{\text{ex}} = \frac{d(P_o/h\nu)}{d[(I - I_t)/e]} \tag{7.8-9}$$

Using (7.8-8) we obtain

$$\begin{aligned}\eta_{ex}^{-1} &= \eta_i^{-1}\frac{\alpha l + \ln(1/R)}{\ln(1/R)} \\ &= \eta_i^{-1}\left[\frac{\alpha l}{\ln(1/R)} + 1\right]\end{aligned} \qquad \textbf{(7.8-10)}$$

This relation is used to determine η_i from the experimentally measured dependence of η_{ex} on l. At 77°K, η_i in GaAs $\simeq$ 0.7–1.

Power efficiency of injection lasers. If the voltage applied to a diode is V_{appl}, the electric power input is $V_{appl}I$. The efficiency of the laser in converting electrical input to laser output is thus

$$\eta = \frac{P_o}{VI} = \eta_i \frac{(I - I_t)}{I}\left(\frac{h\nu}{eV_{appl}}\right)\frac{\ln(1/R)}{\alpha l + \ln(1/R)} \qquad \textbf{(7.8-11)}$$

From Figure 7-21, $eV_{appl} \simeq h\nu$ (in practice the small voltage drop in the diode bulk resistance makes eV_{appl} slightly larger than $h\nu$); therefore, well above threshold ($I \gg I_t$), where optimum coupling (see Section 6.5) dictates that $(1/l)\ln(1/R) \gg \alpha$, η approaches η_i. Since η_i in most lasers is high (0.7–1 in GaAs), the injection laser possesses the highest power efficiency of all the laser types.

The unique features of low power—electrical excitation, small size, and compatibility for interfacing with semiconductor electronics and optical fibers make the injection laser an extremely important component in optical technology.

7.9 Organic-Dye Lasers

Many organic dyes (that is, organic compounds that absorb strongly in certain visible-wavelength regions) also exhibit efficient luminescence, which often spans a large wavelength region in the visible portion of the spectrum. This last property makes it possible to obtain an appreciable tuning range from dye lasers; see References [42]–[44].

A schematic representation of an organic dye molecule (such as rhodamine 6G, for example) is shown in Figure 7-27.

State S_0 is the ground state. S_1, S_2, T_1, and T_2 are excited electronic states—that is, states in which one ground-state electron is elevated to an excited orbit. Typical energy separation, such as S_0–S_1 is about 20,000 cm^{-1}. In a singlet (S) state, the magnetic spin of the excited electron is antiparallel to the spin of the remaining molecule. In a triplet (T) state, the spins are parallel. Singlet → triplet, or triplet → singlet transitions thus involve a spin flip and are far less likely than transitions between two singlet or between two triplet states.

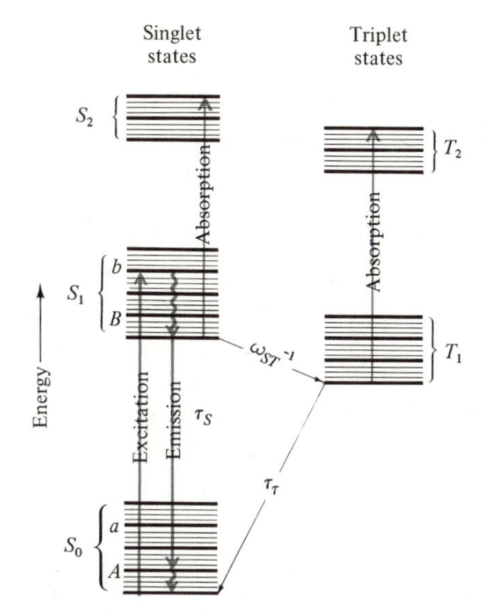

Figure 7-27 Schematic representation of the energy levels of an organic dye molecule. The heavy horizontal lines represent vibrational states and the lighter lines represent the rotational fine structure. Excitation and laser emission are represented by the transitions $A \rightarrow b$ and $B \rightarrow a$, respectively.

Transitions between two singlet states or between two triplet states, which are spin-allowed (that is, they do not involve a spin flip), give rise to intense absorption and fluorescence. The characteristic color of organic dyes is due to the $S_0 \rightarrow S_1$ absorption.

The singlet and triplet states, in turn, are split further into vibrational levels shown as heavy horizontal lines in Figure 7-27. These correspond to the quantized vibrational states of the organic molecule, as discussed in detail in Section 7.6. Typical energy separation between two adjacent vibrational levels within a given singlet or triplet state is about 1500 cm^{-1}. The fine splitting shown corresponds to rotational levels[11] whose spacing is about 15 cm^{-1}.

In the process of pumping the laser, the molecule is first excited, by absorbing a pump photon, into a rotational–vibrational state b within S_1. This is followed by a very fast decay to the bottom of the S_1 group, with the excess energy taken up by the vibrational and rotational energy of the molecules. Most of the excited molecules will then decay spontaneously to state a, emitting a photon of energy $\nu = (E_B - E_a)/h$. The lifetime for this process is τ_S.

There is, however, a small probability, approximately $\omega_{ST}\tau_S$, that an excited molecule will decay instead to the triplet state T_1, where ω_{ST} is the rate per molecule for undergoing an $S_1 \rightarrow T_1$ transition. Since this is a spin-forbidden transition, its rate is usually much smaller than the spontaneous decay rate τ_S^{-1}, so that $\omega_{ST}\tau_S \ll 1$. The lifetime τ_T for decay of T_1 to the ground state is relatively long (since this too is a spin-forbidden

[11] A transition between two adjacent rotational levels involves a change in the total angular momentum of the molecule about some axis.

transition) and may vary from 10^{-7} to 10^{-3} second, depending on the experimental conditions [45]. Owing to its relatively long lifetime, the triplet state T_1 acts as a trap for excited molecules. The absorption of molecules due to a $T_1 \rightarrow T_2$ transition is spin-allowed and is therefore very strong. If the wavelength region of this absorption coincides with that of the laser emission [at $\nu \simeq (E_B - E_a)/h$], an accumulation of molecules in T_1 increases the laser losses and at some critical value quenches the laser oscillation. For this reason, many organic-dye lasers operate only on a pulsed basis. In these cases fast-rise-time pump pulses—often derived from another laser [43]—cause a buildup of the S_1 population with oscillation taking place until an appreciable buildup of the T_1 population occurs.

Another basic property of molecules is that the peak of the absorption spectrum usually occurs at shorter wavelengths than the peak of the corresponding emission spectrum. This is illustrated in Figure 7-28, which shows the absorption and emission spectra of rhodamine 6G which when dissolved in H_2O is used as a CW laser medium [47]. Laser oscillation occurring near the peak of the emission curve is thus absorbed weakly. But for this fortunate circumstance, laser action involving electronic transitions in molecules would not be possible.

Typical excitation and oscillation waveforms of a dye laser are shown in Figure 7-29. The possibility quenching of the laser action by triplet state absorption is evident.

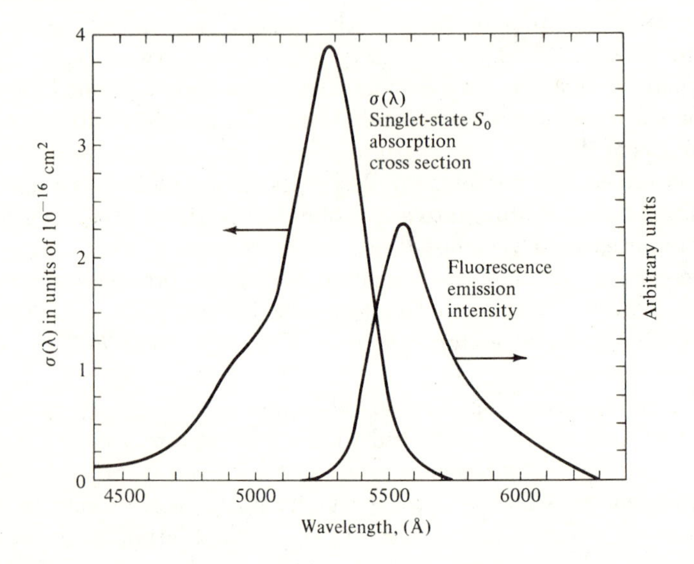

Figure 7-28 Singlet-state absorption and fluorescence spectra of rhodamine 6G obtained from measurements with a 10^{-4} molar ethanol solution of the dye. (After Reference [45].)

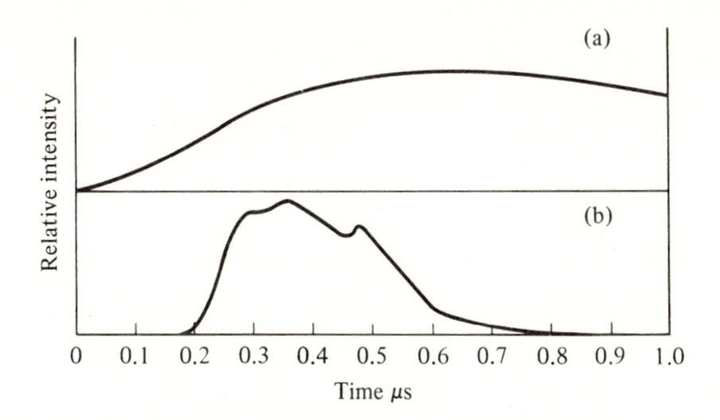

Figure 7-29 (**a**) Flashlamp pulse produced by a linear xenon flashlamp in a low-inductance circuit. (**b**) Laser pulse from a 10^{-3} molar solution of rhodamine 6G in methanol. (After Reference [44].)

A list of some common laser dyes is given in Table 7-2.

The broad fluorescence spectrum of the organic dyes suggests a broad tunability range for lasers using them as the active material. The spectrum in Figure 7-28, as an example, corresponds to a width of $\Delta\nu \simeq 1000 \text{ cm}^{-1}$. One elegant solution for realizing this tuning range consists of replacing one of the laser mirrors with a diffraction grating, as shown in Figure 7-30. A diffraction grating has the property that (for a given order) an incident beam will be reflected back *exactly* along the direction of incidence, provided

$$2d \cos\theta = m\lambda \qquad m = 1, 2, \cdots \tag{7.9-1}$$

where d is the ruling distance, θ is the angle between the propagation direction and its projection on the grating surface, λ is the optical wavelength in the medium next to the grating, and m is the order of diffraction. This type of operation of a grating is usually referred to as the Littrow arrangement. When a grating is used as one of the laser mirrors, it is clear that the oscillation wavelength will be that which satisfies (7.9-1), since other wavelengths are not reflected along the axis of the optical resonator and

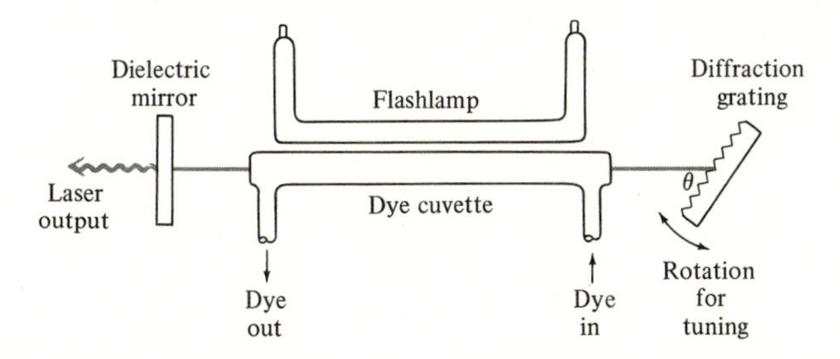

Figure 7-30 A typical pulsed dye laser experimental setup employing a linear flashlamp and a wavelength-selecting diffraction-grating reflector.

Table 7-2 MOLECULAR STRUCTURE, LASER WAVELENGTH, AND SOLVENTS FOR SOME LASER DYES (AFTER REFERENCE [45])

Dye	*Structure*	*Solvent*	*Wavelength*
Acridine red	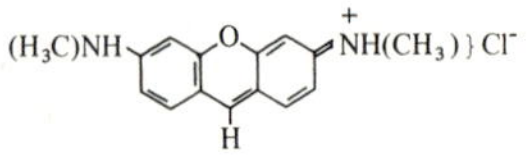 $(H_3C)NH$ … $\overset{+}{N}H(CH_3)\}Cl^-$; O; H	EtOH	Red 600–630 nm
Puronin B	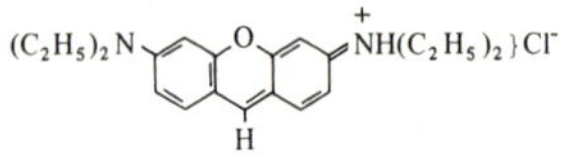 $(C_2H_5)_2N$ … $\overset{+}{N}H(C_2H_5)_2\}Cl^-$; O; H	MeOH H_2O	Yellow
Rhodamine 6G	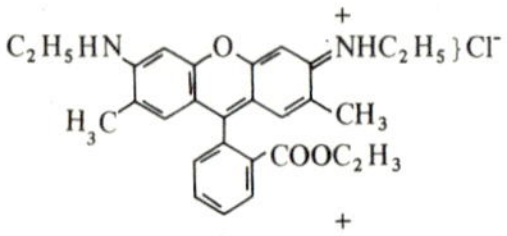 C_2H_5HN … $\overset{+}{N}HC_2H_5\}Cl^-$; O; H_3C; CH_3; $COOC_2H_3$	EtOH MeOH H_2O DMSO Polymethyl-methacrylate	Yellow 570–610 nm
Rhodamine B	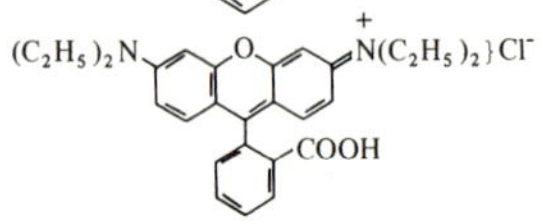 $(C_2H_5)_2N$ … $\overset{+}{N}(C_2H_5)_2\}Cl^-$; O; COOH	EtOH MeOH Polymethyl-methacrylate	Red 605–635 nm
Na-fluorescein	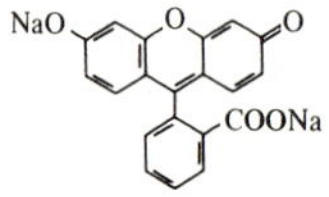 NaO … O; O; COONa	EtOH H_2O	Green 530–560 nm
2, 7-Dichloro-fluorescein	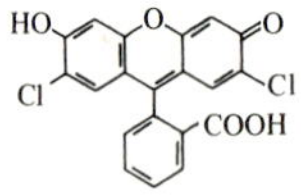 HO … O; O; Cl; Cl; COOH	EtOH	Green 530-560nm
7-Hydroxycoumarin	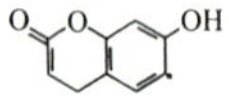 O; O; OH	H_2O (pH ~ 9)	Blue 450–470 nm
4-Methylumbelliferone	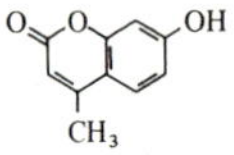 O; O; OH; CH_3	H_2O (pH ~ 9)	Blue 450–470 nm
Esculin	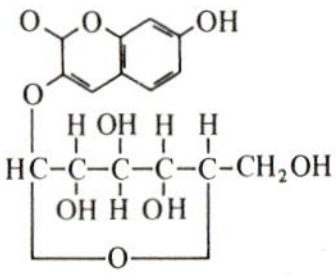 O; O; OH; O; H OH H H; HC–C–C–C–C–CH_2OH; OH H OH; O	H_2O (pH ~ 9)	Blue 450–470 nm
7-Diethylamino-4-Methylcoumarin	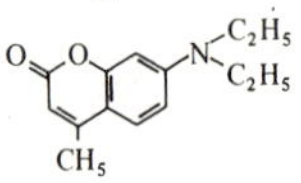 O; O; N(C_2H_5)(C_2H_5); CH_5	EtOH	Blue
Acetamidopyrene-trisuifonate	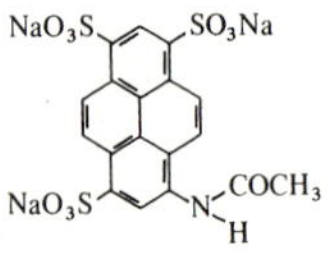 NaO_3S; SO_3Na; NaO_3S; N($COCH_3$)(H)	MeOH H_2O	Green-Yellow
Pyrylium salt	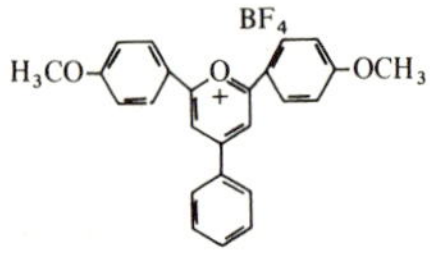 BF_4; H_3CO … $\overset{+}{O}$ … OCH_3	MeOH	Green

will consequently "see" a very lossy (low-Q) resonator. The tuning (wavelength selection) is thus achieved by a rotation of the grating. It follows also that any other means of introducing a controlled, wavelength-dependent loss into the optical resonator can be used for tuning the output.

7.10 High-Pressure Operation of Gas Lasers

Consider a laser medium with an inversion density of ΔN atoms/m^3 at some transition with energy spacing near $h\nu_0$. If this medium is to be used as an amplifier of a pulsed signal at ν_0, then the maximum energy that can be extracted by the signal, through stimulated emission, is $\sim\Delta N h\nu_0$ joules per unit volume of the laser medium. It would follow straightforwardly that to increase the energy gain (= energy out/energy in) of the amplifier we need to increase the inversion density ΔN which, according to (6.4-5), can be done by stronger pumping.

Unfortunately, a mere increase in the pumping strength will increase, according to (5.6-10), the unsaturated gain $\gamma_0(\nu_0)$ of the medium, which will lead at some point to parasitic oscillation off spurious reflections or to energy depletion by amplification of the spontaneous emission [48].

One way around this problem in gas lasers is to increase the density (and pressure) of the amplifying medium. The increase in molecular density causes a proportionate decrease in molecular collision time τ which, according to (5.1-8), causes the transition linewidth $\Delta\nu$ to increase. At a given inversion, this would cause, according to (5.6-10), a reduction in the gain (recall here that $g(\nu_0) = (\Delta\nu)^{-1}$). Alternatively, if the maximum tolerable gain is γ_{max}, the reduction in gain due to increased pressure makes it possible to increase the inversion ΔN (by increased pumping) relative to its low-pressure value, until the maximum allowable gain γ_{max} is achieved. This, as discussed above, leads to increased stored energy density which can be "milked" by the signal pulse.

Let us look, somewhat more formally, at the problem of operating a continuous gas laser oscillator at increased pressures. Much of the work in this field was done on CO_2 lasers so that the following discussion will refer to this particular system, although the considerations are quite general.

The transition linewidth of the mixture of CO_2 and other gases used in CO_2 lasers can be written according to (5.1-8) as

$$\Delta\nu = \Delta\nu_D + \sum_i \frac{1}{\pi\tau_i} \tag{7.10-1}$$

where $\Delta\nu_D$ is the Doppler linewidth (5.1-15) and τ_i is the mean collision lifetime of a CO_2 molecule with a molecule of the ith molecular species

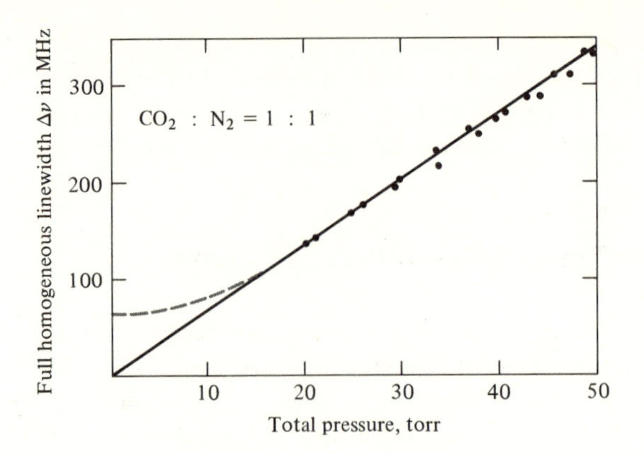

Figure 7-31 The 10.6-μm transition linewidth versus pressure for a gas mixture with equal partial pressures of CO_2 and N_2 at 300 °K.

(N_2, He, and so on) present in the mixture.

For a large range of pressures, τ_i^{-1} is proportional to the pressure [49] so that once $\sum_i (\pi\tau_i)^{-1} > \Delta\nu_D$, the transition linewidth $\Delta\nu$ is essentially proportional to pressure. This region is referred to as the pressure broadened regime and is illustrated by Figure 7-31.

Consider now the problem of maintaining the laser oscillation in a high pressure discharge. First, to achieve a given gain (which is equal to the resonator loss) we need, according to (5.6-10), to increase the inversion density by an amount proportional to the pressure P in order to compensate for the increase of $\Delta\nu$.[12] Second, since the lifetime t_2 in the upper laser level varies as P^{-1}, the pumping power per molecule increases, according to (6.3-4), as P. The result is that the pumping power, for a given gain, increases as P^2. It follows that the output power, along with the excitation power, increases with P^2. This conclusion follows more formally, from (6.5-10) for the power output

$$P_0 = \frac{8\pi n^2 h\nu \Delta\nu A}{\lambda^2(t_2/t_{\text{spont}})} T \left(\frac{g_0}{L_i + T} - 1\right)$$

since $\Delta\nu \propto P$ and $t_2 \propto 1/P$.

The increase of power with pressure is seen in Figure 7-32. The roll-off near $P = 150$ torr reflects the reduction in gain at the higher pressures. A more fundamental measure of the pressure effects is the variation of the saturation intensity (5.6-9)

$$I_s = \frac{8\pi n^2 \Delta\nu h\nu}{(t_2/t_{\text{spont}})\lambda^2}$$

[12] Recall here that in (5.6-10) $g(\nu_0) = (\Delta\nu)^{-1}$.

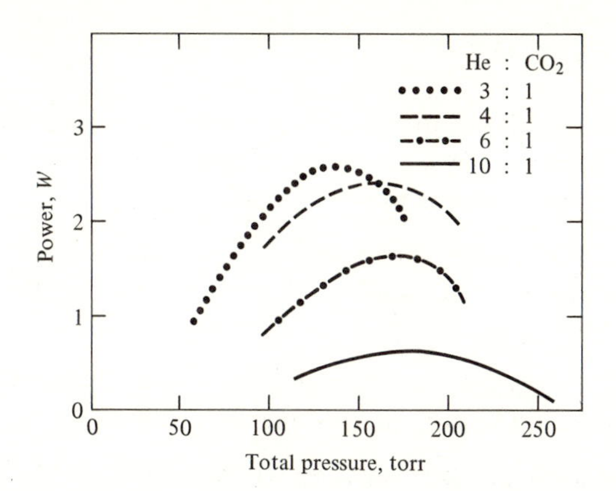

Figure 7-32 Output power versus total pressure under optimum pumping for He:CO_2 mixtures. (After Reference [50].)

that, for the reasons given above, should increase as P^2. Experimental data of I_s versus P is shown in Figure 7-33.

High energy pulsed operation of CO_2 lasers (Reference [51]) at atmospheric pressure has been responsible for large and simple lasers suitable for many industrial uses.

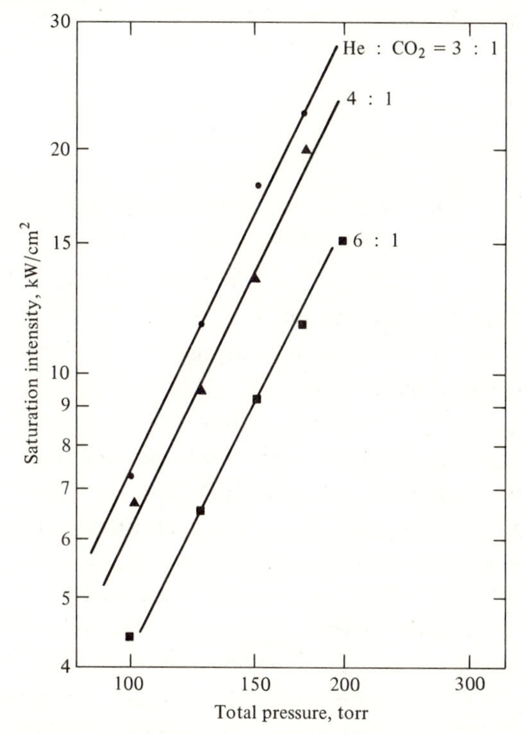

Figure 7-33 Measured saturation intensity versus pressure in a CO_2 laser. (After Reference [50].)

PROBLEMS

7-1 Show how the distributed loss constant α of an injection laser can be determined from two measurements of the threshold current taken at two different values of mirror reflectivity R.

7-2 Show that if the mode thickness d of an injection laser is smaller than t, the thickness of the active region, the gain constant is given by (7.8-2) with d replaced by t.

7-3 Derive the expression relating the absorption cross section at ν in a given $a \rightarrow b$ transition to the spontaneous $b \rightarrow a$ lifetime.

7-4 Derive condition (7.9-1) for the Littrow arrangement of a diffraction grating for which the reflection is parallel to the direction of incidence.

7-5 **a.** Estimate the exponential gain coefficient $\gamma(\nu_0)$ of a 10^{-4} molar solution of rhodamine 6G in ethanol by assuming the peak emission cross section to be comparable to the peak absorption cross section. Use the data of Figure 7-26.
b. Estimate the spontaneous lifetime for an $S_1 \rightarrow S_0$ transition.
c. Estimate the CW pump power threshold assuming 50 percent absorption of pump and 100 percent pumping quantum efficiency.

REFERENCES

[1] Maiman, T. H., "Stimulated optical radiation in ruby masers," Nature, vol. 187, p. 493, 1960.

[2] Maiman, T. H., "Optical and microwave-optical experiments in ruby," *Phys. Rev. Letters*, vol. 4, p. 564, 1960.

[3] Cronemeyer, D. C., "Optical absorption characteristics of pink ruby," *J. Opt. Soc. Am.*, vol. 56, p. 1703, 1966.

[4] Schawlow, A. L., "Fine structure and properties of chromium fluorescence," in *Advances in Quantum Electronics*, J. R. Singer, ed. New York: Columbia University Press, p. 53, 1961.

[5] Yariv, A., "Energy and power considerations in injection and optically pumped lasers," *Proc. IEEE*, vol. 51, p. 1723, 1963.

[6] Geusic, J. E., H. M. Marcos, and L. G. Van Uitert, "Laser oscillations in Nd-doped yttrium aluminum, yttrium gallium and gadolinium garnets," *Appl. Phys. Letters*, vol. 4, p. 182, 1964.

[7] Kushida, T., H. M. Marcos, and J. E. Geusic, "Laser transition cross section and fluorescence branching ratio for Nd^{3+} in yttrium aluminum garnet," *Phys. Rev.*, vol. 167, p. 1289, 1968.

[8] Snitzer, E., and C. G. Young, "Glass lasers," in *Lasers*, vol. 2, A. K. Levine, ed. New York: Marcel Dekker, Inc., p. 191, 1968.

[9] Javan, A., W. R. Bennett, Jr., and D. R. Herriott, "Population inversion and continuous optical maser oscillation in a gas discharge containing a He–Ne mixture," *Phys. Rev. Letters*, vol. 6, p. 106, 1961.

[10] White, A. D., and J. D. Rigden, "Simultaneous gas maser action in the visible and infrared," *Proc. IRE*, vol. 50, p. 2366, 1962.

[11] Bennett, W. R., "Gaseous optical masers," *Appl. Optics, Suppl. 1, Optical Masers*, p. 24, 1962.

[12] Patel, C. K. N., "Interpretation of CO_2 optical maser experiments," *Phys. Rev. Letters*, vol. 12, p. 588, 1964; also, "Continuous-wave laser action on vibrational rotational transitions of CO_2," *Phys. Rev.*, vol. 136, p. A1187, 1964.

[13] Herzberg, G. H., *Spectra of Diatomic Molecules*. Princeton, N.J.: Van Nostrand, 1963.

[14] Patel, C. K. N., "High power CO_2 lasers," *Sci. Am.*, vol. 219, p. 22, Aug. 1968.

[15] Bridges, W. B., "Laser oscillation in singly ionized argon in the visible spectrum," *Appl. Phys. Letters*, vol. 4, p. 128, 1964.

[16] Gordon, E. I., E. F. Labuda, and W. B. Bridges, "Continuous visible laser action in singly ionized argon, krypton and xenon," *Appl. Phys. Letters*, vol. 4, p. 178, 1964.

[17] Hall, R. N., G. E. Fenner, J. D. Kingsley, T. J. Soltys, and R. O. Carlson, "Coherent light emission from GaAs junctions," *Phys. Rev. Letters*, vol. 9, pp. 366–367, 1962.

[18] Nathan, M. I., W. P. Dumke, G. Burns, F. H. Dills, and G. Lasher, "Stimulated emission of radiation from GaAs *p-n* junctions," *Appl. Phys. Letters*, vol. 1, pp. 62–64, 1962.

[19] Quist, T. M., R. J. Keyes, W. E. Krag, B. Lax, A. L. McWhorter, R. H. Rediker, and H. J. Zeiger, "Semiconductor maser of GaAs," *Appl. Phys. Letters*, vol. 1, p. 91, 1962.

[20] Kittel, C., *Introduction to Solid State Physics*, 3d ed. New York: Wiley, 1967.

[21] Bernard, M. G., and G. Duraffourg, "Laser conditions in semiconductors," *Phys. Status Solidi*, vol. 1, p. 699, 1961.

[22] Yariv, A., and R. C. C. Leite, "Dielectric waveguide mode of light propagation in *p-n* junctions," *Appl. Phys. Letters*, vol. 2, p. 55, 1963.

[23] Anderson, W. W., "Mode confinement in junction lasers," *IEEE J. Quantum Electron.*, vol. QE-1, p. 228, 1965.

[24] Stern, F., in *Radiative Recombinations in Semiconductors* (Proc. 7th Int. Conf. on the Physics of Semiconductors). New York: Academic Press, and Paris: Dunod, p. 165, 1964.

[25] See, for example, R. A. Smith, *Semiconductors*. New York: Cambridge, p. 216, 1959.

[26] Phelan, R. J., A. R. Calawa, R. H. Rediker, R. J. Keyes, and B. Lax, "Infrared InSb laser diode in high magnetic fields," *Appl. Phys. Letters*, vol. 3, p. 143, 1963.

[27] Weiser, K., and R. S. Levitt, "Radiative recombination from indium phosphide in *p-n* junctions," *Bull. Am. Phys. Soc.*, vol. 8, p. 29, 1963.

[28] Melngailis, I., "Maser action in InAs diodes," *Appl. Phys. Letters*, vol. 2, p. 176, 1963.

[29] Melngailis, I., and R. H. Rediker, "Properties of InAs lasers," *J. Appl. Phys.*, vol. 37, p. 899, 1966.

[30] Phelan, R. J., A. R. Calawa, R. H. Rediker, R. J. Keyes, and B. Lax, "Infrared InSb laser in high magnetic fields," *Appl. Phys. Letters*, vol. 3, p. 143, 1963.

[31] Butler, J. F., A. R. Calawa, R. J. Phelan, Jr., A. J. Strauss, and R. H. Rediker, "PbSe diode laser," *Solid State Commun.*, vol. 2, p. 303, 1964.

[32] Butler, J. F., A. R. Calawa, R. J. Phelan, Jr., T. C. Harman, A. J. Strauss, and R. H. Rediker, "PbTe diode laser," *Appl. Phys. Letters*, vol. 5, p. 75, 1964.

[33] Holnyak, N., Jr., and S. F. Bevacqua, "Coherent (visible) light emission from $Ga(As_{1-x}P_x)$ junctions," *Appl. Phys. Letters*, vol. 1, p. 82, 1962.

[34] Melngailis, I., A. J. Strauss, and R. H. Rediker, "Semiconductor diode masers of $(In_xGa_{1-x})As$," *Proc. IEEE*, vol. 51, p. 1154, 1963.

[35] Alexander, F. B., V. R. Bird, D. R. Carpenter, G. W. Manley, P. S. McDermott, J. R. Peloke, H. F. Quinn, R. J. Riley, and L. R. Yetter, "Spontaneous and stimulated infrared emission from indium phosphide arsenide diodes," *Appl. Phys. Letters*, vol. 4, p. 13, 1964.

[36] Calawa, A. R., and I. Melngailis, "Infrared radiation from GaSb diodes," *Bull. Am. Phys. Soc.*, vol. 8, p. 29, 1963.

[37] Butler, J. F., and T. C. Harman, "Long wavelength infrared $Pb_{1-x}Sn_xTe$ diode lasers," *Appl. Phys. Letters*, vol. 12, p. 347, 1968.

[38] Alferov, Z. I., V. M. Andreev, V. I. Korolkov, E. L. Portnoi, and D. N. Tretyakov, "Coherent radiation of epitaxial heterojunction structures in the AlAs–GaAs system," *Sov. Phys. Semiconductors*, vol. 2, p. 1289, 1969.

[39] Hayashi, J., M. B. Panish, and P. W. Foy, "A low-threshold room-temperature injection laser," *IEEE J. Quant. Elect.*, vol. 5, p. 211, 1969.

[40] Kressel, H., and H. Nelson, "Close confinement gallium arsenide *p-n* junction laser with reduced optical loss at room temperature," *RCA Review*, vol. 30, p. 106, 1969.

[41] Hayashi, I., M. B. Panish, and F. K. Rinehart, "GaAs–$Ga_{1-x}Al_xAs$ double heterostructure injection lasers," *J. Appl. Phys.*, vol. 42, p. 1929, 1971. Also, Hayashi, I., M. B. Panish, P. W. Foy, and S. Sumski, "Junction lasers which operate continuously at room temperature," *Appl. Phys. Letters*, vol. 17, p. 109, 1970.

[42] Stockman, D. L., W. R. Mallory, and K. F. Tittel, "Stimulated emission in aromatic organic compounds," *Proc. IEEE*, vol. 52, p. 318, 1964.

[43] Sorokin, P. P., and J. R. Lankard, "Stimulated emission observed from an organic dye, chloroaluminum phtalocyanine," *IBM J. Res. Develop.*, vol. 10, p. 162, 1966.

[44] Schafer, F. P., W. Schmidt, and J. Volze, "Organic dye solution laser," *Appl. Phys. Letters*, vol. 9, p. 306, 1966.

[45] Snavely, B. B., "Flashlamp-excited dye lasers," *Proc. IEEE*, vol. 57, p. 1374, 1969.

[46] Soffer, B. H., and B. B. McFarland, "Continuously tunable, narrow band organic dye lasers," *Appl. Phys. Letters*, vol. 10, p. 266, 1967.

[47] Peterson, O. G., Tuccio, S. A., and B. B. Snavely, "CW operation of an organic dye laser," *Appl. Phys. Letters*, vol. 17, 1970.

[48] Yariv, A., *Quantum Electronics*, 2d Ed. New York: Wiley, 1975.

[49] Taylor, R. L., and S. Bitterman, "Survey of vibrational and relaxation data for processes important in the CO_2–N_2 laser system," *Rev. Mod. Phys.*, vol. 41, p. 26, 1969.

[50] Abrams, R. L., and W. B. Bridges, "Characteristics of sealed-off waveguide CO_2 lasers," *IEEE J. Quant. Elect.*, vol. QE-9, p. 940, 1973.

[51] Beaulieu, J. A., "High peak power gas lasers," *Proc. IEEE*, vol. 59, p. 667, 1971.

CHAPTER

8

Second-Harmonic Generation and Parametric Oscillation

8.0 Introduction

In Chapter 1 we considered the propagation of electromagnetic radiation in linear media in which the polarization is proportional to the electric field that induces it. In this chapter we consider some of the consequences of the nonlinear dielectric properties of certain classes of crystals in which, in addition to the linear response, a field produces a polarization proportional to the square of the field.

The nonlinear response can give rise to exchange of energy between a number of electromagnetic fields of different frequencies. Two of the most important applications of this phenomenon are: (1) second-harmonic generation in which part of the energy of an optical wave of frequency ω propagating through a crystal is converted to that of a wave at 2ω; (2) parametric oscillation in which a strong pump wave at ω_3 causes the simultaneous generation in a nonlinear crystal of radiation at ω_1 and ω_2, where $\omega_3 = \omega_1 + \omega_2$. These will be treated in detail in this chapter.

8.1 On the Physical Origin of Nonlinear Polarization

The optical polarization of dielectric crystals is due mostly to the outer, loosely bound valence electrons that are displaced by the optical field. Denoting the electron deviation from the equilibrium position by x and the density of electrons by N, the polarization p is given by

$$p(t) = -Nex(t)$$

In symmetric crystals the potential energy of an electron must reflect the crystal symmetry, so that, using a one-dimensional analog, it can be written as

$$V(x) = \frac{m}{2}\omega_0^2 x^2 + \frac{m}{4}Bx^4 + \cdots \tag{8.1-1}$$

where ω_0^2 and B are constants[1] and m is the electron mass. Because of the symmetry $V(x)$ contains only even powers of x, so $V(-x) = V(x)$. The restoring force on an electron is

$$F = -\frac{\partial V}{\partial x} = -m\omega_0^2 x - mBx^3 \tag{8.1-2}$$

and is zero at the equilibrium position $x = 0$.

The linear polarization of crystals in which the polarization is proportional to the electric field is accounted for by the first term in (8.1-1). To see this, consider a "low" frequency electric field $E(t)$—that is, a field whose Fourier components are at frequencies small compared to ω_0. The excursion $x(t)$ caused by this field is found by equating the total force on the electron to zero[2]

$$-eE(t) - m\omega_0^2 x(t) = 0$$

so that

$$x(t) = -\frac{e}{m\omega_0^2}E(t) \tag{8.1-3}$$

thus resulting in a polarization which is instantaneously proportional to the field.

Now in an asymmetric crystal in which the condition $V(x) = V(-x)$ is no longer fulfilled, the potential function can contain odd powers, and thus

$$V(x) = \frac{m\omega_0^2}{2}x^2 + \frac{m}{3}Dx^3 + \cdots \tag{8.1-4}$$

---

[1] The constant ω_0 was found in Section 5.4 to correspond to the resonance frequency of the electronic oscillator.

[2] The "low" frequency assumption makes it possible to neglect the acceleration term $m\, d^2x/dt^2$ in the force equation.

which corresponds to a restoring force on the electron

$$F = -\frac{\partial V(x)}{\partial x} = -(m\omega_0{}^2 x + mDx^2 + \cdots) \tag{8.1-5}$$

An examination of (8.1-5) reveals that a positive excursion ($x > 0$) results in a larger restoring force, assuming $D > 0$, than does the same excursion in the opposite direction. It follows immediately that if the electric force on the electron is positive ($E < 0$), the induced polarization is smaller than when the field direction is reversed. This situation is depicted in Figure 8-1.

Next consider an alternating electric field at an (optical) frequency ω applied to the crystal. In a linear crystal the induced polarization will be proportional, at any moment, to the field, resulting in a polarization oscillating at ω as shown in Figure 8-2(a). In a nonlinear crystal we can use Figure 8-1(b) to obtain the induced polarization corresponding to a given field and then plot it (vertically) as in Figure 8-2(b). The result is

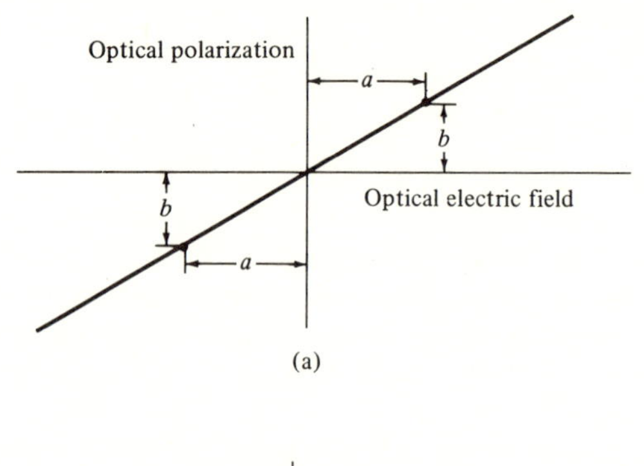

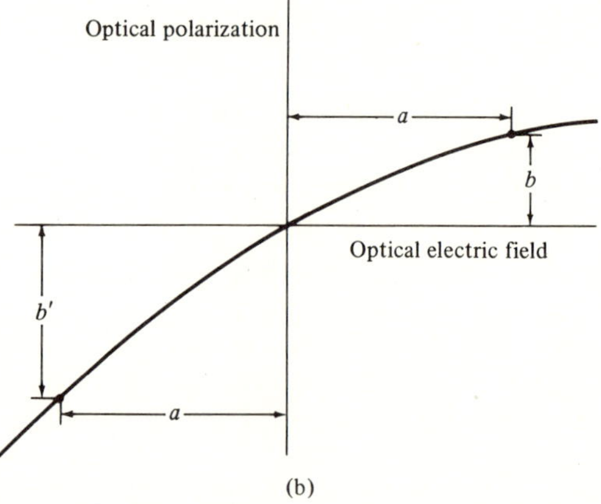

Figure 8-1 Relation between induced polarization and the electric field causing it; (**a**) in a linear dielectric and (**b**) in a crystal lacking inversion symmetry.

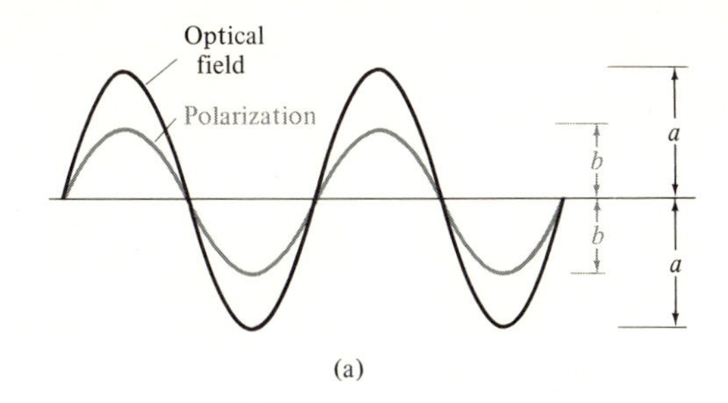

(a)

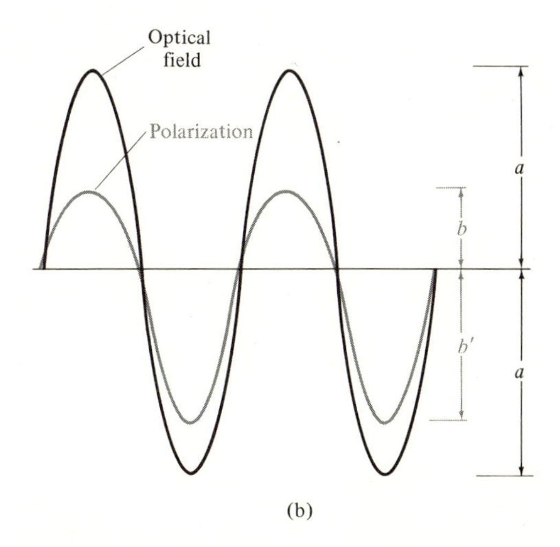

(b)

Figure 8-2 An applied sinusoidal electric field and the resulting polarization; (**a**) in a linear crystal and (**b**) in a crystal lacking inversion symmetry.

a polarization wave in which the stiffer restoring force at $x > 0$ results in positive peaks (b), which are smaller than the negative ones (b'). A Fourier analysis of the nonlinear polarization wave in Figure 8-2(b) shows that it contains the second harmonic of ω as well as an average (dc) term. The average, fundamental, and second-harmonic components are plotted in Figure 8-3.

To relate the nonlinear polarization formally to the inducing field, we use Equation (8.1-5) for the restoring force and take the driving electric field as $E^{(\omega)} \cos \omega t$. The equation of motion of the electron $F = m\ddot{x}$ is then

$$\frac{d^2x(t)}{dt^2} + \sigma \frac{dx(t)}{dt} + {\omega_0}^2 x(t) + Dx^2(t) = -\frac{eE^{(\omega)}}{2m}(e^{i\omega t} + e^{-i\omega t}) \qquad \textbf{(8.1-6)}$$

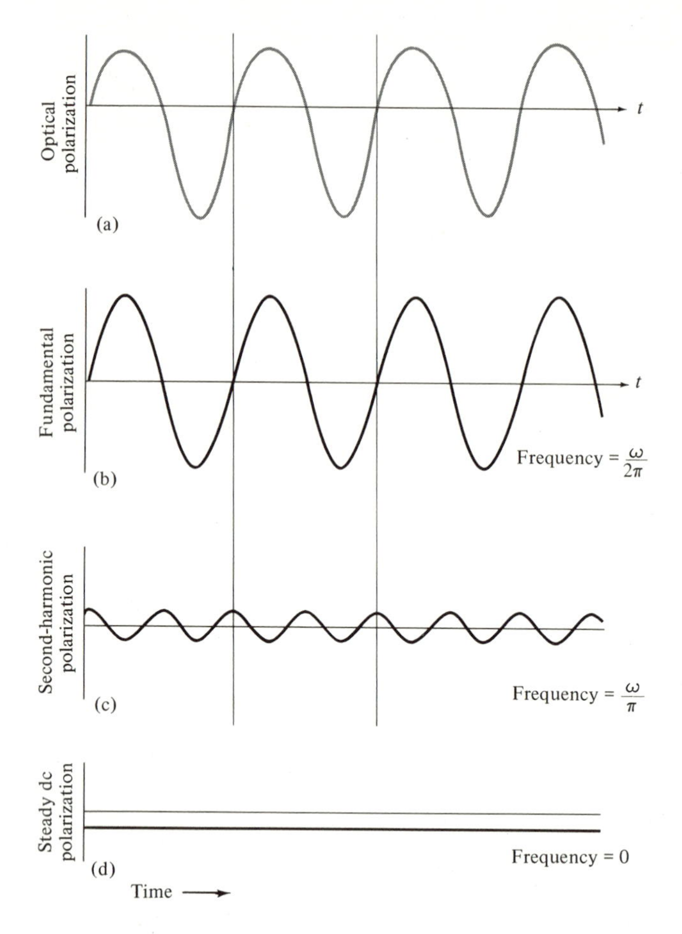

Figure 8-3 Analysis of the nonlinear polarization wave (**a**) of Figure 8.2(**b**) shows that it contains components oscillating at (**b**) the same frequency (ω) as the wave inducing it, (**c**) twice that frequency (2ω), and (**d**) an average (dc) negative component.

where, as in (5.4-1), we account for the losses by a frictional force $-m\sigma\dot{x}$. An inspection of (8.1-6) shows that the term Dx^2 gives rise to a component oscillating at 2ω, so we assume the solution for $x(t)$ in the form[3]

$$x(t) = \tfrac{1}{2}(q_1 e^{i\omega t} + q_2 e^{2i\omega t} + \text{c.c.}) \tag{8.1-7}$$

where c.c. stands for "complex conjugate."

[3] Here we must use the real form of $x(t)$ instead of the complex one since, as discussed in Section 1.1, the differential equation involves x^2.

Substituting the last expression into (8.1-6) gives

$$-\frac{\omega^2}{2}(q_1e^{i\omega t} + \text{c.c.}) - 2\omega^2(q_2e^{2i\omega t} + \text{c.c.}) + \frac{i\omega\sigma}{2}(q_1e^{i\omega t} - \text{c.c.})$$

$$+ i\omega\sigma(q_2e^{2i\omega t} - \text{c.c.}) + \frac{\omega_0^{\ 2}}{2}(q_1e^{i\omega t} + q_2e^{2i\omega t} + \text{c.c.})$$

$$+ \frac{D}{4}(q_1^{\ 2}e^{2i\omega t} + q_2^{\ 2}e^{4i\omega t} + q_1q_1^* + 2q_1q_2e^{3i\omega t}$$

$$+ 2q_1q_2^*e^{-i\omega t} + q_2q_2^* + \text{c.c.}) = \frac{-eE^{(\omega)}}{2m}(e^{i\omega t} + \text{c.c.}) \tag{8.1-8}$$

If (8.1-8) is to be valid for all times t, the coefficients of $e^{\pm i\omega t}$ and $e^{\pm 2i\omega t}$ on both sides of the equation must be equal. Equating first the coefficients of $e^{i\omega t}$, assuming that $|Dq_2| \ll [(\omega_0^2 - \omega^2)^2 + \omega^2\sigma^2]^{1/2}$, gives

$$q_1 = -\frac{eE^{(\omega)}}{m}\frac{1}{(\omega_0^{\ 2} - \omega^2) + i\omega\sigma} \tag{8.1-9}$$

The polarization at ω is related to the electronic deviation at ω by

$$p^{(\omega)}(t) = -\frac{Ne}{2}(q_1e^{i\omega t} + \text{c.c.})$$

$$\equiv \frac{\epsilon_0}{2}[\chi(\omega)E^{(\omega)}e^{i\omega t} + \text{c.c.}] \tag{8.1-10}$$

where $\chi(\omega)$ is thus the linear susceptibility defined by (5.4-8). By using (8.1-9) in (8.1-10) and solving for $\chi(\omega)$, we obtain

$$\chi(\omega) = \frac{Ne^2}{m\epsilon_0[(\omega_0^{\ 2} - \omega^2) + i\omega\sigma]} \tag{8.1-11}$$

We now proceed to solve for the amplitude q_2 of the electronic motion at 2ω. Equating the coefficients of $e^{2i\omega t}$ on both sides of (8.1-8) leads to

$$q_2(-4\omega^2 + 2i\omega\sigma + \omega_0^{\ 2}) = -\tfrac{1}{2}Dq_1^{\ 2}$$

and, after substituting the solution (8.1-9) for q_1, we obtain

$$q_2 = \frac{-De^2[E^{(\omega)}]^2}{2m^2[(\omega_0^{\ 2} - \omega^2) + i\omega\sigma]^2[\omega_0^{\ 2} - 4\omega^2 + 2i\omega\sigma]} \tag{8.1-12}$$

In a manner similar to (8.1-10), the nonlinear polarization at 2ω is

$$p^{(2\omega)}(t) = -\frac{Ne}{2}(q_2 e^{2i\omega t} + \text{c.c.})$$

$$\equiv \tfrac{1}{2}\{d^{(2\omega)}[E^{(\omega)}]^2 e^{2i\omega t} + \text{c.c.}\} \tag{8.1-13}$$

The second of equations (8.1-13) defines the *nonlinear optical coefficient* $d^{(2\omega)}$. If we denote the complex amplitude of the polarization as $P^{(2\omega)}$ we have, from (8.1-13),

$$p^{(2\omega)}(t) = \tfrac{1}{2}[P^{(2\omega)}e^{2i\omega t} + \text{c.c.}]$$

and (8.1-14)

$$P^{(2\omega)} = d^{(2\omega)}E^{(\omega)}E^{(\omega)}$$

that is, $d^{(2\omega)}$ is the ratio of the (complex) amplitude of the polarization at 2ω to the square of the fundamental amplitude. Substituting (8.1-12) for q_2 in (8.1-13), then solving for $d^{(2\omega)}$, results in

$$d^{(2\omega)} = \frac{DNe^3}{2m^2[(\omega_0^2 - \omega^2) + i\omega\sigma]^2[\omega_0^2 - 4\omega^2 + 2i\omega\sigma)]} \tag{8.1-15}$$

Using (8.1-11) we can rewrite (8.1-15) as

$$d^{(2\omega)} = \frac{mD[\chi^{(\omega)}]^2\chi^{(2\omega)}\epsilon_0{}^3}{2N^2e^3} \tag{8.1-16}$$

Equation (8.1-16) is of importance since it relates the nonlinear optical coefficient d to the linear optical susceptibilities χ and to the anharmonic coefficient D. Estimates based on this relation are quite successful in predicting the size of the coefficient d in a large variety of crystals; see References [1] and [2].

Relation (8.1-14) is scalar. In reality the second harmonic polarization along, say, the x direction, is related to the electric field at ω by

$$\begin{aligned} P_x^{(2\omega)} &= d_{xxx}^{(2\omega)}E_x^{(\omega)}E_x^{(\omega)} + d_{xyy}^{(2\omega)}E_y^{(\omega)}E_y^{(\omega)} + d_{xzz}^{(2\omega)}E_z^{(\omega)}E_z^{(\omega)} \\ &+ 2d_{xzy}^{(2\omega)}E_z^{(\omega)}E_y^{(\omega)} + 2d_{xzx}^{(2\omega)}E_z^{(\omega)}E_x^{(\omega)} + 2d_{xxy}^{(2\omega)}E_x^{(\omega)}E_y^{(\omega)} \end{aligned} \tag{8.1-17}$$

Similar relations give $P_y{}^{(2\omega)}$ and $P_z{}^{(2\omega)}$. Considerations of crystal symmetry reduce the number of nonvanishing $d_{ijk}{}^{(2\omega)}$ coefficients—or, in certain cases to be discussed in the following, cause them to vanish altogether. Table 8-1 lists the nonlinear coefficients of a number of crystals.

Crystals are usually divided into two main groups, depending on

Table 8-1 THE NONLINEAR OPTICAL COEFFICIENTS OF A NUMBER OF CRYSTALS*

CRYSTAL	$d_{ijk}^{(2\omega)}$ IN UNITS OF $1/9 \times 10^{-22}$ (mks)
$LiIO_3$	$d_{31} = d_{311} = 6 \pm 1$
$NH_4H_2PO_4$	$d_{36} = d_{312} = 0.45$
(ADP)	$d_{14} = d_{123} = 0.5 \pm 0.02$
KH_2PO_4	$d_{36} = d_{312} = 0.45 \pm 0.03$
(KDP)	$d_{14} = d_{123} = 0.45 \pm 0.03$
KD_2PO_4	$d_{36} = d_{312} = 0.42 \pm 0.02$
	$d_{14} = d_{123} = 0.42 \pm 0.02$
KH_2ASO_4	$d_{36} = d_{312} = 0.48 \pm 0.03$
	$d_{14} = d_{123} = 0.51 \pm 0.03$
Quartz	$d_{11} = d_{111} = 0.37 \pm 0.02$
$AlPO_4$	$d_{11} = d_{111} = 0.38 \pm 0.03$
ZnO	$d_{33} = d_{333} = 6.5 \pm 0.2$
	$d_{31} = d_{311} = 1.95 \pm 0.2$
	$d_{15} = d_{113} = 2.1 \pm 0.2$
CdS	$d_{33} = d_{333} = 28.6 \pm 2$
	$d_{31} = d_{311} = 14.5 \pm 1$
	$d_{15} = d_{113} = 16 \pm 3$
GaP	$d_{14} = d_{123} = 80 \pm 14$
GaAs	$d_{14} = d_{123} = 107 \pm 30$
$BaTiO_3$	$d_{33} = d_{333} = 6.4 \pm 0.5$
	$d_{31} = d_{311} = 18 \pm 2$
	$d_{15} = d_{113} = 17 \pm 2$
$LiNbO_3$	$d_{31} = d_{311} = 4.76 \pm 0.5$
	$d_{22} = d_{222} = 2.3 \pm 1.0$
Te	$d_{11} = d_{111} = 730 \pm 230$
Se	$d_{11} = d_{111} = 130 \pm 30$
$Ba_2NaNb_5O_{15}$	$d_{33} = d_{333} = 10.4 \pm 0.7$
	$d_{32} = d_{322} = 7.4 \pm 0.7$
Ag_3AsS_3	$d_{22} = d_{222} = 22.5$
(Proustite)	$d_{36} = d_{312} = 13.5$

* These coefficients are defined by (8.1-17) with $1 = x$, $2 = y$, $3 = z$.

whether the crystal structure remains unchanged upon inversion (that is, replacing the coordinate **r** by $-$**r**) or not. Crystals belonging to the first group are called centrosymmetric, whereas crystals of the second group are called noncentrosymmetric [3]. In Figure 8-4 we show the crystal structure of NaCl, a centrosymmetric crystal; an example of a crystal lacking inversion symmetry (noncentrosymmetric) is provided by crystals of the ZnS (zinc blende) class such as GaAs, CdTe, and others. The crystal structure of ZnS is shown in Figure 8-5. The lack of inversion symmetry is evident in the projection of the atomic positions given by Figure 8-6.

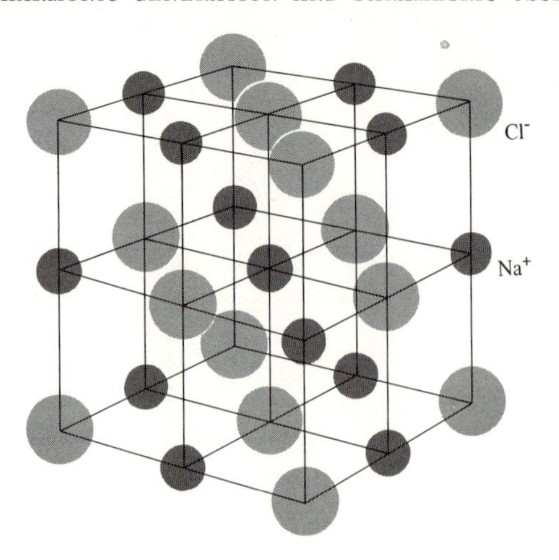

Figure 8-4 The crystal structure of NaCl. The crystal is centrosymmetric, since an inversion of any ion about the central Na^+ ion, as an example, leaves the crystal structure unchanged.

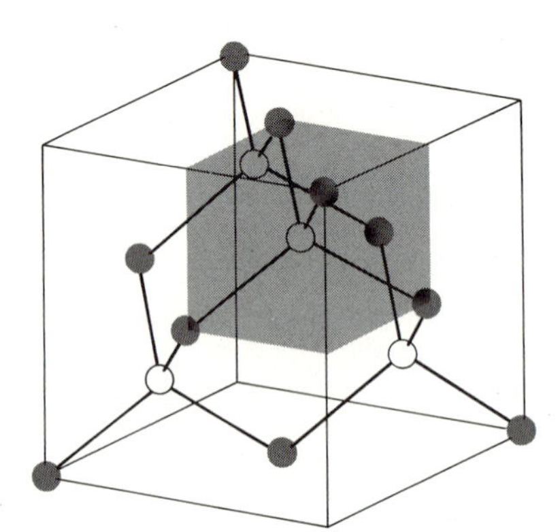

Figure 8-5 The crystal structure of cubic zinc sulfide.

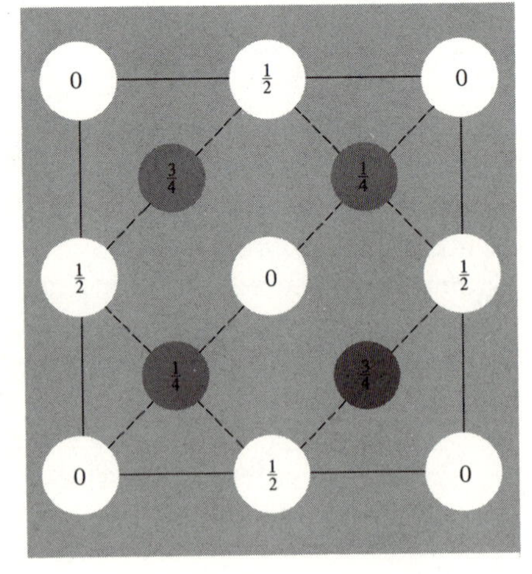

Figure 8-6 The atomic positions in the unit cell of ZnS projected on a cube face. The fractions denote height above base in units of a cube edge. The dark spheres correspond to zinc (or sulfur) atoms and are situated on a face-centered cubic (fcc) lattice, and the white spheres correspond to sulfur (or zinc) atoms and are situated on another fcc lattice displaced by ($\frac{1}{4}$, $\frac{1}{4}$, $\frac{1}{4}$) from the first one. Note the lack of inversion symmetry.

In crystals possessing an inversion symmetry all the nonlinear optical coefficients $d_{ijk}^{(2\omega)}$ must be zero. This follows directly from the relation

$$P_i^{(2\omega)} = \sum_{j,k=x,y,z} d_{ijk}^{(2\omega)} E_j^{(\omega)} E_k^{(\omega)} \tag{8.1-18}$$

which is a compact notation for relation (8.1-17). Let us reverse the direction of the electric field so that in (8.1-18) $E_j^{(\omega)}$ becomes $-E_j^{(\omega)}$ and $E_k^{(\omega)}$ becomes $-E_k^{(\omega)}$. Since the crystal is centrosymmetric, the reversed field "sees" a crystal identical to the original one so that the polarization produced by it must bear the same relationship to the field as originally; that is, the new polarization is $-P_i^{(2\omega)}$. Since the new polarization and the electric field causing it are still related by (8.1-18), we have

$$-P_i^{(2\omega)} = \sum_{j,k} d_{ijk}^{(2\omega)} (-E_j^{(\omega)})(-E_k^{(\omega)}) \tag{8.1-19}$$

Equations (8.1-18) and (8.1-19) can hold simultaneously only if the coefficients $d_{ijk}^{(2\omega)}$ are all zero. We may thus summarize: *In crystals possessing an inversion symmetry there is no second-harmonic generation.*

In the following work we will ignore the vector nature of the optical nonlinearity and use the scalar form (8.1-14). Vectorial considerations are important in determining the interaction strength and are discussed in detail in [11].

8.2 Formalism of Wave Propagation in Nonlinear Media

In this section we derive the equations governing the propagation of electromagnetic waves in nonlinear media. These equations will then be used to describe second-harmonic generation and parametric oscillation.

The starting point is Maxwell's equations (1.2-1), (1.2-2):

$$\begin{aligned} \nabla \times \mathbf{h} &= \mathbf{i} + \frac{\partial \mathbf{d}}{\partial t} \\ \nabla \times \mathbf{e} &= -\mu \frac{\partial \mathbf{h}}{\partial t} \end{aligned} \tag{8.2-1}$$

and

$$\begin{aligned} \mathbf{d} &= \epsilon_0 \mathbf{e} + \mathbf{p} \\ \mathbf{i} &= \sigma \mathbf{e} \end{aligned} \tag{8.2-2}$$

where σ is the conductivity. If we separate the total polarization $\mathbf{p}$ into its linear and nonlinear portions according to

$$\mathbf{p} = \epsilon_0 \chi_e \mathbf{e} + \mathbf{p}_{NL} \tag{8.2-3}$$

the first of equations (8.2-1) becomes

$$\nabla \times \mathbf{h} = \sigma \mathbf{e} + \epsilon \frac{\partial \mathbf{e}}{\partial t} + \frac{\partial}{\partial t} \mathbf{p}_{NL} \tag{8.2-4}$$

with $\epsilon \equiv \epsilon_0(1 + \chi_e)$. Taking the curl of both sides of the second of (8.2-1), using the vector identity

$$\nabla \times \nabla \times \mathbf{e} = \nabla\nabla \cdot \mathbf{e} - \nabla^2\mathbf{e},$$

From (8.2-4) and taking $\nabla \cdot \mathbf{e} = 0$, we get

$$\nabla^2\mathbf{e} = \mu\sigma \frac{\partial \mathbf{e}}{\partial t} + \mu\epsilon \frac{\partial^2 \mathbf{e}}{\partial t^2} + \mu \frac{\partial^2}{\partial t^2} \mathbf{p}_{NL} \tag{8.2-5}$$

Next we go over to a scalar notation and rewrite (8.2-5) as

$$\nabla^2 e = \mu\sigma \frac{\partial e}{\partial t} + \mu\epsilon \frac{\partial^2 e}{\partial t^2} + \mu \frac{\partial^2}{\partial t^2} p_{NL}(\mathbf{r}, t) \tag{8.2-6}$$

where we assumed, for simplicity, that $\mathbf{p}_{NL}$ is parallel to $\mathbf{e}$. Let us limit our consideration to a field made up of three plane waves propagating in the z direction with frequencies ω_1, ω_2, and ω_3 according to

$$\begin{aligned} e^{(\omega_1)}(z, t) &= \tfrac{1}{2}[E_1(z)e^{i(\omega_1 t - k_1 z)} + \text{c.c.}] \\ e^{(\omega_2)}(z, t) &= \tfrac{1}{2}[E_2(z)e^{i(\omega_2 t - k_2 z)} + \text{c.c.}] \\ e^{(\omega_3)}(z, t) &= \tfrac{1}{2}[E_3(z)e^{i(\omega_3 t - k_3 z)} + \text{c.c.}] \end{aligned} \tag{8.2-7}$$

Then the total instantaneous field is

$$e = e^{(\omega_1)}(z, t) + e^{(\omega_2)}(z, t) + e^{(\omega_3)}(z, t) \tag{8.2-8}$$

Next we substitute (8.2-8), using (8.2-7), into the wave equation (8.2-6) and separate the resulting equation into three equations, each containing only terms oscillating at one of the three frequencies. The expression involving $p_{NL}(\mathbf{r}, t)$ in (8.2-6) will give rise, according to (8.1-14), to

$$\tfrac{1}{2}\mu d \frac{\partial^2}{\partial t^2} E_1 E_2 e^{i[(\omega_1 + \omega_2)t - (k_1 + k_2)z]}$$

or

$$\tfrac{1}{2}\mu d \frac{\partial^2}{\partial t^2} E_3 E_2^* e^{i[(\omega_3 - \omega_2)t - (k_3 - k_2)z]}$$

These oscillate at the new frequencies $(\omega_1 + \omega_2)$ and $(\omega_3 - \omega_2)$ and, in general being nonsynchronous, will not be able to drive the oscillation at ω_1, ω_2, or ω_3. An exception to the last statement is the case when

$$\omega_3 = \omega_1 + \omega_2 \tag{8.2-9}$$

In this case the term

$$\mu d \frac{\partial^2}{\partial t^2} E_1 E_2 e^{i[(\omega_1 + \omega_2)t - (k_1 + k_2)z]}$$

oscillates at $\omega_1 + \omega_2 = \omega_3$ and can thus act as a source for the wave at ω_3. In physical terms, we have power flow from the fields at ω_1 and ω_2

into that at ω_3, or vice versa. Assuming that (8.2-9) holds, we return to (8.2-6) and, writing it for the oscillation at ω_1, obtain

$$\nabla^2 e^{(\omega_1)} = \mu\sigma_1 \frac{\partial e^{(\omega_1)}}{\partial t} + \mu\epsilon \frac{\partial^2 e^{(\omega_1)}}{\partial t^2} + \mu_0 d \frac{\partial^2}{\partial t^2} \times \left[\frac{E_3(z)E_2^*(z)}{2} e^{i[(\omega_3-\omega_2)t-(k_3-k_2)z]} + \text{c.c}\right] \quad \textbf{(8.2-10)}$$

Next we observe that, in view of (8.2-7),

$$\nabla^2 e^{(\omega_1)} = \frac{1}{2}\frac{\partial^2}{\partial z^2}[E_1(z)e^{i(\omega_1 t - k_1 z)} + \text{c.c.}]$$
$$= -\frac{1}{2}\left[k_1{}^2 E_1(z) + 2ik_1 \frac{dE_1(z)}{dz}\right] e^{i(\omega_1 t - k_1 z)} + \text{c.c.}$$

where we assumed that

$$\left|k_1 \frac{dE_1(z)}{dz}\right| \gg \left|\frac{d^2 E_1(z)}{dz^2}\right| \quad \textbf{(8.2-11)}$$

If we use (8.2-9) and the last result in (8.2-10), and take $\partial/\partial t = i\omega_i$, we obtain

$$-\frac{1}{2}\left[k_1{}^2 E_1(z) + 2ik_1 \frac{dE_1(z)}{dz}\right] e^{i(\omega_1 t - k_1 z)} + \text{c.c.}$$
$$= [i\omega_1\mu_0\sigma_1 - \omega_1{}^2\mu\epsilon]\left[\frac{E_1(z)}{2} e^{i(\omega_1 t - k_1 z)}\right] + \text{c.c.}$$
$$- \left[\frac{\omega_1{}^2\mu d}{2} E_3(z)E_2^*(z)e^{i[\omega_1 t-(k_3-k_2)z]} + \text{c.c.}\right] \quad \textbf{(8.2-12)}$$

Recognizing that $k_1{}^2 = \omega_1{}^2\mu\epsilon$, we can rewrite (8.2-12) after multiplying all the terms by

$$\frac{i}{k_1}\exp(-i\omega_1 t + ik_1 z)$$

as

$$\frac{dE_1}{dz} = -\frac{\sigma_1}{2}\sqrt{\frac{\mu}{\epsilon_1}}E_1 - \frac{i\omega_1}{2}\sqrt{\frac{\mu}{\epsilon_1}}dE_3E_2{}^*e^{-i(k_3-k_2-k_1)z}$$

and, similarly,

$$\frac{dE_2{}^*}{dz} = -\frac{\sigma_2}{2}\sqrt{\frac{\mu}{\epsilon_2}}E_2{}^* + \frac{i\omega_2}{2}\sqrt{\frac{\mu}{\epsilon_2}}dE_1E_3{}^*e^{-i(k_1-k_3+k_2)z}$$
$$\frac{dE_3}{dz} = -\frac{\sigma_3}{2}\sqrt{\frac{\mu}{\epsilon_3}}E_3 - \frac{i\omega_3}{2}\sqrt{\frac{\mu}{\epsilon_2}}dE_1E_2e^{-i(k_1+k_2-k_3)z} \quad \textbf{(8.2-13)}$$

for the fields at ω_2 and ω_3. These are the basic equations describing nonlinear parametric interactions [4]. We notice that they are coupled to each other via the nonlinear constant d.

8.3 Optical Second-Harmonic Generation

The first experiment in nonlinear optics [5] consisted of generating the second harmonic ($\lambda = 0.3470\ \mu$m) of a ruby laser beam ($\lambda = 0.694\ \mu$m) that was focused on a quartz crystal. The experimental arrangement is depicted in Figure 8-7. The conversion efficiency of this first experiment ($\sim 10^{-8}$) was improved by methods to be described below to a point where about 30 percent conversion has been observed in a single pass through a few centimeters length of a nonlinear crystal. This technique is finding important applications in generating short-wave radiation from longer-wave lasers.

In the case of second-harmonic generation, two of the three fields that figure in (8.2-13) are of the same frequency. We may thus put $\omega_1 = \omega_2 = \omega$, for which case the first two equations are the complex conjugate of one another and we need to consider only one of them. We take the input field at ω to correspond to E_1 in (8.2-13) and the second-harmonic field to E_3, and we put $\omega_3 = \omega_1 + \omega_2 = 2\omega$, neglecting the absorption, so $\sigma_{1,2,3} = 0$. The last equation becomes

$$\frac{dE^{(2\omega)}}{dz} = -i\omega\sqrt{\frac{\mu}{\epsilon}}\,d[E^{(\omega)}(z)]^2 e^{i(\Delta k)z} \tag{8.3-1}$$

where

$$\Delta k \equiv k_3 - 2k_1 = k^{(2\omega)} - 2k^{(\omega)} \tag{8.3-2}$$

To simplify the analysis further, we may assume that the depletion of the input wave at ω due to conversion of its power to 2ω is negligible. Under those conditions, which apply in the majority of the experimental situations, we can take $E^{(\omega)}(z) =$ constant in (8.3-1) and neglect its dependence on z. Assuming no input at 2ω—that is, $E^{(2\omega)}(0) = 0$—we obtain from

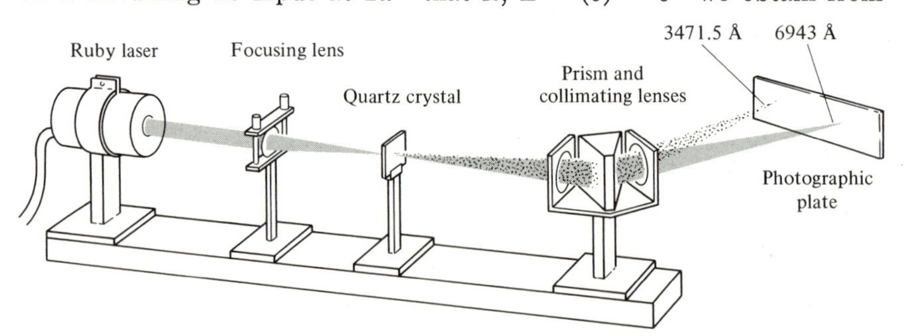

Figure 8-7 Arrangement used in first experimental demonstration of second-harmonic generation [5]. Ruby laser beam at $\lambda_0 = 0.694\ \mu$m is focused on a quartz crystal, causing generation of a (weak) beam at $\lambda_0/2 = 0.347\ \mu$m. The two beams are then separated by a prism and detected on a photographic plate.

(8.3-1) by integration the output field at the end of a crystal of length l:

$$E^{(2\omega)}(l) = -i\omega\sqrt{\frac{\mu}{\epsilon}}\, d[E^{(\omega)}]^2 \frac{e^{i\Delta k l} - 1}{i\Delta k}$$

The output intensity is proportional to

$$E^{(2\omega)}(l)E^{(2\omega)*}(l) = \left(\frac{\mu}{\epsilon_0}\right)\frac{\omega^2 d^2}{n^2}[E^{(\omega)}]^4 l^2 \frac{\sin^2(\Delta k l/2)}{(\Delta k l/2)^2} \tag{8.3-3}$$

Here we used $\epsilon/\epsilon_0 = n^2$, where n is the index of refraction. If the input beam is confined to a cross section $A(m^2)$, then, according to (1.3-26), the power per unit area (intensity) is related to the field by

$$I \equiv \frac{P_{2\omega}}{A} = \frac{1}{2}\sqrt{\frac{\epsilon}{\mu}}\,|E^{(2\omega)}|^2 \tag{8.3-4}$$

and (8.3-3) can be written as

$$\eta_{SHG} \equiv \frac{P_{2\omega}}{P_\omega} = 2\left(\frac{\mu}{\epsilon_0}\right)^{3/2} \frac{\omega^2 d^2 l^2}{n^3} \frac{\sin^2(\Delta k l/2)}{(\Delta k l/2)^2} \frac{P_\omega}{A} \tag{8.3-5}$$

for the conversion efficiency from ω to 2ω. We notice that the conversion efficiency is proportional to the intensity P_ω/A of the fundamental beam.

Phase-matching in second-harmonic generation. According to (8.3-5), a prerequisite for efficient second-harmonic generation is that $\Delta k = 0$—or, using (8.3-2),

$$k^{(2\omega)} = 2k^{(\omega)} \tag{8.3-6}$$

If $\Delta k \neq 0$, the second-harmonic power generated at some plane, say z_1, having propagated to some other plane (z_2), is not in phase with the second-harmonic wave generated at z_2. This results in the interference described by the factor

$$\frac{\sin^2(\Delta k l/2)}{(\Delta k l/2)^2}$$

in (8.3-5). Two adjacent peaks of this spatial interference pattern are separated by the so-called "coherence length"

$$l_c = \frac{2\pi}{\Delta k} = \frac{2\pi}{k^{(2\omega)} - 2k^{(\omega)}} \tag{8.3-7}$$

The coherence length l_c is thus a *measure* of the *maximum crystal length that is useful in producing the second-harmonic power*. Under ordinary circumstances it may be no larger than 10^{-2} cm. This is because the index of refraction n_ω normally increases with ω so Δk is given by

$$\Delta k = k^{(2\omega)} - 2k^{(\omega)} = \frac{2\omega}{c}[n^{2\omega} - n^{\omega}] \tag{8.3-8}$$

where we used the relation $k^{(\omega)} = \omega n^{\omega}/c$. The coherence length is thus

$$l_c = \frac{\pi c}{\omega[n^{2\omega} - n^{\omega}]} = \frac{\lambda}{2[n^{2\omega} - n^{\omega}]} \tag{8.3-9}$$

where λ is the free-space wavelength of the fundamental beam. If we take a typical value of $\lambda = 1\ \mu\text{m}$ and $n^{2\omega} - n^{\omega} \simeq 10^{-2}$, we get $l_c \simeq 100\ \mu\text{m}$. If l_c were to increase from 100 μm to 2 cm, as an example, according to (8.3-5) the second-harmonic power would go up by a factor of 4×10^4.

The technique that is used widely (see [6] and [7]) to satisfy the *phase-matching* requirement $\Delta k = 0$ takes advantage of the natural birefringence of anisotropic crystals, which was discussed in Section 1.4. Using the relation $k^{(\omega)} = \omega\sqrt{\mu\epsilon_0}n^{\omega}$, (8.3-6) becomes

$$n^{2\omega} = n^{\omega} \tag{8.3-10}$$

so the indices of refraction at the fundamental and second-harmonic frequencies must be equal. In normally dispersive materials the index of the ordinary wave or the extraordinary wave along a given direction increases with ω, as can be seen from Table 8-2. This makes it impossible to satisfy (8.3-10) when both the ω and 2ω beams are of the same type—that is, when both are extraordinary or ordinary. We can, however, under certain circumstances, satisfy (8.3-10) by making the two waves be of

Table 8-2 INDEX OF REFRACTION DISPERSION DATA OF KH_2PO_4. (AFTER REFERENCE [8])

	INDEX	
WAVELENGTH, μm	n_o (ORDINARY RAY)	n_e (EXTRAORDINARY RAY)
0.2000	1.622630	1.563913
0.3000	1.545570	1.498153
0.4000	1.524481	1.480244
0.5000	1.514928	1.472486
0.6000	1.509274	1.468267
0.7000	1.505235	1.465601
0.8000	1.501924	1.463708
0.9000	1.498930	1.462234
1.0000	1.496044	1.460993
1.1000	1.493147	1.459884
1.2000	1.490169	1.458845
1.3000	1.487064	1.457838
1.4000	1.483803	1.456838
1.5000	1.480363	1.455829
1.6000	1.476729	1.454797
1.7000	1.472890	1.453735
1.8000	1.468834	1.452636
1.9000	1.464555	1.451495
2.0000	1.460044	1.450308

different type. To illustrate the point, consider the dependence of the index of refraction of the extraordinary wave in a uniaxial crystal on the angle θ between the propagation direction and the crystal optic (z) axis. It is given by (1.4-12) as

$$\frac{1}{n_e^2(\theta)} = \frac{\cos^2\theta}{n_o^2} + \frac{\sin^2\theta}{n_e^2} \tag{8.3-11}$$

If $n_e^{2\omega} < n_o^{\omega}$, there exists an angle θ_m at which $n_e^{2\omega}(\theta_m) = n_o^{\omega}$; so if the fundamental beam (at ω) is launched along θ_m as an ordinary ray, the second-harmonic beam will be generated along the *same direction* as an extraordinary ray. The situation is illustrated by Figure 8-8. The angle θ_m is determined by the intersection between the sphere (shown as a circle in the figure) correspondending to the index surface of the ordinary beam at ω with the index surface of the extraordinary ray which gives $n_e^{2\omega}(\theta)$. The angle θ_m for negative uniaxial crystals—that is, crystals in which $n_e^{\omega} < n_o^{\omega}$—is that satisfying $n_e^{2\omega}(\theta_m) = n_o^{\omega}$ or, using (8.3-11),

$$\frac{\cos^2\theta_m}{(n_o^{2\omega})^2} + \frac{\sin^2\theta_m}{(n_e^{2\omega})^2} = \frac{1}{(n_o^{\omega})^2} \tag{8.3-12}$$

and, solving for θ_m,

$$\sin^2\theta_m = \frac{(n_o^{\omega})^{-2} - (n_o^{2\omega})^{-2}}{(n_e^{2\omega})^{-2} - (n_o^{2\omega})^{-2}} \tag{8.3-13}$$

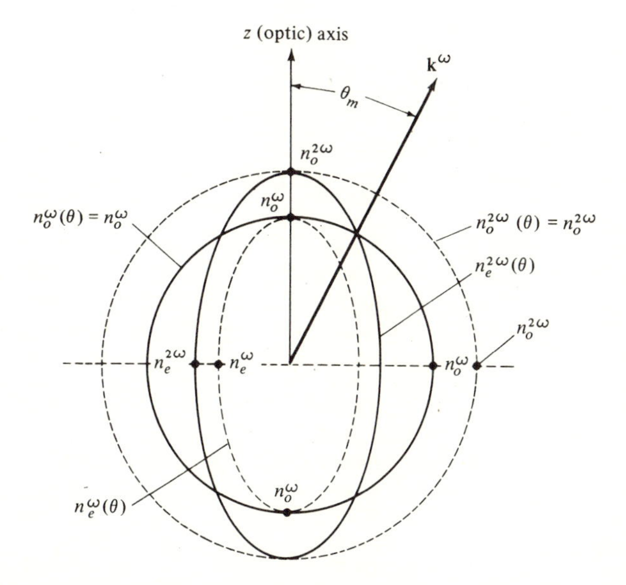

Figure 8-8 Normal (index) surfaces for the ordinary and extraordinary rays in a negative ($n_e < n_o$) uniaxial crystal. If $n_e^{2\omega} < n_o^{\omega}$, the condition $n_e^{2\omega}(\theta) = n_o^{\omega}$ is satisfied at $\theta = \theta_m$. The eccentricities shown are vastly exaggerated.

Numerical example—second-harmonic generation. Consider the problem of second-harmonic generation using the output of a pulsed ruby laser ($\lambda_0 = 0.6940\,\mu$m) in a KH_2PO_4 crystal (KDP) under the following conditions:

$$l = 1 \text{ cm}$$

$$P_\omega/A = 10^8 \text{ W/cm}^2$$

The appropriate d coefficient is, according to Table 8-1, $d = 5 \times 10^{-24}$ mks units. Using these data in (8.3-5) and assuming $\Delta k = 0$ gives a conversion efficiency of

$$\frac{P_{(\lambda_0=0.347\mu\text{m})}}{P_{(\lambda_0=0.694\mu\text{m})}} \simeq 15 \text{ percent}$$

The angle θ_m between the z axis and the direction of propagation for which $\Delta k = 0$ is given by (8.3-13). The appropriate indices are taken from Table 8-2, and are

$$n_e(\lambda = 0.694\ \mu\text{m}) = 1.465 \qquad n_e(\lambda = 0.347\ \mu\text{m}) = 1.487$$

$$n_o(\lambda = 0.694\ \mu\text{m}) = 1.505 \qquad n_o(\lambda = 0.347\ \mu\text{m}) = 1.534$$

Substituting the foregoing data into (8.3-13) gives

$$\theta_m = 51°$$

To obtain phase-matching along this direction, the fundamental beam in the crystal must be polarized as appropriate to an ordinary ray in accordance with the discussion following (8.3-11).

We conclude from this example that very large intensities are needed to obtain high-efficiency second-harmonic generation. This efficiency will, according to (8.3-5), increase as the square of the nonlinear optical coefficient d and will consequently improve as new materials are developed. Another approach is to take advantage of the dependence of η_{SHG} on P_ω/A and to place the nonlinear crystal inside the laser resonator where the energy flux P_ω/A can be made very large.[4] This approach has been used successfully [9] and it will be discussed in considerable detail further in this chapter.

Experimental verification of phase-matching. According to (8.3-5) if the phase-matching condition $\Delta k = 0$ is violated, the output power is reduced by a factor

$$F = \frac{\sin^2(\Delta kl/2)}{(\Delta kl/2)^2} \tag{8.3-14}$$

[4] The one-way power flow inside the optical resonator P_i is related to the power output P_e as $P_i = P_e/(1 - R)$, where R is the reflectivity.

from its (maximum) phase-matched value. The phase mismatch $\Delta kl/2$ is given according to (8.3-8) by

$$\frac{\Delta kl}{2} = \frac{\omega l}{c}\,[n_e^{2\omega}(\theta) - n_o^{\omega}] \tag{8.3-15}$$

and is thus a function of θ. If we use (8.3-11) to expand $n_e^{2\omega}(\theta)$ as a Taylor series near $\theta \simeq \theta_m$, retain the first two terms only, and assume perfect phase-matching at $\theta = \theta_m$ so $n_e^{2\omega}(\theta_m) = n_o^{\omega}$, we obtain

$$\begin{aligned}\Delta k(\theta)l &= -\frac{2\omega l}{c}\sin(2\theta_m)\,\frac{(n_e^{2\omega})^{-2} - (n_o^{2\omega})^{-2}}{2(n_o^{\omega})^{-3}}\,(\theta - \theta_m) \\ &\equiv 2\beta(\theta - \theta_m)\end{aligned} \tag{8.3-16}$$

where β, as defined by (8.3-16), is a constant depending on $n_e^{2\omega}$, $n_o^{2\omega}$, n_o^{ω}, ω, and l. If we plot the output power at 2ω as a function of θ we would expect, according to (8.3-5) and (8.3-16), to find it varying as

$$P_{2\omega}(\theta) \propto \frac{\sin^2[\beta(\theta - \theta_m)]}{[\beta(\theta - \theta_m)]^2} \tag{8.3-17}$$

Figure 8-9 shows an experimental plot of $P_{2\omega}(\theta)$ as well as a plot of (8.3-17).

Second harmonic generation with focused Gaussian beams. The analysis of second harmonic generation leading to (8.3-5) is based on a plane wave model. In practice one uses Gaussian beams which are focused so as to reach their minimum radius (waist) inside the crystal. A typical situation is depicted in Figure 8-10. The incident Gaussian beam is charac-

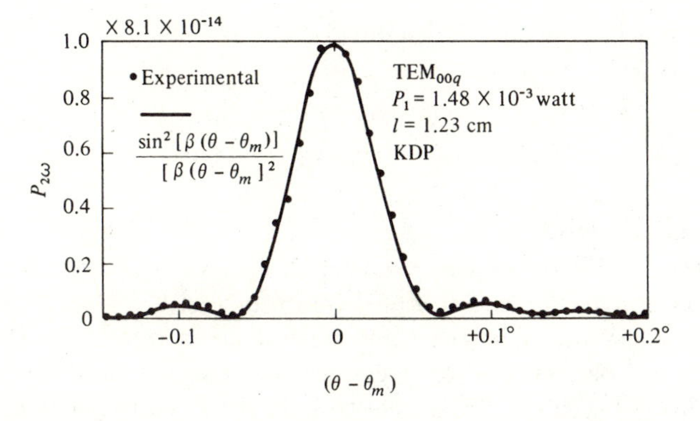

Figure 8-9 Variation of the second-harmonic power $P_{2\omega}$ with the angular departure $(\theta - \theta_m)$ from the phase-matching angle. (After Reference [10].)

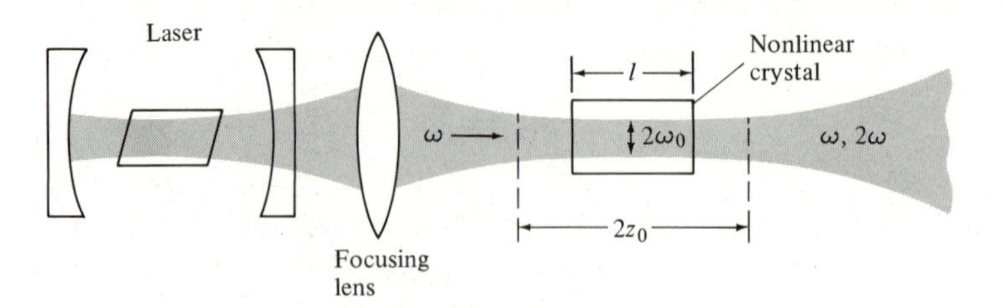

Figure 8-10 Second harmonic generation with a focused Gaussian beam.

terized by confocal parameter z_0, which according to (3.2-11) is the distance from the beam waist in which the beam "area" $\pi\omega^2$ is double that of the waist. We recall that $z_0 = \pi\omega_0^2 n/\lambda$, where ω_0 is the minimum beam radius (waist). If $z_0 \gg l$ (l is the crystal length), the beam area, hence the intensity, of the incident wave is nearly independent of z within the crystal, and we may apply the plane wave result (8.3-3) to write

$$|E^{(2\omega)}(r)|^2 = \frac{\mu}{\epsilon}\omega^2 d^2 |E^{(\omega)}(r)|^4 l^2 \frac{\sin^2(\Delta kl/2)}{(\Delta kl/2)^2} \tag{8.3-18}$$

where $E^{(\omega)}(r)$ is taken as

$$E^{(\omega)}(r) \cong E_0 e^{-r^2/\omega_0^2} \tag{8.3-19}$$

as appropriate to a fundamental Gaussian beam. Using

$$P^{(\omega)} = \frac{1}{2}\sqrt{\frac{\epsilon}{\mu}}\int_{\text{cross section}} |E^{(\omega)}|^2\,dx\,dy = \sqrt{\frac{\epsilon}{\mu}}E_0^2\left(\frac{\pi\omega_0^2}{4}\right)$$

as well as (8.3-19), we obtain, by integrating (8.3-18),

$$\frac{P^{(2\omega)}}{P^{(\omega)}} = 2\left(\frac{\mu}{\epsilon_0}\right)^{3/2}\frac{\omega^2 d^2 l^2}{n^3}\left(\frac{P^{(\omega)}}{\pi\omega_0^2}\right)\frac{\sin^2(\Delta kl/2)}{(\Delta kl/2)^2} \tag{8.3-20}$$

where we used $(n^{\omega})^2 n^{2\omega} \equiv n^3$.

Equation (8.3-20) is identical to (8.3-5). We must recall, however, that it was derived for a Gaussian beam input with $z_0 \gg l$. According to (8.3-20) in a crystal of length l and with a given input $P^{(\omega)}$ the output power $P^{(2\omega)}$ can be increased by decreasing ω_0. This is indeed the case until z_0 $(= \pi\omega_0^2 n/\lambda)$ becomes comparable to l. Further reduction of ω_0 (and z_0) will lead to a situation in which the beam begins to spread appreciably within the crystal, thus leading to a reduced intensity and a reduced second harmonic generation. It is thus reasonable to focus the beam until $l = 2z_0$. At this point $\omega_0^2 = \lambda l/2\pi n$, which is referred to as confocal focusing, and (8.3-20) becomes

$$\left.\frac{P^{(2\omega)}}{P^{(\omega)}}\right|_{\text{confocal focusing}} = \frac{2}{\pi c}\left(\frac{\mu}{\epsilon_0}\right)^{3/2}\frac{\omega^2 d^2 l}{n^2}P^{(\omega)}\frac{\sin^2(\Delta kl/2)}{(\Delta kl/2)^2} \tag{8.3-21}$$

A more exact analysis of second harmonic generation with focused Gaussian beam shows that the maximum conversion efficiency is approximately 20 percent higher than the confocal result (8.3-21).

The main difference between (8.3-21) and the plane wave result (8.3-5) is that the conversion efficiency in this case increases as l instead of l^2. This reflects the fact that a longer crystal entails the use of a larger beam spot size ω_0 so as to keep $z_0 \approx l/2$, which reduces the intensity of the fundamental beam.

Example—optimum focusing. Consider second harmonic conversion under confocal focusing conditions, in KH_2PO_4 from $\lambda = 1\ \mu m$ to $\lambda = 0.5\ \mu m$. Using $l = 1$ cm, $d_{eff} = 3.6 \times 10^{24}$ MKS, $n = 1.5$, we obtain from (8.3-21)

$$\frac{P^{(2\omega)}}{P^{(\omega)}} = 4.3 \times 10^{-5} P^{(\omega)}$$

8.4 Second-Harmonic Generation Inside the Laser Resonator

According to the numerical example of Section 8.3 we need to use large power densities at the fundamental frequency ω to obtain appreciable conversion from ω to 2ω in typical nonlinear optical crystals. These power densities are not usually available from continuous (CW) lasers. The situation is altered, however, if the nonlinear crystal is placed within the laser resonator. The intensity (one-way power per unit area in watts per square meter) inside the resonator exceeds its value outside a mirror by $(1 - R)^{-1}$, where R is the mirror reflectivity. If $R \simeq 1$, the enhancement is very large and since the second-harmonic conversion efficiency is, according to (8.3-5), proportional to the intensity, we may expect a far more efficient conversion inside the resonator. We will show below that under the proper conditions we can extract the *total available power* of the laser at 2ω instead of at ω and in that sense obtain 100 percent conversion efficiency. In order to appreciate the last statement, consider as an example the case of a (CW) laser in which the maximum power output, at a given pumping rate, is available when the output mirror has a (optimal) transmission of 5 percent.

The output mirror is next replaced with one having 100 percent reflection at ω and a nonlinear crystal is placed inside the laser resonator. If with the crystal inside the conversion efficiency from ω to 2ω in a *single pass* is 5 percent, the laser is loaded optimally as in the previous case except that the coupling is attributable to loss of power caused by second-harmonic generation instead of by the output mirror. It follows that the power generated at 2ω is the same as that coupled previously through the mirror and that the total available power of a laser can thus be converted to the second harmonic.

An experimental setup similar to the one used in the first internal second-harmonic generation experiment [9] is shown in Figure 8-11. The Nd^{3+}:YAG laser (see Chapter 7 for a description of this laser) emits a (fundamental) wave at $\lambda_0 = 1.06\,\mu\text{m}$. The mirrors are, as nearly as possible, totally reflecting at $\lambda_0 = 1.06\,\mu\text{m}$. A Ba_2NaNbO_{15} crystal is used to generate the second harmonic at $\lambda_0 = 0.53\,\mu\text{m}$. The latter is coupled through the mirror—which, ideally, transmits all the radiation at this wavelength.

In the mathematical treatment of internal second-harmonic generation that follows we use the results of the analysis of optimum power coupling in laser oscillators of Section 6.5.

The mirror transmission T_{opt} that results in the maximum power output from a laser oscillator is given by (6.5-11) as

$$T_{\text{opt}} = \sqrt{g_0 L_i} - L_i \tag{8.4-1}$$

where L_i is the residual (that is, unavoidable) fractional intensity loss per pass and g_0 is the fractional unsaturated gain per pass.[5] The useful power output under optimum coupling is, according to (6.5-12),

$$P_o = I_s A(\sqrt{g_0} - \sqrt{L_i})^2 \tag{8.4-2}$$

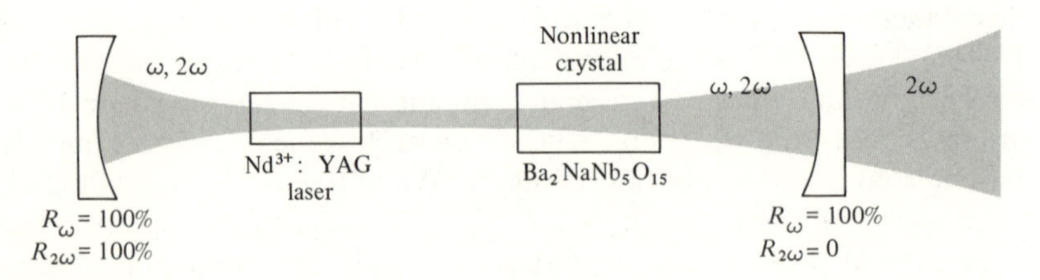

Figure 8-11 Typical setup for second-harmonic conversion inside a laser resonator. (After Reference [9].)

where the saturation intensity of the laser transition I_sA was given by (5.6-9) as[6]

$$I_s = \frac{8\pi n^2 h\nu\Delta\nu}{\lambda^2(t_2/t_{\text{spont}})} \tag{8.4-3}$$

In the present problem the conversion from ω to 2ω can be considered, as far as the ω oscillation is concerned, just as another loss mechanism. We may think of it as due to a mirror with a transmission T' taken as equal to the conversion efficiency (from ω to 2ω) per pass, which, according to (8.3-5), is

$$T' \equiv \frac{P_{2\omega}}{P_\omega} = 2\left(\frac{\mu_0}{\epsilon_0}\right)^{3/2} \frac{\omega^2 d^2 l^2}{n^3} \left[\frac{\sin^2(\Delta kl/2)}{(\Delta kl/2)^2}\right] \frac{P_\omega}{A} \tag{8.4-4}$$

where d is the crystal nonlinear coefficient, l its length, A its cross-sectional area, Δk the wave-vector mismatch, and P_ω the one-way traveling power *inside* the laser. We can rewrite T' in the form

$$T' = \kappa P_\omega \tag{8.4-5}$$

where the value of the constant κ is evident from Equation (8.4-4). The equivalent mirror transmission T' is thus proportional to the power.

Using the last result in (8.4-1) we find immediately that at optimum conversion the product κP_ω must have the value

$$(\kappa P_\omega)_{\text{opt}} = \sqrt{g_0 L_i} - L_i \tag{8.4-6}$$

The total loss per pass seen by the fundamental beam is the sum of the conversion loss (κP_ω) and the residual losses, which, under optimum coupling, becomes

$$L_{\text{opt}} = L_i + (\kappa P_\omega)_{\text{opt}} = \sqrt{g_0 L_i} \tag{8.4-7}$$

Our next problem is to find the internal power P_ω at optimum coupling so that using (8.4-4) we may calculate the second-harmonic power. We start with the expression (6.5-6) for the total power P_e extracted from the

---

[5] We may recall here that the residual losses include all loss mechanisms except those representing useful power coupling. The unsaturated gain g_0 is that exercised by a very weak wave and represents the maximum available gain at a given pumping strength.

[6] I_s is, according to (5.6-8) (and putting $g(\nu)^{-1} = \Delta\nu$) the optical intensity (watts per square meter) that reduces the inversion, hence the gain, to one half its zero intensity (unsaturated) value.

laser atoms and replace the loss L by its optimum value (8.4-7) to obtain

$$(P_e)_{\text{opt}} = P_s\left[\frac{g_0}{L_{\text{opt}}} - 1\right] = P_s\left[\sqrt{\frac{g_0}{L_i}} - 1\right]$$

$$= \frac{8\pi n^3 h\Delta\nu V}{\lambda^3 (t_c)_{\text{opt}}}\left(\frac{t_{\text{spont}}}{t_2}\right)\left[\sqrt{\frac{g_0}{L_i}} - 1\right] = L_{\text{opt}} I_s A\left[\sqrt{\frac{g_0}{L_i}} - 1\right] \quad \text{(8.4-8)}$$

where to get the last equality we used relation (4.7-2)

$$t_c = \frac{nl}{cL}$$

to relate the resonator decay time t_c to the loss per pass L. The fraction of the total power P_e emitted by the atoms that is available as useful output is T'/L. This power is also given by the product $P_\omega T'$ of the one-way internal power P_ω and the fraction T' of this power that is converted per pass. Equating these two forms gives

$$P_\omega = \frac{P_e}{L}$$

and using (8.4-8) we get

$$(P_\omega)_{\text{opt}} = I_s A\left[\sqrt{\frac{g_0}{L_i}} - 1\right] \quad \text{(8.4-9)}$$

for the one-way fundamental power inside the laser under optimum coupling conditions. The amount of second-harmonic power generated under optimum coupling is

$$(P_{2\omega})_{\text{opt}} = (\kappa P_\omega)_{\text{opt}}(P_\omega)_{\text{opt}}$$

which, through the use of (8.4-6) and (8.4-9), results in

$$(P_{2\omega})_{\text{opt}} = I_s A(\sqrt{g_0} - \sqrt{L_i})^2 \quad \text{(8.4-10)}$$

This is the same expression as the one previously obtained in (6.5-12) for the maximum available power output from a laser oscillator.

The nonlinear coupling constant κ was defined by (8.4-4) and (8.4-5) as

$$\kappa = 2\left(\frac{\mu_0}{\epsilon_0}\right)^{3/2}\frac{\omega^2 d^2 l^2}{n^3 A}\left[\frac{\sin^2(\Delta kl/2)}{(\Delta kl/2)^2}\right] \quad \text{(8.4-11)}$$

Its value under optimum coupling can be derived from (8.4-6) and (8.4-9) and is

$$\kappa_{\text{opt}} = \frac{(\kappa P_\omega)_{\text{opt}}}{(P_\omega)_{\text{opt}}} = \frac{L_i}{I_s A} \quad \text{(8.4-12)}$$

and is thus *independent of the pumping strength.*[7] It follows that once κ is adjusted to its optimum value L_i/I_sA, it remains optimal at any pumping

[7] We recall here that the pumping strength in our analysis is represented by the unsaturated gain g_0.

level. This is quite different from the case of optimum coupling in ordinary lasers, in which optimum mirror transmission was found [see (6.5-11)] to depend on the pumping strength.

In closing we may note that apart from its dependence on the crystal length l, the nonlinear coefficient d, and the beam cross section A, κ depends also on the phase mismatch Δkl. Since Δk was shown in (8.3-15) to depend on the direction of propagation in the crystal, we can use the crystal orientation as a means of varying κ.

Numerical example—internal second-harmonic generation. Consider the problem of designing an internal second harmonic generator of the type illustrated in Figure 8-11. The Nd^{3+}:YAG laser is assumed to have the following characteristics:

$$\lambda_0 = 1.06\ \mu\text{m} = 1.06 \times 10^{-6}\ \text{meter}$$

$$\Delta\nu = 1.35 \times 10^{11}\ \text{Hz}$$

Beam diameter (averaged over entire resonator length) = 2 mm

$$L_i = \text{internal loss per pass} = 2 \times 10^{-2}$$

$$n = 1.5$$

The crystal used for second-harmonic generation is $BaNaNb_5O_{15}$, whose second-harmonic coefficient (see Table 8-1) is $d \simeq 1.1 \times 10^{-22}$ MKS units.

Our problem is to calculate the length l of the nonlinear crystal that results in a full conversion of the optimally available fundamental power into the second harmonic at $\lambda = 0.53\ \mu$m. The crystal is assumed to be oriented at the phase-matching angle, so $\Delta k = k^{2\omega} - 2k^{\omega} = 0$.

The optimum coupling parameter is given by (8.4-12) as $\kappa_{\text{opt}} = L_i/I_sA$, where I_s is the saturation intensity defined by (8.4-3). Using the foregoing data in (8.4-3) gives

$$I_sA = 2\ \text{watts}$$

which, taking $L_i = 2 \times 10^{-2}$, yields

$$\kappa_{\text{opt}} = 10^{-2}$$

Next we use the definition (8.4-11)

$$\kappa = 2\left(\frac{\mu_0}{\epsilon_0}\right)^{3/2} \frac{\omega^2 d^2 l^2}{n^3 A}$$

where we put $\Delta k = 0$ and take the beam diameter at the crystal as 50 μm. (The crystal can be placed near a beam waist so the diameter is a minimum.) Equating the last expression to $\kappa_{\text{opt}} = 10^{-2}$ using the numerical data given above, and solving for the crystal length, results in

$$l_{\text{opt}} = 0.804\ \text{cm}$$

8.5 Photon Model of Second-Harmonic Generation

A very useful point of view and one that follows directly from the quantum mechanical analysis of nonlinear optical processes [11] is based on the photon model illustrated in Figure 8-12. According to this picture, the basic process of second-harmonic generation can be viewed as an annihilation of two photons at ω and a simultaneous creation of a photon at 2ω. Recalling that a photon has an energy $\hbar\omega$ and a momentum $\hbar\mathbf{k}$, it follows that if the fundamental conversion process is to conserve momentum as well as energy that

$$\mathbf{k}^{(2\omega)} = 2\mathbf{k}^{(\omega)} \tag{8.5-1}$$

which is a generalization to three dimensions of the condition $\Delta k = 0$ shown in Section 8.3 to lead to maximum second-harmonic generation.

8.6 Parametric Amplification

Optical parametric amplification in its simplest form involves the transfer of power from a "pump" wave at ω_3 to waves at frequencies ω_1 and ω_2, where $\omega_3 = \omega_1 + \omega_2$. It is fundamentally similar to the case of second-harmonic generation treated in Section 8.3. The only difference is in the direction of power flow. In second-harmonic generation, power is fed from the low-frequency optical field at ω to the field at 2ω. In parametric

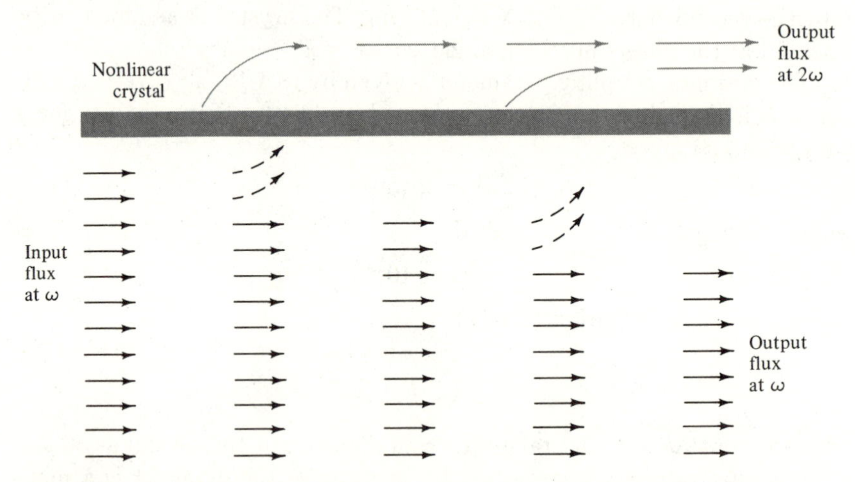

Figure 8-12 Schematic representation of the process of second-harmonic generation. Input photons (each arrow represents one photon) at ω are "annihilated" by the nonlinear crystal in pairs, with a new photon at 2ω being created for each annihilated pair. (Note that in reality both ω and 2ω occupy the same space inside the crystal.)

amplification, power flow is from the high-frequency field (ω_3) to the low-frequency fields at ω_1 and ω_2. In the special case where $\omega_1 = \omega_2$ we have the exact reverse of second-harmonic generation. This is the case of the so-called "degenerate parametric amplification."

Before we embark on a detailed analysis of the optical case it may be worthwhile to review some of the low-frequency beginnings of parametric oscillation.

Consider a classical nondriven oscillator whose equation of motion is given by

$$\frac{d^2v}{dt^2} + \kappa \frac{dv}{dt} + \omega_0^2 v = 0 \tag{8.6-1}$$

The variable v may correspond to the excursion of a mass M, which is connected to a spring with a constant ω_0^2M, or to the voltage across a parallel RLC circuit, in which case $\omega_0^2 = (LC)^{-1}$ and $\kappa = (RC)^{-1}$. The solution of (8.6-1) is

$$v(t) = v(0) \exp\left[-\frac{\kappa t}{2}\right] \exp\left[\pm i\sqrt{\omega_0^2 - \frac{\kappa^2}{4}}\, t\right] \tag{8.6-2}$$

that is, a damped sinusoid.

In 1883 Lord Rayleigh [12], investigating parasitic resonances in pipe organs, considered the consequences of the following equation

$$\frac{d^2v}{dt^2} + \kappa \frac{dv}{dt} + (\omega_0^2 + 2\alpha \sin \omega_p t)v = 0 \tag{8.6-3}$$

This equation may describe an oscillator in which an energy storage parameter (mass or spring constant in the mechanical oscillator, L or C in the RLC oscillator) is modulated at a frequency ω_p. As an example consider the case of the RLC circuit shown in Figure 8-13, in which the capacitance is modulated according to

$$C = C_0\left(1 - \frac{\Delta C}{C_0} \sin \omega_p t\right) \tag{8.6-4}$$

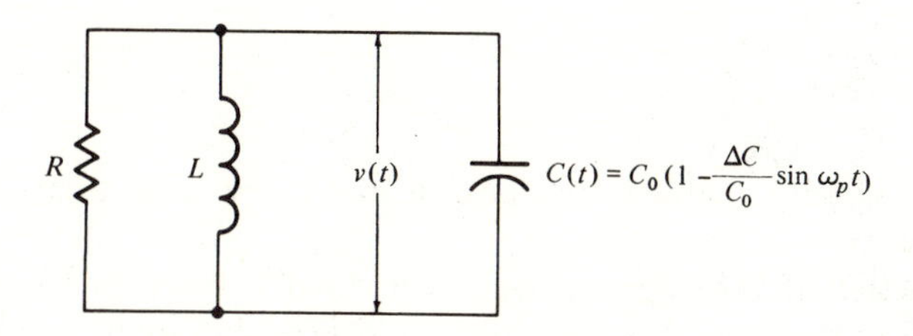

Figure 8-13 A degenerate parametric oscillator with a sinusoidally modulated capacitance.

The equation of the voltage across the *RLC* circuit is given by (8.6-1) with $\omega_0{}^2 = (LC)^{-1}$.

Using (8.6-4) and assuming $\Delta C \ll C_0$ (8.6-1) becomes

$$\frac{d^2v}{dt^2} + \kappa\frac{dv}{dt} + \frac{1}{LC_0}\left(1 + \frac{\Delta C}{C_0}\sin\omega_p t\right)v = 0 \tag{8.6-5}$$

which, if we make the identification

$$\omega_0{}^2 = \frac{1}{LC_0}, \qquad \alpha = \frac{\omega_0{}^2\Delta C}{2C_0} \tag{8.6-6}$$

is identical to (8.6-3).

The most important feature of the parametrically driven oscillator described by (8.6-3) is that it is capable of *sustained oscillation* at ω_0. To show this let us assume a solution

$$v = a\cos[\omega t + \phi] \tag{8.6-7}$$

Expanding $\sin\omega_p t$ in (8.6-3) in terms of exponentials, substituting (8.6-7), and neglecting nonsynchronous terms oscillating at $(\omega_p + \omega)$ leads to

$$(\omega_0{}^2 - \omega^2)e^{i(\omega t+\phi)} + i\omega\kappa e^{i(\omega t+\phi)} - i\alpha e^{i[(\omega_p-\omega)t-\phi]} = 0 \tag{8.6-8}$$

From (8.6-8) it follows that steady-state oscillation is possible if

$$\begin{gathered}\omega_p = 2\omega \quad (\text{so that } \omega_p - \omega = \omega) \\ \omega = \omega_0 \qquad \phi = 0 \qquad \alpha = \omega\kappa\end{gathered} \tag{8.6-9}$$

or, in words:

The pump frequency ω_p is twice the oscillation frequency ω_0. The oscillation phase[8] is $\phi = 0$ and the strength of the pumping α must satisfy $\alpha = \omega\kappa$. The last condition is referred to as the "start-oscillation" condition or "threshold condition," since it gives the pumping strength (α) needed to overcome the losses (κ) at the oscillation threshold. In the case of the *RLC* circuit, whose capacitance is pumped according to (8.6-4), the threshold oscillation condition $\alpha = \omega\kappa$ can be written with the aid of (8.6-6) as

$$\frac{\Delta C}{2C_0} = \frac{\kappa}{\omega_0} = \frac{1}{Q} \tag{8.6-10}$$

where the quality factor $Q = \omega_0 RC$ is related to the decay rate κ by $\kappa = \omega_0/Q$.

In practice, if the capacitance of the circuit shown in Figure 8-13 is modulated so that condition (8.6-10) is satisfied, the circuit will break into

[8] The phase ϕ is relative to that of the pump oscillation as given by (8.6-4).

spontaneous oscillation at a frequency $\omega_0 = \omega_p/2$. This constitutes a transfer of energy from ω_p to $\omega_p/2$.

The physical nature of this transfer may become clearer if we consider the time behavior of the voltage $v(t)$, the charge $q(t)$, and the capacitance $C(t)$ as illustrated in Figure 8-14.

$C(t)$ is a parallel-plate capacitor whose capacitance is periodically varied. Assume first that $C(t)$ is varied as in Figure 8-14(a) by pulling the capacitor plates apart and pushing them together again [$C \propto$ (plate separation)$^{-1}$]. At the same time the circuit is caused to oscillate so that the charge $q(t)$ on the capacitor plates varies as in Figure 8-14(b). Now, according to Figure 8-14(a), when the charge on the plates is a maximum, the plates are pulled apart slightly. The charge cannot change instantaneously, but since work must be done (against the Coulomb attraction of the opposite charges on the capacitor plates) to separate the plates, energy is fed into the capacitor and appears as a sudden increase in the voltage ($v = q/C$, $\mathcal{E} = \frac{1}{2}q^2/C$), as in Figure 8-14(c). One quarter of a period later, the charge and thus the field between the plates is zero and the plates can be returned to their original position with no energy expenditure. At the end of half a cycle, the charge has reversed sign and is again a maximum, so the plates are pulled apart once more. This process is then repeated many times, causing the total voltage to increase twice in each oscillation cycle. In this way, energy at *twice* the resonant frequency is pumped into the circuit where it appears as an increase in energy of the resonant frequency.

There are two noteworthy features to this degenerate oscillator. First, the frequency of the pump *must* be very nearly twice the resonant frequency of the oscillator for gain to occur, in agreement with the previous conclusions; see (8.6-9). In addition, the phase of the pump relative to the charge on the capacitor plates must be chosen properly. Consider the case where $C(t) = C_0 \pm (\Delta C) \sin (2\omega_0 t)$, as in Figure 8-14(d). If we take the minus sign, which corresponds to the $\phi = 0°$ curve, then energy is continuously fed *into* the system as described above. If, however, the pumping phase is inverted (that is, the plus sign), then the capacitor plates are pushed together when the charge is a maximum, thus performing work, giving up energy, and decreasing the total voltage. Any initial oscillations that may be present will be damped out. The phase condition ($\phi = 0$) agrees with the second of (8.6-9).

To make a connection between the lumped-circuit parametric oscillator and the optical nonlinearity discussed in (8.1-14) we show that the (time) modulation of a capacitance at some frequency ω_p which was shown to give rise to oscillation at $\omega_p/2$ is formally equivalent to appyling a field at ω_p to a nonlinear dielectric in which the polarization p and the electric field e are related by

$$p = \epsilon_0 \chi e + de^2 \qquad \textbf{(8.6-11)}$$

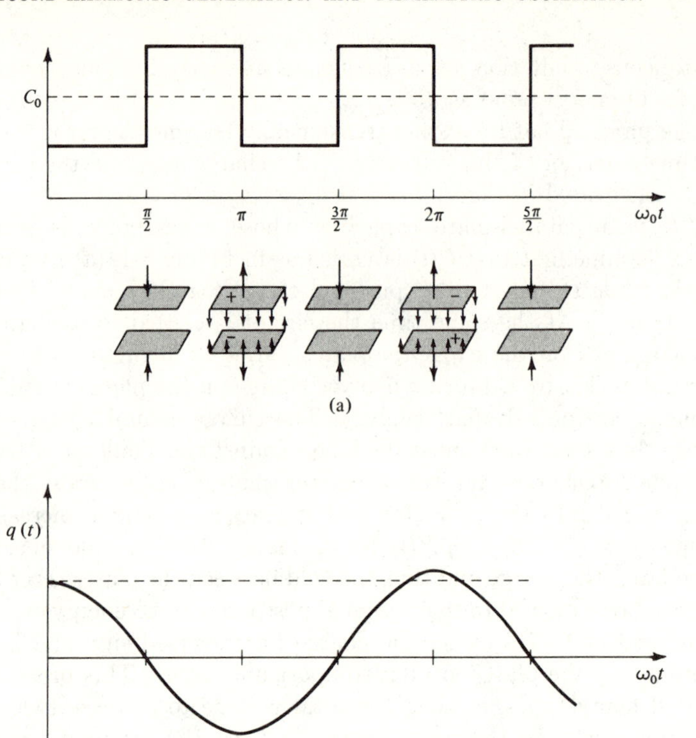

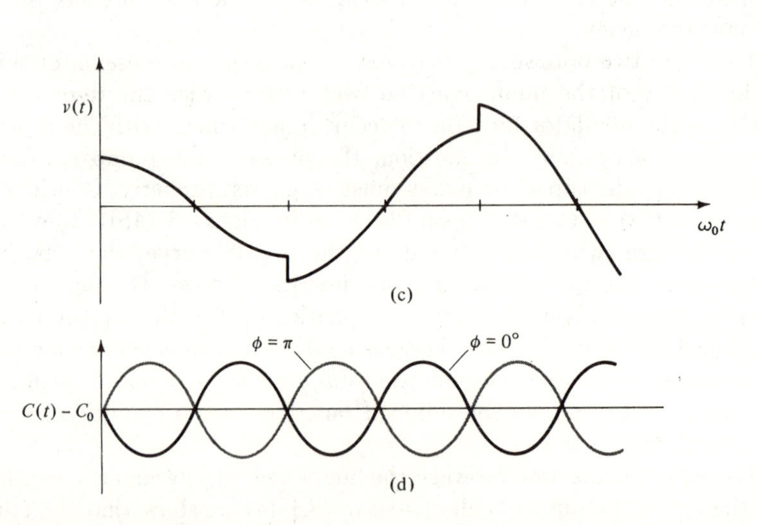

Figure 8-14 Physical model of a capacitively pumped parametric oscillator. (**a**) Square-wave capacitance variation at twice the circuit oscillation frequency. (Also shown is the motion of the capacitor plates, the charge, and the forces on the plates.) (**b**) The charge on one of the capacitor plates. (**c**) The voltage across the circuit. (**d**) Variation of the capacitance $C(t)$ at two phases relative to that of the charge.

This can be done by considering a parallel-plate capacitance of area A and separation s which is filled with a medium whose polarization is given by (8.6-11). Using the relations[9]

$$d(t) = \epsilon_0 e(t) + p(t) = \epsilon e(t) \tag{8.6-12}$$

the dielectric constant ϵ can be written as

$$\epsilon = \epsilon_0(1 + \chi) + de$$

and the capacitance $C = \epsilon A/s$ as

$$C = \frac{\epsilon_0(1+\chi)A}{s} + \frac{Ad}{s}e \tag{8.6-13}$$

If the electric field is given by

$$e = -E_0 \sin \omega_p t$$

the capacitance becomes

$$C = \frac{\epsilon_0(1+\chi)A}{s} - \frac{AdE_0}{s}\sin \omega_p t \tag{8.6-14}$$

which is of a form identical to (8.6-4). It follows that the two points of view used to describe parametric processes—the one represented by (8.6-4), in which an energy-storage parameter is modulated, and that in which the electric (or magnetic) response is nonlinear, as in (8.6-11)—are equivalent.

We return now to the basic nonlinear parametric equations (8.2-13) to analyze the case of optical parametric amplification. We find it convenient to introduce a new field variable, defined by

$$A_l \equiv \sqrt{\frac{n_l}{\omega_l}}\, E_l \qquad l = 1, 2, 3 \tag{8.6-15}$$

so that the power flow per unit area at ω_l is given by

$$\frac{P_l}{A} = \frac{1}{2}\sqrt{\frac{\epsilon_0}{\mu_0}}\, n_l|E_l|^2 = \frac{1}{2}\sqrt{\frac{\epsilon_0}{\mu_0}}\, \omega_l|A_l|^2 \tag{8.6-16}$$

where n_l is the index of refraction at ω_l. The power flow P_l/A per unit area is related to the flux N_l (photons per square meter per second) by

$$\frac{P_l}{A} = N_l \hbar\omega_l = \frac{1}{2}\sqrt{\frac{\epsilon_0}{\mu_0}}\, |A_l|^2\omega_l \tag{8.6-17}$$

---

[9] The electric displacement $d(t)$ should not be confused with the nonlinear constant d in (8.6-11).

so that $|A_l|^2$ is proportional to the photon flux at ω_l. The equations of motion (8.2-13) for the A_l variables become

$$\frac{dA_1}{dz} = -\frac{1}{2}\alpha_1 A_1 - \frac{i}{2}\lambda A_2^* A_3 e^{-i(\Delta k)z}$$
$$\frac{dA_2^*}{dz} = -\frac{1}{2}\alpha_2 A_2^* + \frac{i}{2}\lambda A_1 A_3^* e^{i(\Delta k)z} \tag{8.6-18}$$
$$\frac{dA_3}{dz} = -\frac{1}{2}\alpha_3 A_3 - \frac{i}{2}\lambda A_1 A_2 e^{i(\Delta k)z}$$

where

$$\Delta k \equiv k_3 - (k_1 + k_2)$$
$$\lambda \equiv d\sqrt{\left(\frac{\mu}{\epsilon_0}\right)\frac{\omega_1\omega_2\omega_3}{n_1 n_2 n_3}} \tag{8.6-19}$$
$$\alpha_l \equiv \sigma_l\sqrt{\frac{\mu}{\epsilon_l}} \qquad l = 1, 2, 3$$

The advantage of using the A_l instead of E_l is now apparent since, unlike (8.2-13), relations (8.6-18) involve a single coupling parameter λ.

We will now use (8.6-18) to solve for the field variables $A_1(z)$, $A_2(z)$, and $A_3(z)$ for the case in which three waves with amplitudes $A_1(0)$, $A_2(0)$, and $A_3(0)$ at frequencies ω_1, ω_2, and ω_3, respectively, are incident on a nonlinear crystal at $z = 0$. We take $\omega_3 = \omega_1 + \omega_2$, $\alpha_1 = \alpha_2 = \alpha_3 = 0$ (no losses), and $\Delta k = k_3 - k_1 - k_2 = 0$. In addition, we assume that $\omega_1|A_1(z)|^2$ and $\omega_2|A_2(z)|^2$ remain small compared to $\omega_3 A_3(0)^2$ throughout the interaction region. This last condition, in view of (8.6-17), is equivalent to assuming that the power drained off the "pump" (at ω_3) by the "signal" (ω_1) and idler (ω_2) is negligible compared to the input power at ω_3. This enables us to view $A_3(z)$ as a constant. With the assumptions stated above, equations (8.6-18) become

$$\frac{dA_1}{dz} = -\frac{ig}{2}A_2^* \qquad \frac{dA_2^*}{dz} = \frac{ig}{2}A_1 \tag{8.6-20}$$

where

$$g \equiv \lambda A_3(0) = \sqrt{\left(\frac{\mu}{\epsilon_0}\right)\frac{\omega_1\omega_2}{n_1 n_2}}\, dE_3(0) \tag{8.6-21}$$

The solution of the coupled equations (8.6-20) subject to the initial conditions $A_1(z = 0) \equiv A_1(0)$, $A_2(z = 0) \equiv A_2(0)$, $A_3(0) = A_3^*(0)$ is

$$A_1(z) = A_1(0)\cosh\frac{g}{2}z - iA_2^*(0)\sinh\frac{g}{2}z$$
$$A_2^*(z) = A_2^*(0)\cosh\frac{g}{2}z + iA_1(0)\sinh\frac{g}{2}z \tag{8.6-22}$$

Equations (8.6-22) describe the growth of the signal and idler waves under phase-matching conditions. In the case of parametric amplification the input will consist of the pump (ω_3) wave and one of the other two fields, say ω_1. In this case $A_2(0) = 0$, and using the relation $N_i \propto A_i A_i^*$ for the photon flux we obtain from (8.6-22)

$$N_1(z) \propto A_1^*(z)A_1(z) = |A_1(0)|^2 \cosh^2 \frac{gz}{2} \underset{gz \gg 1}{\longrightarrow} \frac{|A_1(0)|^2}{4} e^{gz}$$

$$N_2(z) \propto A_2^*(z)A_2(z) = |A_1(0)|^2 \sinh^2 \frac{gz}{2} \underset{gz \gg 1}{\longrightarrow} \frac{|A_1(0)|^2}{4} e^{gz} \tag{8.6-23}$$

Thus, for $gz \gg 1$, the photon fluxes at ω_1 and ω_2 grow exponentially. If we limit our attention to the wave at ω_1, it undergoes an amplification by a factor

$$\frac{A_1^*(z)A_1(z)}{A_1^*(0)A_1(0)} \underset{gz \gg 1}{=} \tfrac{1}{4}e^{gz} \tag{8.6-24}$$

Numerical example—parametric amplification. The magnitude of the gain coefficient g available in a traveling-wave parametric interaction is estimated for the following case involving the use of a $LiNbO_3$ crystal.

$$d_{311} = 5 \times 10^{-23}$$

$$\nu_1 \cong \nu_2 = 3 \times 10^{14}$$

$$P_3 = (\text{pump power}) = 5 \times 10^6 \text{ W/cm}^2$$

$$n_1 \simeq n_2 = 2.2$$

Converting P_3 to $|E_3|^2$ with the use of (8.6-16), and then substituting in (8.6-21), yields

$$g = 0.7 \text{ cm}^{-1}$$

This shows that traveling-wave parametric amplification is not expected to lead to large values of gain except for extremely large pump-power densities. The main attraction of the parametric amplification just described is probably in giving rise to parametric oscillation, which will be described in Section 8.8.

8.7 Phase-Matching in Parametric Amplification

In the preceding section the analysis of parametric amplification assumed that the phase-matching condition

$$k_3 = k_1 + k_2 \tag{8.7-1}$$

is satisfied. It is important to determine the consequences of violating this condition. We start with equations (8.6-18) taking the loss coefficients $\alpha_1 = \alpha_2 = 0$:

$$\frac{dA_1}{dz} = -i\frac{g}{2}A_2^* e^{-i(\Delta k)z}$$
$$\frac{dA_2^*}{dz} = +i\frac{g}{2}A_1 e^{i(\Delta k)z} \tag{8.7-2}$$

The solution of (8.7-2) is facilitated by the substitution

$$A_1(z) = m_1 e^{[s-i(\Delta k/2)]z}$$
$$A_2^*(z) = m_2 e^{[s+i(\Delta k/2)]z} \tag{8.7-3}$$

where m_1 and m_2 are coefficients independent of z. The exponential growth constant s is to be determined. Substitution of (8.7-3) in (8.7-2) leads to

$$\left(s - i\frac{\Delta k}{2}\right)m_1 + i\frac{g}{2}m_2 = 0$$
$$-i\frac{g}{2}m_1 + \left(s + i\frac{\Delta k}{2}\right)m_2 = 0 \tag{8.7-4}$$

By equating the determinant of the coefficients of m_1 and m_2 in (8.7-4) to zero we obtain the two solutions

$$s_\pm = \pm\tfrac{1}{2}\sqrt{g^2 - (\Delta k)^2} \equiv \pm b \tag{8.7-5}$$

The general solution of (8.7-2) is the sum of the two independent solutions

$$A_1(z) = m_1^+ e^{[s_+ - i(\Delta k/2)]z} + m_1^- e^{[s_- - i(\Delta k/2)]z}$$
$$A_2^*(z) = m_2^+ e^{[s_+ + i(\Delta k/2)]z} + m_2^- e^{[s_- + i(\Delta k/2)]z} \tag{8.7-6}$$

The coefficients m_1^+, m_1^-, m_2^+, m_2^- are next determined by requiring that at $z = 0$ the solution (8.7-6) agree with the input amplitudes $A_1(0)$ and $A_2^*(0)$. This leads straightforwardly to the result

$$A_1(z)e^{i(\Delta k/2)z} = A_1(0)\left[\cosh(bz) + \frac{i(\Delta k)}{2b}\sinh(bz)\right] - i\frac{g}{2b}A_2^*(0)\sinh(bz)$$
$$A_2^*(z)e^{-i(\Delta k/2)z} = A_2^*(0)\left[\cosh(bz) - \frac{i(\Delta k)}{2b}\sinh(bz)\right] + i\frac{g}{2b}A_1(0)\sinh(bz) \tag{8.7-7}$$

The last result reduces, as it should, to (8.6-22) if we put $\Delta k = 0$.

The most noteworthy feature of (8.7-5) and (8.7-7) is that the exponential gain coefficient b is a function of Δk and that unless

$$g \geqq \Delta k \tag{8.7-8}$$

no sustained growth of the signal (A_1) and idler (A_2) waves is possible, since in this case the cosh and sinh functions in (8.7-7) become

$$\sin \{\tfrac{1}{2}[(\Delta k)^2 - g^2]^{1/2}z\}$$

$$\cos \{\tfrac{1}{2}[(\Delta k)^2 - g^2]^{1/2}z\}$$

respectively, and the energies at ω_1 and ω_2 oscillate as a function of the distance z.

The problem of phase-matching in parametric amplification is fundamentally the same as that in second-harmonic generation. Instead of satisfying the condition (8.3-6), $k^{2\omega} = 2k^{\omega}$, we have, according to (8.7-1), to satisfy the condition

$$k_3 = k_1 + k_2$$

This is done, as in second-harmonic generation, by using the dependence of the phase velocity of the extraordinary wave in axial crystals on the direction of propagation. In a negative uniaxial crystal ($n_e < n_0$) we can, as an example, choose the signal and idler waves as ordinary while the pump at ω_3 is applied as an extraordinary wave. Using (8.3-11) and the relation $k^{\omega} = (\omega/c)n^{\omega}$ the phase-matching condition (8.7-1) is satisfied when all three waves propagate at an angle θ_m to the z (optic) axis where

$$n_e^{\omega_3}(\theta_m) = \left[\left(\frac{\cos\theta_m}{n_o^{\omega_3}}\right)^2 + \left(\frac{\sin\theta_m}{n_e^{\omega_3}}\right)^2\right]^{-1/2} = \frac{\omega_1}{\omega_3}\, n_o^{\omega_1} + \frac{\omega_2}{\omega_3}\, n_o^{\omega_2} \tag{8.7-9}$$

8.8 Parametric Oscillation[10]

In the two preceding sections we have demonstrated that a pump wave at ω_3 can cause a simultaneous amplification in a nonlinear medium of "signal" and "idler" waves at frequencies ω_1 and ω_2, respectively, where $\omega_3 = \omega_1 + \omega_2$. If the nonlinear crystal is placed within an optical resonator (as shown in Figure 8-15) that provides resonances for the signal or idler waves (or both), the parametric gain will, at some threshold pumping intensity, cause a simultaneous oscillation at the signal and idler frequencies. The threshold pumping corresponds to the point at which the parametric gain just balances the losses of the signal and idler waves. This is the physical basis of the optical parametric oscillator. Its practical importance

[10] See References [13]–[15].

derives from its ability to convert the power output of the pump laser to power at the signal and idler frequencies which, as will be shown below, can be tuned continuously over large ranges.

To analyze this situation we return to (8.6-18). We take $\Delta k = 0$ and neglect the depletion of the pump waves, so $A_3(z) = A_3(0)$. The result is

$$\begin{aligned} \frac{dA_1}{dz} &= -\frac{1}{2}\alpha_1 A_1 - i\frac{g}{2}A_2^* \\ \frac{dA_2^*}{dz} &= -\frac{1}{2}\alpha_2 A_2^* + i\frac{g}{2}A_1 \end{aligned} \tag{8.8-1}$$

where, as in (8.6-21),

$$\begin{aligned} g &\equiv \sqrt{\left(\frac{\mu_0}{\epsilon_0}\right)\frac{\omega_1\omega_2}{n_1 n_2}}\, dE_3(0) \\ \alpha_{1,2} &\equiv \sigma_{1,2}\sqrt{\frac{\mu_0}{\epsilon_{1,2}}} \end{aligned} \tag{8.8-2}$$

Equations (8.8-1) describe traveling-wave parametric interaction. We will use them to describe the interaction inside a resonator such as the one shown in Figure 8-15. This procedure seems plausible if we think of propagation inside an optical resonator as a folded optical path. The magnitude of the spatial distributed loss constants α_1 and α_2 must then be chosen so that they account for the actual losses in the resonator. The latter will include losses caused by the less than perfect reflection at the mirrors, as well as distributed loss in the nonlinear crystal and that due to diffraction.[11]

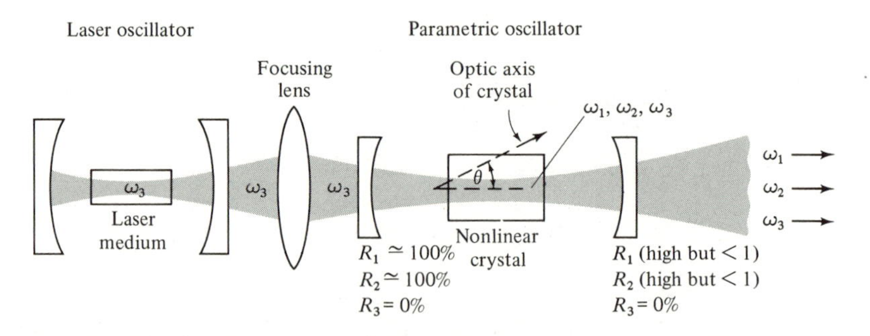

Figure 8-15 Schematic diagram of an optical parametric oscillator in which the laser output at ω_3 is used as the pump, giving rise to oscillations at ω_1 and ω_2 (where $\omega_3 = \omega_1 + \omega_2$) in an optical cavity that contains the nonlinear crystal and resonates at ω_1 and ω_2.

[11] The effective loss constant α_i is chosen so that $\exp(-\alpha_i l)$ is the total attenuation in intensity per resonator pass at ω_i, where l is the crystal length.

If the parametric gain is sufficiently high to overcome the losses, steady-state oscillation results. When this is the case,

$$\frac{dA_1}{dz} = \frac{dA_2^*}{dz} = 0 \tag{8.8-3}$$

and thus the power gained via the parametric interaction just balances the losses.

Putting $d/dz = 0$ in (8.8-1) gives

$$\begin{aligned} -\frac{\alpha_1}{2} A_1 - i\frac{g}{2} A_2^* &= 0 \\ i\frac{g}{2} A_1 - \frac{\alpha_2}{2} A_2^* &= 0 \end{aligned} \tag{8.8-4}$$

The condition for nontrivial solutions for A_1 and A_2^* is that the determinant at (8.8-4) vanish; that is,

$$\det \begin{vmatrix} -\frac{\alpha_1}{2} & -i\frac{g}{2} \\ i\frac{g}{2} & -\frac{\alpha_2}{2} \end{vmatrix} = 0$$

and, therefore,

$$g^2 = \alpha_1\alpha_2 \tag{8.8-5}$$

This is the *threshold condition* for *parametric oscillation.*

If we choose to express the mode losses at ω_1 and ω_2 by the quality factors Q_1 and Q_2, respectively, we have[12]

$$\alpha_i = \frac{\omega_i n_i}{Q_i c} \tag{8.8-6}$$

By the use of (8.8-2), condition (8.8-5) can be written as

$$\frac{d(E_3)_t}{\sqrt{\epsilon_1\epsilon_2}} = \frac{1}{\sqrt{Q_1 Q_2}} \tag{8.8-7}$$

where $(E_3)_t$ is the value of E_3 at threshold. This relation can be shown to be formally analogous to that obtained in (8.6-10) for the lumped-circuit parametric oscillator. According to (8.6-14), $\Delta C/C_0 = dE/\epsilon$; therefore, apart from a factor of two, if we put $Q_1 = Q_2$ and $\epsilon_1 = \epsilon_2$, (8.8.7) is the same as (8.6-10).

Another useful form of the threshold relation results from representing the quality factor Q in terms of the (effective) mirror reflectivities as in (4.7-3). If, furthermore, we express E_3 in terms of the power flow per unit area according to

$$E_3^2 = 2\frac{P_3}{A}\sqrt{\frac{\mu_0}{\epsilon_0 n_3^2}}$$

[12] This relation follows from recognizing that the temporal decay rate $\sigma = \omega/Q$ is related to α by $\sigma = \alpha c = \alpha c/n$.

we can rewrite (8.8-7) as

$$\left(\frac{P_3}{A}\right)_t = \frac{1}{2}\left(\frac{\epsilon_0}{\mu}\right)^{3/2} \frac{n_1 n_2 n_3 (1 - R_1)(1 - R_2)}{\omega_1 \omega_2 l^2 d^2} \tag{8.8-8}$$

where l is the length of the nonlinear crystal.

Numerical example—parametric oscillation threshold. Let us estimate the threshold pump requirement P_3/A (watts per square centimeter) of a parametric oscillator of the kind shown in Figure 8-15, which utilizes an $LiNbO_3$ crystal. We use the following set of parameters:

$$(1 - R_1) = (1 - R_2) = 2 \times 10^{-2} \text{ (that is, loss per pass at } \omega_1 \text{ and } \omega_2 = 2 \text{ percent)}$$

$$(\lambda)_1 = (\lambda)_2 = 1\ \mu\text{m}$$

$$l = 5 \text{ cm}$$

$$n_1 = n_2 = n_3 = 1.5$$

$$d_{311}(\text{LiNbO}_3) = 5 \times 10^{-23}$$

Substitution in (8.8-8) yields

$$\left(\frac{P_3}{A}\right)_t \cong 50 \text{ watts/cm}^2$$

This is a modest amount of power so that the example helps us appreciate the attractiveness of optical parametric oscillation as a means for generating coherent optical frequency at new optical frequencies.

8.9 Frequency Tuning in Parametric Oscillation

We have shown above that the pair of signals (ω_1) and idler frequencies that are caused to oscillate by parametric pumping at ω_3 satisfy the condition $k_3 = k_1 + k_2$. Using $k_i = \omega_i n_i / c$ we can write it as

$$\omega_3 n_3 = \omega_1 n_1 + \omega_2 n_2 \tag{8.9-1}$$

In a crystal the indices of refraction generally depend as shown in Section 8-3, on the frequency, crystal orientation (if the wave is extraordinary), electric field (in electrooptic crystals), and on the temperature. If, as an example, we change the crystal orientation in the oscillator shown in Figure 8-15, the oscillation frequencies ω_1 and ω_2 will change so as to compensate for the change in indices, and thus condition (8.9-1) will be satisfied at the new frequencies.

To be specific, we consider the case of a parametric oscillator pumped by an extraordinary beam at a fixed frequency ω_3. The signal (ω_1) and idler (ω_2) are ordinary waves. At some crystal orientation θ_0 the oscillation takes place at frequencies ω_{10} and ω_{20}. Let the indices of refraction at ω_{10}, ω_{20}, and ω_{30} under those conditions be n_{10}, n_{20}, and n_{30}, respectively. We want to find the change in ω_1 and ω_2 due to a small change $\Delta\theta$ in the crystal orientation.

From (8.9-1) we have, at $\theta = \theta_0$,

$$\omega_3 n_{30} = \omega_{10} n_{10} + \omega_{20} n_{20} \tag{8.9-2}$$

After the crystal orientation has been changed from θ_0 to $\theta_0 + \Delta\theta$, the following changes occur:

$$n_{30} \rightarrow n_{30} + \Delta n_3$$

$$n_{10} \rightarrow n_{10} + \Delta n_1$$

$$n_{20} \rightarrow n_{20} + \Delta n_2$$

$$\omega_{10} \rightarrow \omega_{10} + \Delta\omega_1$$

Since $\omega_1 + \omega_2 = \omega_3 =$ constant,

$$\omega_{20} \rightarrow \omega_{20} + \Delta\omega_2 = \omega_{20} - \Delta\omega_1$$

that is, $\Delta\omega_2 = -\Delta\omega_1$. Since (8.9-1) must be satisfied at $\theta = \theta_0 + \Delta\theta$, we have

$$\omega_3(n_{30} + \Delta n_3) = (\omega_{10} + \Delta\omega_1)(n_{10} + \Delta n_1) + (\omega_{20} - \Delta\omega_1)(n_{20} + \Delta n_2)$$

Neglecting the second-order terms $\Delta n_1 \Delta\omega_1$ and $\Delta n_2 \Delta\omega_1$ and using (8.9-2), we obtain

$$\Delta\omega_1 \Big|_{\substack{\omega_1 \simeq \omega_{10} \\ \omega_2 \simeq \omega_{20}}} = \frac{\omega_3 \Delta n_3 - \omega_{10}\Delta n_1 - \omega_{20}\Delta n_2}{n_{10} - n_{20}} \tag{8.9-3}$$

According to our starting hypotheses the pump is an extraordinary ray; therefore, according to (1.4-12), its index depends on the orientation θ, giving

$$\Delta n_3 = \frac{\partial n_3}{\partial\theta}\Big|_{\theta_0} \Delta\theta \tag{8.9-4}$$

The signal and idler are ordinary rays, so their indices depend on the frequencies but not on the direction. It follows that

$$\Delta n_1 = \frac{\partial n_1}{\partial\omega_1}\Big|_{\omega_{10}} \Delta\omega_1$$

$$\Delta n_2 = \frac{\partial n_2}{\partial\omega_2}\Big|_{\omega_{20}} \Delta\omega_2 \tag{8.9-5}$$

Using the last two equations in (8.9-3) results in

$$\frac{\partial\omega_1}{\partial\theta} = \frac{\omega_3(\partial n_3/\partial\theta)}{(n_{10} - n_{20}) + [\omega_{10}(\partial n_1/\partial\omega_1) - \omega_{20}(\partial n_2/\partial\omega_2)]} \tag{8.9-6}$$

for the rate of change of the oscillation frequency with respect to the crystal orientation. Using (1.4-12) and the relation $d(1/x^2) = -(2/x^3)\,dx$, we obtain

$$\frac{\partial n_3}{\partial \theta} = -\frac{{n_3}^3}{2}\sin(2\theta)\left[\left(\frac{1}{n_e^{\omega_3}}\right)^2 - \left(\frac{1}{n_o^{\omega_3}}\right)^2\right]$$

which, when substituted in (8.9-6), gives

$$\frac{\partial \omega_1}{\partial \theta} = \frac{-\frac{1}{2}\,\omega_3 {n_{30}}^3\left[\left(\frac{1}{n_e^{\omega_3}}\right)^2 - \left(\frac{1}{n_o^{\omega_3}}\right)^2\right]\sin(2\theta)}{(n_{10} - n_{20}) + \left(\omega_{10}\frac{\partial n_1}{\partial \omega_1} - \omega_{20}\frac{\partial n_2}{\partial \omega_2}\right)} \tag{8.9-7}$$

An experimental curve showing the dependence of the signal and idler frequencies on θ in $NH_4H_2PO_4$ (ADP) is shown in Figure 8-16. Also shown is a theoretical curve based on a quadratic approximation of (8.9-7), which was plotted using the dispersion (that is, n versus ω) data of ADP; see Reference [16].

Reasoning similar to that used to derive the angle-tuning expression (8.9-7) can be applied to determine the dependence of the oscillation

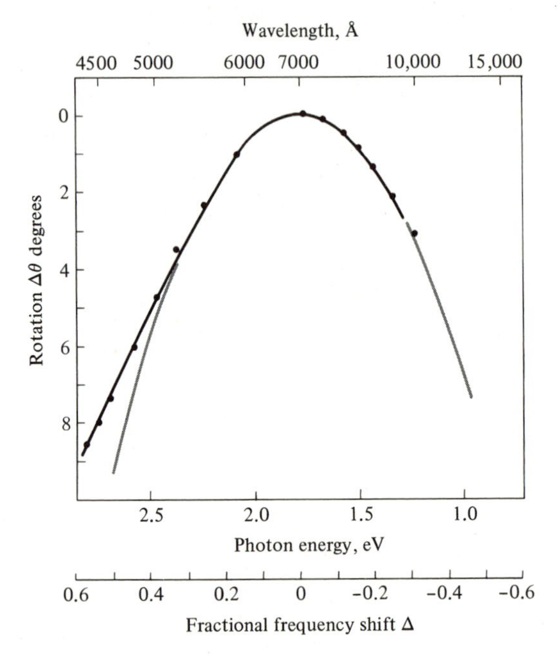

Figure 8-16 Dependence of the signal (ω_1) frequency on the angle between the pump propagation direction and the optic axis of the ADP crystal. The angle θ is measured with respect to the angle for which $\omega_1 = \omega_3/2$. $\Delta = (\omega_1 - \omega_3/2)/(\omega_3/2)$. (After Reference [16].)

frequency on temperature. Here we need to know the dependence of the various indices on temperature. This is discussed further in Problem 8-6. An experimental temperature-tuning curve is shown in Figure 8-17.

8.10 Power Output and Pump Saturation in Optical Parametric Oscillators

In the treatment of the laser oscillator in Section 6.5 we showed that in the steady state the gain could not exceed the threshold value regardless of the intensity of the pump. A closely related phenomenon exists in the case of parametric oscillation. The pump field E_3 gives rise to amplification of the signal and idler waves. When E_3 reaches its critical (threshold) value given by (8.8-7) the gain just equals the losses and the device is on the threshold of oscillation. If the pump field E_3 is increased beyond its threshold value the gain can no longer follow it and must be "clamped" at its threshold value. This follows from the fact that if the gain constant g exceeds its threshold value (8.8-5), a steady state is no longer possible and

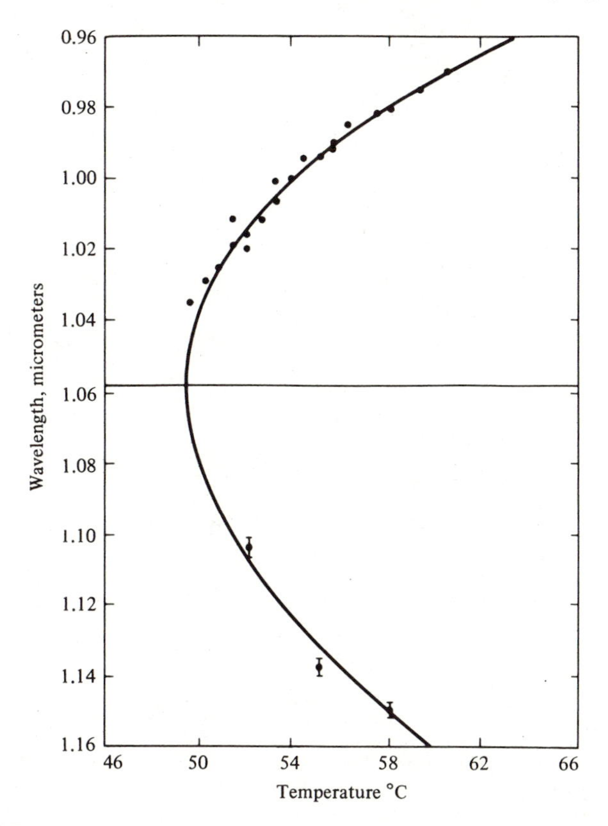

Figure 8-17 Signal and idler wavelength as a function of the temperature of the oscillator crystal. (After Reference [14].)

the signal and idler intensities will increase with time. Since the gain g is proportional to the pump field E_3, it follows that above threshold the *pump field inside* the optical resonator must saturate at its level just prior to oscillation. As power is conserved it follows that any additional pump power input must be diverted into power at the signal and idler fields. Since $\omega_3 = \omega_1 + \omega_2$, it follows that for each input pump photon above threshold we generate one photon at the signal (ω_1) and one at the idler (ω_2) frequencies, so [17]

$$\frac{P_1}{\omega_1} = \frac{P_2}{\omega_2} = \frac{(P_3)_t}{\omega_3}\left[\frac{P_3}{(P_3)_t} - 1\right] \tag{8.10-1}$$

The last argument shows that in principle the parametric oscillator can attain high efficiencies. This requires operation well above threshold, and thus $P_3/(P_3)_t \gg 1$. These considerations are borne out by actual experiments [18].

Figure 8-18 shows experimental confirmation of the phenomenon of

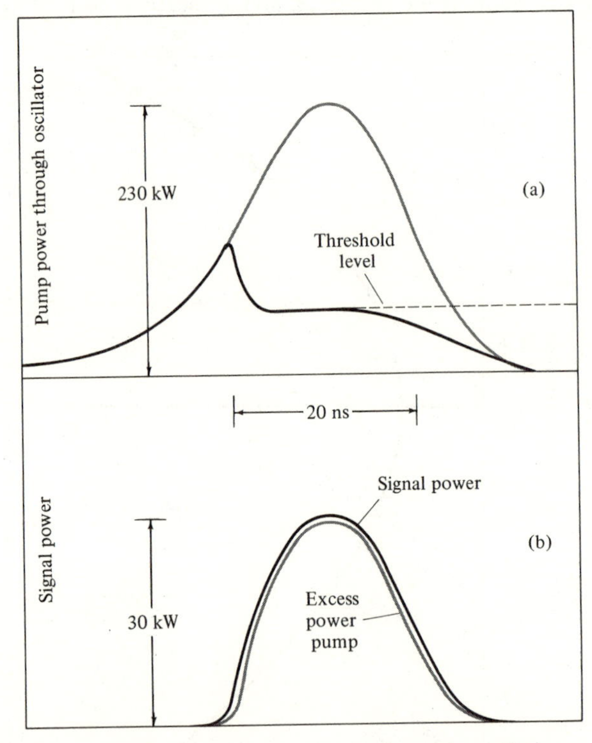

Figure 8-18 Power levels and pumping in a parametric oscillator. (**a**) Waveforms of P_3, the pump power passing through the oscillator. The gray waveform was obtained when the crystal was rotated so that oscillation did not occur; the solid waveform was obtained when oscillation took place. (**b**) Signal power and excess pump power. The gray waveform is the normalized difference between the waveforms in (**a**). (After Reference [18].)

pump saturation; see References [17] and [20]. After a transient buildup the pump intensity inside the resonator settles down to its threshold value.

Figure 8-18(b) shows that the signal power is proportional to the excess (above threshold) pump input power. This is in agreement with Equation (8.10-1).

8.11 Frequency Up-Conversion

Parametric interactions in a crystal can be used to convert a signal from a "low" frequency ω_1 to a "high" frequency ω_3 by mixing it with a strong laser beam at ω_2, where

$$\omega_1 + \omega_2 = \omega_3 \tag{8.11-1}$$

Using the quantum mechanical photon picture described in Section 8-5 we can consider the basic process taking place in frequency up-conversion as one in which a signal (ω_1) photon and a pump (ω_2) photon are annihilated while, simultaneously, one photon at ω_3 is generated; see References [11] and [21]–[24]. Since a photon energy is $\hbar\omega$, conservation of energy dictates that $\omega_3 = \omega_1 + \omega_2$ and, in a manner similar to (8.5-1), the conservation of momentum leads to the relationship

$$\mathbf{k}_3 = \mathbf{k}_1 + \mathbf{k}_2 \tag{8.11-2}$$

between the wave vectors at the three frequencies. This point of view also suggests that the number of output photons at ω_3 cannot exceed the input number of photons at ω_1.

The experimental situation is demonstrated by Figure 8-19. The ω_1 and ω_2 beams are combined in a partially transmissive mirror (or prism), so they traverse together (in near parallelism) the length l of a crystal possessing nonlinear optical characteristics.

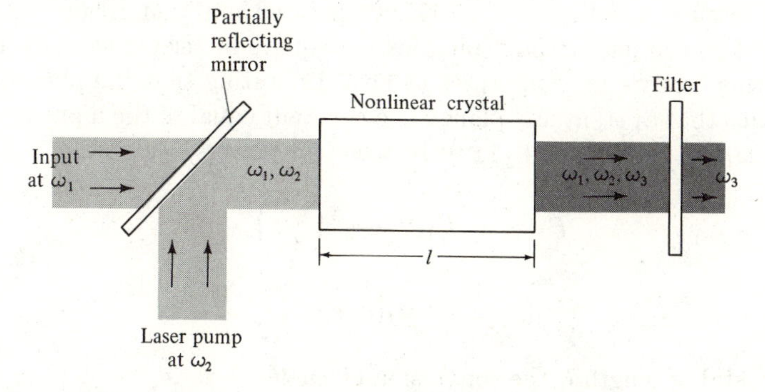

Figure 8-19 Parametric up-conversion in which a signal at ω_1 and a strong laser beam at ω_2 combine in a nonlinear crystal to generate a beam at the sum frequency $\omega_3 = \omega_1 + \omega_2$.

The analysis of frequency up-conversion starts with Equations (8.6-18). Assuming negligible depletion of the pump wave A_2, no losses ($\alpha = 0$) at ω_1 and ω_3 and taking $\Delta k = 0$ we can write the first and third of these equations as

$$\frac{dA_1}{dz} = -i\frac{g}{2}A_3 \qquad \frac{dA_3}{dz} = -i\frac{g}{2}A_1 \tag{8.11-3}$$

where, using (8.6-15) and (8.6-19) and choosing without loss of generality the pump phase as zero so that $A_2(0) = A_2^*(0)$,

$$g \equiv \sqrt{\frac{\omega_1\omega_3}{n_1 n_3}\left(\frac{\mu_0}{\epsilon_0}\right)}\, dE_2 \tag{8.11-4}$$

where E_2 is the amplitude of the electric field of the pump laser. Taking the input waves with (complex) amplitudes $A_1(0)$ and $A_3(0)$, the general solution of (8.11-3) is

$$\begin{aligned} A_1(z) &= A_1(0)\cos\left(\frac{g}{2}z\right) - iA_3(0)\sin\left(\frac{g}{2}z\right) \\ A_3(z) &= A_3(0)\cos\left(\frac{g}{2}z\right) - iA_1(0)\sin\left(\frac{g}{2}z\right) \end{aligned} \tag{8.11-5}$$

In the case of a single (low) frequency input at ω_1, we have $A_3(0) = 0$. In this case,

$$\begin{aligned} |A_1(z)|^2 &= |A_1(0)|^2\cos^2\left(\frac{g}{2}z\right) \\ |A_3(z)|^2 &= |A_1(0)|^2\sin^2\left(\frac{g}{2}z\right) \end{aligned} \tag{8.11-6}$$

therefore,

$$|A_1(z)|^2 + |A_3(z)|^2 = |A_1(0)|^2$$

In the discussion following (8.6-17) we pointed out that $|A_l(z)|^2$ is proportional to the photon flux (photons per square meter per second) at ω_l. Using this fact we may interpret (8.11-6) as stating that the photon flux at ω_1 plus that at ω_3 at any plane z is a constant equal to the input ($z = 0$) flux at ω_1. If we rewrite (8.11-6) in terms of powers, we obtain

$$\begin{aligned} P_1(z) &= P_1(0)\cos^2\left(\frac{g}{2}z\right) \\ P_3(z) &= \frac{\omega_3}{\omega_1}P_1(0)\sin^2\left(\frac{g}{2}z\right) \end{aligned} \tag{8.11-7}$$

In a crystal of length l, the conversion efficiency is thus

$$\frac{P_3(l)}{P_1(0)} = \frac{\omega_3}{\omega_1}\sin^2\left(\frac{g}{2}l\right) \tag{8.11-8}$$

and can have a maximum value of ω_3/ω_1, corresponding to the case in which all the input (ω_1) photons are converted to ω_3 photons.

In most practical situations the conversion efficiency is small (see the following numerical example) so using $\sin x \simeq x$ for $x \ll 1$, we get

$$\frac{P_3(l)}{P_1(0)} \simeq \frac{\omega_3}{\omega_1}\left(\frac{g^2l^2}{4}\right)$$

which, by the use of (8.11-4) and (8.6-16), can be written as

$$\frac{P_3(l)}{P_1(0)} \simeq \frac{\omega_3{}^2l^2d^2}{2n_1n_2n_3}\left(\frac{\mu_0}{\epsilon_0}\right)^{3/2}\left(\frac{P_2}{A}\right) \qquad \textbf{(8.11-9)}$$

where A is the cross-sectional area of the interaction region.

Numerical example—frequency up-conversion. The main practical interest in parametric frequency up-conversion stems from the fact that it offers a means of detecting infrared radiation (a region where detectors are either inefficient, very slow, or require cooling to cryogenic temperatures) by converting the frequency into the visible or near-visible part of the spectrum. The radiation can then be detected by means of efficient and fast detectors such as photomultipliers or photodiodes; see References [22]–[25].

As an example of this application, consider the problem of up-converting a 10.6-μm signal, originating in a CO_2 laser to 0.96 μm by mixing it with the 1.06-μm output of an Nd^{3+}:YAG laser. The nonlinear crystal chosen for this application has to have low losses at 1.06 μm and 10.6 μm, as well as at 0.96 μm. In addition, its birefringence has to be such as to make phase matching possible. The crystal proustite (Ag_3AsS_3) listed in Table 8-1 meets these requirements [25].

Using the data

$$\frac{P_{1.06\ \mu\text{m}}}{A} = 10^4\ \text{W/cm}^2 = 10^8\ \text{W/m}^2$$

$$l = 10^{-2}\ \text{meter}$$

$n_1 \simeq n_2 \simeq n_3 = 2.6$ (an average number based on the data of Reference [25])

$d_{\text{eff}} = 1.1 \times 10^{-22}$ (taken conservatively as a little less than half the value given in Table 8.1 for d_{22})

we obtain, from (8.11-9),

$$\frac{P_{\lambda=0.96\ \mu\text{m}}(l = 1\ \text{cm})}{P_{\lambda=10.6\ \mu\text{m}}(l = 0)} = 6 \times 10^{-4}$$

indicating a useful amount of conversion efficiency.

PROBLEMS

8-1 Show that if θ_m is the phase-matching angle for an ordinary wave at ω and an extraordinary wave at 2ω, then

$$\Delta k(\theta)l|_{\theta \simeq \theta_m} = -\frac{2\omega l}{c_0} \sin(2\theta_m) \frac{(n_e^{2\omega})^{-2} - (n_o^{2\omega})^{-2}}{2(n_o^{\omega})^{-3}} (\theta - \theta_m)$$

8-2 Derive the expression for the phase-matching angle of a parametric amplifier using KDP in which two of the waves are extraordinary while the third is ordinary. Which of the three waves (that is, signal, idler, or pump) would you choose as ordinary? Can this type of phase-matching be accomplished with $\omega_3 = 10{,}000 \text{ cm}^{-1}$, $\omega_1 = \omega_2 = 5000 \text{ cm}^{-1}$? If so, what is θ_m?

8-3 Show that Equations (8.6-22) are consistent with the fact that the increases in the photon flux at ω_1 and ω_2 are identical—that is, that $A_1^*(z)A_1(z) - A_1^*(0)A_1(0) = A_2^*(z)A_2(z) - A_2^*(0)A_2(0)$.

8-4 Complete the missing steps in the derivation of Equation (8.7-7).

8-5 Show that the voltage $v(t)$ across an open-circuited parallel RLC circuit obeys

$$\frac{d^2v}{dt^2} + \frac{1}{RC}\frac{dv}{dt} + \frac{1}{LC}v = 0$$

and is thus of the form of Equation (8.6-1).

8-6 Consider a parametric oscillator setup such as that shown in Figure 8-15. The crystal orientation angle is θ, its temperature is T, and the signal and idler frequencies are ω_{10} and ω_{20}, respectively, with $\omega_{10} + \omega_{20} = \omega_3$. Show that a small temperature change ΔT causes the signal frequency to change by

$$\Delta\omega_1 = \Delta T \times \left\{\omega_3\left[\cos^2\theta\left(\frac{n_e^{\omega_3}(\theta)}{n_o^{\omega_3}}\right)^3 \frac{\partial n_o^{\omega_3}}{\partial T} + \sin^2\theta\left(\frac{n_e^{\omega_3}(\theta)}{n_e^{\omega_3}}\right)^3 \frac{\partial n_e^{\omega_3}}{\partial T}\right] - \omega_{10}\frac{\partial n_o^{\omega_1}}{\partial T} - \omega_{20}\frac{\partial n_o^{\omega_2}}{\partial T}\right\} \times \frac{1}{n_{10} - n_{20}}$$

The pump is taken as an extraordinary ray, whereas the signal and idler are ordinary. [*Hint:* The starting point is Equation (8.9-3), which is valid regardless of the nature of the perturbation.]

8-7 Using the published dispersion data of proustite (Reference [25]), calculate the maximum angular deviation of the input beam at ν_1 (from parallelism with the pump beam at ν_2) that results in a reduction by a factor of 2 in the conversion efficiency. Take $\lambda_1 = 10.6\ \mu\text{m}$, $\lambda_2 = 1.06\ \mu\text{m}$,

$\lambda_3 = 0.964\ \mu$m. [*Hint:* A proper choice must be made for the polarizations at ω_1, ω_2, and ω_3 so that phase-matching can be achieved along some angle.] The maximum angular deviation is that for which

$$\frac{\sin^2 [\Delta k(\theta)l/2]}{[\Delta k(\theta)l/2]^2} = \frac{1}{2}$$

where, at the phase-matching angle θ_m, $\Delta k(\theta_m) = 0$. Approximate the dispersion data by a Taylor-series expansion about the nominal ($\Delta k = 0$) frequencies.

8-8 Using the dispersion data of Reference [25], discuss what happens to phase-matching in an upconversion experiment due to a deviation of the input frequency from the nominal ($\Delta k = 0$) ν_{10} value. Derive an expression for the spectral width of the output in the case where the input spectral density (power per unit frequency) in the vicinity of ν_{10} is uniform. [*Hint:* Use a Taylor-series expansion of the dispersion data about the phase-matching ($\Delta k = 0$) frequencies to obtain an expression for $\Delta k(\nu_3)$.] Define the output spectral width as twice the frequency deviation at which the output is half its maximum ($\Delta k = 0$) value.

REFERENCES

[1] Miller, R. C., "Optical second harmonic generation in piezoelectric crystals," *Appl. Phys. Letters*, vol. 5, p. 17, 1964.

[2] Garret, C. G. B., and F. N. H. Robinson, "Miller's phenomenological rule for computing nonlinear susceptibilities," *IEEE J. Quantum Electron.*, vol. QE-2, p. 328, 1966.

[3] See, for example, J. F. Nye, *Physical Properties of Crystals*. New York: Oxford, 1957.

[4] Armstrong, J. A., N. Bloembergen, J. Ducuing, and P. S. Pershan, "Interactions between light waves in a nonlinear dielectric," *Phys. Rev.*, vol. 127, p. 1918, 1962.

[5] Franken, P. A., A. E. Hill, C. W. Peters, and G. Weinreich, "Generation of optical harmonics," *Phys. Rev. Letters*, vol. 7, p. 118, 1961.

[6] Maker, P. D., R. W. Terhune, M. Nisenoff, and C. M. Savage, "Effects of dispersion and focusing on the production of optical harmonics," *Phys. Rev. Letters*, vol. 8, p. 21, 1962.

[7] Giordmaine, J. A., "Mixing of light beams in crystals," *Phys. Rev. Letters*, vol. 8, p. 19, 1962.

[8] Zernike, F., Jr., "Refractive indices of ammonium dihydrogen phosphate and potassium dihydrogen phosphate between 2000 Å and 1.5 μ," *J. Opt. Soc. Am.*, vol. 54, p. 1215, 1964.

[9] Geusic, J. E., H. J. Levinstein, S. Singh, R. G. Smith, and L. G. Van Uitert, "Continuous 0.53-μm solid-state source using $Ba_2NaNb_5O_{15}$," *IEEE J. Quantum Electron.*, vol. QE-4, p. 352, 1968.

[10] Ashkin, A., G. D. Boyd, and J. M. Dziedzic, "Observation of continuous second harmonic generation with gas lasers," *Phys. Rev. Letters*, vol. 11, p. 14, 1963.

[11] Yariv, A., *Quantum Electronics,* 2d Ed. New York: Wiley, 1975.

[12] Lord, Rayleigh, "On maintained vibrations," *Phil. Mag.*, vol. 15, ser. 5, pt. I, p. 229, 1883.

[13] Parametric amplification was first demonstrated by C. C. Wang and G. W. Racette, "Measurement of parametric gain accompanying optical difference frequency generation," *Appl. Phys. Letters*, vol. 6, p. 169, 1965.

[14] The first demonstration of optical parametric oscillation is that of J. A. Giordmaine and R. C. Miller, "Tunable optical parametric oscillation in $LiNbO_3$ at optical frequencies," *Phys. Rev. Letters*, vol. 14, p. 973, 1965.

[15] Some of the early theoretical analyses of optical parametric oscillation are attributable to R. H. Kingston, "Parametric amplification and oscillation at optical frequencies," *Proc. IRE*, vol. 50, p. 472, 1962, and N. M. Kroll, "Parametric amplification in spatially extended media and applications to the design of tunable oscillators at optical frequencies," *Phys. Rev.*, vol. 127, p. 1207, 1962.

[16] Magde, D., and H. Mahr, "Study in ammonium dihydrogen phosphate of spontaneous parametric interaction tunable from 4400 to 16000 Å," *Phys. Rev. Letters*, vol. 18, p. 905, 1967.

[17] Yariv, A., and W. H. Louisell, "Theory of the optical parametric oscillator," *IEEE J. Quantum Electron.*, vol. QE-2, p. 418, 1966.

[18] Bjorkholm, J. E., "Efficient optical parametric oscillation using doubly and singly resonant cavities," *Appl. Phys. Letters*, vol. 13, p. 53, 1968.

[19] Kreuzer, L. B., "High-efficiency optical parametric oscillation and power limiting in $LiNbO_3$," *Appl. Phys. Letters*, vol. 13, p. 57, 1968.

[20] Siegman, A. E., "Nonlinear optical effects: An optical power limiter," *Appl. Opt.*, vol. 1, p. 739, 1962.

[21] Louisell, W. H., A. Yariv, and A. E. Siegman, "Quantum fluctuations and noise in parametric processes," *Phys. Rev.*, vol. 124, p. 1646, 1961.

[22] Johnson, F. M., and J. A. Durado, "Frequency up-conversion," *Laser Focus*, vol. 3, p. 31, 1967.

[23] Midwinter, J. E., and J. Warner, "Up-conversion of near infrared to visible radiation in lithium-meta-niobate," *J. Appl. Phys.*, vol. 38, p. 519, 1967.

[24] Warner, J., "Photomultiplier detection of 10.6 μ radiation using optical up-conversion in proustite," *Appl. Phys. Letters*, vol. 12, p. 222, 1968.

[25] Hulme, K. F., O. Jones, P. H. Davies, and M. V. Hobden, "Synthetic proustite (Ag_3AsS_3): A new material for optical mixing," *Appl. Phys. Letters*, vol. 10, p. 133, 1967.

CHAPTER

9

Electrooptic Modulation of Laser Beams

9.0 Introduction

In Chapter 1 we treated the propagation of electromagnetic waves in anisotropic crystal media. It was shown how the properties of the propagating wave can be determined from the index ellipsoid surface.

In this chapter we consider the problem of propagation of optical radiation in crystals in the presence of an applied electric field. We find that in certain types of crystals it is possible to effect a change in the index of refraction which is proportional to the field. This is the linear electrooptic effect. It affords a convenient and widely used means of controlling the intensity or phase of the propagating radiation. This modulation is used in an ever expanding number of applications including: the impression of information onto optical beams, Q-switching of lasers (Sec. 6.7) for generation of giant optical pulses, mode locking, and optical beam deflection. Some of these applications will be discussed further in this chapter. Modulation and deflection of laser beams by acoustic beams are considered in Chapter 12.

9.1 Electrooptic Effect

In Chapter 1 we found that, given a direction in a crystal, in general two possible linearly polarized modes exist; the so-called rays of propagation. Each mode possesses a unique direction of polarization (that is, direction of **D**) and a corresponding index of refraction (that is, a velocity of propagation). The mutually orthogonal polarization directions and the indices of the two rays are found most easily by using the index ellipsoid

$$\frac{x^2}{n_x^2} + \frac{y^2}{n_y^2} + \frac{z^2}{n_z^2} = 1 \tag{9.1-1}$$

where the directions x, y, and z are the principal dielectric axes—that is, the directions in the crystal along which **D** and **E** are parallel. The existence of two "ordinary" and "extraordinary" rays with different indices of refraction is called birefringence.

The linear electrooptic effect is the change in the indices of the ordinary and extraordinary rays that is caused by and is proportional to an applied electric field. This effect exists only in crystals that do not possess inversion symmetry.[1] This statement can be justified as follows: Assume that in a crystal possessing an inversion symmetry, the application of an electric field E along some direction causes a change $\Delta n_1 = sE$ in the index, where s is a constant characterizing the linear electrooptic effect. If the direction of the field is reversed, the change in the index is given by $\Delta n_2 = s(-E)$, but because of the inversion symmetry the two directions are physically equivalent, so $\Delta n_1 = \Delta n_2$. This requires that $s = -s$, which is possible only for $s = 0$, so no linear electrooptic effect can exist. The division of all crystal classes into those that do and those that do not possess an inversion symmetry is an elementary consideration in crystallography and this information is widely tabulated [1].

Since the propagation characteristics in crystals are fully described by means of the index ellipsoid (9.1-1), the effect of an electric field on the propagation is expressed most conveniently by giving the changes in the constants $1/n_x^2$, $1/n_y^2$, $1/n_z^2$ of the index ellipsoid.

Following convention [2], we take the equation of the index ellipsoid in the presence of an electric field as

$$\left(\frac{1}{n^2}\right)_1 x^2 + \left(\frac{1}{n^2}\right)_2 y^2 + \left(\frac{1}{n^2}\right)_3 z^2 + 2\left(\frac{1}{n^2}\right)_4 yz + 2\left(\frac{1}{n^2}\right)_5 xz + 2\left(\frac{1}{n^2}\right)_6 xy = 1 \tag{9.1-2}$$

[1] If a crystal contains points such that inversion (replacing each atom at **r** by one at $-$**r**, with **r** being the position vector relative to the point) about any one of these points leaves the crystal structure invariant, the crystal is said to possess inversion symmetry.

If we choose x, y, and z to be parallel to the principal dielectric axes of the crystal, then with zero applied field, Equation (9.1-2) must reduce to (9.1-1); therefore,

$$\left(\frac{1}{n^2}\right)_1\bigg|_{E=0} = \frac{1}{n_x^2} \qquad \left(\frac{1}{n^2}\right)_2\bigg|_{E=0} = \frac{1}{n_y^2}$$

$$\left(\frac{1}{n^2}\right)_3\bigg|_{E=0} = \frac{1}{n_z^2} \qquad \left(\frac{1}{n^2}\right)_4\bigg|_{E=0} = \left(\frac{1}{n^2}\right)_5\bigg|_{E=0} = \left(\frac{1}{n^2}\right)_6\bigg|_{E=0} = 0$$

The linear change in the coefficients

$$\left(\frac{1}{n^2}\right)_i \qquad i = 1, \cdots, 6$$

due to an arbitrary electric field $\mathbf{E}(E_x, E_y, E_z)$ is defined by

$$\Delta\left(\frac{1}{n^2}\right)_i = \sum_{j=1}^{3} r_{ij}E_j \tag{9.1-3}$$

where in the summation over j we use the convention $1 = x$, $2 = y$, $3 = z$. Equation (9.1-3) can be expressed in a matrix form as

$$\begin{vmatrix} \Delta\left(\frac{1}{n^2}\right)_1 \\ \Delta\left(\frac{1}{n^2}\right)_2 \\ \Delta\left(\frac{1}{n^2}\right)_3 \\ \Delta\left(\frac{1}{n^2}\right)_4 \\ \Delta\left(\frac{1}{n^2}\right)_5 \\ \Delta\left(\frac{1}{n^2}\right)_6 \end{vmatrix} = \begin{vmatrix} r_{11} & r_{12} & r_{13} \\ r_{21} & r_{22} & r_{23} \\ r_{31} & r_{32} & r_{33} \\ r_{41} & r_{42} & r_{43} \\ r_{51} & r_{52} & r_{53} \\ r_{61} & r_{62} & r_{63} \end{vmatrix} \begin{vmatrix} E_1 \\ E_2 \\ E_3 \end{vmatrix} \tag{9.1-4}$$

where, using the rules for matrix multiplication, we have, for example,

$$\Delta\left(\frac{1}{n^2}\right)_6 = r_{61}E_1 + r_{62}E_2 + r_{63}E_3$$

The 6×3 matrix with elements r_{ij} is called the electrooptic tensor. We have shown in Section 9.1 that in crystals possessing an inversion symmetry (centrosymmetric), $r_{ij} = 0$. The form, but not the magnitude, of the tensor r_{ij} can be derived from symmetry considerations [1], which dictate which of the 18 r_{ij} coefficients are zero, as well as the relationships that exist between the remaining coefficients. In Table 9-1 we give the form of the electrooptic tensor for all the noncentrosymmetric crystal classes. The electrooptic coefficients of some crystals are given in Table 9-2.

Table 9-1 THE FORM OF THE ELECTROOPTIC TENSOR FOR ALL CRYSTAL SYMMETRY CLASSES

Symbols:

- • zero element
- ● nonzero element
- ●—● equal nonzero elements
- ●—○ equal nonzero elements, but opposite in sign

The symbol at the upper left corner of each tensor is the conventional symmetry group designation.

Centrosymmetric—All elements zero

Triclinic

$$\begin{pmatrix} \bullet & \bullet & \bullet \\ \bullet & \bullet & \bullet \\ \bullet & \bullet & \bullet \\ \bullet & \bullet & \bullet \\ \bullet & \bullet & \bullet \\ \bullet & \bullet & \bullet \end{pmatrix}$$

Monoclinic

2 (parallel to x_2)

$$\begin{pmatrix} \cdot & \bullet & \cdot \\ \cdot & \bullet & \cdot \\ \cdot & \bullet & \cdot \\ \bullet & \cdot & \bullet \\ \cdot & \bullet & \cdot \\ \bullet & \cdot & \bullet \end{pmatrix}$$

(parallel to x_3)

$$\begin{pmatrix} \cdot & \cdot & \bullet \\ \cdot & \cdot & \bullet \\ \cdot & \cdot & \bullet \\ \bullet & \bullet & \cdot \\ \bullet & \bullet & \cdot \\ \cdot & \cdot & \bullet \end{pmatrix}$$

m (perpendicular to x_2)

$$\begin{pmatrix} \bullet & \cdot & \bullet \\ \bullet & \cdot & \bullet \\ \bullet & \cdot & \bullet \\ \cdot & \bullet & \cdot \\ \bullet & \cdot & \bullet \\ \cdot & \bullet & \cdot \end{pmatrix}$$

(perpendicular to x_3)

$$\begin{pmatrix} \bullet & \bullet & \cdot \\ \bullet & \bullet & \cdot \\ \bullet & \bullet & \cdot \\ \cdot & \cdot & \bullet \\ \cdot & \cdot & \bullet \\ \bullet & \bullet & \cdot \end{pmatrix}$$

Orthorhombic

222

$$\begin{pmatrix} \cdot & \cdot & \cdot \\ \cdot & \cdot & \cdot \\ \cdot & \cdot & \cdot \\ \bullet & \cdot & \cdot \\ \cdot & \bullet & \cdot \\ \cdot & \cdot & \bullet \end{pmatrix}$$

$mm2$

$$\begin{pmatrix} \cdot & \cdot & \bullet \\ \cdot & \cdot & \bullet \\ \cdot & \cdot & \bullet \\ \cdot & \bullet & \cdot \\ \bullet & \cdot & \cdot \\ \cdot & \cdot & \cdot \end{pmatrix}$$

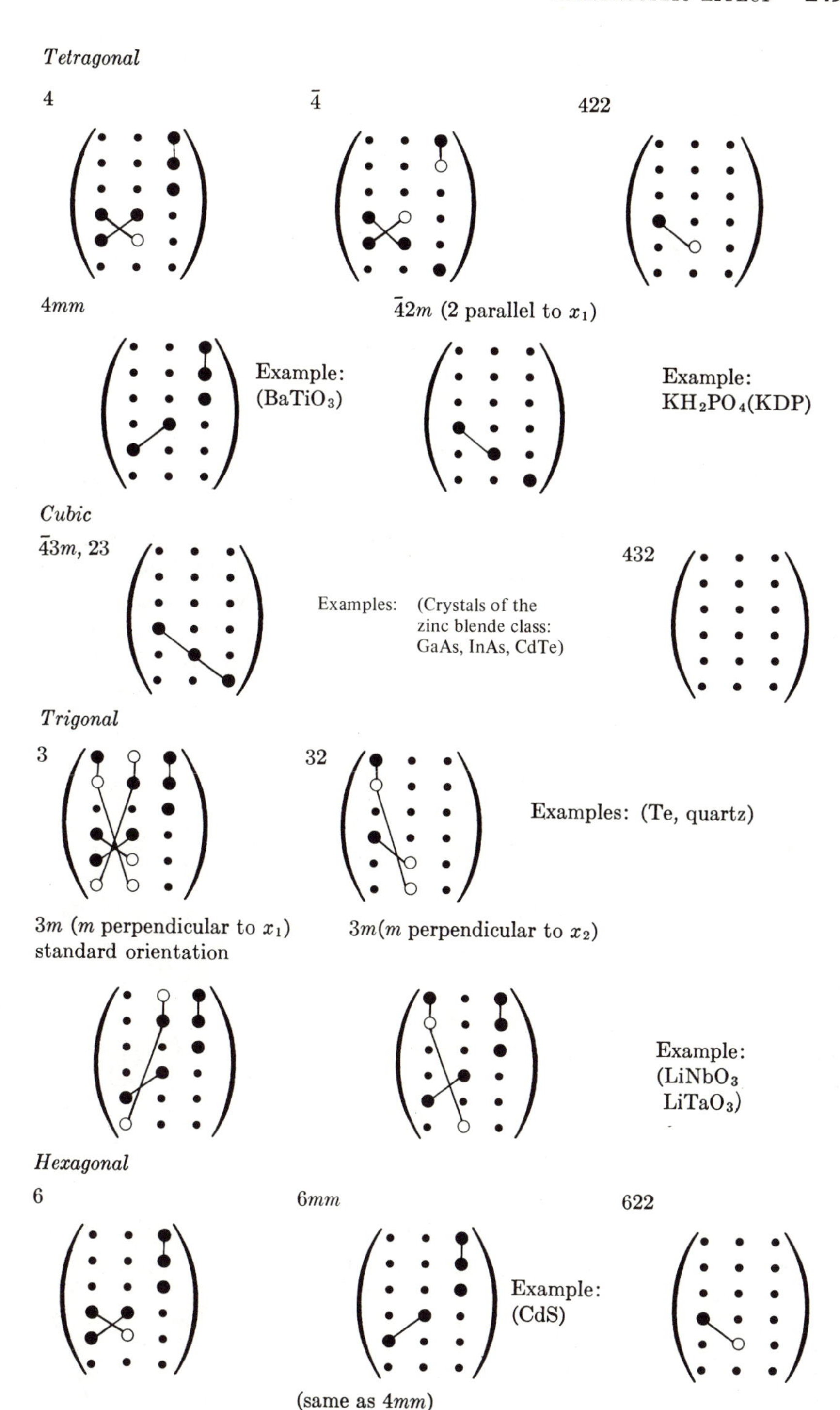

Tetragonal
4
$\bar{4}$
422
4mm
Example: (BaTiO3)
$\bar{4}2m$ (2 parallel to x_1)
Example: KH2PO4(KDP)
Cubic
$\bar{4}3m$, 23
Examples: (Crystals of the zinc blende class: GaAs, InAs, CdTe)
432
Trigonal
3
32
Examples: (Te, quartz)
3m (m perpendicular to x_1) standard orientation
3m(m perpendicular to x_2)
Example: (LiNbO3 LiTaO3)
Hexagonal
6
6mm
Example: (CdS)
622
(same as 4mm)

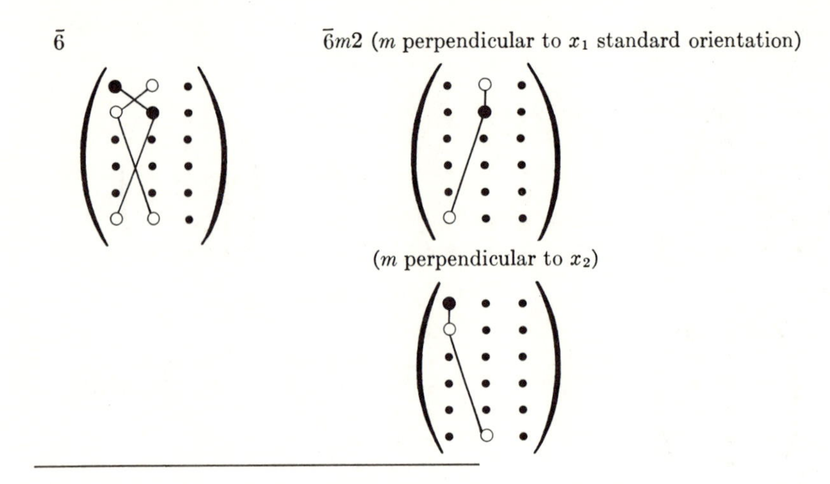

Example: the electrooptic effect in KH_2PO_4. Consider the specific example of a crystal of potassium dihydrogen phosphate (KH_2PO_4) also known as KDP. The crystal has a fourfold axis of symmetry,[2] which by strict convention is taken as the z (optic) axis, as well as two mutually orthogonal twofold axes of symmetry that lie in the plane normal to z. These are designated as the x and y axes. The symmetry group[3] of this crystal is $\bar{4}2m$. Using Table 9-1, we take the electrooptic tensor in the form of

$$r_{ij} = \left| \begin{matrix} 0 & 0 & 0 \\ 0 & 0 & 0 \\ 0 & 0 & 0 \\ r_{41} & 0 & 0 \\ 0 & r_{41} & 0 \\ 0 & 0 & r_{63} \end{matrix} \right| \tag{9.1-5}$$

so the only nonvanishing elements are $r_{41} = r_{52}$ and r_{63}. Using (9.1-1), (9.1-4), and (9.1-5), we obtain the equation of the index ellipsoid in the presence of a field $\mathbf{E}(E_x, E_y, E_z)$ as

$$\frac{x^2}{n_o^2} + \frac{y^2}{n_o^2} + \frac{z^2}{n_e^2} + 2r_{41}E_x yz + 2r_{41}E_y xz + 2r_{63}E_z xy = 1 \tag{9.1-6}$$

where the constants involved in the first three terms do not depend on the field and, since the crystal is uniaxial, are taken as $n_x = n_y = n_o$, $n_z = n_e$.

---

[2] That is, a rotation by $2\pi/4$ about this axis leaves the crystal structure invariant.

[3] The significance of the symmetry group symbols and a listing of most known crystals and their symmetry groups is to be found in any basic book on crystallography.

Table 9-2 SOME ELECTROOPTIC MATERIALS AND THEIR PROPERTIES [2]

MATERIAL	ROOM TEMPERATURE ELECTROOPTIC COEFFICIENTS IN UNITS OF 10^{-12} m/V	INDEX OF REFRACTION*	$n_0^3 r$, IN UNITS OF 10^{-12} m/V	ϵ/ϵ_0 (ROOM TEMPERATURE)	POINT-GROUP SYMMETRY
KDP (KH_2PO_4)	$r_{41} = 8.6$ $r_{63} = 10.6$	$n_o = 1.51$ $n_e = 1.47$	29 34	$\epsilon \parallel c = 20$ $\epsilon \perp c = 45$	$\bar{4}2m$
KD_2PO_4	$r_{63} = 23.6$	~ 1.50	80	$\epsilon \parallel c \sim 50$ at 24°C	$\bar{4}2m$
ADP ($NH_4H_2PO_4$)	$r_{41} = 28$ $r_{63} = 8.5$	$n_o = 1.52$ $n_e = 1.48$	95 27	$\epsilon \parallel c = 12$	$\bar{4}2m$
Quartz	$r_{41} = 0.2$ $r_{63} = 0.93$	$n_o = 1.54$ $n_e = 1.55$	0.7 3.4	$\epsilon \parallel c \sim 4.3$ $\epsilon \perp c \sim 4.3$	32
CuCl	$r_{41} = 6.1$	$n_o = 1.97$	47	7.5	$\bar{4}3m$
ZnS	$r_{41} = 2.0$	$n_o = 2.37$	27	~ 10	$\bar{4}3m$
GaAs at 10.6 μm	$r_{41} = 1.6$	$n_o = 3.34$	59	11.5	$\bar{4}3m$
ZnTe at 10.6 μm CdTe at 10.6 μm	$r_{41} = 3.9$ $r_{41} = 6.8$	$n_o = 2.79$ $n_o = 2.6$	85 120	7.3	$\bar{4}3m$ $\bar{4}3m$
$LiNbO_3$	$r_{33} = 30.8$ $r_{13} = 8.6$ $r_{22} = 3.4$ $r_{42} = 28$	$n_o = 2.29$ $n_e = 2.20$	$n_e^3 r_{33} = 328$ $n_o^3 r_{22} = 37$ $\frac{1}{2}(n_e^3 r_{33} - n_o^3 r_{13}) = 112$	$\epsilon \perp c = 98$ $\epsilon \parallel c = 50$	$3m$
GaP	$r_{41} = 0.97$	$n_o = 3.31$	$n_o^3 r_{41} = 29$		$\bar{4}3m$
$LiTaO_3$ (30°C)	$r_{33} = 30.3$ $r_{13} = 5.7$	$n_o = 2.175$ $n_e = 2.180$	$n_e^3 r_{33} = 314$	$\epsilon \parallel c = 43$	$3m$
$BaTiO_3$ (30°C)	$r_{33} = 23$ $r_{13} = 8.0$ $r_{42} = 820$	$n_o = 2.437$ $n_e = 2.365$	$n_e^3 r_{33} = 334$	$\epsilon \perp c = 4300$ $\epsilon \parallel c = 106$	$4mm$

* Typical value near 5500 Å

We thus find that the application of an electric field causes the appearance of "mixed" terms in the equation of the index ellipsoid. These are the terms with xy, xz, and yz. This means that the major axes of the ellipsoid, with a field applied, are no longer parallel to the x, y, and z axes. It becomes necessary, then, to find the directions and magnitudes of the new axes, in the presence of **E**, so that we may determine the effect of the field on the propagation. To be specific we choose the direction of the applied field parallel to the z axis, so (9.1-6) becomes

$$\frac{x^2 + y^2}{n_o^2} + \frac{z^2}{n_e^2} + 2r_{63}E_z xy = 1 \tag{9.1-7}$$

The problem is one of finding a new coordinate system—x', y', z'—in which the equation of the ellipsoid (9.1-7) contains no mixed terms; that is, it is of the form

$$\frac{x'^2}{n_{x'}^2} + \frac{y'^2}{n_{y'}^2} + \frac{z'^2}{n_{z'}^2} = 1 \tag{9.1-8}$$

x', y', and z' are then the directions of the major axes of the ellipsoid in the presence of an external field applied parallel to z. The length of the major axes of the ellipsoid is, according to (9.1-8), $2n_{x'}$, $2n_{y'}$, and $2n_{z'}$ and these will, in general, depend on the applied field.

In the case of (9.1-7) it is clear from inspection that in order to put it in a diagonal form we need to choose a coordinate system x', y', z', where z' is parallel to z, and because of the symmetry of (9.1-7) in x and y, x' and y' are related to x and y by a 45° rotation, as shown in Figure 9-1. The transformation relations from x, y to x', y' are thus

$$x = x' \cos 45° - y' \sin 45°$$
$$y = x' \sin 45° + y' \cos 45°$$

which, upon substitution in (9.1-7), yield

$$\left(\frac{1}{n_o^2} + r_{63}E_z\right) x'^2 + \left(\frac{1}{n_o^2} - r_{63}E_z\right) y'^2 + \frac{z^2}{n_e^2} = 1 \tag{9.1-9}$$

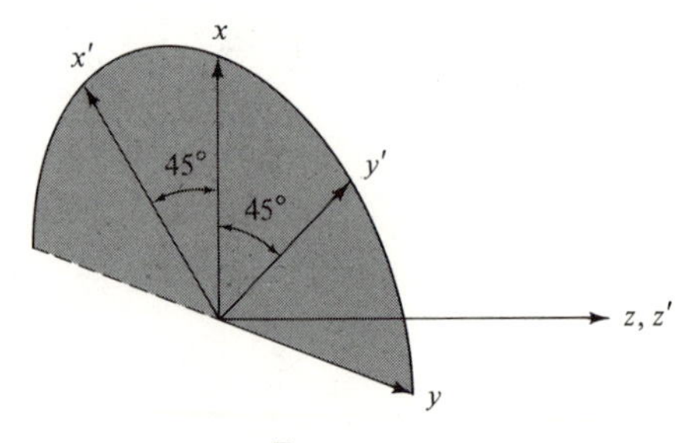

Figure 9-1 The x, y, and z axes of $\bar{4}2m$ crystals (such as KH_2PO_4) and the x', y', and z' axes, where z is the fourfold optic axis and x and y are the twofold axes of crystals with $\bar{4}2m$ symmetry.

Equation (9.1-9) shows that x', y', and z are indeed the principal axes of the ellipsoid when a field is applied along the z direction. According to (9.1-9), the length of the x' axis of the ellipsoid is $2n_{x'}$, where

$$\frac{1}{n_{x'}^2} = \frac{1}{n_o^2} + r_{63}E_z$$

which, assuming $r_{63}E_z \ll n_o^{-2}$ and using the differential relation $dn = -(n^3/2)\,d(1/n^2)$, gives

$$n_{x'} = n_o - \frac{n_o^3}{2} r_{63}E_z \tag{9.1-10}$$

and, similarly,

$$n_{y'} = n_o + \frac{n_o^3}{2} r_{63}E_z \tag{9.1-11}$$

$$n_z = n_e \tag{9.1-12}$$

9.2 Electrooptic Retardation

The index ellipsoid for KDP with **E** applied parallel to z is shown in Figure 9-2. If we consider propagation along the z direction, then according to the procedure described in Section 1.4 we need to determine the ellipse formed by the intersection of the plane $z = 0$ (in general, the plane that contains the origin and is normal to the propagation direction) and the ellipsoid. The equation of this ellipse is obtained from (9.1-9) by putting $z = 0$ and is

$$\left(\frac{1}{n_o^2} + r_{63}E_z\right)x'^2 + \left(\frac{1}{n_o^2} - r_{63}E_z\right)y'^2 = 1 \tag{9.2-1}$$

One quadrant of the ellipse is shown shaded in Figure 9-2, along with its minor and major axes, which in this case coincide with x' and y', respec-

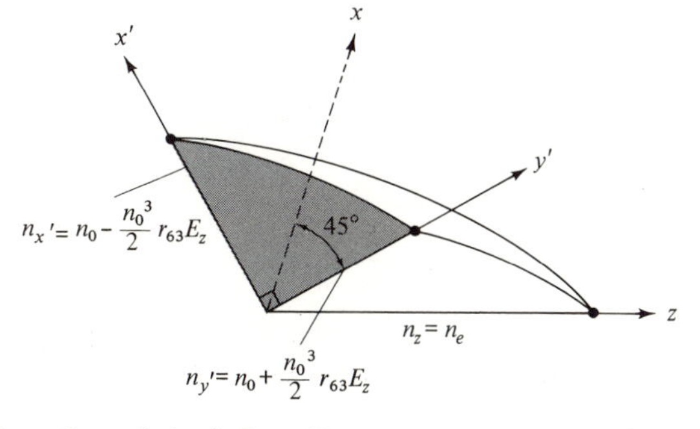

Figure 9-2 A section of the index ellipsoid of KDP, showing the principal dielectric axes x', y', and z due to an electric field applied along the z axis. The directions x' and y' are defined by Figure 9.1.

tively. It follows from Section 1.4 that the two allowed directions of polarization are x' and y' and that their indices of refraction are $n_{x'}$ and $n_{y'}$, which are given by (9.1-10) and (9.1-11).

We are now in a position to take up the concept of retardation. We consider an optical field which is incident normally on the $x'y'$ plane with its **E** vector along the x direction. We can resolve the optical field at $z = 0$ (input plane) into two mutually orthogonal components polarized along x' and y'. The x' component propagates as

$$e_{x'} = Ae^{i[\omega t-(\omega/c)n_{x'}z]}$$

which, using (9.1-10), becomes

$$e_{x'} = Ae^{i\{\omega t-(\omega/c)[n_o-(n_o{}^3/2)r_{63}E_z]z\}} \tag{9.2-2}$$

while the y' component is given by

$$e_{y'} = Ae^{i\{\omega t-(\omega/c_0)[n_o+(n_o{}^3/2)r_{63}E_z]z\}} \tag{9.2-3}$$

The phase difference at the output plane $z = l$ between the two components is called the *retardation*. It is given by the difference of the exponents in (9.2-2) and (9.2-3) and is equal to

$$\Gamma = \phi_{x'} - \phi_{y'} = \frac{\omega n_o{}^3 r_{63} V}{c} \tag{9.2-4}$$

where $V = E_z l$ and $\phi_{x'} = (\omega n_{x'}/c)l$.

Figure 9-3 shows $E_{x'}(z)$ and $E_{y'}(z)$ at some moment in time. Also shown are the curves traversed by the tip of the optical field vector at various points along the path. At $z = 0$, the retardation is $\Gamma = 0$ and the field is linearly polarized along x. At point e, $\Gamma = \pi/2$; thus, omitting a common phase factor, we have

$$\begin{aligned} e_{x'} &= A \cos \omega t \\ e_{y'} &= A \cos\left(\omega t - \frac{\pi}{2}\right) = A \sin \omega t \end{aligned} \tag{9.2-5}$$

and the electric field vector is circularly polarized in the clockwise sense as shown in the figure. At point i, $\Gamma = \pi$ and thus

$$\begin{aligned} e_{x'} &= A \cos \omega t \\ e_{y'} &= A \cos(\omega t - \pi) = -A \cos \omega t \end{aligned}$$

and the radiation is again linearly polarized, but this time along the y direction—that is, at 90° to its input direction of polarization.

The retardation as given by (9.2-4) can also be written as

$$\Gamma = \pi \frac{E_z l}{V_\pi} = \pi \frac{V}{V_\pi} \tag{9.2-6}$$

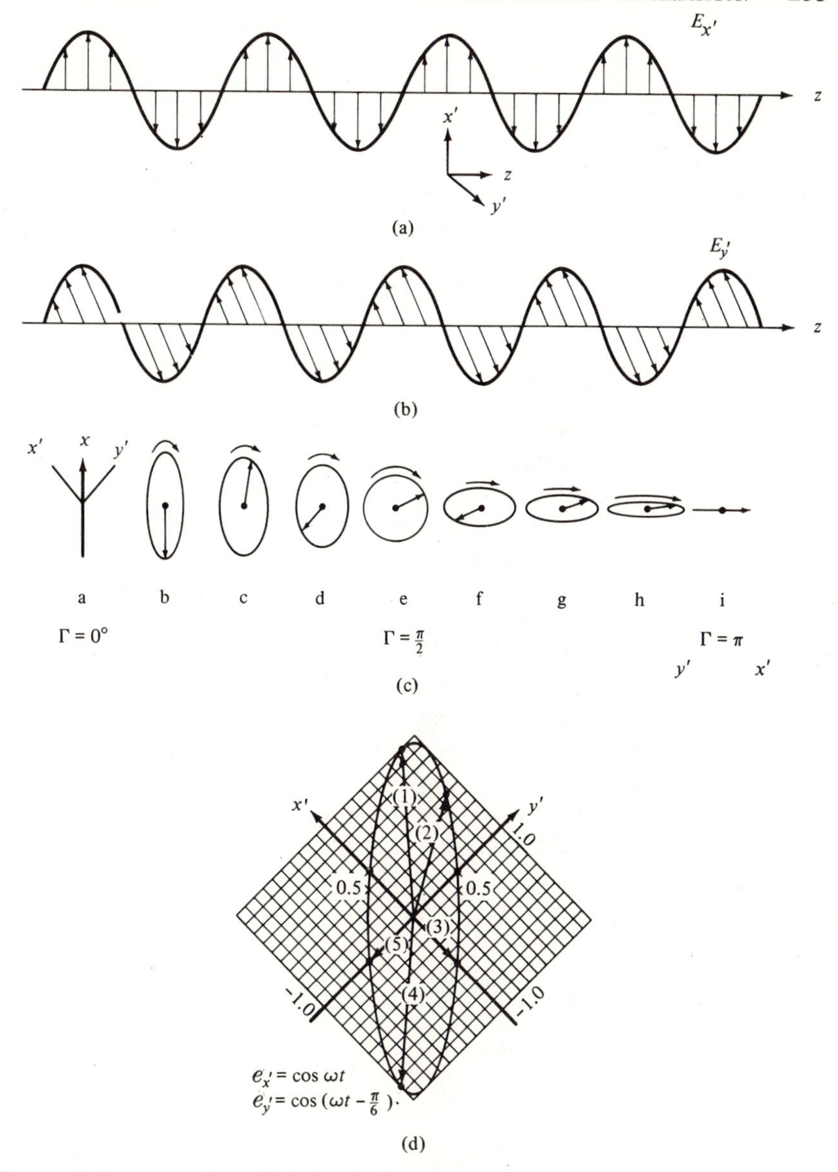

Figure 9-3 An optical field that is linearly polarized along x is incident on an electrooptic crystal having its electrically induced principal axes along x' and y'. (This is the case in KH_2PO_4 when an electric field is applied along its z axis.) (**a**) The component $e_{x'}$ at some time t as a function of the position z along the crystal. (**b**) $e_{y'}$ as a function of z at the same value of t as in (**a**). (**c**) The ellipses in the x'-y' plane traversed by the tip of the optical electric field at various points (a through i) along the crystal during one optical cycle. The arrow shows the instantaneous field vector at time t, while the curved arrow gives the sense in which the ellipse is traversed. (**d**) A plot of the polarization ellipse due to two orthogonal components $e_{x'} = \cos \omega t$ and $e_{y'} = \cos (\omega t - \pi/6)$. Also shown are the instantaneous field vectors at (1) $\omega t = 0^0$, (2) $\omega t = 60°$, (3) $\omega t = 120°$, (4) $\omega t = 210°$, and (5) $\omega t = 270°$.

where V_π, the voltage yielding a retardation $\Gamma = \pi$,[4] is

$$V_\pi = \frac{\lambda}{2n_o^3 r_{63}} \tag{9.2-7}$$

where $\lambda = 2\pi c/\omega$ is the free space wavelength. Using, as an example, the value of r_{63} for ADP as given in Table 9-2, we obtain from (9.2-7)

$$(V_\pi)_{\text{ADP}} = 10{,}000 \text{ volts} \quad \text{at} \quad \lambda = 0.5\ \mu\text{m}$$

9.3 Electrooptic Amplitude Modulation

An examination of Figure 9-3 reveals that the electrically induced birefringence causes a wave launched at $z = 0$ with its polarization along x to acquire a y polarization,. which grows with distance at the expense of the x component until at point i, at which $\Gamma = \pi$, the polarization becomes parallel to y. If point i corresponds to the output plane of the crystal and if one inserts at this point a polarizer at right angles to the input polarization—that is, one that allows only E_y to pass—then with the field on, the optical beam passes through unattenuated, whereas with the field off ($\Gamma = 0$), the output beam is blocked off completely by the crossed output polarizer. This control of the optical energy flow serves as the basis of the electrooptic amplitude modulation of light.

A typical arrangement of an electrooptic amplitude modulator is shown in Figure 9-4. It consists of an electrooptic crystal placed between two crossed polarizers, which, in turn, are at an angle of 45° with respect to

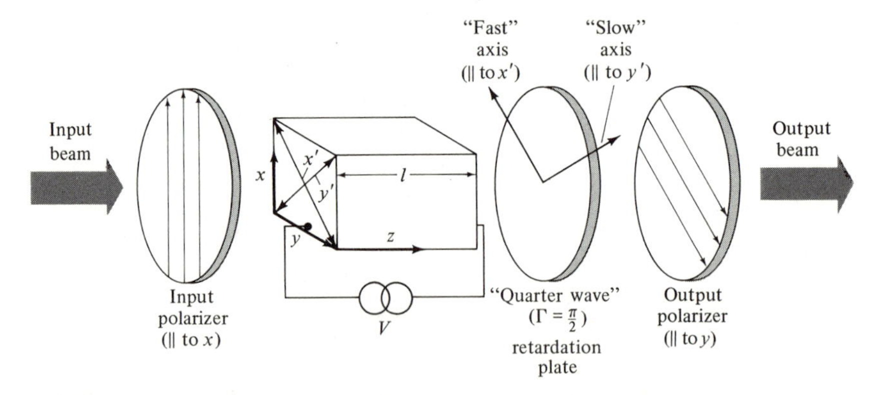

Figure 9-4 A typical electrooptic amplitude modulator. The total retardation Γ is the sum of the fixed retardation bias ($\Gamma_B = \pi/2$) introduced by the "quarter-wave" plate and that attributable to the electrooptic crystal.

[4] V_π is referred to as the "half-wave" voltage since, as can be seen in Figure 9.3c(i), it causes the two waves that are polarized along x' and y' to acquire a relative spatial displacement of $\Delta z = \lambda/2$, where λ is the optical wavelength.

the electrically induced birefringent axes x' and y'. To be specific, we show how this arrangement is achieved using a KDP crystal. Also included in the optical path is a naturally birefringent crystal that introduces a fixed retardation, so the total retardation Γ is the sum of the retardation due to this crystal and the electrically induced one. The incident field is parallel to x at the input face of the crystal, thus having equal-in-phase components along x' and y' which we take as

$$e_{x'} = A \cos \omega t$$
$$e_{y'} = A \cos \omega t$$

or, using the complex amplitude notation,

$$E_{x'}(0) = A$$
$$E_{y'}(0) = A$$

The incident intensity is thus[5]

$$I_i \propto \mathbf{E} \cdot \mathbf{E}^* = |E_{x'}(0)|^2 + |E_{y'}(0)|^2 = 2A^2 \tag{9.3-1}$$

Upon emerging from the output face $z = l$, the x' and y' components have acquired, according to (9.2-4), a relative phase shift (retardation) of Γ radians, so we may take them as

$$\begin{aligned} E_{x'}(l) &= A \\ E_{y'}(l) &= Ae^{-i\Gamma} \end{aligned} \tag{9.3-2}$$

The total (complex) field emerging from the output polarizer is the sum of the y components of $E_{x'}(l)$ and $E_{y'}(l)$

$$(E_y)_o = \frac{A}{\sqrt{2}} (e^{-i\Gamma} - 1) \tag{9.3-3}$$

which corresponds to an output intensity

$$\begin{aligned} I_o &\propto [(E_y)_o(E_y^*)_o] \\ &= \frac{A^2}{2} [(e^{-i\Gamma} - 1)(e^{i\Gamma} - 1)] = 2A^2 \sin^2 \frac{\Gamma}{2} \end{aligned}$$

where the proportionality constant is the same as in (9.3-1). The ratio of the output intensity to the input is thus

$$\frac{I_o}{I_i} = \sin^2 \frac{\Gamma}{2} = \sin^2 \left[\left(\frac{\pi}{2} \right) \frac{V}{V_\pi} \right] \tag{9.3-4}$$

[5] We recall here that the time average of the product of two harmonic fields Re $[Be^{i\omega t}]$ and Re $[Ce^{i\omega t}]$ is equal to $\frac{1}{2}$ Re $[BC^*]$.

The second equality in (9.3-4) was obtained from (9.2-6). The transmission factor (I_o/I_i) is plotted in Figure 9-5 against the applied voltage.

The process of amplitude modulation of an optical signal is also illustrated in Figure 9-5. The modulator is usually biased[6] with a fixed retardation $\Gamma = \pi/2$ to the 50 percent transmission point. A small sinuosidal modulation voltage would then cause a nearly sinusoidal modulation of the transmitted intensity as shown.

To treat the situation depicted by Figure 9-5 mathematically, we take

$$\Gamma = \frac{\pi}{2} + \Gamma_m \sin \omega_m t \tag{9.3-5}$$

where the retardation bias is taken as $\pi/2$, and Γ_m is related to the amplitude V_m of the modulation voltage $V_m \sin \omega_m t$ by (9.2-6); thus, $\Gamma_m = \pi(V_m/V_\pi)$.

Using (9.3-4) we obtain

$$\frac{I_o}{I_i} = \sin^2\left(\frac{\pi}{4} + \frac{\Gamma_m}{2} \sin \omega_m t\right) \tag{9.3-6}$$

$$= \tfrac{1}{2}[1 + \sin(\Gamma_m \sin \omega_m t)] \tag{9.3-7}$$

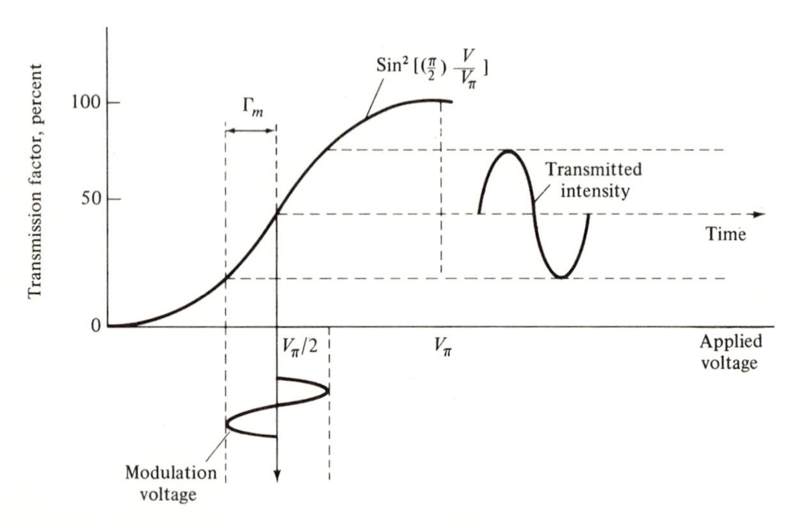

Figure 9-5 Transmission factor of a cross-polarized electrooptic modulator as a function of an applied voltage. The modulator is biased to the point $\Gamma = \pi/2$, which results in a 50 percent intensity transmission. A small applied sinusoidal voltage modulates the transmitted intensity about the bias point.

---

[6] This bias can be achieved by applying a voltage $V = V_\pi/2$ or, more conveniently, by using a naturally birefringent crystal as in Figure 9-4 to introduce a phase difference (retardation) of $\pi/2$ between the x' and y' components.

which, for $\Gamma_m \ll 1$, becomes

$$\frac{I_o}{I_i} \simeq \frac{1}{2}[1 + \Gamma_m \sin \omega_m t] \tag{9.3-8}$$

so that the intensity modulation is a linear replica of the modulating voltage $V_m \sin \omega_m t$. If the condition $\Gamma_m \ll 1$ is not fulfilled, it follows from Figure 9-5 or from (9.3-7) that the intensity variation is distorted and will contain an appreciable amount of the higher (odd) harmonics. The dependence of the distortion on Γ_m is discussed further in Problem 9.3.

In Figure 9-6 we show how some information signal $f(t)$ (the electric output of a phonograph stylus in this case) can be impressed electrooptically as an amplitude modulation on a laser beam and subsequently be recovered by an optical detector. The details of the optical detection are considered in Chapter 11.

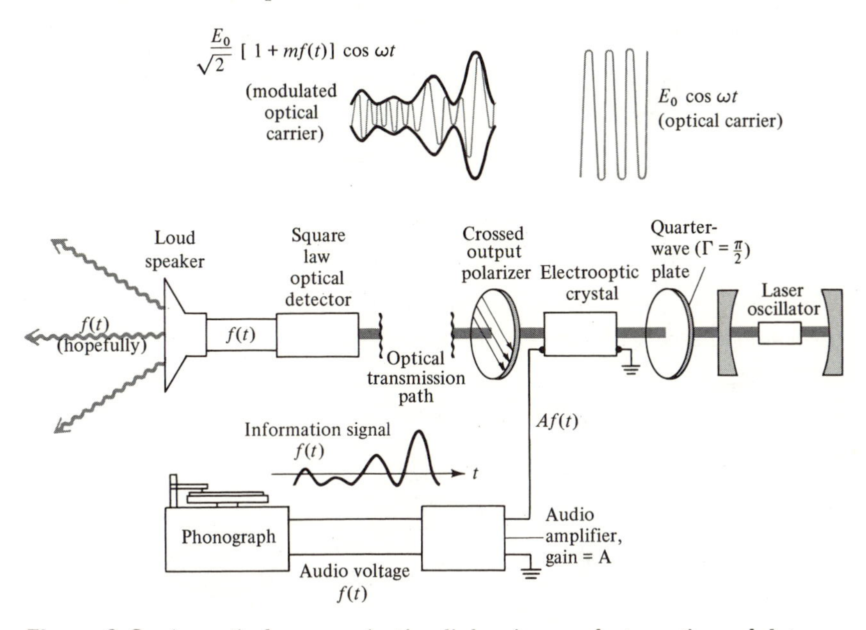

Figure 9-6 An optical communication link using an electrooptic modulator.

9.4 Phase Modulation of Light

In the preceding section we saw how the modulation of the state of polarization, from linear to elliptic, of an optical beam by means of the electrooptic effect can be converted, using polarizers, to intensity modulation. Here we consider the situation depicted by Figure 9-7, in which, instead of there being equal components along the induced birefringent axes (x' and y' in Figure 9-4), the incident beam is polarized parallel to one of them, x'

say. In this case the application of the electric field does not change the state of polarization, but merely changes the output phase by

$$\Delta\phi'_x = -\frac{\omega l}{c}\,\Delta n_{x'}$$

where, from (9.1-10),

$$\Delta\phi'_x = -\frac{\omega n_o^3 r_{63}}{2c} E_z l \tag{9.4-1}$$

If the bias field is sinusoidal and is taken as

$$E_z = E_m \sin \omega_m t \tag{9.4-2}$$

then an incident optical field which, at the input ($z = 0$) face of the crystal varies as $e_{\text{in}} = A \cos \omega t$, will emerge according to (9.2-2) as

$$e_{\text{out}} = A \cos\left[\omega t - \frac{\omega}{c}\left(n_o - \frac{n_o^3}{2} r_{63} E_m \sin \omega_m t\right) l\right]$$

where l is the length of the crystal. Dropping the constant phase factor, which is of no consequence here, we rewrite the last equation as

$$e_{\text{out}} = A \cos\left[\omega t + \delta \sin \omega_m t\right] \tag{9.4-3}$$

where

$$\delta = \frac{\omega n_o^3 r_{63} E_m l}{2c} = \frac{\pi n_o^3 r_{63} E_m l}{\lambda} \tag{9.4-4}$$

is referred to as the phase modulation index. The optical field is thus phase-modulated with a modulation index δ. Is we use the Bessel function identities

$$\cos(\delta \sin \omega_m t) = J_0(\delta) + 2J_2(\delta)\cos 2\omega_m t + 2J_4(\delta)\cos 4\omega_m t + \cdots$$

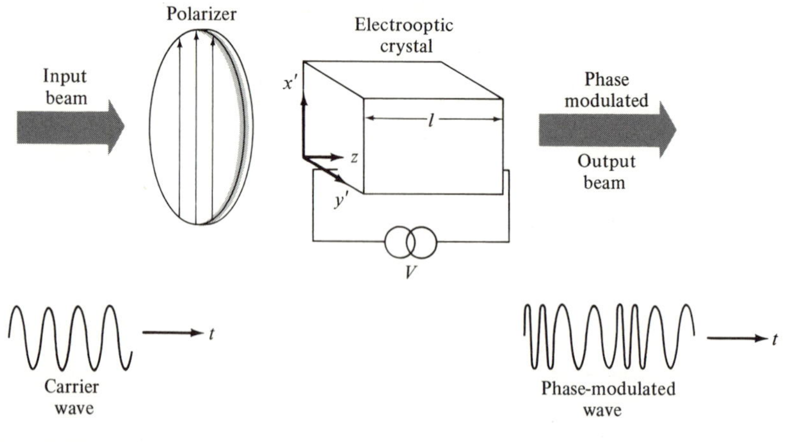

Figure 9-7 An electrooptic phase modulator. The crystal orientation and applied directions are appropriate to KDP. The optical polarization is parallel to an electrically induced principal dielectric axis (x').

and

$$\sin(\delta \sin \omega_m t) = 2J_1(\delta) \sin \omega_m t + 2J_3(\delta) \sin 3\omega_m t + \cdots \tag{9.4-5}$$

we can rewrite (9.4-3) as

$$\begin{aligned} e_{\text{out}} = A[&J_0(\delta) \cos \omega t + J_1(\delta) \cos (\omega + \omega_m)t \\ &- J_1(\delta) \cos (\omega - \omega_m)t + J_2(\delta) \cos (\omega + 2\omega_m)t \\ &+ J_2(\delta) \cos (\omega - 2\omega_m)t + J_3(\delta) \cos (\omega + 3\omega_m)t \\ &- J_3(\delta) \cos (\omega - 3\omega_m)t + J_4(\delta) \cos (\omega + 4\omega_m)t \\ &- J_4(\delta) \cos (\omega - 4\omega_m)t + \cdots] \end{aligned} \tag{9.4-6}$$

which form gives the distribution of energy in the sidebands as a function of the modulation index δ. We note that, for $\delta = 0$, $J_0(0) = 1$ and $J_n(\delta) = 0$, $n \neq 0$. Another point of interest is that the phase modulation index δ as given by (9.4-4) is one half the retardation Γ as given by (9.2-4).

9.5 Transverse Electrooptic Modulators

In the examples of electrooptic retardation discussed in the two preceding sections, the electric field was applied along the direction of light propagation. This is the so-called longitudinal mode of modulation. A more desirable mode of operation is the transverse one, in which the field is applied normal to the direction of propagation. The reason is that in this case the field electrodes do not interfere with the optical beam, and the retardation, being proportional to the product of the field times the crystal length, can be increased by the use of longer crystals. In the longitudinal case the retardation, according to (9.2-4), is proportional to $E_z l = V$ and is independent of the crystal length l. Figures 9-1 and 9-2 suggest how transverse retardation can be obtained using a KDP crystal with the actual arrangement shown in Figure 9-8. The light propagates along y' and its polarization is in the x'–z plane at 45° from the z axis. The retardation, with a field applied along z, is, from (9.1-10) and (9.1-12),

$$\begin{aligned} \Gamma &= \phi_z - \phi_{x'} \\ &= \frac{\omega l}{c}\left[(n_o - n_e) - \frac{n_o^3}{2} r_{63} \left(\frac{V}{d}\right)\right] \end{aligned} \tag{9.5-1}$$

where d is the crystal dimension along the direction of the applied field. We note that Γ contains a term that does not depend on the applied voltage. This point will be discussed in Problem 9-2. A detailed example of transverse electrooptic modulation using $\bar{4}3m$, cubic zinc-blende type crystals is given in Appendix C.

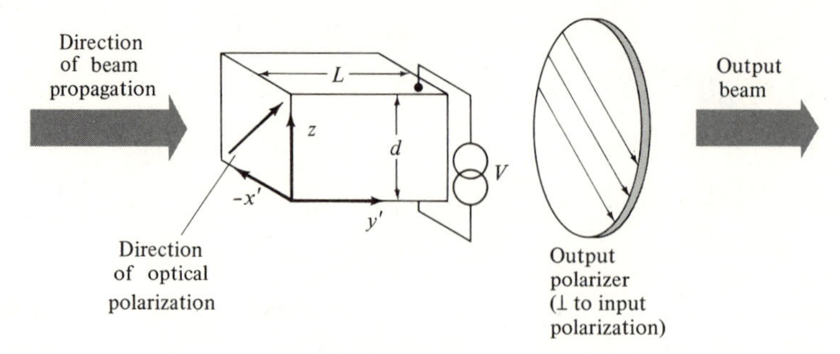

Figure 9-8 A transverse electrooptic amplitude modulator using a KH_2PO_4 (KDP) crystal in which the field is applied normal to the direction of propagation.

9.6 High-Frequency Modulation Considerations

In the examples considered in the three preceding sections, we derived expressions for the retardation caused by electric fields of low frequencies. In many practical situations the modulation signal is often at very high frequencies and, in order to utilize the wide frequency spectrum available with lasers, may occupy a large bandwidth. In this section we consider some of the basic factors limiting the highest usable modulation frequencies in a number of typical experimental situations.

Consider first the situation described by Figure 9-9. The electrooptic crystal is placed between two electrodes with a modulation field containing frequencies near $\omega_0/2\pi$ applied to it. R_s is the internal resistance of the modulation source and C represents the parallel-plate capacitance due to the electrooptic crystal. If $R_s > (\omega_0 c)^{-1}$, most of the modulation voltage drop is across R_s and is thus wasted, since it does not contribute to the retardation. This can be remedied by resonating the crystal capacitance with an inductance L, where $\omega_0{}^2 = (LC)^{-1}$, as shown in Figure 9-9. In addition, a shunting resistance R_L is used so that at $\omega = \omega_0$ the impedance

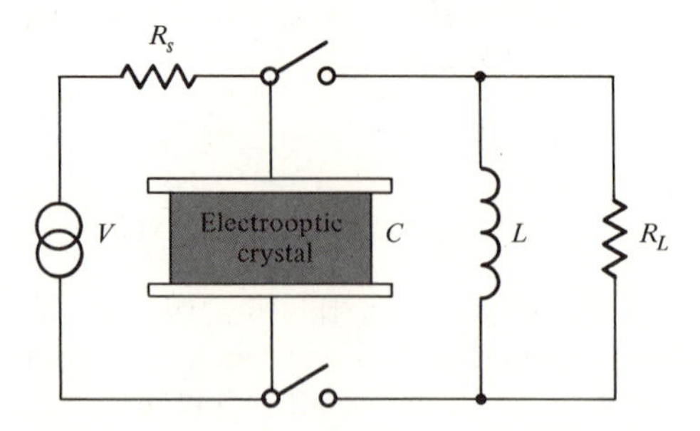

Figure 9-9 Equivalent circuit of an electrooptic modulation crystal in a parallel-plate configuration.

of the parallel RLC circuit is R_L, which is chosen to be larger than R_s so most of the modulation voltage appears across the crystal. The resonant circuit has a finite bandwidth—that is, its impedance is high only over a frequency interval $\Delta\omega/2\pi \simeq 1/2\pi R_L C$ (centered on ω_0). Therefore, the maximum modulation bandwidth (the frequency spectrum occupied by the modulation signal) must be less than

$$\frac{\Delta\omega}{2\pi} \simeq \frac{1}{2\pi R_L C} \tag{9.6-1}$$

if the modulation field is to be a faithful replica of the modulation signal.

In practice, the size of the modulation bandwith $\Delta\omega/2\pi$ is dictated by the specific application. In addition, one requires a certain peak retardation Γ_m. Using (9.2-4) to relate Γ_m to the peak modulation voltage $V_m = (E_z)_m l$ we can show, with the aid of (9.6-1), that the power $V_m^2/2R_L$ needed in KDP-type crystals to obtain a peak retardation Γ_m is related to the modulation bandwidth $\Delta\nu = \Delta\omega/2\pi$ as

$$P = \frac{\Gamma_m^2 \lambda^2 A \epsilon \Delta\nu}{4\pi l n_0^6 r_{63}^2} \tag{9.6-2}$$

where $n_0 l$ is the length of the optical path in the crystal, A is the cross-sectional area of the crystal normal to l, and ϵ is the dielectric constant at the modulation frequency ω_0.

Transit-time limitations to high-frequency electrooptic modulation. According to (9.2-4) the electrooptic retardation due to a field E can be written as

$$\Gamma = aEl \tag{9.6-3}$$

where $a = \omega n_0^3 r_{63}/c$ and l is the length of the optical path in the crystal. If the field E changes appreciably during the transit time $\tau_d = nl/c$ of light through the crystal, we must replace (9.6-3) by

$$\Gamma(t) = a\int_0^l e(t')\,dz = a\,\frac{c}{n}\int_{t-\tau_d}^{t} E(t')\,dt' \tag{9.6-4}$$

where c is the velocity of light and $e(t')$ is the instantaneous electric field. In the second integral we replace integration over z by integration over time, recognizing that the portion of the wave which reaches the output face $z = l$ at time t entered the crystal at time $t - \tau_d$. We also assumed that at any given moment the field $e(t)$ has the same value throughout the crystal.

Taking $e(t')$ as a sinusoid

$$e(t') = E_m e^{i\omega_m t'}$$

we obtain from (9.6-4)

$$\Gamma(t) = a\frac{c}{n}E_m \int_{t-\tau_d}^{t} e^{i\omega_m t'}\, dt'$$

$$= \Gamma_0 \left[\frac{1 - e^{-i\omega_m\tau_d}}{i\omega_m\tau_d}\right] e^{i\omega_m t} \qquad \textbf{(9.6-5)}$$

where $\Gamma_0 = a(c/n)\tau_d E_m = alE_m$ is the peak retardation, which obtains when $\omega_m\tau_d \ll 1$. The factor

$$r = \frac{1 - e^{-i\omega_m\tau_d}}{i\omega_m\tau_d} \qquad \textbf{(9.6-6)}$$

gives the decrease in peak retardation resulting from the finite transit time. For $r \simeq 1$ (that is, no reduction), the condition $\omega_m\tau_d \ll 1$ must be satisfied, so the transit time must be small compared to the shortest modulation period. The factor r is plotted in Figure 11-17.

If, somewhat arbitrarily, we take the highest useful modulation frequency as that for which $\omega_m\tau_d = \pi/2$ (at this point, according to Figure 11-17, $|r| = 0.9$) and we use the relation $\tau_d = nl/c$, we obtain

$$(\nu_m)_{\max} = \frac{c}{4nl} \qquad \textbf{(9.6-7)}$$

which, using a KDP crystal ($n \simeq 1.5$) and a length $l = 1$ cm, yields $(\nu_m)_{\max} = 5 \times 10^9$ Hz.

Traveling-wave modulators. One method that can, in principle, overcome the transit-time limitation, involves applying the modulation signal in the form of a traveling wave [3], as shown in Figure 9-10. If the optical and modulation field phase velocities are equal to each other, then a portion of an optical wavefront will exercise the same instantaneous

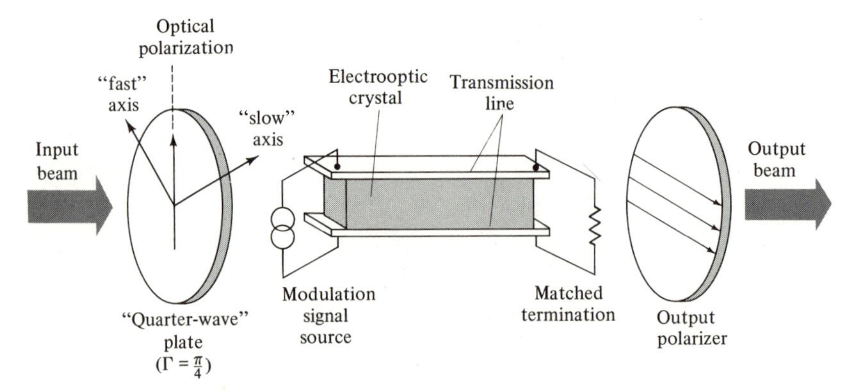

Figure 9-10 A traveling-wave electrooptic modulator.

electric field, which corresponds to the field it encounters at the entrance face, as it propagates through the crystal and the transit-time problem discussed above is eliminated. This form of modulation can be used only in the transverse geometry was discussed in the preceding section, since the RF field in most propagating structures is predominantly transverse.

Consider an element of the optical wavefront that *enters* the crystal at $z = 0$ at time t. The position z of this element at some later time t' is

$$z(t') = \frac{c}{n}(t' - t) \tag{9.6-8}$$

where $c = c_0/n$ is the optical phase velocity. The retardation exercised by this element is given similarly to (9.6-4) by

$$\Gamma(t) = ac\int_t^{t+\tau d} e[t', z(t')]\, dt' \tag{9.6-9}$$

where $e(t', z(t'))$ is the instantaneous modulation field as seen by an observer traveling with the phase front. Taking the traveling modulation field as

$$e(t', z) = E_m e^{i[\omega_m t' - k_m z]}$$

we obtain, using (9.6-8),

$$e[t', z(t')] = E_m e^{i[\omega_m t' - k_m c(t' - t)]} \tag{9.6-10}$$

Recalling that $k_m = \omega_m/c_m$, where c_m is the phase velocity of the modulation field, we substitute (9.6-10) in (9.6-9) and, carrying out the simple integration, obtain

$$\Gamma(t) = \Gamma_0 e^{i\omega_m t}\left[\frac{e^{i\omega_m \tau_d(1 - c/nc_m)} - 1}{i\omega_m \tau_d(1 - c/nc_m)}\right] \tag{9.6-11}$$

where $\Gamma_0 = alE_m = ac\tau_d E_m$ is the retardation that would result from a dc field equal to E_m.

The reduction factor

$$r = \frac{e^{i\omega_m \tau_d(1 - c/nc_m)} - 1}{i\omega_m \tau_d(1 - c/nc_m)} \tag{9.6-12}$$

is of the same form as that of the lumped-constant modulator (9.6-6) except that τ_d is replaced by $\tau_d(1 - c/nc_m)$. If the two phase velocities are made equal so that $c/n = c_m$, then $r = 1$ and maximum retardation is obtained *regardless* of the crystal length.

The maximum useful modulation frequency is taken, as in the treatment leading to (9.6-7), as that for which $\omega_m \tau_d(1 - c/nc_m) = \pi/2$, yielding

$$(\nu_m)_{\max} = \frac{c}{4ln(1 - c/nc_m)} \tag{9.6-13}$$

which, upon comparison with (9.6-7), shows an increase in the frequency limit or useful crystal length of $(1 - c/nc_m)^{-1}$. The problem of designing traveling wave electrooptic modulators is considered in References [4]–[6].

9.7 Electrooptic Beam Deflection

The electrooptic effect is also used to deflect light beams [7]. The operation of such a beam deflector is shown in Figure 9-11. Imagine an optical wavefront incident on a crystal in which the optical path length depends on the transverse position x. This could be achieved by having the velocity of propagation—that is, the index of refraction n—depend on x, as in Figure 9-11. Taking the index variation to be a linear function of x, the upper ray A "sees" an index $n + \Delta n$ and hence traverses the crystal in a time

$$T_A = \frac{l}{c}(n + \Delta n)$$

The lower portion of the wavefront (that is, ray B) "sees" an index n and has a transit time

$$T_B = \frac{l}{c} n$$

The difference in transit times results in a lag of ray A with respect to B of

$$\Delta y = \frac{c}{n}(T_A - T_B) = l \frac{\Delta n}{n}$$

which corresponds to a deflection of the beam-propagation axis, as measured inside the crystal, at the output face of

$$\theta' = -\frac{\Delta y}{D} = -\frac{l\,\Delta n}{Dn} = -\frac{l}{n}\frac{dn}{dx} \tag{9.7-1}$$

where we replaced $\Delta n/D$ by dn/dx. The external deflection angle θ, measured with respect to the horizontal axis, is related to θ' by Snell's law

$$\frac{\sin\theta}{\sin\theta'} = n$$

which, using (9.7-1) and assuming $\sin\theta \simeq \theta \ll 1$ yields

$$\theta = \theta' n = -l\frac{\Delta n}{D} = -l\frac{dn}{dx} \tag{9.7-2}$$

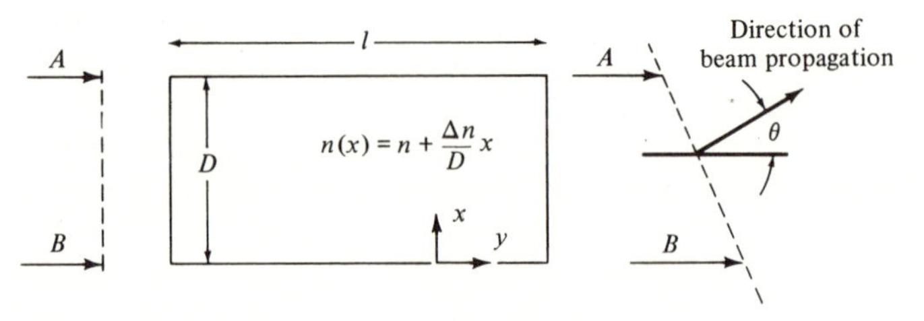

Figure 9-11 Schematic diagram of a beam deflector. The index of refraction varies linearly in the x direction as $n(x) = n_o + ax$. Ray B "gains" on ray A in passing through the crystal axis, thus causing a tilting of the wavefront by θ.

A simple realization of such a deflector using a KH_2PO_4(KDP) crystal is shown in Figure 9-12. It consists of two KDP prisms with edges along the x', y', and z directions.[7] The two prisms have their z axes opposite to one another, but are otherwise similarly oriented. The electric field is applied parallel to the z direction and the light propagates in the y' direction with its polarization along x'. For this case the index of refraction "seen" by ray A, which propagates entirely in the upper prism, is given by (9.1-10) as

$$n_A = n_o - \frac{n_o^3}{2} r_{63} E_z$$

while in the lower prism the sign of the electric field with respect to the z axis is reversed so that

$$n_B = n_o + \frac{n_o^3}{2} r_{63} E_z$$

Using (9.7-2) with $\Delta n = n_A - n_B$ the deflection angle is given by

$$\theta = \frac{l}{D} n_o^3 r_{63} E_z \tag{9.7-3}$$

According to (3.2-18), every optical beam has a finite, far-field divergence angle which we call θ_{beam}. It is clear that a fundamental figure of merit for the deflector is not the angle of deflection θ that can be changed by a lens, but the factor N by which θ exceeds θ_{beam}. If one were, as an example, to focus the output beam, then N would correspond to the

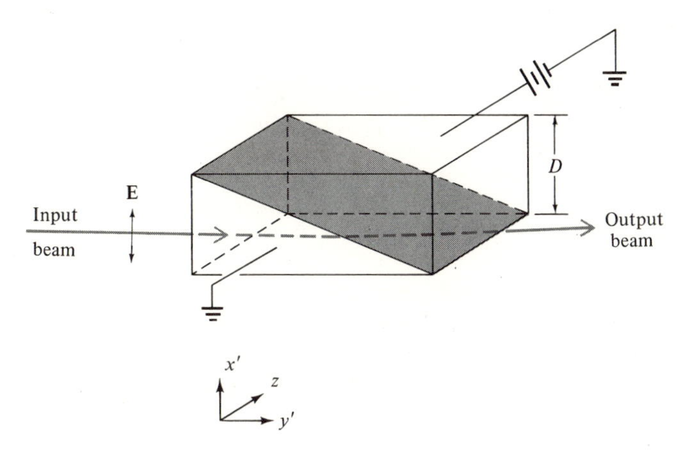

Figure 9-12 Double-prism KDP beam deflector. Upper and lower prisms have their z axes reversed with respect to each other. The deflection field is applied parallel to z.

---

[7] These are the principal axes of the index ellipsoid when an electric field is applied along the z direction as described in Section 9.1 and in the example of Section 9.5.

number of resolvable spots that can be displayed in the focal plane using fields with a magnitude up to E_z.

To get an expression for N we assume that the crystal is placed astride the "waist" of a Gaussian (fundamental) beam with a spot size ω_0. According to (3.2-18) the far-field diffraction angle in air is

$$\theta_{\text{beam}} = \frac{\lambda}{\pi\omega_0}$$

Such a beam can be passed through a crystal with height $D = 2\omega_0$ so that, using (9.7-3), the number of resolvable spots is

$$N = \frac{\theta}{\theta_{\text{beam}}} = \frac{\pi l n_o^3 r_{63}}{2\lambda} E_z \tag{9.7-4}$$

It follows directly from (9.7-4), the details being left as a problem, that an electric field that induces a birefringent retardation (in a distance l) $\Delta\Gamma = \pi$ will yield $N \simeq 1$. Therefore, fundamentally, the electrooptic extinction of a beam, which according to (9.3-4) requires $\Gamma = \pi$, is equivalent to a deflection by one spot diameter.

The deflection of an optical beam by diffraction from a sound wave is discussed in Chapter 12. Electrooptic modulation in thin dielectric waveguides ([8]–[10]) is discussed in Chapter 14.

PROBLEMS

9-1 Derive the equations of the nine ellipses traced by the optical field vector as shown in Figure 9-3(c), as a function of the retardation Γ.

9-2 Discuss the consequence of the field-independent retardation $(\omega L/c_0)(n_0 - n_e)$ in Equation (9.5-1) on an amplitude modulator such as that shown in Figure 9.4.

9-3 Use the Bessel-function expansion of sin $[a \sin x]$ to express (9.3-7) in terms of the harmonics of the modulation frequency ω_m. Plot the ratio of the third harmonic $(3\omega_m)$ of the output intensity to the fundamental as a function of Γ_m. What is the maximum allowed Γ_m if this ratio is not to exceed 10^{-2}? (*Answer:* $\Gamma_m < 0.5$).

9-4 Show that, if a phase-modulated optical wave is incident on a square-law detector, the output contains no alternating currents.

9-5 Using References [4] and [5], design a partially loaded KDP phase modulator that operates at $\nu_m = 10^9$ Hz and yields a peak phase excursion of $\delta = \pi/3$. What is the modulation power?

9-6 Derive the expression [similar to Equation (9.6-2)] for the modulation power of a transverse $\bar{4}3m$ crystal electrooptic modulator of the type described in the Appendix C.

9-7 Derive an expression for the modulation power requirement (corresponding to Equation (9.6-2)) for a GaAs transverse modulator.

9-8 a. Show that if a ray propagates at an angle $\theta(\ll 1)$ to the z axis in the arrangement of Figure 9-4, it exercises a birefringent contribution to the retardation.

$$\Delta\Gamma_{\text{birefringent}} = \frac{\omega l}{2c} n_0 \left(\frac{n_0{}^2}{n_e{}^2} - 1\right) \theta^2$$

which corresponds to a change in index

$$n_0 - n_e(\theta) = \frac{n_0 \theta^2}{2} \left(\frac{n_0{}^2}{n_e{}^2} - 1\right)$$

b. Derive an approximate expression for the maximum allowable beam spreading angle for which $\Delta\Gamma_{\text{birefringent}}$ does not interfere with the operation of the modulator. *Answer:*

$$\theta < \left[\lambda / 4 l n_0 \left(\frac{n_0{}^2}{n_e{}^2} - 1\right)\right]^{1/2}$$

REFERENCES

[1] See, for example, J. F. Nye, *Physical Properties of Crystals*. New York: Oxford, 1957, p. 123.

[2] See, for example, A. Yariv, *Quantum Electronics*. New York, Wiley, 1967, Chap. 18.

[3] Peters, L. C., "Gigacycle bandwidth coherent light traveling-wave phase modulators," *Proc. IEEE*, vol. 51, p. 147, 1963.

[4] Rigrod, W. W., and I. P. Kaminow, "Wide-band microwave light modulation," *Proc. IEEE*, vol. 51, p. 137, 1963.

[5] Kaminow, I. P., and J. Lin, "Propagation characteristics of partially loaded two-conductor transmission lines for broadband light modulators," *Proc. IEEE*, vol. 51, p. 132, 1963.

[6] White, R. M., and C. E. Enderby, "Electro-optical modulators employing intermittent interaction," *Proc. IEEE*, vol. 51, p. 214, 1963.

[7] Fowler, V. J., and J. Schlafer, "A survey of laser beam deflection techniques," *Proc. IEEE*, vol. 54, p. 1437, 1966.

[8] Hall, D., A. Yariv, and E. Garmire, "Observation of propagation cutoff and its control in thin optical waveguides," *Appl. Phys. Lett.*, vol. 17, p. 127, 1970.

[9] Hall, D., A. Yariv, and E. Garmire "Optical guiding and electro-optic modulation in GaAs epitaxial layers," *Opt. Commun.*, vol. 1, p. 403, 1970.

[10] Hammer, J. M., and W. Phillips, "Low-loss single mode optical waveguide and efficient high-speed modulators of $LiNb_x Ta_{1-x}O_3$ on $LiTaO_3$," *Appl. Phys. Lett.*, vol. 24, p. 545, 1974.

CHAPTER

10

Noise in Optical Detection and Generation

10.0 Introduction

In this chapter we study the effect of noise in a number of important physical processes. We will take the term noise to represent random electromagnetic fields occupying the same spectral region as that occupied by some "signal." The effect of noise will be considered in the following cases.

1. *Measurement of optical power.* In this case the noise causes fluctuations in the measurement, thus placing a lower limit on the smallest amount of power that can be measured.

2. *Linewidth of laser oscillators.* The presence of incoherent spontaneous emission power will be found to be the cause for a finite amount of spectral line broadening in the output of single-mode laser oscillators. This broadening manifests itself as a limited coherence time.

3. *Optical communication system.* We will consider the case of an optical communication system using a binary pulse code modulation in which the information is carried by means of a string of 1 and 0 pulses. The presence of noise will be shown to lead to a certain probability that any given pulse in the reconstructed train pulse is in error.

In this chapter we consider optical detectors utilizing light-generated charge carriers. These include the photomultiplier, the photoconductive detector, the *p-n* junction photodiode, and the avalanche photodiode. These detectors are the main ones used in the field of quantum electronics, because they combine high sensitivity with very short response times. Other types of detectors, such as bolometers, Golay cells, and thermocouples, whose operation depends on temperature changes induced by the absorbed radiation, will not be discussed.[1]

Two types of noise will be discussed in detail. The first type is thermal (Johnson) noise, which represents noise power generated by thermally agitated charge carriers. The expression for this noise will be derived by using the conventional thermodynamic treatment as well as by a statistical analysis of a particular model in which the physical origin of the noise is more apparent. The second type, shot noise (or generation-recombination noise in photoconductive detectors), is attributable to the random way in which electrons are emitted or generated in the process of interacting with a radiation field. This noise exists even at zero temperature, where thermal agitation or generation of carriers can be neglected. In this case it results from the randomness with which carriers are generated by the *very signal that is measured.* Detection in the limit of signal-generated shot noise is called quantum-limited detection, since the corresponding sensitivity is that allowed by the uncertainty principle in quantum mechanics. This point will be brought out in the next chapter.

10.1 Limitations Due to Noise Power

Measurement of optical power. Consider the problem of measuring an optical signal field

$$v_S(t) = V_S \cos \omega t \tag{10.1-1}$$

in the presence of a noise field. The instantaneous noise field which adds to that of the signal can be taken as the sum of an in-phase component and a quadrature component according to

$$v_N(t) = V_{NC}(t) \cos \omega t + V_{NS}(t) \sin \omega t \tag{10.1-2}$$

[1] The interested reader will find a good description of these devices in Reference [6].

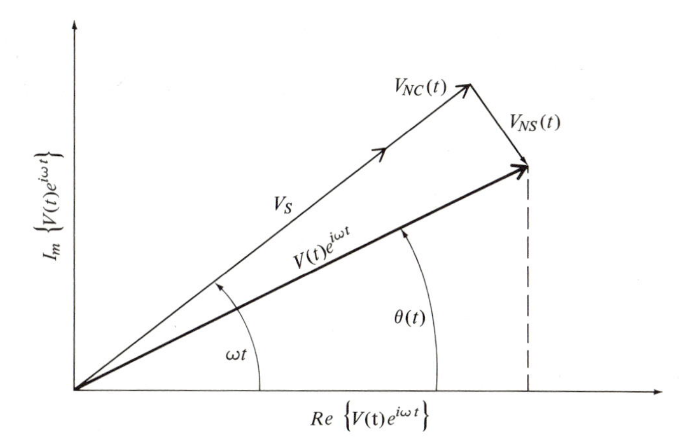

Figure 10-1 A phasor diagram showing the total (signal plus noise) field phasor $V(t)$ at time t. The instantaneous field is given by the horizontal projection of $V(t) \exp(i\omega t)$.

where $V_{NC}(t)$ and $V_{NS}(t)$ are slowly [compared to $\exp(i\omega t)$] varying random uncorrelated quantities with a zero mean. The total field at the detector $v(t) = v_S(t) + v_N(t)$ can be written as

$$v(t) = \text{Re}\,\{[V_S + V_{NC}(t) - iV_{NS}(t)]e^{i\omega t}\} \tag{10.1-3}$$

$$\equiv \text{Re}\,[V(t)e^{i\omega t}] \tag{10.1-4}$$

The total (signal plus noise) field phasor $V(t)$ is shown in Figure 10-1.

In most situations of interest to optical detection the sources of noise are due to the concerted action of a large number of independent agents. In this case the central limit theorem of statistics [1] tells us that the probability function for finding $V_{NC}(t)$ at time t between V_{NC} and $V_{NC} + dV_{NC}$ is described by a Gaussian

$$p(V_{NC})\,dV_{NC} = \frac{1}{\sqrt{2\pi}\sigma}\,e^{-V_{NC}^2/2\sigma^2}\,dV_{NC} \tag{10.1-5}$$

and by a similar expression in which V_{NS} replaces V_{NC} for $p(V_{NS})$. Since $V_{NC}(t)$ has a unity probability of having some value between $-\infty$ and ∞, it follows that

$$\int_{-\infty}^{\infty} p(V_{NC})\,dV_{NC} = 1 \tag{10.1-6}$$

It follows from (10.1-5) that $\bar{V}_{NC}$, the ensemble average[2] (denoted by a horizontal bar) of V_{NC}, is zero,[3] whereas the mean square value is

$$\overline{V_{NC}^2} = \overline{V_{NS}^2} = \int_{-\infty}^{\infty} V_{NC}^2 p(V_{NC})\, dV_{NC} = \sigma^2 \tag{10.1-7}$$

The power in $v(t)$ is obtained using (1.1-12) as

$$\begin{aligned} P(t) &= [V(t)e^{i\omega t}][V^*(t)e^{-i\omega t}] \\ &= V_S^2 + 2V_S V_{NC} + V_{NC}^2 + V_{NS}^2 \end{aligned} \tag{10.1-8}$$

The ensemble average (or *long* time average) of $P(t)$ is

$$\bar{P} \equiv \overline{P(t)} = V_S^2 + \overline{V_{NC}^2} + \overline{V_{NS}^2} = V_S^2 + 2\sigma^2 \tag{10.1-8a}$$

where use has been made of the fact that $\bar{V}_{NC} = 0$ and of (10.1-7).

The physical significance of the time-varying power $P(t)$ and its long time (or ensemble) average $\bar{P}$ is illustrated by Figure 10-2.

It is clear from the fluctuating nature of $P(t)$ that any measurement of this power is subject to an uncertainty which is due to the random nature of V_{NC} and V_{NS} in (10.1-8). As a measure of the uncertainty in power measurement, we may reasonably take the root mean square (rms) power deviation

$$\Delta P \equiv [\overline{(P(t) - \bar{P})^2}]^{1/2}$$

Using (10.1-8) and (10.1-8a) we obtain after some algebra

$$\Delta P = [4V_S^2\overline{V_{NC}^2} + 2\overline{V_{NC}^4} - 2\overline{V_{NS}^2}\,\overline{V_{NC}^2}]^{1/2} \tag{10.1-9}$$

[2] The ensemble average $\overline{A(t)}$ of a quantity $A(t)$ is obtained by measuring A simultaneously at time t in a very large number of systems that, *to the best of our knowledge*, are identical. Mathematically,

$$\overline{A(t)} = \lim_{N\to\infty} \left[\frac{1}{N}\sum_{n=1}^{N} A_n(t)\right]$$

where $A_n(t)$ denotes the observation in the nth system. In a truly random phenomenon, the time averaging and ensemble averaging lead to the same result, so the ensemble average is independent of the time t in which it is performed and can also be obtained from

$$\bar{A} = \int_{-\infty}^{\infty} Ap(A)\, dA$$

where $p(A)$ is the probability function, in the sense of (10.1-5), of the variable A.

[3] The reason for $V_{NC}(t) = 0$ can be appreciated from Figure 10-1. $V_{NC}(t)$ has an equal probability of being in phase with V_S as of being out phase, thus averaging out to zero.

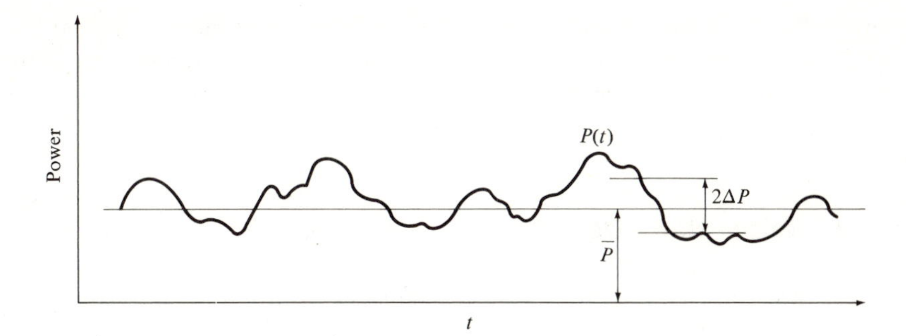

Figure 10-2 The intermingling of noise power with that of a signal causes the total power to fluctuate. The rms fluctuation ΔP limits the accuracy of power measurements.

Using (10.1-5) we obtain

$$\overline{V_{NC}^4} = \int_{-\infty}^{\infty} V_{NC}^4 p(V_{NC})\, dV_{NC} = 3\sigma^4 \tag{10.1-10}$$

so that using $\overline{V_{NC}^2} = \overline{V_{NS}^2} = \sigma^2$ in (10.1-9) results in

$$\begin{aligned} \Delta P &= 2\sigma(V_S^2 + \sigma^2)^{1/2} \\ &= 2\sigma(P_S + \sigma^2)^{1/2} \end{aligned} \tag{10.1-11}$$

where according to (10.1-8) we may associate $P_S = V_S^2$ with the signal power that is, the power that would be measured if V_{NC} and V_{NS} were, hypothetically, rendered zero.

A question of practical importance involves the minimum signal power that can be measured in the presence of noise. We may, somewhat arbitrarily, take this power P_{limit} to be that at which the uncertainty ΔP becomes equal to the signal power P_S. At this point we have from (10.1-11)

$$P_{\text{limit}} = 2\sigma(P_{\text{limit}} + \sigma^2)^{1/2}$$

or after solving for P_{limit}

$$P_{\text{limit}} = 2\sigma^2(1 + \sqrt{2}) = P_N(1 + \sqrt{2}) \tag{10.1-12}$$

where $P_N = 2\sigma^2 = \bar{V}_{NC}^2 + \bar{V}_{NS}^2$ is the noise power. Widespread convention chooses to define the minimum detectable signal power as equal to P_N instead of $2.414P_N$, as obtained above. This simplification is understandable, since our choice of the limit of detectability $\Delta P = P_S$ was somewhat arbitrary. In any case the main conclusion to remember is that near the limit of detectivity, the rms power fluctuation is comparable to the noise power. The next task, which will be taken up in this chapter and in Chapter 11, is to find out the main sources of noise power and consequently ways to minimize them. Before tackling this task, however, we need to develop some mathematical tools for dealing with random processes.

10.2 Noise—Basic Definitions and Theorems

A real function $v(t)$ and its Fourier transform $V(\omega)$ are related by

$$V(\omega) = \int_{-\infty}^{\infty} v(t)e^{-i\omega t}\, dt \tag{10.2-1}$$

and

$$v(t) = \frac{1}{2\pi}\int_{-\infty}^{\infty} V(\omega)e^{i\omega t}\, d\omega \tag{10.2-2}$$

In the process of measuring a signal $v(t)$ we are not in a position to use the infinite time interval needed, according to (10.2-1), to evaluate $V(\omega)$. If the time duration of the measurement is T we may consider the function $v(t)$ to be zero for $t \leqslant -T/2$ and $t \geqslant T/2$ and, instead of (10.2-1), get

$$V_T(\omega) = \int_{-T/2}^{T/2} v(t)e^{-i\omega t}\, dt \tag{10.2-3}$$

Since $v(t)$ is real, it follows that

$$V_T(\omega) = V_T^*(-\omega) \tag{10.2-4}$$

T is usually called the resolution or integration time of the system.

Let us evaluate the average power P associated with $v(t)$. Taking the instantaneous power as $v^2(t)$,[4] we obtain

$$P = \frac{1}{T}\int_{-T/2}^{T/2} v^2(t)\, dt = \frac{1}{2\pi T}\int_{-T/2}^{T/2} \left\{v(t)\left[\int_{-\infty}^{\infty} V_T(\omega)e^{i\omega t}\, d\omega\right]\right\} dt \tag{10.2-5}$$

Using (10.2-3) and (10.2-4) in the last equation and interchanging the order of integration leads to

$$P = \frac{1}{2\pi T}\int_{-\infty}^{\infty} |V_T(\omega)|^2\, d\omega \tag{10.2-6}$$

[4] It may be convenient for this purpose to think of $v(t)$ as the voltage across a one-ohm resistance.

or

$$P = \frac{1}{\pi T} \int_0^\infty |V_T(\omega)|^2 \, d\omega \tag{10.2-7}$$

where we used

$$\lim_{T \to \infty} (2\pi)^{-1} \int_{-T/2}^{T/2} dt \exp [i(\omega - \omega')t] = \delta(\omega - \omega')$$

If we define the *spectral density function* $S_T(\omega)$ of $v(t)$ by

$$S_T(\omega) \equiv \frac{|V_T(\omega)|^2}{\pi T} \tag{10.2-8}$$

then, according to (10.2-7), $S_T(\omega)\, d\omega$ is the portion of the average power of $v(t)$ that is due to frequency components between ω and $\omega + d\omega$. According to this physical interpretation we may measure $S_T(\omega)$ by separating the spectrum of $v(t)$ into its various frequency classes as shown in Figure 10-3 and then measuring the power output $S_T(\omega_i)\, \Delta\omega_i$ of each of the filters [2].

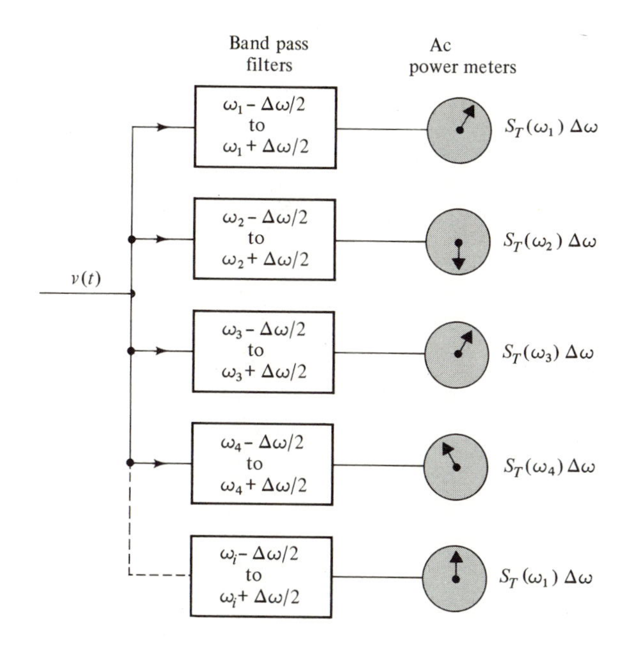

Figure 10-3 Diagram illustrating how the spectral density function $S_T(\omega)$ of a signal $v(t)$ can be obtained by measuring the power due to different frequency intervals.

10.3 The Spectral Density Function of a Train of Randomly Occurring Events

Consider a time-dependent random variable $i(t)$ made up of a very large number of individual events $f(t - t_i)$, which occur at random times t_i.[5] An observation of $i(t)$ during a period T will yield

$$i_T(t) = \sum_{i=1}^{N_T} f(t - t_i) \qquad 0 \leqslant t \leqslant T \tag{10.3-1}$$

where N_T is the total number of events occurring in T. Typical examples of a random function $i(t)$ are provided by the thermionic emission current emitted by a hot cathode (under temperature-limited conditions), or the electron current caused by photoemission from a surface. In these cases $f(t - t_i)$ represents the current resulting from a single electron emission occurring at t_i.

The Fourier transform of $i_T(t)$ is given according to (10.2-3) by

$$I_T(\omega) = \sum_{i=1}^{N_T} F_i(\omega) \tag{10.3-2}$$

where $F_i(\omega)$ is the Fourier transform[6] of $f(t - t_i)$

$$\begin{aligned} F_i(\omega) &= \int_{-\infty}^{\infty} f(t - t_i)e^{-i\omega t}\,dt = e^{-i\omega t_i}\int_{-\infty}^{\infty} f(t)e^{-i\omega t}\,dt \\ &= e^{-i\omega t_i}F(\omega) \end{aligned} \tag{10.3-3}$$

From (10.3-2) and (10.3-3) we obtain

$$\begin{aligned} |I_T(\omega)|^2 &= |F(\omega)|^2 \sum_{i=1}^{N_T}\sum_{j=1}^{N_T} e^{-i\omega(t_i - t_j)} \\ &= |F(\omega)|^2\left[N_T + \sum_{i\neq j}^{N_T}\sum_{j}^{N_T} e^{i\omega(t_j - t_i)}\right] \end{aligned} \tag{10.3-4}$$

If we take the average of (10.3-4) over an ensemble of a very large number of physically identical systems, the second term on the right side of (10.3-4) can be neglected in comparison to N_T since the times t_i are random.

---

[5] This means that the *a priori* probability that a given event will occur in any time interval is distributed uniformly over the interval, or equivalently, that the probability $p(n)$ for n events to occur in an observation period T is given by the Poisson distribution function [2]

$$p(n) = \frac{[(\bar{n})^n e^{-\bar{n}}]}{n!}$$

where $\bar{n}$ is the average number of events occurring in T.

[6] We assume that the individual event $f(t - t_i)$ is over in a time short compared to the observation period T, so the integration limits can be taken as $-\infty$ to ∞ instead of 0 to T.

This results in

$$\overline{|I_T(\omega)|^2} = \overline{N}_T|F(\omega)|^2 \equiv \overline{N}T|F(\omega)|^2 \tag{10.3-5}$$

where the horizontal bar denotes ensemble averaging, and where $\overline{N}$ is the average rate at which the events occur so that $\overline{N}_T = \overline{N}T$. The spectral density function $S_T(\omega)$ of the function $i_T(t)$ is given according to (10.2-8) and (10.3-5) as

$$S(\omega) = \frac{\overline{N}|F(\omega)|^2}{\pi} \tag{10.3-6}$$

In practice, one uses more often the spectral density function $S(\nu)$ defined so that the average power due to frequencies between ν and $\nu + d\nu$ is equal to $S(\nu)\,d\nu$. It follows then, that $S(\nu)\,d\nu = S(\omega)\,d\omega$; thus, since $\omega = 2\pi\nu$,

$$S(\nu) = 2\overline{N}|F(2\pi\nu)|^2 \tag{10.3-7}$$

The last result is known as Carson's theorem and its usefulness will be demonstrated in the following sections where we employ it in deriving the spectral density function associated with a number of different physical processes related to optical detection.

Equation (10.3-7) was derived for the case in which the individual events $f(t - t_i)$ were displaced in time but were, otherwise, identical. There are physical situations in which the individual events may depend on one or more additional parameters. Denoting the parameter (or group of parameters) as α we can clearly single out the subclass of events $f_\alpha(t - t_i)$ whose α is nearly the same and use (10.3-7) to obtain directly

$$S_\alpha(\nu) = 2\overline{N}(\alpha)|F_\alpha(2\pi\nu)|^2\Delta\alpha \tag{10.3-8}$$

for the contribution of this subclass of events to $S(\nu)$. $F_\alpha(\omega)$ is the Fourier transform of $f_\alpha(t)$ and thus $\overline{N}(\alpha)\Delta\alpha$ is the average number of events per second whose α parameter falls between α and $\alpha + \Delta\alpha$.

$$\int_{-\infty}^{\infty} \bar{N}(\alpha)\,d\alpha = \bar{N}$$

The probability distribution function for α is $p(\alpha) = \overline{N}(\alpha)/\overline{N}$; therefore,

$$\int_{-\infty}^{\infty} p(\alpha)\,d\alpha = \frac{1}{\overline{N}}\int_{-\infty}^{\infty} \overline{N}(\alpha)\,d\alpha = 1 \tag{10.3-9}$$

Summing (10.3-8) over all classes α and weighting each class by the probability $p(\alpha)\Delta\alpha$ of its occurrence, we obtain

$$\begin{aligned} S(\nu) &= \sum_\alpha S_\alpha(\nu) = 2\sum_\alpha \overline{N}(\alpha)|F_\alpha(2\pi\nu)|^2\Delta\alpha \\ &= 2\overline{N}\sum_\alpha |F_\alpha(2\pi\nu)|^2 p(\alpha)\Delta\alpha \\ &= 2\overline{N}\int_{-\infty}^{\infty} |F_\alpha(2\pi\nu)|^2 p(\alpha)\,d\alpha = 2\overline{N}\overline{|F(2\pi\nu)|^2} \end{aligned} \tag{10.3-10}$$

where the bar denotes averaging over α. Equation (10.3-10) is thus the extension of (10.3-7) to the case of events whose characterization involves, in addition to their time t_i, some added parameters. We will use it further in this chapter to derive the noise spectrum of photoconductive detectors in which case α is the lifetime of the excited photocarriers.

10.4 Shot Noise [3]

Let us consider the spectral density function of current arising from random generation and flow of mobile charge carriers. This current is identified with "shot noise." To be specific, we consider the case illustrated in Figure 10-4, in which electrons are released at random into the vacuum from electrode A to be collected at electrode B, which is maintained at a slight positive potential relative to A.

The average rate $\overline{N}$ of electron emission from A is $\overline{N} = \overline{I}/e$, where $\overline{I}$ is the average current and the electronic charge is taken as $-e$. The current pulse due to a single electron as observed in the external circuit is

$$i_e(t) = \frac{ev(t)}{d} \tag{10.4-1}$$

where $v(t)$ is the instantaneous velocity and d is the separation between A and B. To prove (10.4-1), consider the case in which the moving electron is replaced by a thin sheet of a very large area and of total charge $-e$ moving between the plates, as illustrated in Figure 10-5.

It is a simple matter to show (see Problem 10-1), using the relation $\nabla \cdot \mathbf{E} = \rho/\epsilon$, that the charge induced by the moving sheet on the left electrode is

$$Q_1 = \frac{e(d - x)}{d} \tag{10.4-2}$$

and that on the right electrode is

$$Q_2 = \frac{ex}{d} \tag{10.4-3}$$

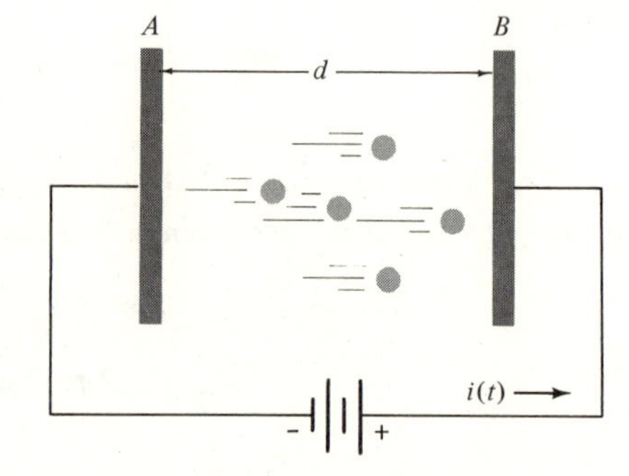

Figure 10-4 Random electron flow between two electrodes. This basic configuration is used in the derivation of shot noise.

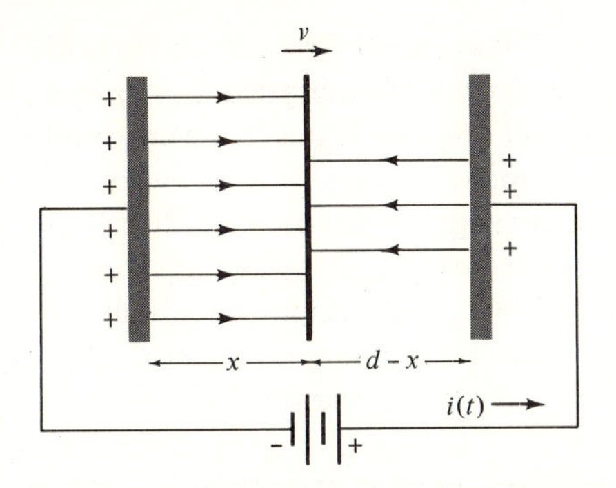

Figure 10-5 Induced charges and field lines due to a thin charge layer between the electrodes.

where x is the position of the charged sheet measured from the left electrode. The current in the external circuit due to a single electron is thus

$$i_e(t) = \frac{dQ_2}{dt} = \frac{e}{d}\frac{dx}{dt} = \frac{e}{d}v(t) \tag{10.4-4}$$

in agreement with (10.4-1).

The Fourier transform of a single current pulse is

$$F(\omega) = \frac{e}{d}\int_0^{t_a} v(t)e^{-i\omega t}\,dt \tag{10.4-5}$$

where t_a is the arrival time of an electron emitted at $t = 0$. If the transit time of an electron is sufficiently small that, at the frequency of interest ω,

$$\omega t_a \ll 1 \tag{10.4-6}$$

we can replace exp $(-i\omega t)$ in (10.4-5) by unity and obtain

$$F(\omega) = \frac{e}{d}\int_0^{t_a} \frac{dx}{dt}\,dt = e \tag{10.4-7}$$

since $x(t_a)$ is, by definition, equal to d. Using (10.4-7) in (10.3-7) and recalling that $\bar{I} = e\bar{N}$ gives

$$S(\nu) = 2\bar{N}e^2 = 2e\bar{I} \tag{10.4-8}$$

The power (in the sense of 10.2-5) in the frequency interval ν to $\nu + \Delta\nu$ associated with the current is, according to the discussion following (10.2-8), given by $S(\nu)\,\Delta\nu$. It is convenient to represent this power by an *equivalent noise generator* at ν with a mean-square current amplitude

$$\overline{i_N{}^2(\nu)} \equiv S(\nu)\,\Delta\nu = 2e\bar{I}\Delta\nu \tag{10.4-9}$$

The noise mechanism described above is referred to as shot noise.

It is interesting to note that e in (10.4-9) is the charge of the particle responsible for the current flow. If, hypothetically, these carriers had a charge of $2e$, then at the *same average current* $\bar{I}$ the shot noise power would double. Conversely, shot noise would disappear if the magnitude of an individual charge tended to zero. This is a reflection of the fact that shot noise is caused by fluctuations in the current that are due to the discreteness of the charge carriers and to the random electronic emission (for which the number of electrons emitted per unit time obey Poisson statistics [2]). The ratio of the fluctuations to the average current decreases with increasing number of events.[7]

Another point to remember is that, in spite of the appearance of $\bar{I}$ on the right side of (10.4-9), $i_N{}^2(\nu)$ represents an alternating current with frequencies near ν.

10.5 Johnson Noise

Johnson, or Nyquist, noise describes the fluctuations in the voltage across a dissipative circuit element; see References [4] and [5]. These fluctuations are most often caused by the thermal motion of the charge carriers.[8] The charge neutrality of an electrical resistance is satisfied when we consider the whole volume, but locally the random thermal motion of the carriers sets up fluctuating charge gradients and, correspondingly, a fluctuating (ac) voltage. If we now connect a second resistance across the first one, the thermally induced voltage described above will give rise to a current and hence to a power transfer to the second resistor.[9] This is the so-called Johnson noise, whose derivation follows.

Consider the case illustrated in Figure 10-6 of a transmission line connected between two similar resistances R, which are maintained at the same temperature T. We choose the resistance R to be equal to the characteristic impedance Z_0 of the lines, so that no reflection can take place at the ends. The transmission line can support traveling voltage waves of the form

$$v(t) = A \cos(\omega t \pm kz) \tag{10.5-1}$$

where $k = 2\pi/\lambda$ and the phase velocity is $c = \omega/k$.

---

[7] More precisely, for Poisson statistics we have [1]

$$\frac{[\overline{(\Delta N)^2}]^{1/2}}{\overline{N}} = \frac{1}{(\overline{N})^{1/2}}$$

where N is the number of events in an observation time, $\overline{N}$ is the average value of N, and $(\Delta N)^2 \equiv (N - \overline{N})^2$.

[8] We use the word "carriers" rather than "electrons" to include cases of ionic conduction or conduction by holes.

[9] The same argument applies to the second resistor, so at thermal equilibrium the net power leaving each resistor is zero.

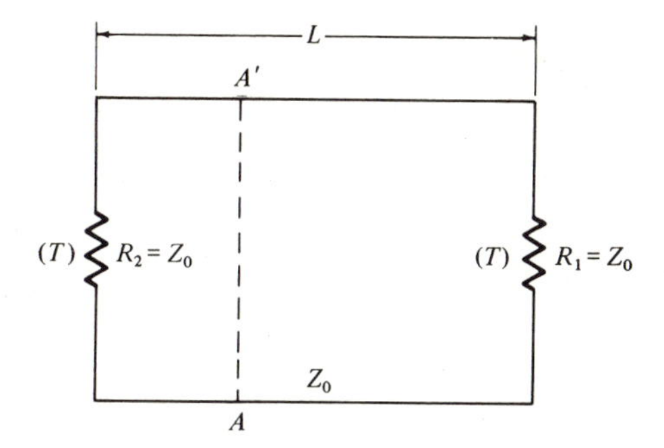

Figure 10-6 Lossless transmission line of characteristic impedance Z_0 connected between two matched loads ($R = Z_0$) at temperature T.

For simplicity we require that the allowed solutions be periodic in the distance L,[10] so if we extend the solution outside the limits $0 \leqslant z \leqslant L$ · we obtain

$$v(t) = A \cos [\omega t \pm k(z + L)] = A \cos [\omega t \pm kz]$$

This condition is fulfilled when

$$kL = 2m\pi \qquad m = 1, 2, 3, \cdots \tag{10.5-2}$$

Therefore, two adjacent modes differ in their value of k by

$$\Delta k = \frac{2\pi}{L} \tag{10.5-3}$$

and the number of modes having their k values somewhere between zero and $+k$ is[11]

$$N_k = \frac{kL}{2\pi} \tag{10.5-4}$$

or, using $k = 2\pi\nu/c$, we obtain

$$N(\nu) = \frac{\nu L}{c}$$

for the number of positively traveling modes having their frequencies between zero and ν.

The number of modes per unit frequency interval is

$$p(\nu) = \frac{dN(\nu)}{d\nu} = \frac{L}{c} \tag{10.5-5}$$

---

[10] This, seemingly arbitrary, type of boundary condition is used extensively in similar situations in thermodynamics to derive the blackbody radiation density, or in solid-state physics to derive the density of electronic states in crystals.

[11] Negative k values correspond, according to (10.5-1), to waves traveling in the $-z$ direction. Our bookkeeping is thus limited to modes carrying power in the $+z$ direction.

Consider the power flowing in the $+z$ direction across some arbitrary plane, $A - A'$ say. It is clear that due to the lack of reflections this power must originate in R_2. Since the power is carried by the electromagnetic modes of the system, we have

$$\text{power} = (\text{energy/distance})(\text{velocity of energy})$$

We find, taking the velocity of light as c, that the power P due to frequencies between ν and $\nu + \Delta\nu$ is given by

$$P = \left(\frac{1}{L}\right)\begin{pmatrix}\text{number of modes between}\\ \nu \text{ and } \nu + \Delta\nu\end{pmatrix}(\text{energy per mode})(c)$$
$$= \left(\frac{1}{L}\right)\left(\frac{L}{c}\Delta\nu\right)\left(\frac{h\nu}{e^{h\nu/kT} - 1}\right)(c)$$

or

$$P = \frac{h\nu\Delta\nu}{e^{h\nu/kT} - 1} \approx kT\Delta\nu \qquad (kT \gg h\nu) \tag{10.5-6}$$

where we used the fact that in thermal equilibrium the energy of a mode is given by [7]

$$\mathcal{E} = \frac{h\nu}{e^{h\nu/kT} - 1} \tag{10.5-7}$$

An equal amount of noise power is, of course, generated in the right resistor and is dissipated in the left one, so in thermal equilibrium the net power crossing any plane is zero.

The power given by (10.5-6) represents the maximum noise power available from the resistance, since it is delivered to a matched load. If the load connected across R has a resistance different from R the noise power delivered is less than that given by (10.5-6). The noise-power bookkeeping is done correctly if the resistance R appearing in a circuit is replaced by either one of the following two equivalent circuits: a noise generator in series with R with mean-square voltage amplitude

$$\overline{v_N^2} = \frac{4h\nu R\Delta\nu}{e^{h\nu/kT} - 1} \underset{kT \gg h\nu}{\simeq} 4kTR\Delta\nu \tag{10.5-8}$$

or a noise current generator of mean square value

$$\overline{i_N^2} = \frac{4h\nu\Delta\nu}{R(e^{h\nu/kT} - 1)} \underset{kT \gg h\nu}{\simeq} \frac{4kT\Delta\nu}{R} \tag{10.5-9}$$

in parallel with R. The noise representations of the resistor are shown in Figure 10-7. There are numerous other derivations of the formula for Johnson noise. For derivations using lumped-circuit concepts and an antenna example, the reader is referred to References [6] and [7], respectively.

Statistical derivation of Johnson noise. The derivation of Johnson noise leading to (10.5-6) leans heavily on thermodynamic and statistical mechanics considerations. It may be instructive to obtain this result using a

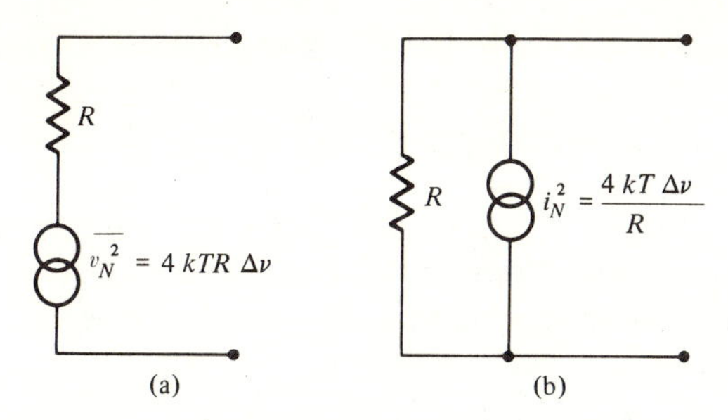

Figure 10-7 (**a**) Voltage and (**b**) current noise equivalent circuits of a resistance.

physical model for a resistance and applying the mathematical tools developed in this chapter. The model used is shown in Figure 10-8.

The resistor consists of a medium of volume $V = Ad$, which contains N free electrons per unit volume. In addition, there are N positively charged ions, which preserve the (average) charge neutrality. The electrons move about randomly with an average kinetic energy per electron of

$$\overline{E} = \tfrac{3}{2}kT = \tfrac{1}{2}m(\overline{v_x^{\,2}} + \overline{v_y^{\,2}} + \overline{v_z^{\,2}}) \tag{10.5-10}$$

where $\overline{v_x^{\,2}} = \overline{v_y^{\,2}} = \overline{v_z^{\,2}}$ refer to thermal averages. A variety of scattering mechanisms including electron–electron, electron–ion, and electron–phonon collisions act to interrupt the electron motion at an average rate of τ_0^{-1} times per second. τ_0 is thus the mean scattering time. These scattering mechanisms are responsible for the electrical resistance and give rise to a conductivity[12]

$$\sigma = \frac{Ne^2\tau_0}{m} \tag{10.5-11}$$

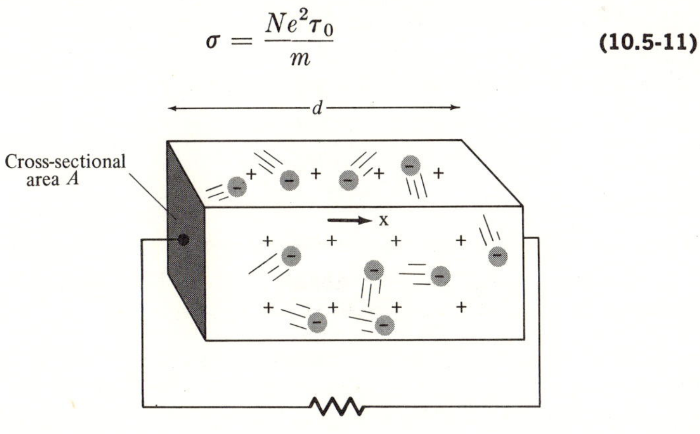

Figure 10-8 Model of a resistance used in deriving the Johnson-noise formula.

[12] The derivation of (10.5-11) can be found in any introductory book on solid-state physics.

where m is the mass of the electron.[13] The sample resistance is thus

$$R = \frac{d}{\sigma A} = \frac{md}{Ne^2\tau_0 A} \tag{10.5-12}$$

We apply next the results of Section 10-3 to the problem and choose as our basic single event the current pulse $i_e(t)$ in the external circuit due to the motion of *one* electron between two successive scattering events. Using (10.4-1), we write

$$i_e(t) = \begin{cases} \dfrac{ev_x}{d} & 0 \leqslant t \leqslant \tau \\ 0 & \text{otherwise} \end{cases} \tag{10.5-13}$$

where v_x is the x component of the velocity (assumed constant) and where τ is the scattering time of the electron under observation. Taking the Fourier transform of $i_e(t)$, we have

$$I_e(\omega, \tau, v_x) = \int_0^\tau i_e(t)e^{-i\omega t}\,dt = \frac{ev_x}{-i\omega d}[e^{-i\omega\tau} - 1] \tag{10.5-14}$$

from which

$$|I_e(\omega, \tau, v_x)|^2 = \frac{e^2 v_x^2}{\omega^2 d^2}[2 - e^{i\omega\tau} - e^{-i\omega\tau}] \tag{10.5-15}$$

According to (10.3-10) we need to average $|I_e(\omega, \tau, v_x)|^2$ over the parameters τ and v_x. We assume that τ and v_x are independent variables—that is, that the probability function

$$p(\alpha) = p(\tau, v_x) = g(\tau)f(v_x)$$

is the product of the individual probabilities [1]—and take $g(\tau)$ as[14]

$$g(\tau) = \frac{1}{\tau_0}e^{-\tau/\tau_0} \tag{10.5-16}$$

---

[13] In a semiconductor we use the effective mass of the charge carrier.

[14] If the collision probability per carrier per unit times is $1/\tau_0$ and $q(t)$ is the probability that an electron *has not* collided by time t, we have:

$$q'(t) = -q(t)\frac{1}{\tau_0} \Rightarrow q(t) = e^{-t/\tau_0}$$

Taking $g(\tau)\,d\tau$ as the probability that a collision will occur between τ and $\tau + d\tau$, it follows that

$$q(t) = 1 - \int_0^t g(t')\,dt'$$

and thus

$$g(t) = -\frac{dq}{dt} = \frac{1}{\tau_0}e^{-t/\tau_0}$$

as in (10.5-16).

and, performing the averaging over τ, obtain

$$\overline{|I_e(\omega, v_x)|^2} = \int_0^\infty g(\tau)|I_e(\omega, v_x, \tau)|^2\, d\tau = \frac{2e^2 v_x^2 \tau_0^2}{d^2(1 + \omega^2\tau_0^2)} \qquad \textbf{(10.5-17)}$$

The second averaging over v_x^2 is particularly simple, since it results in the replacement of v_x^2 in (10.5-17) by its average $\overline{v_x^2}$, which, for a sample at thermal equilibrium, is given according to (10.5-10) by $\overline{v_x^2} = kT/m$. The final result is then

$$\overline{|I_e(\omega)|^2} = \frac{2e^2\tau_0^2 kT}{md^2(1 + \omega^2\tau_0^2)} \qquad \textbf{(10.5-18)}$$

The average number of scattering events per second $\overline{N}$ is equal to the total number of electrons NV divided by the mean scattering time τ_0

$$\overline{N} = \frac{NV}{\tau_0} \qquad \textbf{(10.5-19)}$$

thus, from (10.3-10), we obtain

$$S(\nu) = 2\bar{N}\,\overline{|I_e(\omega)|^2} = \frac{4NVe^2\tau_0 kT}{md^2(1 + \omega^2\tau_0^2)}$$

and, after using (10.5-12) and limiting ourselves as in (10.4-6) to frequencies where $\omega\tau_0 \ll 1$, we get

$$\overline{i_N^2}(\nu) \equiv S(\nu)\Delta\nu = \frac{4kT\Delta\nu}{R} \qquad \textbf{(10.5-20)}$$

in agreement with (10.5-9).

10.6 Spontaneous Emission Noise in Laser Oscillators

Another type of noise that plays an important role in quantum electronics is that of spontaneous emission in laser oscillators and amplifiers. As shown in Chapter 5, a necessary condition for laser amplification is that the atomic population of a pair of levels 1 and 2 be inverted. If $E_2 > E_1$, gain occurs when $N_2 > N_1$. Assume that an optical wave with frequency $\nu \simeq (E_2 - E_1)/h$ is propagating through an inverted population medium. This wave will grow coherently due to the effect of stimulated emission. In addition, its radiation will be contaminated by noise radiation caused by spontaneous emission from level 2 to level 1. Some of the radiation emitted by the spontaneous emission will propagate very nearly along the same direction as that of the stimulated emission and cannot be separated from it. This has two main consequences. First, the laser output has a finite spectral width. This effect is described in this section. Second, the signal-to-noise ratio achievable at the output of laser amplifiers [7] is limited because of the intermingling of spontaneous emission noise power with that of the amplified signal. (See Figure 10-9.)

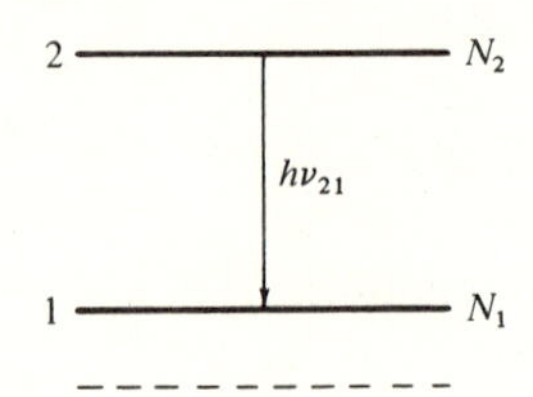

Figure 10-9 An atomic transition with $N_2 > N_1$ providing gain for laser oscillation.

Returning to the case of a laser oscillator, we represent it by an *RLC* circuit, as shown in Figure 10-10. The presence of the laser medium with negative loss (that is, gain) is accounted for by including a negative conductance $-G_m$ while the ordinary loss mechanisms described in Chapter 6 are represented by the positive conductance G_0. The noise generator associated with the losses G_0 is given according to (10.5-9) as

$$\overline{i_N{}^2} = \frac{4\hbar\omega G_0(\Delta\omega/2\pi)}{e^{\hbar\omega/kT} - 1}$$

where T is the actual temperature of the losses. Spontaneous emission is represented by a similar expression[15]

$$(\overline{i_N{}^2})_{\substack{\text{spont.}\\\text{emission}}} = \frac{4\hbar\omega(-G_m)(\Delta\omega/2\pi)}{e^{\hbar\omega/kT_m} - 1} \tag{10.6-1}$$

where the term $(-G_m)$ represents negative losses and T_m is a temperature determined by the population ratio according to

$$\frac{N_2}{N_1} = e^{-\hbar\omega/kT_m} \tag{10.6-2}$$

Since $N_2 > N_1$, then $T_m < 0$, $(\overline{i_N{}^2})$ in (10.6-1) is positive definite.

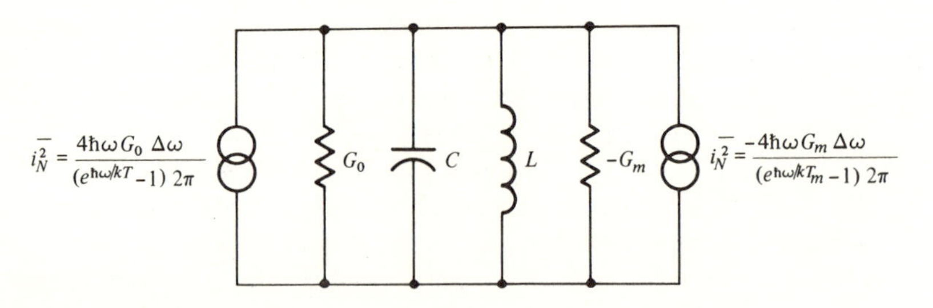

Figure 10-10 Equivalent circuit of a laser oscillator.

[15] The 2π factor appearing in the denominators of $\overline{i_N{}^2}$ is due to the fact that here we use $\overline{i_N{}^2}(\omega)$ instead of $\overline{i_N{}^2}(\nu)$ with

$$\overline{i_N{}^2}(\omega)\Delta\omega = \overline{i_N{}^2}(\nu)\Delta\nu$$

Although a detailed justification of (10.6-1) is outside the scope of the present treatment, a strong case for its plausibility can be made by noting that since $G_m \propto N_2 - N_1$, $(\overline{i_N^2})$ in (10.6-1) can be written, using (10.6-2), as[16]

$$(\overline{i_N^2})_{\substack{\text{spont.}\\ \text{emission}}} \propto \frac{-4\hbar\omega\Delta\omega(N_2 - N_1)}{(N_1/N_2) - 1} = 4\hbar\omega\Delta\omega N_2 \tag{10.6-3}$$

and is thus proportional to N_2. This makes sense, since spontaneous emission power is due to $2 \rightarrow 1$ transitions and should consequently be proportional to N_2.

Returning to the equivalent circuit, its quality factor Q is given by

$$Q^{-1} = \frac{G_0 - G_m}{\omega_0 C} = \frac{1}{Q_0} - \frac{1}{Q_m} \tag{10.6-4}$$

where $\omega_0^2 = (LC)^{-1}$. The circuit impedance is

$$\begin{aligned} Z(\omega) &= \frac{1}{(G_0 - G_m) + (1/i\omega L) + i\omega C} \\ &= \frac{i\omega}{C} \frac{1}{(i\omega\omega_0/Q) + (\omega_0^2 - \omega^2)} \end{aligned} \tag{10.6-5}$$

so the voltage across this impedance due to a current source with a complex amplitude $I(\omega)$ is

$$V(\omega) = \frac{i}{C} \frac{I(\omega)}{[(\omega_0^2 - \omega^2)/\omega] + (i\omega_0/Q)} \tag{10.6-6}$$

which, near $\omega = \omega_0$, becomes

$$\overline{|V(\omega)|^2} = \frac{1}{4C^2} \frac{\overline{|I(\omega)|^2}}{(\omega_0 - \omega)^2 + (\omega_0^2/4Q^2)} \tag{10.6-7}$$

The current sources driving the resonant circuit are those shown in Figure 10-10; since they are not correlated, we may take $|I(\omega)|^2$ as the sum of their mean-square values

$$\overline{|I(\omega)|^2} = 4\hbar\omega \left[\frac{G_m N_2}{N_2 - N_1} + \frac{G_0}{e^{\hbar\omega/kT} - 1}\right] \frac{d\omega}{2\pi} \tag{10.6-8}$$

---

[16] The proportionality of G_m to $N_2 - N_1$ can be justified by noting that in the equivalent circuit (Figure 10.10) the stimulated emission power is given by $v^2 G_m$ where v is the voltage. Using the field approach, this power is proportional to $E^2(N_2 - N_1)$ where E is the field amplitude. Since v is proportional to E, G_m is proportional to $N_2 - N_1$.

where in the first term inside the square brackets we used (10.6-2). In the optical region, $\lambda = 1\ \mu$m say, and for $T = 300°$K we have $\hbar\omega/kT \simeq 50$; thus, since near oscillation $G_m \simeq G_0$, we may neglect the thermal (Johnson) noise term in (10.6-8), thereby obtaining

$$\overline{|V(\omega)|^2}_{\omega \simeq \omega_0} = \frac{\hbar G_m}{2\pi C^2}\left(\frac{N_2}{N_2 - N_1}\right)\frac{\omega\, d\omega}{(\omega_0 - \omega)^2 + (\omega_0^2/4Q^2)} \tag{10.6-9}$$

Equation (10.6-9) represents the spectral distribution of the laser output. If we subject the output to high-resolution spectral analysis we should, according to (10.6-9), measure a linewidth

$$\Delta\omega = \frac{\omega_0}{Q} \tag{10.6-10}$$

between the half-intensity points. The trouble is that, though correct, (10.6-10) is not of much use in practice. The reason is that according to (10.6-4), Q^{-1} is equal to the difference of two nearly equal quantities neither of which is known with high enough accuracy. We can avoid this difficulty by showing that Q is related to the laser power output, and thus $\Delta\omega$ may be expressed in terms of the power.

The total power extracted from the atoms comprising the laser is

$$\begin{aligned} P &= G_0 \int_0^\infty \frac{\overline{|V(\omega)|^2}}{d\omega}\, d\omega \\ &= \frac{\hbar G_m G_0}{2\pi C^2}\left(\frac{N_2}{N_2 - N_1}\right)\int_0^\infty \frac{\omega\, d\omega}{(\omega_0 - \omega)^2 + (\omega_0/2Q)^2} \end{aligned} \tag{10.6-11}$$

Since the integrand peaks sharply near $\omega \simeq \omega_0$, we may replace ω in the numerator of (10.6-11) by ω_0 and after integration obtain

$$P = \frac{\hbar G_m G_0 Q}{C^2}\left(\frac{N_2}{N_2 - N_1}\right) \tag{10.6-12}$$

which is the desired result linking P to Q. In a laser oscillator the gain very nearly equals the loss, or in our notation, $G_m \simeq G_0$. Using this result in (10.6-12), we obtain

$$Q = \frac{C^2}{\hbar G_0^2}\left(\frac{N_2 - N_1}{N_2}\right) P$$

which, when substituted in (10.6-10), yields

$$\Delta\nu = \frac{2\pi h\nu_0(\Delta\nu_{1/2})^2}{P}\left(\frac{N_2}{N_2 - N_1}\right) \tag{10.6-13}$$

where $\Delta\nu_{1/2}$ is the full width of the passive cavity resonance given in (4.7-6) as $\Delta\nu_{1/2} = \nu_0/Q_0 = (1/2\pi)(G_0/C)$.

Numerical example. Consider a laser oscillator with the following characteristics

$$\nu = 4.73 \times 10^{14}\ \text{Hz}$$

$$l = 100\ \text{cm}$$

$$\text{Loss} = 1\ \text{percent per pass (Loss} \equiv 1 - R)$$

$$P = 1\ \text{mW}$$

$$N_2 \gg N_1$$

These numbers are typical of low-power laboratory He-Ne lasers. From (4.6-6) we get

$$\Delta\nu_{1/2} = \frac{1}{2\pi t_c} = \frac{c(1 - R)}{2\pi n l} \approx 5 \times 10^5$$

Using the foregoing data in (10.6-13) gives

$$\Delta\nu \simeq 2 \times 10^{-3}\ \text{Hz}$$

for the spectral width of the laser output. We must emphasize, however, that $\Delta\nu$ as given by (10.6-13) represents a theoretical limit and does not necessarily correspond to the value commonly observed in the laboratory. The output of operational laser is broadened mostly by thermal and acoustic fluctuations in the optical resonator length, which cause the resonance frequencies to shift about rapidly. An experimental determination of the limiting $\Delta\nu$ requires great care in acoustic isolation and thermal stabilization; see References [11] and [12]. An observed value of $\Delta\nu \sim 10^3$ Hz reported in [11] in a 1-mW He-Ne laser is still limited by vibrations and thermal fluctuations.

Figure 10-11 shows the output spectrum of a $Pb_{0.88}Sn_{0.12}Te$ injection laser at 10.6 μm [13]. The narrowing of the spectrum from $\Delta\nu = 1.75$ MHz to $\Delta\nu = 0.75$ MHz is consistent with the inverse dependence on P predicted by (10.6-13).

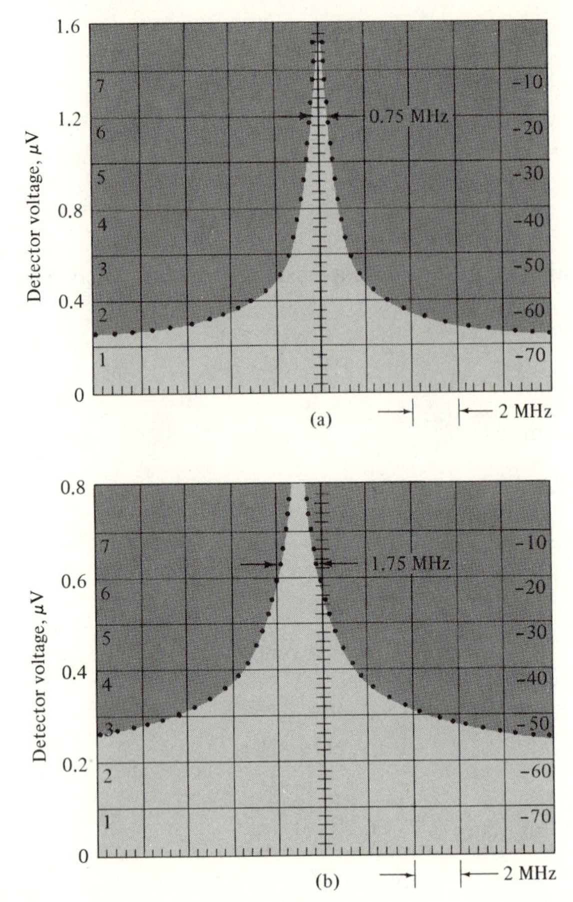

Figure 10-11 Spectrum analyzer display of beat note between low-power diode laser mode and P(16) transition of a CO_2 gas laser, corresponding to diode laser current of 865 mA in (**a**) and 845 mA in (**b**). Center frequency is 92 MHz. The dotted curves correspond to a Lorentzian lineshape, with the indicated half-power linewidths. (After Reference [13].)

10.7 Error Probability in a Binary Pulse Code Modulation System

The simplicity and reliability of digital processing by integrated electronic circuits has made it increasingly attractive to transmit information in the form of binary pulse trains. In the case of optical communication systems the analog data to be transmitted are coded into a train of 1 and 0 electrical pulses so that each pulse carries one bit of information. The electrical signal thus generated is impressed by means of a modulator on an optical beam, resulting in an optical train pulse. The optical signal having propagated through air or an optical fiber, is detected in the receiving end, thus yielding an electrical train of pulses.

Now, ideally, the reconstructed train of electrical pulses should be an exact replica of (or, more generally, constitute an exact analog of) the input train. The intermingling of noise at the detector output with the signal makes this perfect reconstruction impossible. A figure of merit used to describe the "quality" of the reconstructed signal is the *error probability*, which is defined as the probability that any given pulse in the detected train does not agree with the corresponding pulse in the input train.

Figure 10-12 shows part of a pulse sequence containing three "1" pulses and two "0" pulses. An ideal noiseless detection should yield the sequence [Figure 10-12(a)] where the pulse height (say in amperes) is i_S. The presence of noise, however, introduces random fluctuations so that the detected signal may appear as in Figure 10-12(b).

A threshold decision circuit is usually employed which samples the signal [Figure 10-12(b)] once each period yielding a 1 pulse if the sample exceeds a predetermined value ki_S ($k < 1$), and a 0 pulse if the measured sample is smaller than ki_S [14]. In the case shown in Figure 10-12(b) the choice of the indicated threshold value will lead to a correct reconstruction of all pulses except the last one, where a negative noise fluctuation has conspired to keep the pulse below the threshold value.

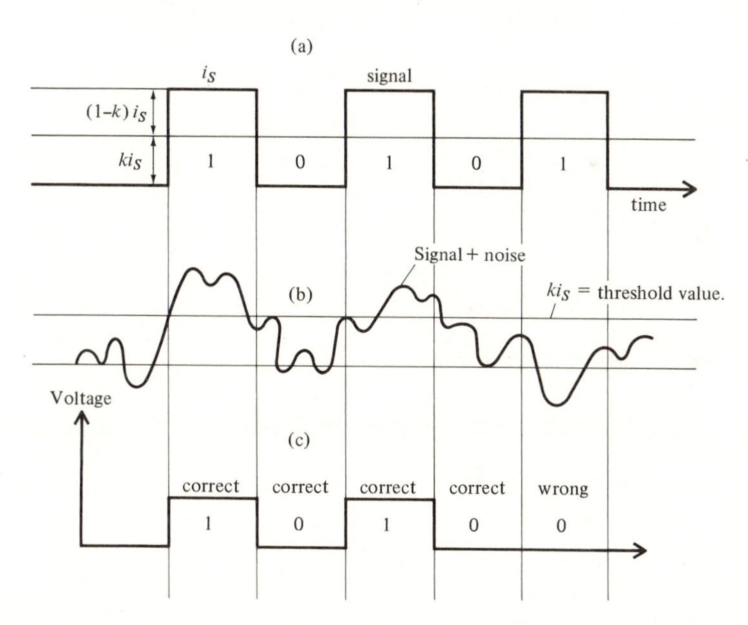

Figure 10-12 An ideal noiseless pulse train (**a**) is contaminated by noise as in (**b**). A reconstruction using a threshold decision level kI_S leads to (**c**). Note that the reconstruction of the last "1" pulse is in error because of a large negative noise fluctuation.

If a given pulse is a "1," then an erroneous reconstruction would result if during the sampling the noise current i_N is negative and such that

$$i_N < -i_S(1-k)$$

since in this case $i_S + i_N$ is smaller than the threshold value ki_S and "0" will result. In a like manner the reconstruction of a "0" pulse will be in error if

$$i_N > ki_S$$

On the average half the pulses are 0 and half are 1 so that the probability of a wrong reconstruction of any given pulse is

$$P_e = \tfrac{1}{2}[\text{Probability that } i_N < -i_S(1-k) + \text{Probability that } i_N > ki_S] \tag{10.7-1}$$

If the noise current i_N is a random Gaussian variable which is the case in most applications, we can use (10.1-5) to evaluate the error probability. In this case σ is the root mean square (rms) value of the noise current i_N, so that $\sigma^2 = \overline{i_N^2}$ is the mean square noise current, as derived in Sections 10.4, 10.5, and 10.6. To simplify the result, let us choose $k = \frac{1}{2}$. Using the fact that, according to (10.1-5), $p(i_N) = p(-i_N)$ (10.7-1) becomes

$$P_e = \text{probability that } i_N > \frac{i_S}{2}$$

$$= \int_{i_s/2}^{\infty} p(i_N)\, di_N = \frac{1}{\sigma\sqrt{2\pi}} \int_{i_s/2}^{\infty} e^{-(i_N^2/2\sigma^2)}\, di_N \tag{10.7-2}$$

$$= \frac{1}{\sqrt{\pi}} \int_{i_s/2\sqrt{2}\sigma}^{\infty} e^{-\xi^2}\, d\xi \tag{10.7-3}$$

Using the definition of the error function

$$\text{erf } z = \frac{2}{\sqrt{\pi}} \int_0^z e^{-\xi^2}\, d\xi$$

we can write (10.7-3) as

$$P_e = \frac{1}{2}\left[1 - \text{erf}\,\frac{i_S}{2\sqrt{2}\sigma}\right]$$
$$\equiv \frac{1}{2}\,\text{erfc}\,\frac{i_S}{2\sqrt{2}\sigma} = \tfrac{1}{2}\,\text{erfc}\,\frac{i_S}{2\sqrt{2}\langle i_N\rangle} \tag{10.7-4}$$

where $\langle i_N\rangle \equiv \sigma = (\overline{i_N^2})^{1/2}$ is the rms noise current.

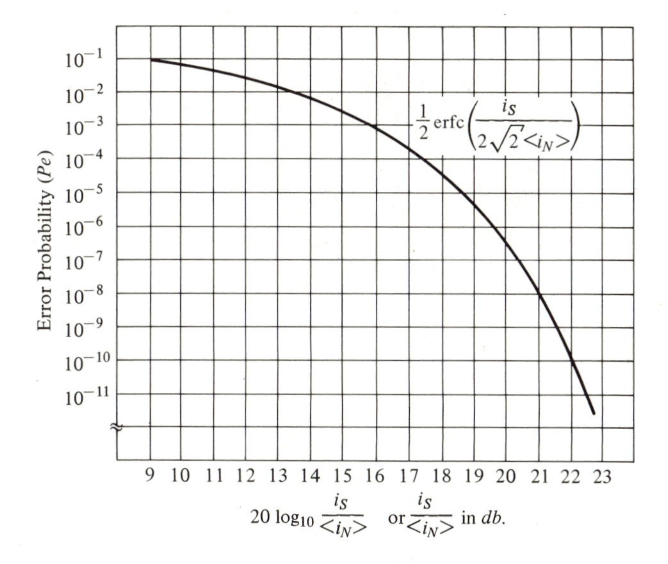

Figure 10-13 Plot of Equation (10.11-4) for the error probability as a function of the (peak) signal to noise current ratio at the detector output.

A theoretical plot of P_e as a function of the (peak) signal to noise ratio $i_S/\langle i_N \rangle$ is shown in Figure 10-13. We recall that S represents the electrical signal power at the detector output and not the optical power. It is interesting to note the extremely small error probabilities resulting from even moderate signal to noise power ratios. As an example, $P_e = 10^{-9}$ when $i_S/\langle i_N \rangle = 11.89$ (21.5 db).

Experimental measurement of error probability in a detected optical pulse train is described by Reference [15]. Other pertinent discussions are to be found in References [16] and [17].

A detailed example using the results of this section in designing a binary optical fiber communication system appears at the end of Chapter 11.

PROBLEMS

10-1 Derive Equations (10.4-3) and (10.4-4). (*Hint:* Apply the relation

$$\int_s \mathbf{D} \cdot \mathbf{n}\, ds = \int_v \rho\, dv$$

to a differential volume containing the charge sheet.)

10-2 Derive the shot noise formula without making the restriction (Equation 10.4-6) $\omega t_a \gg 1$. Assume the carriers move between the electrodes at a constant velocity.

10-3 Derive Equation (10.5-11).

10-4 Complete the missing steps in the derivation of Equation (10.5-20).

10-5 Estimate the scattering time τ_0 of carriers in copper at $T = 300°K$ using a tabulated value for its conductivity. At what frequencies is the condition $\omega\tau_0 \ll 1$ violated?

10-6 Repeat Problem 10-5 for a material with a carrier density of 10^{22} cm^{-3} and $\sigma = 10^{-5}$ (ohm-cm)$^{-1}$.

10-7 What is the change $\Delta\nu$ in the resonant frequency of a laser whose cavity length changes by Δl?

10-8 a. Estimate the frequency smearing $\Delta\nu$ of a laser in which fused-quartz rods are used to determine the length of the optical cavity in an environment where the temperature stability is $\pm 0.5°K$. [*Caution:* Do not forget the dependence of n on T.]

b. What temperature stability is needed to reduce $\Delta\nu$ to less than 10^3 Hz?

10-9 Derive expression (10.5-9), $\overline{i_N^2(\omega)} = 4kT\Delta\nu/R$, for the Johnson noise by considering a high-Q parallel RLC circuit which is shunted by a current source of mean-square amplitude $\overline{i_N^2(\omega)}$. The magnitude of $\overline{i_N^2(\omega)}$ is to be chosen so that the resulting excitation of the circuit corresponds to a stored electromagnetic energy of kT. [*Hint:* Since the magnetic and electric energies are equal, then

$$kT = C\overline{v^2(t)} = C\int_0^\infty \frac{\overline{V_N^2(\omega)}}{d\omega}\,d\omega$$

where $\overline{V_N^2(\omega)} = \overline{i_N^2(\omega)}|Z(\omega)|^2$. Also assume that $\overline{i_N^2(\omega)}/\Delta\nu$ is independent of frequency.]

10-10 Derive and plot the error probability as a function of $i_S/\langle i_N\rangle$ for (**a**) $k = 0.75$, (**b**) $k = 0.25$.

REFERENCES

[1] The basic concepts of noise theory used in this chapter can be found, for example, in W. B. Davenport and W. L. Root, *An Introduction to the Theory of Random Signals and Noise.* New York: McGraw-Hill, 1958.

[2] Bennett, W. R., "Methods of solving noise problems," *Proc. IRE*, vol. 44, p. 609, 1956.

[3] The classic reference to this topic is: S. O. Rice, "Mathematical analysis of random noise," *Bell System Tech. J.*, vol. 23, p. 282, 1944; vol. 24, p. 46, 1945.

[4] Johnson, J. B., "Thermal agitation of electricity in conductors," *Phys. Rev.*, vol. 32, p. 97, 1928.

[5] Nyquist, H., "Thermal agitation of electric charge in conductors," *Phys. Rev.*, vol. 32, p. 110, 1928.

[6] Smith, R. A., F. A. Jones, and R. P. Chasmar, *The Detection and Measurement of Infrared Radiation*. New York: Oxford, 1968.

[7] Yariv, A., *Quantum Electronics*. 2d Ed. New York: Wiley, 1975.

[8] Gordon, J. P., H. J. Zeiger, and C. H. Townes, "The maser—New type of microwave amplifier, frequency standard and spectrometer," *Phys. Rev.*, vol. 99, p. 1264, 1955.

[9] Gordon, E. I., "Optical maser oscillators and noise," *Bell System Tech. J.*, vol. 43, p. 507, 1964.

[10] Grivet, P. A., and A. Blaquiere, *Optical Masers*. New York: Polytechnic Press, 1963, p. 69.

[11] Jaseja, T. J., A. Javan, and C. H. Townes, "Frequency stability of He-Ne masers and measurements of length," *Phys. Rev. Letters*, vol. 10, p. 165, 1963.

[12] Egorov, Y. P., "Measurements of natural line width of the emission of a gas laser with coupled modes," *JETP Letters*, vol. 8, p. 320, 1968.

[13] Hinkley, E. D., and C. Freed, "Direct observation of the Lorentzian lineshape as limited by quantum phase noise in a laser above threshold," *Phys. Rev. Letters*, vol. 23, p. 277, 1969.

[14] Bennett, W. R., and J. R. Davey, *Data Transmission*, New York: McGraw-Hill, 1965, p. 100.

[15] Goell, J. E., "A 274 Mb/s optical repeater experiment employing a GaAs laser," *Proc. IEEE*, vol. 61, p. 1504, 1973.

[16] Personick, S. D., "Receiver design for digital fiber optic communication systems," *Bell Syst. Tech. J.*, vol. 52, p. 843, 1973.

[17] Miller, S. E., T. Li, and E. A. J. Marcatili, "Toward optic fiber transmission systems—devices and systems considerations," *Proc. IEEE*, vol. 61, p. 1726, 1973.

CHAPTER

11

Detection of Optical Radiation

11.0 Introduction

The detection of optical radiation is often accomplished by converting the radiant energy into an electric signal whose intensity is measured by conventional techniques. Some of the physical mechanisms that may be involved in this conversion include

1. the generation of mobile charge carriers in solid-state photoconductive detectors,
2. changing through absorption the temperature of thermocouples, thus causing a change in the junction voltage, and
3. the release by the photoelectric effect of free electrons from photoemissive surfaces.

In this chapter we consider in some detail the operation of four of the most important detectors:

1. the photomultiplier,
2. the photoconductive detector,
3. the photodiode, and
4. the avalanche photodiode.

The limiting sensitivity of each is discussed and compared to the theoretical limit. We will find that by use of the heterodyne mode of detection the theoretical limit of sensitivity may be approached.

11.1 Optically Induced Transition Rates

A common feature of all the optical detection schemes discussed in this chapter is that the electric signal is proportional to the rate at which electrons are excited by the optical field. This excitation involves a transition of the electron from some initial bound state, say a, to a final state (or a group of states) b in which it is free to move and contribute to the current flow. For example, in an n-type photoconductive detector, state a corresponds to electrons in the filled valence band or localized donor impurity atoms, while state b corresponds to electrons in the conduction band. The two levels involved are shown schematically in Figure 11-1. A photon of energy $h\nu$ is absorbed in the process of exciting an electron from a "bound" state a to a "free" state b in which the electron can contribute to the current flow.

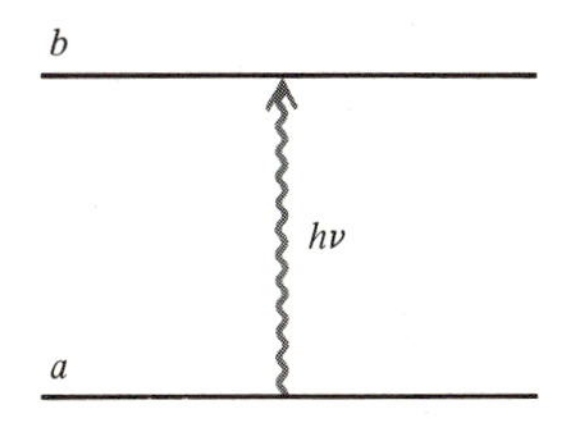

Figure 11-1 Most high speed optical detectors depend on absorption of photons of energy $h\nu$ accompanied by a simultaneous transition of an electron (or hole) from a quantum state of low mobility (a) to one of higher mobility (b).

An important point to understand before proceeding with the analysis of different detection schemes is the manner of relating the transition rate per electron from state a to b to the intensity of the optical field. This rate is derived by quantum mechanical considerations.[1] In our case it can be stated in the following form: Given a nearly sinusoidal[2] optical field

$$e(t) = \tfrac{1}{2}[E(t)e^{i\omega_0 t} + E^*(t)e^{-i\omega_0 t}] \equiv \text{Re}\,[V(t)] \qquad \textbf{(11.1-1)}$$

where $V(t) = E(t)\exp(i\omega_0 t)$,[3] the transition rate per electron induced by this field is proportional to $V(t)V^*(t)$. Denoting the transition rate as $W_{a\to b}$, we have

$$W_{a\to b} \propto V(t)V^*(t) \qquad \textbf{(11.1-2)}$$

We can easily show that $V(t)V^*(t)$ is equal to twice the average value of $e^2(t)$, where the averaging is performed over a few optical periods.

---

[1] More specifically, from first order time-dependent perturbation theory; see for example, Reference [1].

[2] By "nearly sinusoidal" we mean a field where $E(t)$ varies slowly compared to $\exp(i\omega_0 t)$ or, equivalently, where the Fourier spectrum of $E(t)$ occupies a bandwidth that is small compared to ω_0. Under these conditions the variation of the amplitude $E(t)$ during a few optical periods can be neglected.

[3] $V(t)$ is referred to as the "analytic signal" of $e(t)$. See Problem 1.1.

To illustrate the power of this seemingly simple result, consider the problem of determining the transition rate due to a field

$$e(t) = E_0 \cos(\omega_0(t) + \phi_0) + E_1 \cos(\omega_1 t + \phi_1) \tag{11.1-3}$$

taking $\omega_1 - \omega_0 \equiv \omega \ll \omega_0$. We can rewrite (11.1-3) as

$$\begin{aligned} e(t) &= \mathrm{Re}\,(E_0 e^{i(\omega_0 t + \phi_0)} + E_1 e^{i(\omega_1 t + \phi_1)}) \\ &= \mathrm{Re}\,[(E_0 e^{i\phi_0} + E_1 e^{i(\omega t + \phi_1)}) e^{i\omega_0 t}] \end{aligned} \tag{11.1-4}$$

and, using (11.1-1), identify $V(t)$ as

$$V(t) = [E_0 e^{i\phi_0} + E_1 e^{i(\omega t + \phi_1)}] e^{i\omega_0 t}$$

thus, using (11.1-2), we obtain

$$\begin{aligned} W_{a \to b} &\propto (E_0 e^{i\phi_0} + E_1 e^{i(\omega t + \phi_1)})(E_0 e^{-i\phi_0} + E_1 e^{-i(\omega t + \phi_1)}) \\ &= E_0^2 + E_1^2 + 2E_0 E_1 \cos(\omega t + \phi_1 - \phi_0) \end{aligned} \tag{11.1-5}$$

This shows that the transition rate has, in addition to a constant term $E_0^2 + E_1^2$, a component oscillating at the difference frequency ω with a phase equal to the difference of the two original phases. This coherent "beating" effect forms the basis of the heterodyne detection scheme, which is discussed in detail in Section 11.4.

11.2 Photomultiplier

The photomultiplier, one of the most common optical detectors, is used to measure radiation in the near ultraviolet, visible, and near infrared regions of the spectrum. Because of its inherent high current amplification and low noise, the photomultiplier is one of the most sensitive instruments devised by man and under optimal operation—which involves long integration time, cooling of the photocathode, and pulse-height discrimination—has been used to detect power levels as low as about 10^{-19} watts [2].

A schematic diagram of a conventional photomultiplier is shown in Figure 11–2. It consists of a photocathode (C) and a series of electrodes, called dynodes, which are labeled 1 through 8. The dynodes are kept at progressively higher potentials with respect to the cathode, with a typical potential difference between adjacent dynodes of 100 volts. The last electrode (A), the anode, is used to collect the electrons. The whole assembly is contained within a vacuum envelope in order to reduce the possibility of electronic collisions with gas molecules.

The photocathode is the most crucial part of the photomultiplier, since it converts the incident optical radiation to electronic current and thus determines the wavelength-response characteristics of the detector and, as will be seen, its limiting sensitivity. The photocathode consists of materials with low surface work functions. Compounds involving Ag-O-Cs and

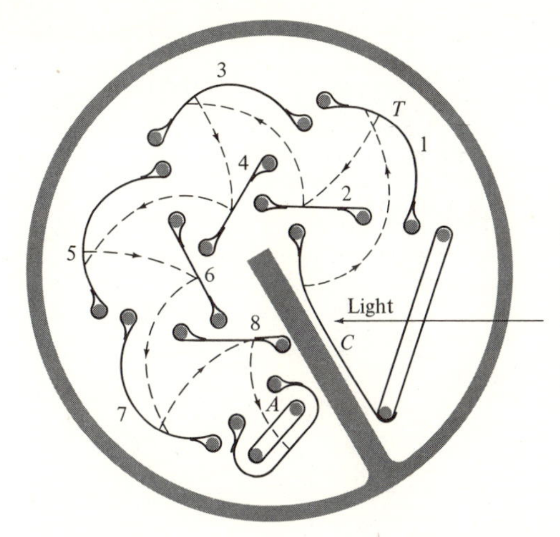

Figure 11-2 Photocathode and focusing dynode configuration of a typical commercial photomultiplier. C = cathode; 1–8 = secondary-emission dynodes; A = collecting anode. (After Reference [3].)

Sb-Cs are often used; see References [2] and [3]. These compounds possess work functions as low as 1.5 eV as compared to 4.5 eV in typical metals. As can be seen in Figure 11-3, this makes it possible to detect photons with longer wavelengths. It follows from the figure that the low-frequency detection limit corresponds to $h\nu = \phi$. At present the lowest-work-function materials make possible photoemission at wavelengths as long as 1–1.1 μm.

Spectral response curves of a number of commercial photocathodes are shown in Figure 11-4. The quantum efficiency (or quantum yield as it is often called) is defined as the number of electrons released per incident photon.

The electrons that are emitted from the photocathode are focused electrostatically and accelerated toward the first dynode arriving with a kinetic energy of, typically, about 100 eV. Secondary emission from dynode surfaces causes a multiplication of the initial current. This process repeats itself at each dynode until the initial current emitted by the photocathode is amplified by a very large factor. If the average secondary emission multiplication at each dynode is δ (that is, δ secondary electrons for each incident one) and the number of dynodes is N, the total current multipli-

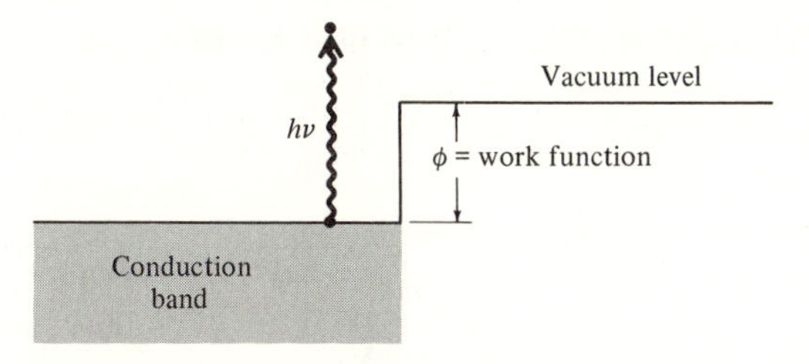

Figure 11-3 Photomultiplier photocathode. The vacuum level corresponds to the energy of an electron at rest an infinite distance from the cathode. The work function ϕ is the minimum energy required to lift an electron from the metal into the vacuum level, so only photons with $h\nu > \phi$ can be detected.

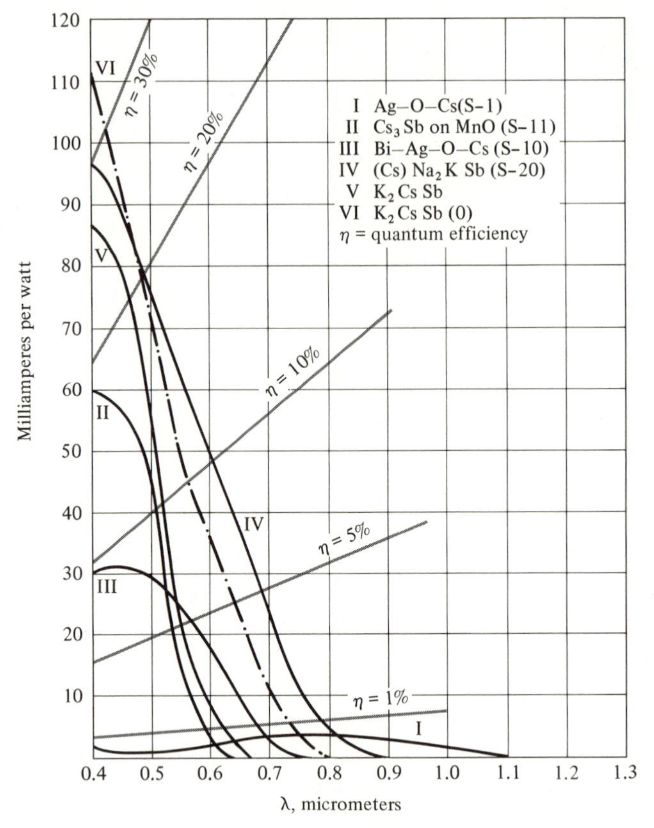

Figure 11-4 Photoresponse versus wavelength characteristics and quantum efficiency of a number of commercial photocathodes. (After Reference [3], p. 228.)

cation between the cathode and anode is

$$G = \delta^N$$

which for typical values[4] of $\delta = 5$ and $N = 9$ gives $G \simeq 2 \times 10^6$.

11.3 Noise Mechanisms in Photomultipliers

The random fluctuations observed in the photomultiplier output are due to:

1. Cathode shot noise, given according to (10.4-9) by

$$\overline{(i_{N1}{}^2)} = G^2 2e(\bar{i}_c + i_d)\Delta\nu \qquad \textbf{(11.3-1)}$$

[4] The value of δ depends on the voltage V between dynodes and values of $\delta \simeq 10$ can be obtained (for $V \simeq 400$ volts). In commercial tubes, values of $\delta \simeq 5$, achievable with $V \simeq 100$ volts, are commonly used.

where $\bar{i}_c$ is the average current emitted by the photocathode due to the signal power which is incident on it. The current i_d is the so-called "dark current," which is due to random thermal excitation of electrons from the surface as well as to excitation by cosmic rays and radioactive bombardment.

2. Dynode shot noise, which is the shot noise due to the random nature of the secondary emission process at the dynodes. Since current originating at a dynode does not exercise the full gain of the tube, the contribution of all the dynodes to the total shot noise output is smaller by a factor of $\sim\delta^{-1}$ than that of the cathode; since $\delta \simeq 5$ it amounts to a small correction and will be ignored in the following.

3. Johnson noise, which is the thermal noise associated with the output resistance R connected across the anode. Its magnitude is given by (10.5-9) as

$$\overline{(i_{N2}{}^2)} = \frac{4kT\Delta\nu}{R} \tag{11.3-2}$$

Minimum detectable power in photomultipliers—video detection. Photomultipliers are used primarily in one of two ways. In the first, the optical wave to be detected is modulated at some low frequency ω_m before impinging on the photocathode. The signal consists then, of an output current oscillating at ω_m, which, as will be shown below, has an amplitude proportional to the optical intensity. This mode of operation is known as *video*, or straight, detection.

In the second mode of operation, the signal to be detected, whose optical frequency is ω_s, is combined at the photocathode with a much stronger optical wave of frequency $\omega_s + \omega$. The output signal is then a current at the offset frequency ω. This scheme, known as *heterodyne* detection, will be considered in detail in Section 11-4.

The optical signal in the case of video detection may be taken as

$$\begin{aligned} e_s(t) &= E_s(1 + m \cos \omega_m t) \cos \omega_s t \\ &= \mathrm{Re}\,[E_s(1 + m \cos \omega_m t)e^{i\omega_s t}] \end{aligned} \tag{11.3-3}$$

where the factor $(1 + m \cos \omega_m t)$ represents amplitude modulation of the carrier.[5] The photocathode current is given, according to (11.1-2), by

$$\begin{aligned} i_c(t) &\propto [E_s(1 + m \cos \omega_m t)]^2 \\ &= E_s{}^2\left[\left(1 + \frac{m^2}{2}\right) + 2m \cos \omega_m t + \frac{m^2}{2} \cos 2\omega_m t\right] \end{aligned} \tag{11.3-4}$$

To determine the proportionality constant involved in (11.3-4), consider

[5] The amplitude modulation can be due to the information carried by the optical wave or, as an example, to chopping before detection.

the case of $m = 0$. The average photocathode current due to the signal is then[6]

$$\bar{i}_c = \frac{Pe\eta}{h\nu_s} \tag{11.3-5}$$

where $\nu_s = \omega_s/2\pi$, P is the average optical power, and η (the quantum efficiency) is the average number of electrons emitted from the photocathode per incident photon. This number depends on the photon frequency, the photocathode surface, and in practice (see Figure 11-4) is found to approach 0.3. Using (11.3-5), we rewrite (11.3-4) as

$$i_c(t) = \frac{Pe\eta}{h\nu_s}\left[\left(1 + \frac{m^2}{2}\right) + 2m \cos \omega_m t + \frac{m^2}{2} \cos 2\omega_m t\right] \tag{11.3-6}$$

The signal output current at ω_m is

$$i_s = \frac{GPe\eta}{h\nu_s}(2m) \cos \omega_m t \tag{11.3-7}$$

If the output of the detector is limited by filtering to a bandwidth $\Delta\nu$ centered on ω_m, it contains a shot-noise current, which, according to (11.3-1), has a mean-squared amplitude

$$\overline{(i_{N1}{}^2)} = 2G^2e(\bar{i}_c + i_d)\Delta\nu \tag{11.3-8}$$

where $\bar{i}_c$ is the average signal current and i_d is the dark current.

The noise and signal equivalent circuit is shown in Figure 11–5, where for the sake of definiteness we took the modulation index $m = 1$. R represents the output load of the photomultiplier. T_e is chosen so that the term $4kT_e\,\Delta\nu/R$ accounts for the thermal noise of R as well as for the noise generated by the amplifier that follows the photomultiplier.

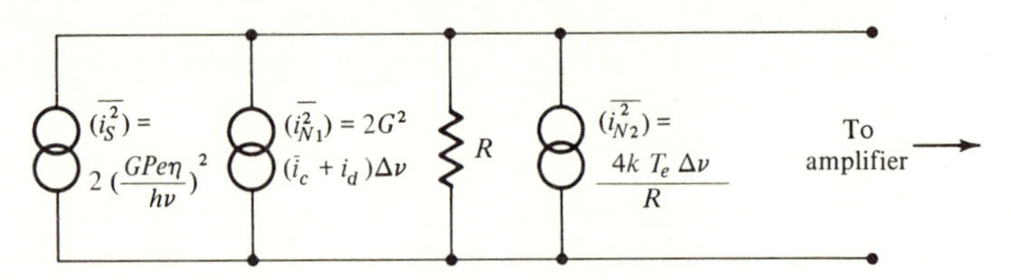

Figure 11-5 Equivalent circuit of a photomultiplier.

The signal-to-noise power ratio at the output is thus

$$\frac{S}{N} = \frac{\overline{i_s^2}}{\overline{(i_{N1}{}^2)} + \overline{(i_{N2}{}^2)}}$$

$$= \frac{2(Pe\eta/h\nu_s)^2G^2}{2G^2e(\bar{i}_c + i_d)\Delta\nu + (4kT_e\Delta\nu/R)} \tag{11.3-9}$$

---

[6] $P/h\nu_s$ is the rate of photon incidence on the photocathode; thus, if it takes $1/\eta$ photons to generate one electron, the average current is given by (11.3-5).

Due to the large current gain ($G \simeq 10^6$) the first term in the denominator of (11.3-9) which represents amplified cathode shot noise is much larger than the thermal and amplifier noise term $4kT_e\,\Delta\nu/R$. Neglecting the term $4kT_e\,\Delta\nu/R$, assuming $i_d \gg \bar{i}_c$, and setting $S/N = 1$, we can solve for the minimum detectable optical power.

$$P_{\min} = \frac{h\nu_s(i_d\Delta\nu)^{1/2}}{\eta e^{1/2}} \tag{11.3-10}$$

Numerical example—sensitivity of photomultiplier. Consider a typical case of detecting an optical signal under the following conditions:

$$\begin{aligned}
\nu_s &= 6 \times 10^{14}\text{ Hz } (\lambda = 0.5\,\mu\text{m})\\
\eta &= 10\text{ percent}\\
\Delta\nu &= 1\text{ Hz}\\
i_d &= 10^{-15}\text{ ampere (a typical value of the dark photocathode current)}
\end{aligned}$$

Substitution in (11.3-10) gives

$$P_{\min} = 3 \times 10^{-16}\text{ watt}$$

The corresponding cathode signal current is $\bar{i}_c \sim 10^{-17}$ ampere, so the assumption $i_d \gg \bar{i}_c$ is justified.

Signal-limited shot noise. If one could, somehow, eliminate the dark current altogether, so that the only contribution to the average photocathode current is $\bar{i}_c$, that is that due to the optical signal, then, using (11.3-5) and (11.3-9) to solve self-consistently for $P_{\min}$,

$$P_{\min} \simeq \frac{h\nu_s\Delta\nu}{\eta} \tag{11.3-11}$$

This corresponds to the quantum limit of optical detection. Its significance will be discussed in the next section. The practical achievement of this limit in video detection is nearly impossible since it depends on near total suppression of the dark current and other extraneous noise sources such as background radiation reaching the photocathode and causing shot noise.

The quantum detection limit (11.3-11) can, however, be achieved in the heterodyne mode of optical detection. This is discussed in the next section.

11.4 Heterodyne Detection with Photomultipliers

In the heterodyne mode of optical detection, the signal to be detected $E_s \cos \omega_s t$ is combined with a second optical field, referred to as the local-oscillator field, $E_L \cos (\omega_s + \omega)t$, shifted in frequency by ω ($\omega \ll \omega_s$). The total field incident on the photocathode is therefore given by

$$e(t) = \text{Re}\,[E_L e^{i(\omega_s+\omega)t} + E_s e^{i\omega_s t}] \equiv \text{Re}\,[V(t)] \qquad \textbf{(11.4-1)}$$

The local-oscillator field originates usually at a laser at the receiving end, so that it can be made very large compared to the signal to be detected. In the following we will assume that

$$E_L \gg E_s \qquad \textbf{(11.4-2)}$$

A schematic diagram of a heterodyne detection scheme is shown in Figure 11-6. The current emitted by the photocathode is given, according to (11.1-2) and (11.4-1), by

$$i_c(t) \propto V(t)V^*(t) = E_L^2 + E_s^2 + 2E_L E_s \cos \omega t$$

which, using (11.4-2) can be written as

$$i_c(t) \equiv aE_L^2\left(1 + \frac{2E_s}{E_L}\cos \omega t\right) = aE_L^2\left(1 + 2\sqrt{\frac{P_s}{P_L}}\cos \omega t\right) \qquad \textbf{(11.4-3)}$$

where P_s and P_L are the signal and local-oscillator powers, respectively. The proportionality constant a in (11.4-3) can be determined as in (11.3-6) by requiring that when $E_s = 0$ the direct current be related to the local-oscillator power P_L by $\bar{i}_c = P_L \eta e / h\nu_L$,[7] so taking $\nu \approx \nu_L$

$$i_c(t) = \frac{P_L e \eta}{h\nu}\left(1 + 2\sqrt{\frac{P_s}{P_L}}\cos \omega t\right) \qquad \textbf{(11.4-4)}$$

The total cathode shot noise is thus

$$\overline{(i_{N1}{}^2)} = 2e\left(i_d + \frac{P_L e \eta}{h\nu}\right)\Delta\nu \qquad \textbf{(11.4-5)}$$

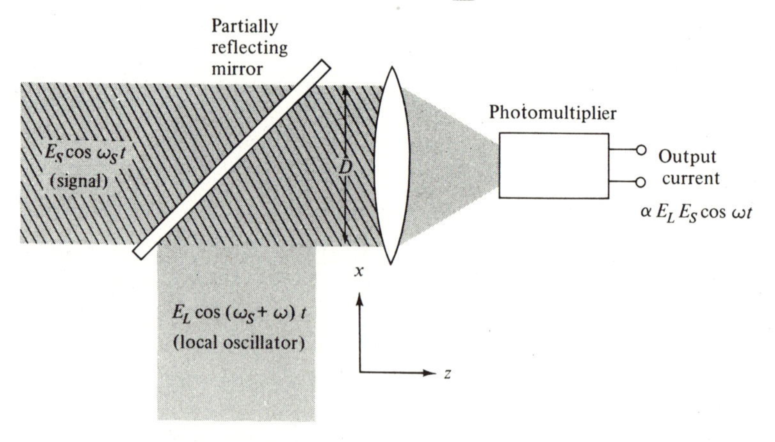

Figure 11-6 Schematic diagram of a heterodyne detector using a photomultiplier.

---

[7] This is just a statement of the fact that each incident photon has a probability η of releasing an electron.

where i_d is the average dark current while $P_L e\eta/h\nu$ is the dc cathode current due to the strong local-oscillator field. The shot-noise current is amplified by G, resulting in an output noise

$$\overline{(i_N{}^2)}_{\text{anode}} = G^2 2e\left(i_d + \frac{P_L e\eta}{h\nu}\right)\Delta\nu \tag{11.4-6}$$

The mean-square signal current at the output is, according to (11.4-4),

$$\overline{(i_s{}^2)}_{\text{anode}} = 2G^2\left(\frac{P_s}{P_L}\right)\left(\frac{P_L e\eta}{h\nu}\right)^2 \tag{11.4-7}$$

The signal-to-noise power ratio at the output is given by

$$\frac{S}{N} = \frac{2G^2(P_s P_L)(e\eta/h\nu)^2}{\{G^2 2e(i_d + (P_L e\eta/h\nu))] + (4kT_e/R)\}\Delta\nu} \tag{11.4-8}$$

where, as in (11.3-9), the last term in the denominator represents the Johnson (thermal) noise generated in the output load, plus the effective input noise of the amplifier following the photomultiplier. The big advantage of the heterodyne detection scheme is now apparent. By increasing P_L the S/N ratio increases until the denominator is dominated by the term $G^2 2eP_L e\eta/h\nu$. This corresponds to the point at which the *shot noise produced by the local oscillator current dwarfs all the other noise contributions.* When this state of affairs prevails, we have, according to (11.4-8),

$$\frac{S}{N} \simeq \frac{P_s}{h\nu\Delta\nu/\eta} \tag{11.4-9}$$

which corresponds to the quantum-limited detection limit. The minimum detectable signal—that is, the signal input power leading to an output signal-to-noise ratio of 1—is thus

$$(P_s)_{\text{min}} = \frac{h\nu\Delta\nu}{\eta} \tag{11.4-10}$$

This power corresponds for $\eta = 1$ to a flux at a rate of one photon per $(\Delta\nu)^{-1}$ seconds—that is, one photon per resolution time of the system.[8]

Numerical example. It is interesting to compare the minimum detectable power for the heterodyne system as given by (11.4-10) with that calculated in the example of Section 11-3 for the video system. Using the same data,

$\nu = 6 \times 10^{14}$ Hz ($\lambda = 0.5\,\mu$m)

$\eta = 10$ percent

$\Delta\nu = 1$ Hz

---

[8] A detection system that is limited in bandwidth to $\Delta\nu$ cannot resolve events in time that are separated by less than $\sim(\Delta\nu)^{-1}$ second. Thus $(\Delta\nu)^{-1}$ is the resolution time of the system.

we obtain

$$(P_s)_{\min} \simeq 4 \times 10^{-18} \text{ watt}$$

to be compared with $P_{\min} \simeq 3 \times 10^{-16}$ watt in the video case.

Limiting sensitivity as a result of the particle nature of light. The quantum limit to optical detection sensitivity is given by (11.4-10) as

$$(P_s)_{\min} = \frac{h\nu\Delta\nu}{\eta} \tag{11.4-11}$$

This limit was shown to be due to the shot noise of the photoemitted current. We may alternatively, attribute this noise to the granularity—that is, the particle nature—of light, according to which the minimum energy increment of an electromagnetic wave at frequency ν is $h\nu$. The power average P of an optical wave can be written as

$$P = \overline{N}h\nu \tag{11.4-12}$$

where $\overline{N}$ is the average number of photons arriving at the photocathode per second. Next assume a hypothetical noiseless pohotomultiplier in which *exactly* one electron is produced for each η^{-1} incident photon. The measurement of P is performed by counting the number of electrons produced during an observation period T and then averaging the result over a large number of similar observations.

The average number of electrons emitted per observation period T is

$$\overline{N_e} = \overline{N}T\eta \tag{11.4-13}$$

which, assuming perfect randomness in the arrival, is equal to the mean-square fluctuation[9]

$$\overline{(\Delta N_e)^2} \equiv \overline{(N_e - \overline{N}_e)^2} = \bar{N}_e = \bar{N}T\eta$$

Taking the minimum detectable number of quanta as that for which the rms fluctuation equals the average value, we get

$$(\overline{N}_{\min}T\eta)^{1/2} = \overline{N}_{\min}T\eta$$

---

[9] This follows from the assumption that the photon arrival is perfectly random, so the probability of having N photons arriving in a given time interval is given by the Poisson law

$$p(N) = (\overline{N})^N e^{-\overline{N}}/N!$$

The mean-square fluctuation is given by

$$\overline{(\Delta N)^2} = \sum_{N=0}^{\infty} p(N)(N - \overline{N})^2 = \overline{N}$$

where

$$\overline{N} = \sum_{0}^{\infty} Np(N)$$

is the average N.

or

$$(\bar{N})_{\min} = \frac{1}{T\eta} \tag{11.4-14}$$

If we convert the last result to power by multiplying it by $h\nu$ and recall that $T^{-1} \simeq \Delta\nu$, where $\Delta\nu$ is the bandwidth of the system, we get

$$(P_s)_{\min} = \frac{h\nu\Delta\nu}{\eta} \tag{11.4-15}$$

in agreement with (11.4-10).

The question as to whether the real limit to sensitivity is imposed by the cathode shot noise or the fluctuations in the incident photon flux is thus academic. An examination of the photocurrent cannot distinguish between these cases.

11.5 Photoconductive Detectors

The operation of photoconductive detectors is illustrated in Figure 11-7. A semiconductor crystal is connected in series with a resistance R and a supply voltage V. The optical field to be detected is incident on and absorbed in the crystal, thereby exciting electrons into the conduction band (or, in p-type semiconductors, holes into the valence band). Such excitation results in a lowering of the resistance R_d of the semiconductor crystal and hence in an increase in the voltage drop across R which, for $\Delta R_d/R_d \ll 1$, is proportional to the incident optical intensity.

To be specific, we show the energy levels involved in one of the more popular semiconductive detectors—mercury-doped germanium [7]. Mercury atoms enter germanium as acceptors with an ionization energy of 0.09 eV. It follows that it takes a photon energy of at least 0.09 eV (that is, a photon with a wavelength shorter than 14 μm) to lift an electron from the top of the valence band and have it trapped by the Hg (acceptor) atom. Usually the germanium crystal contains a smaller density N_D of donor atoms, which at low temperatures find it energetically profitable to lose their valence electrons to one of the far more numerous Hg acceptor atoms, thereby becoming positively ionized and ionizing (negatively) an equal number of acceptors.

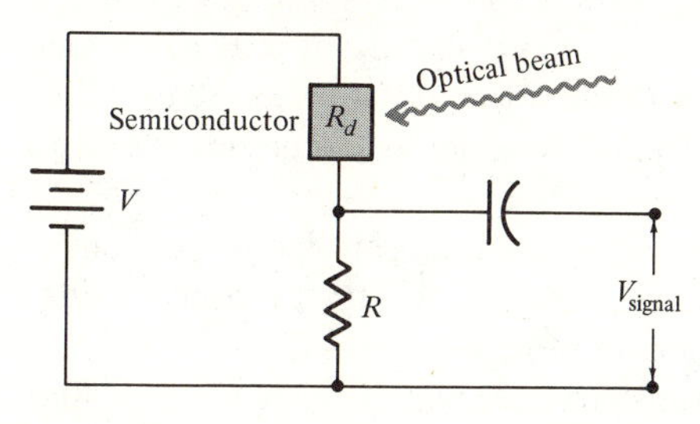

Figure 11-7 Typical biasing circuit of a photoconductive detector.

Since the acceptor density $N_A \gg N_D$, most of the acceptor atoms remain neutrally charged.

An incident photon is absorbed and lifts an electron from the valence band onto an acceptor atom, as shown in process A in Figure 11-8. The electronic deficiency (that is, the hole) thus created is acted upon by the electric field and its drift along the field direction gives rise to the signal current. The contribution of a given hole to the current ends when an electron drops from an ionized acceptor level back into the valence band, thus eliminating the hole as in B. This process is referred to as electron–hole recombination or trapping of a hole by an ionized acceptor atom.

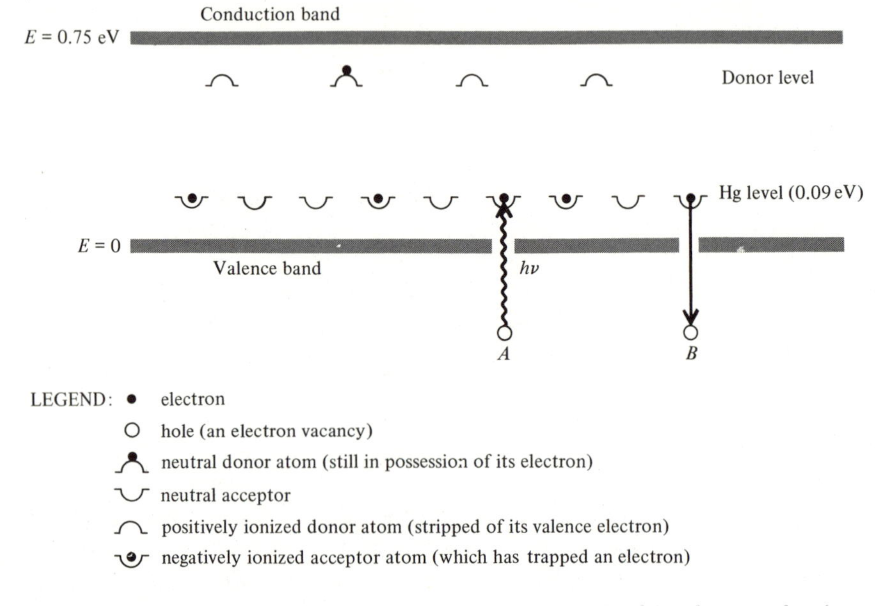

Figure 11-8 Donor and acceptor impurity levels involved in photoconductive semiconductors.

By choosing impurities with lower ionization energies, even lower-energy photons can be detected and indeed photoconductive detectors operate commonly at wavelengths up to $\lambda = 50\ \mu\text{m}$. Cu, as an example, enters into Ge as an acceptor with an ionization energy of 0.04 eV, which would correspond to long-wavelength detection cutoff of $\lambda \simeq 32\ \mu\text{m}$. The response of a number of commercial photoconductive detectors is shown in Figure 11-9.

It is clear from this discussion that the main advantage of photoconductors compared to photomultipliers is their ability to detect long-wavelength radiation, since the creation of mobile carriers does not involve overcoming the large surface potential barrier. On the debit side we find

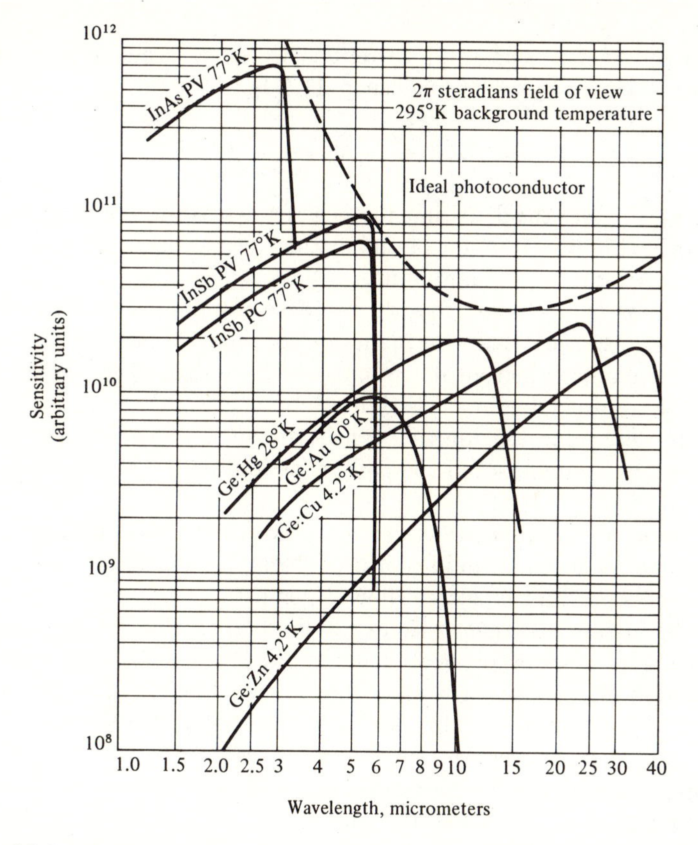

Figure 11-9 Relative sensitivity of a number of commercial photoconductors. (Courtesy Santa Barbara Research Corp.)

the lack of current multiplication and the need to cool the semiconductor so that photoexcitation of carriers will not be masked by thermal excitation.

Consider an optical beam, of power P and frequency ν, which is incident on a photoconductive detector. Taking the probability for excitation of a carrier by an incident photon—the so-called quantum efficiency—as η, the carrier generation rate is $G = P\eta/h\nu$. If the carriers last on the average τ_0 seconds before recombining, the average number of carriers N_c is found by equating the generation rate to the recombination rate (N_c/τ_0), so

$$N_c = G\tau_0 = \frac{P\eta\tau_0}{h\nu} \tag{11.5-1}$$

Each one of these carriers drifts under the electric field influence[10] at a velocity $\bar{v}$ giving rise, according to (10.4-1), to a current in the external circuit of $i_e = e\bar{v}/d$, where d is the length (between electrodes) of the semi-

[10] The drift velocity is equal to μE, where μ is the mobility and E is the electric field.

conductor crystal. The total current is thus the product of i_e and the number of carriers present, or, using (11.5-1),

$$\bar{i} = N_c i_e = \frac{P\eta\tau_0 e\bar{v}}{h\nu d} = \frac{e\eta}{h\nu}\left(\frac{\tau_0}{\tau_d}\right)P \tag{11.5-2}$$

where $\tau_d = d/\bar{v}$ is the drift time for a carrier across the length d. The factor (τ_0/τ_d) is thus the fraction of the crystal length drifted by the average excited carrier before recombining.

Equation (11.5-2) describes the response of a photoconductive detector to a constant optical flux. Our main interest, however, is in the heterodyne mode of photoconductive detection, which, as has been shown in Section 11.4, allows detection sensitivities approaching the quantum limit. In order to determine the limiting sensitivity of photoconductive detectors we need first to understand the noise contribution in these devices.

Generation recombination noise in photoconductive detectors. The principal noise mechanism in cooled photoconductive detectors reflects the randomness inherent in current flow. Even if the incident optical flux were constant in time, the generation of individual carriers by the flux would constitute a random process. This is exactly the type of randomness involved in photoemission, and we may expect, likewise, that the resulting noise will be shot noise. This is almost true except for the fact that in a photoconductive detector a photoexcited carrier lasts τ seconds[11] (its recombination lifetime) before being captured by an ionized impurity. The contribution of the carrier to the charge flow in the external circuit is thus $e(\tau/\tau_d)$, as is evident from inspection of (11.5-2). Since the lifetime τ is not a constant, but must be described statistically, another element of randomness is introduced into the current flow.

Consider a carrier excited by a photon absorption and lasting τ seconds. Its contribution to the external current is, according to (10.4-1)

$$i_e(t) = \begin{cases} \dfrac{e\bar{v}}{d} & 0 \leqslant t \leqslant \tau \\ 0 & \text{otherwise} \end{cases} \tag{11.5-3}$$

which has a Fourier transform

$$I_e(\omega, \tau) = \frac{e\bar{v}}{d}\int_0^\tau e^{-i\omega\tau}\,dt = \frac{e\bar{v}}{d}[1 - e^{-i\omega\tau}] \tag{11.5-4}$$

so that

$$|I_e(\omega, \tau)|^2 = \frac{e^2\bar{v}^2}{\omega^2 d^2}[2 - e^{-i\omega\tau} - e^{i\omega\tau}] \tag{11.5-5}$$

---

[11] The parameter τ_0 appearing in (11.5-2) is the value of τ averaged over a large number of carriers.

According to (10.3-10) we need to average $|I_e(\omega, \tau)|^2$ over τ. This is done in a manner similar to the procedure used in Section 10.5. Taking the probability function[12] $g(\tau) = \tau_0^{-1} \exp(-\tau/\tau_0)$, we average (11.5-5) over all the possible values of τ according to

$$\overline{|I_e(\omega)|^2} = \int_0^\infty |I_e(\omega, \tau)|^2 g(\tau)\,d\tau$$

$$= \frac{2e^2\bar{v}^2\tau_0^2}{d^2(1 + \omega^2\tau_0^2)} \tag{11.5-6}$$

The spectral density function of the current fluctuations is obtained using Carson's theorem (10.3-10) as

$$S(\nu) = 2\bar{N}\,\frac{2e^2(\tau_0^2/\tau_d^2)}{(1 + \omega^2\tau_0^2)} \tag{11.5-7}$$

where we used $\tau_d = d/\bar{v}$ and where $\bar{N}$, the average number of carriers generated per second, can be expressed in terms of the average current $\bar{I}$ by use of the relation[13]

$$\bar{I} = \bar{N}e\frac{\tau_0}{\tau_d} \tag{11.5-8}$$

leading to

$$S(\nu) = \frac{4e\bar{I}(\tau_0/\tau_d)}{1 + 4\pi^2\nu^2\tau_0^2}$$

Therefore, the mean-square current representing the noise power in a frequency interval ν to $\nu + \Delta\nu$ is

$$\overline{i_N^2} \equiv S(\nu)\Delta\nu = \frac{4e\bar{I}(\tau_0/\tau_d)\Delta\nu}{1 + 4\pi^2\nu^2\tau_0^2} \tag{11.5-9}$$

which is the basic result for generation–recombination noise.

Numerical example. To better appreciate the kind of numbers involved in the expression for $\overline{i_N^2}$ we may consider a typical mercury-doped germanium detector operating at 20°K with the following characteristics:

$d = 10^{-1}$ cm

$\tau_0 = 10^{-9}$ second

V (across the length d) $= 10$ volts $\Rightarrow E = 10^2$ V/cm

$\mu = 3 \times 10^4$ cm^2/V-s

---

[12] $g(\tau)\,d\tau$ is the probability that a carrier lasts between τ and $\tau + d\tau$ seconds before recombining.

[13] This relation follows from the fact that the average charge per carrier flowing through the external circuit is $e(\tau_0/\tau_d)$, which, when multiplied by the generation rate $\bar{N}$, gives the current.

The drift velocity is $\bar{v} = \mu E = 3 \times 10^6$ cm/s and $\tau_d = d/\bar{v} \simeq 3.3 \times 10^{-8}$ second, and therefore $\tau_0/\tau_d = 3 \times 10^{-2}$. Thus, on the average, a carrier traverses only 3 percent of the length ($d = 1$ mm) of the sample before recombining. Comparing (11.5-9) to the shot-noise result (10.4-9), we find that for a given average current $\bar{I}$ the generation recombination noise is reduced from the shot-noise value by a factor

$$\frac{\overline{(i_N{}^2)}_{\substack{\text{generation-}\\ \text{recombination}}}}{\overline{(i_N{}^2)}_{\text{shot noise}}} \underset{\omega\tau_0 \ll 1}{=} 2\left(\frac{\tau_0}{\tau_d}\right) \tag{11.5-10}$$

which, in the foregoing example, has a value of about 1/15. Unfortunately, as will be shown subsequently, the reduced noise is accompanied by a reduction by a factor of (τ_0/τ_d) in the magnitude of the signal power, which wipes out the advantage of the low noise.

Heterodyne detection in photoconductors. The situation here is similar to that described by Figure 11-6 in connection with heterodyne detection using photomultipliers. The signal field

$$e_s(t) = E_s \cos \omega_s t$$

is combined with a strong local-oscillator field

$$e_L(t) = E_L \cos (\omega + \omega_s)t \qquad E_L \gg E_s$$

so the total field incident on the photoconductor is

$$e(t) = \text{Re}\,[E_s e^{i\omega_s t} + E_L e^{i(\omega_s+\omega)t}] \equiv \text{Re}\,[V(t)] \tag{11.5-11}$$

The rate at which carriers are generated is taken, following (11.1-2), as $aV(t)V^*(t)$ where a is a constant to be determined. The equation describing the number of excited carriers N_c is thus

$$\frac{dN_c}{dt} = aVV^* - \frac{N_c}{\tau_0} \tag{11.5-12}$$

where τ_0 is the average carrier lifetime, so N_c/τ_0 corresponds to the carrier's decay rate. We assume a solution for $N_c(t)$ that consists of the sum of dc and a sinusoidal component in the form of

$$N_c(t) = N_0 + (N_1 e^{i\omega t} + \text{c.c.}) \tag{11.5-13}$$

where c.c. stands for "complex conjugate."

Substitution in (11.5-12) gives

$$N_c(t) = a\tau_0(E_s{}^2 + E_L{}^2) + a\tau_0\left(\frac{E_s E_L e^{i\omega t}}{1 + i\omega\tau_0} + \text{c.c.}\right) \tag{11.5-14}$$

where we took E_s and E_L as real. The current through the sample is given by the number of carriers per unit length N_c/d times $e\bar{v}$, where $\bar{v}$ is the drift velocity

$$i(t) = \frac{N_c(t)e\bar{v}}{d} \tag{11.5-15}$$

which, using (11.5-14), gives

$$i(t) = \frac{e\bar{v}a\tau_0}{d}\left[E_s^2 + E_L^2 + \frac{2E_sE_L\cos(\omega t - \phi)}{\sqrt{1+\omega^2\tau_0^2}}\right] \tag{11.5-16}$$

where $\phi = \tan^{-1}(\omega\tau_0)$.

The current is thus seen to contain a signal component that oscillates at ω and is proportional to E_s. The constant a in (11.5-16) can be determined by requiring that, when $P_s = 0$, the expression for the direct current predicted by (11.5-16) agree with (11.5-2). This condition is satisfied if we rewrite (11.5-16) as

$$i(t) = \frac{e\eta}{h\nu}\left(\frac{\tau_0}{\tau_d}\right)\left[P_s + P_L + \frac{2\sqrt{P_sP_L}}{\sqrt{1+\omega^2\tau_0^2}}\cos(\omega t - \phi)\right] \tag{11.5-17}$$

where P_s and P_L refer, respectively, to the incident-signal and local-oscillator powers and η, the quantum efficiency, is the number of carriers excited per incident photon. The signal current is thus

$$i_s(t) = \frac{2e\eta}{h\nu}\left(\frac{\tau_0}{\tau_d}\right)\frac{\sqrt{P_sP_L}}{\sqrt{1+\omega^2\tau_0^2}}\cos(\omega t - \phi) \tag{11.5-18}$$

while the dc (average) current is

$$\bar{I} = \frac{e\eta}{h\nu}\left(\frac{\tau_0}{\tau_d}\right)(P_s + P_L) \tag{11.5-19}$$

Since the average current $\bar{I}$ appearing in the expression (11.5-9) for the generation recombination noise is given in this case by

$$\bar{I} = \left(\frac{e\eta}{h\nu}\right)\left(\frac{\tau_0}{\tau_d}\right)P_L \qquad P_L \gg P_S$$

we can, by increasing P_L, increase the noise power $\overline{i_N^2}$ and at the same time, according to (11.5-18), the signal $\overline{i_s^2}$ until the generation recombination noise (11.5-9) is by far the largest contribution to the total output noise. When this condition is satisfied, the signal-to-noise ratio can be written, using (11.5-9), (11.5-18) and (11.5-19) and taking $P_L \gg P_s$, as

$$\frac{S}{N} = \frac{\overline{i_s^2}}{\overline{i_N^2}} = \left[2\left(\frac{e\eta\tau_0}{h\nu\tau_d}\right)^2\frac{P_sP_L}{1+\omega^2\tau_0^2}\right]\bigg/\left[\frac{4e^2\eta(\tau_0/\tau_d)^2P_L\Delta\nu}{(1+\omega^2\tau_0^2)h\nu}\right] = \frac{P_s\eta}{2h\nu\Delta\nu} \tag{11.5-20}$$

The minimum detectable signal—that which leads to a signal to noise ratio of unity—is found by setting the left side of (11.5-20) equal to unity and solving for P_s. It is

$$(P_s)_{\min} = \frac{2h\nu\Delta\nu}{\eta} \tag{11.5-21}$$

which, for the same η, is twice that of the photomultiplier heterodyne detection as given by (11.4-10). In practice, however, η in photoconductive detectors can approach unity, whereas in the best photomultipliers $\eta \simeq 30$ percent.

Numerical example—minimum detectable power of a heterodyne receiver using a photoconductor at 10.6 μm. Assume the following:

$$\lambda = 10.6\ \mu\text{m}$$

$$\Delta\nu = 1\ \text{Hz}$$

$$\eta \simeq 1$$

Substitution in (11.5-21) gives a minimum detectable power of

$$(P_s)_{\min} \simeq 10^{-19}\ \text{watt}$$

Experiments ([8] and [9]) have demonstrated that the theoretical signal-to-noise ratio as given by (11.5-20) can be realized quite closely in practice; see Figure 11-10.

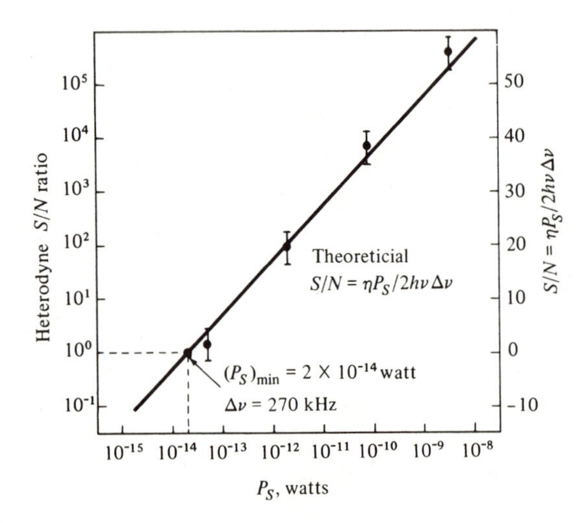

Figure 11-10 Signal-to-noise ratio of heterodyne signal in Ge:Cu detector at a heterodyne frequency of 70 MHz. Data points represent observed values. (After Reference [8].)

11.6 The p-n Junction

Before embarking on a description of the p-n diode detector, we need to understand the operation of the semiconductor p-n junction. Consider the junction illustrated in Figure 11-11. It consists of an abrupt transition from a donor-doped (that is, n-type) region of a semiconductor, where the free-charge carriers are predominantly electrons, to an acceptor-doped (p-type) region, where the carriers are holes. The doping profile—that is, the density of excess donor (in the n region) atoms or acceptor atoms (in the p region)—is shown in Figure 11.11(a). This abrupt transition results usually from diffusing suitable impurity atoms into a substrate of a semiconductor with the opposite type of conductivity. In our slightly idealized abrupt junction we assume that the n region ($x > 0$) has a constant (net) donor density N_D and the p region ($x < 0$) has a constant acceptor density N_A.

The energy-band diagram at zero applied bias is shown in Figure 11.11(b). The top (or bottom) curve can be taken to represent the potential energy of an electron as a function of position x, so the minimum energy needed to take an electron from the n to the p side of the junction is eV_d. Taking the separations of the Fermi level from the respective band edges as ϕ_n and ϕ_p as shown, we have

$$eV_d = E_g - (\phi_n + \phi_p)$$

V_d is referred to as the "built-in" junction potential.

Figure 11-11(c) shows the potential distribution in the junction with an applied reverse bias of magnitude V_a. This leads to a separation of eV_a between the Fermi levels in the p and n regions and causes the potential barrier across the junction to increase from eV_d to $e(V_d + V_a)$. The change of potential between the p and n regions is due to a sweeping of the mobile charge carriers from the region $-l_p < x < l_n$, giving rise to a charge double layer of stationary (ionized) impurity atoms, as shown in Figure 11-11(d).

In the analytical treatment of the problem we assume that in the depletion layer ($-l_p < x < l_n$) the excess impurity atoms are fully ionized and thus, using $\nabla \cdot \mathbf{E} = \rho/\epsilon$ and $\mathbf{E} = -\nabla V$, where V is the potential, we have

$$\frac{d^2V}{dx^2} = \frac{eN_A}{\epsilon} \qquad \text{for } -l_p < x < 0 \tag{11.6-1}$$

and

$$\frac{d^2V}{dx^2} = -\frac{eN_D}{\epsilon} \qquad 0 < x < l_p \tag{11.6-2}$$

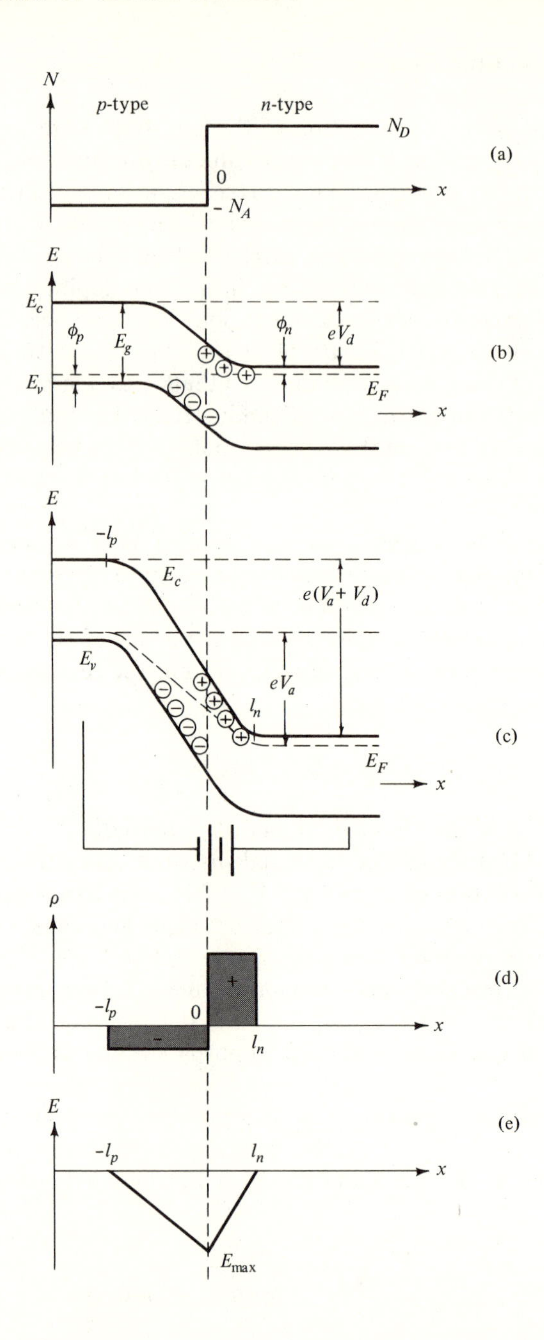

Figure 11-11 The abrupt *p-n* junction. (**a**) Impurity profile. (**b**) Energy-band diagram with zero applied bias. (**c**) Energy-band diagram with reverse applied bias. (**d**) Net charge density in the depletion layer. (**e**) The electric field. The circles in (**b**) and (**c**) represent ionized impurity atoms in the depletion layer.

where the charge of the electron is $-e$ and the dielectric constant is ϵ. The boundary conditions are

$$E = -\frac{dV}{dx} = 0 \text{ at } x = -l_p \text{ and } x = +l_n \tag{11.6-3}$$

$$\frac{dV}{dx} \text{ is continuous at } x = 0 \tag{11.6-4}$$

$$V(l_n) - V(-l_p) = V_d + V_a \tag{11.6-5}$$

The solutions of (11.6-1) and 11.6-2) are

$$V = \frac{e}{2\epsilon} N_A(x^2 + 2l_p x) \qquad \text{for } -l_p < x < 0 \tag{11.6-6}$$

$$V = -\frac{e}{2\epsilon} N_D(x^2 - 2l_n x) \qquad 0 < x < l_n \tag{11.6-7}$$

which, using (11.6-4), gives

$$N_A l_p = N_D l_n \tag{11.6-8}$$

so the double layer contains an equal amount of positive and negative charge.

Condition (11.6-5) gives

$$V_d + V_a = \frac{e}{2\epsilon}(N_D l_n{}^2 + N_A l_p{}^2) \tag{11.6-9}$$

which, together with (11.6-8) leads to

$$l_p = (V_d + V_a)^{1/2}\left(\frac{2\epsilon}{e}\right)^{1/2}\left[\frac{N_D}{N_A(N_A + N_D)}\right]^{1/2} \tag{11.6-10}$$

and

$$l_n = (V_d + V_a)^{1/2}\left(\frac{2\epsilon}{e}\right)^{1/2}\left[\frac{N_A}{N_D(N_A + N_D)}\right]^{1/2} \tag{11.6-11}$$

and, therefore, as before,

$$\frac{l_p}{l_n} = \frac{N_D}{N_A} \tag{11.6-12}$$

Differentiation of (11.6-6) and (11.6-7) yields

$$\begin{aligned} E &= -\frac{e}{\epsilon} N_A(x + l_p) \qquad \text{for } -l_p < x < 0 \\ E &= -\frac{e}{\epsilon} N_D(l_n - x) \qquad 0 < x < l_n \end{aligned} \tag{11.6-13}$$

The field distribution of (11.6-13) is shown in Figure 11-11(e). The maximum field occurs at $x = 0$ and is given by

$$E_{\max} = -2(V_d + V_a)^{1/2}\left(\frac{e}{2\epsilon}\right)^{1/2}\left(\frac{N_D N_A}{N_A + N_D}\right)^{1/2} = -\frac{2(V_d + V_a)}{l_p + l_n} \tag{11.6-14}$$

The presence of a charge $Q = -eN_Al_p$ per unit junction area on the p side and an equal and negative charge on the n side leads to a junction capacitance. The reason is that l_p and l_n depend, according to (11.6-10) and (11.6-11), on the applied voltage V_a, so a change in voltage leads to a change in the charge $eN_Al_p = eN_Dl_n$ and hence to a differential capacitance per unit area,[14] given by

$$\frac{C_d}{\text{area}} \equiv \frac{dQ}{dV_a} = eN_A \frac{dl_p}{dV_a}$$

$$= \left(\frac{\epsilon e}{2}\right)^{1/2} \left(\frac{N_A N_D}{N_A + N_D}\right)^{1/2} \left(\frac{1}{V_a + V_d}\right)^{1/2} \tag{11.6-15}$$

which, using (11.6-10) and (11.6-11), can be shown to be equal to

$$\frac{C_d}{\text{area}} = \frac{\epsilon}{l_p + l_n} \tag{11.6-16}$$

as appropriate to a parallel-plate capacitance of separation $l = l_p + l_n$. The equivalent circuit of a p-n junction is shown in Figure 11-12. The capacitance C_d was discussed above. The diode shunt resistance R_d in back-biased junctions is usually very large ($>10^6$ ohms) compared to the load impedance R_L and can be neglected. The resistance R_s represents ohmic losses in the bulk p and n regions adjacent to the junction.

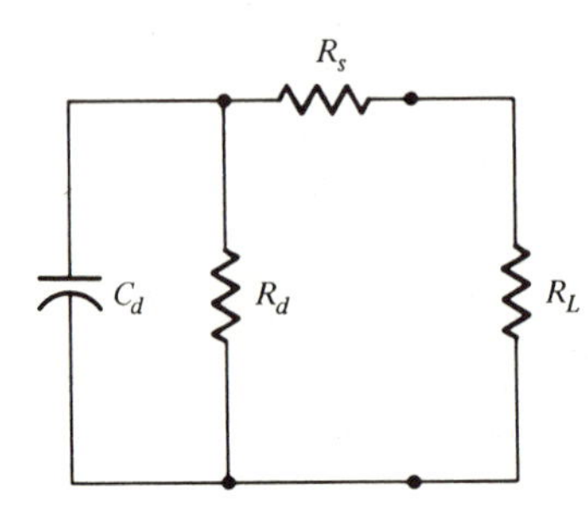

Figure 11-12 Equivalent circuit of a p-n junction. In typical back-biased diodes, $R_d \gg R_s$ and R_L, and $R_L \gg R_s$, so the resistance across the junction can be taken as equal to the load resistance R_L.

11.7 Semiconductor Photodiodes

Semiconductor p-n junctions are used widely for optical detection: see References [10]–[12]. In this role they are referred to as junction photodiodes. The main physical mechanisms involved in junction photodetection are illustrated in Figure 11-13. At A, an incoming photon is absorbed in the p side creating a hole and a free electron. If this takes place within a diffusion length (the distance in which an excess minority concentration

---

[14] The capacitance is defined by $C = Q/V_a$, whereas the differential capacitance $C_d = dQ/dV_a$ is the capacitance "seen" by a small ac voltage when the applied bias is V_a.

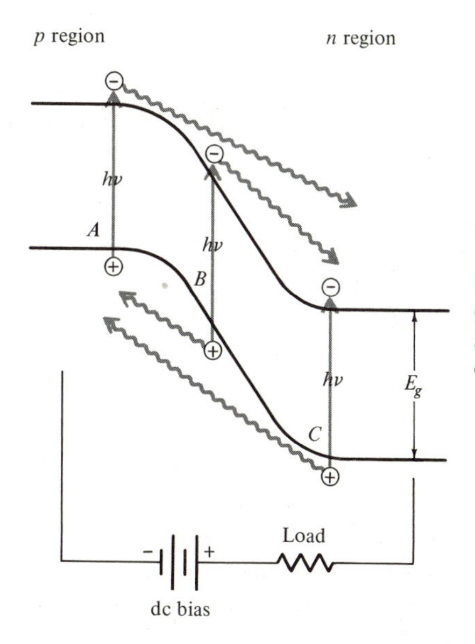

Figure 11-13 The three types of electron–hole pair creation by absorbed photons that contribute to current flow in a *p-n* photodiode.

is reduced to e^{-1} of its peak value, or in physical terms, the average distance a minority carrier traverses before recombining with a carrier of the opposite type) of the depletion layer, the electron will, with high probability, reach the layer boundary and will drift under the field influence across it. An electron traversing the junction contributes a charge e to the current flow in the external circuit, as described in Section 10.4. If the photon is absorbed near the n side of the depletion layer, as shown at C the resulting hole will diffuse to the junction and then drift across it again, giving rise to a flow of charge e in the external load. The photon may also be absorbed in the depletion layer as at B, in which case both the hole and electron which are created drift (in opposite directions) under the field until they reach the p and n sides, respectively. Since in this case each carrier traverses a distance that is less than the full junction width, the contribution of this process to charge flow in the external circuit is, according to (10.4-1) and (10.4-7), e. In practice this last process is the most desirable, since each absorption gives rise to a charge e, and delayed current response caused by finite diffusion time is avoided. As a result, photodiodes often use a *p-i-n* structure in which an intrinsic high resistivity (i) layer is sandwiched between the p and n regions. The potential drop occurs mostly across this layer, which can be made long enough to insure that most of the incident photons are absorbed within it. Typical construction of a *p-i-n* photodiode is shown in Figure 11-14.

It is clear from Figure 11-13 that a photodiode is capable of detecting only radiation with photon energy $h\nu > E_g$, where E_g is the energy gap of the semiconductor. If, on the other hand, $h\nu \gg E_g$, the absorption, which in a semiconductor increases strongly with frequency, will take place

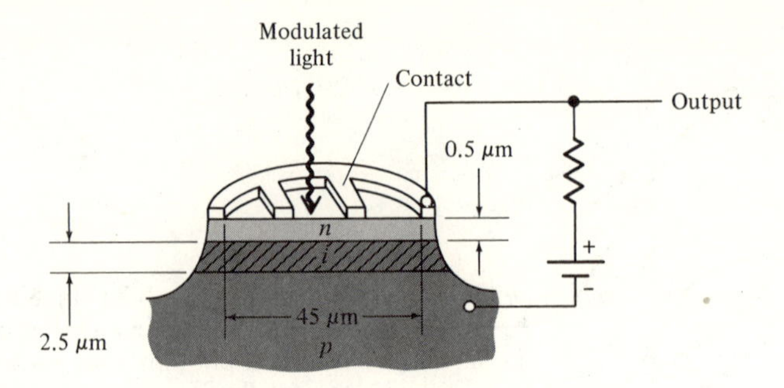

Figure 11-14 A *p-i-n* photodiode. (After Reference [13].)

entirely near the input face (in the n region of Figure 11-14) and the minority carriers generated by absorbed photons will recombine with majority carriers before diffusing to the depletion layer. This event does not contribute to the current flow and, as far as the signal is concerned, is wasted. This is why the photoresponse of diodes drops off when $h\nu > E_g$. Typical frequency response curves of photodiodes are shown in Figure 11-15. The number of carriers flowing in the external circuit per incident photon, the so-called quantum efficiency, is seen to approach 50 percent in Ge.

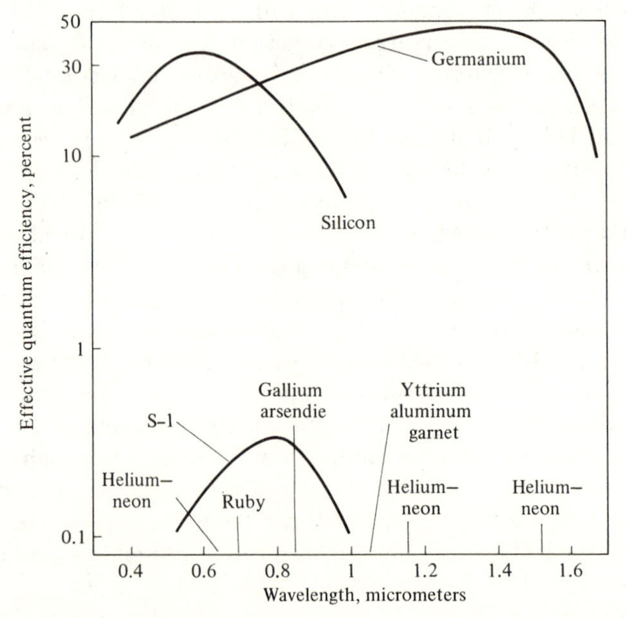

Figure 11-15 Quantum efficiencies for silicon and germanium photodiodes compared with the efficiency of the S-1 photodiode used in a photomultiplier tube. Emission wavelengths for various lasers are also indicated. (After Reference [13].)

Frequency response of photodiodes. One of the major considerations in optical detectors is their frequency response—that is, the ability to respond to variations in the incident intensity such as those caused by high-frequency modulation. The three main mechanisms limiting the frequency response in photodiodes are:

1. The finite diffusion time of carriers produced in the p and n regions. This factor was described in the last section and its effect can be minimized by a proper choice of the length of the depletion layer.
2. The shunting effect of the signal current by the junction capacitance C_d shown in Figure 11-12. This places an upper limit of

$$\omega_m \simeq \frac{1}{R_e C_d} \tag{11.7-1}$$

on the intensity modulation frequency where R_e is the equivalent resistance in parallel with the capacitance C_d.
3. The finite transit time of the carriers drifting across the depletion layer.

To analyze this situation, we assume the slightly idealized case in which the carriers are generated in a *single* plane, say point A in Figure 11-13 and then drift the full width of the depletion layer at a constant velocity v. For high enough electric fields the drift velocity of carriers in semiconductors tends to saturate, so the constant velocity assumption is not very far from reality even for a nonuniform field distribution, such as that shown in Figure 11-11(e), provided the field exceeds its saturation value over most of the depletion layer length. The saturation of the hole velocity in germanium, as an example, is illustrated by the data of Figure 11-16.

The incident optical field is taken as

$$\begin{aligned} e(t) &= E_s(1 + m \cos \omega_m t) \cos \omega t \\ &\equiv \mathrm{Re}\,[V(t)] \end{aligned} \tag{11.7-2}$$

where

$$V(t) \equiv E_s(1 + m \cos \omega_m t) e^{i\omega t} \tag{11.7-3}$$

Thus, the amplitude is modulated at a frequency $\omega_m/2\pi$. Following the discussion of Section 11-1 we take the generation rate $G(t)$; that is, the number of carriers generated per second, as proportional to the average of $e^2(t)$ over a time long compared to the optical period $2\pi/\omega$. This average is equal to $\frac{1}{2}V(t)V^*(t)$, so the generation rate is taken as

$$G(t) = aE_s^2\left[\left(1 + \frac{m^2}{2}\right) + 2m \cos(\omega_m t) + \frac{m^2}{2} \cos(2\omega_m t)\right] \tag{11.7-4}$$

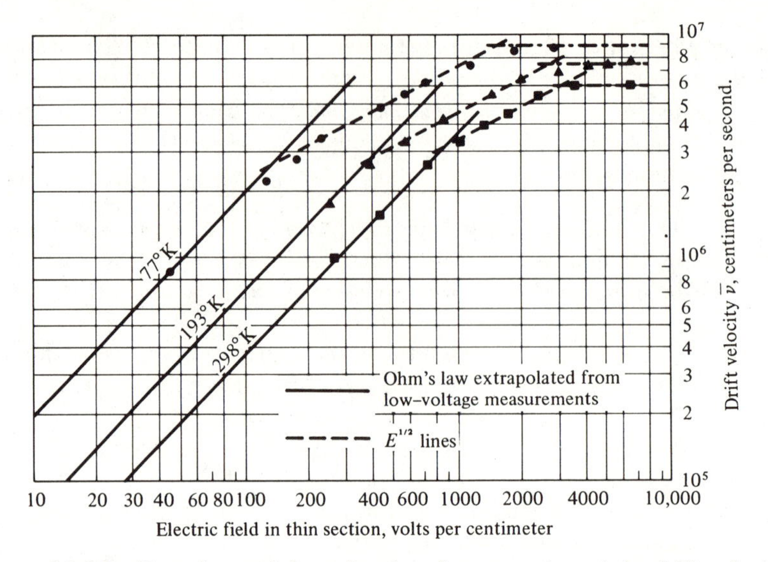

Figure 11-16 Experimental data showing the saturation of the drift velocity of holes in germanium at high electric fields. (After Reference [14].)

where a is a proportionality constant to be determined. Dropping the term involving cos $(2\omega_m t)$ and using complex notation, we rewrite $G(t)$ as

$$G(t) = aE_s{}^2\left[1 + \frac{m^2}{2} + 2me^{i\omega_m t}\right] \tag{11.7-5}$$

A single carrier drifting at a velocity $\bar{v}$ contributes, according to (10.4-1), an instantaneous current

$$i = \frac{e\bar{v}}{d} \tag{11.7-6}$$

to the external circuit, where d is the width of the depletion layer. The current due to carriers generated between t and $t' + dt'$ is thus $(e\bar{v}/d)G(t')\,dt'$ but, since each carrier spends a time $\tau_d = d/\bar{v}$ in transit, the instantaneous current at time t is the sum of contributions of carriers generated between t and $t - \tau_d$

$$i(t) = \frac{e\bar{v}}{d}\int_{t-\tau_d}^{t} G(t')dt' = \frac{e\bar{v}aE_s{}^2}{d}\int_{t-\tau_d}^{t}\left(1 + \frac{m^2}{2} + 2me^{i\omega_m t'}\right)dt'$$

and, after integration,

$$i(t) = \left(1 + \frac{m^2}{2}\right)eaE_s{}^2 + 2meaE_s{}^2\left(\frac{1 - e^{-i\omega_m\tau_d}}{i\omega_m\tau_d}\right)e^{i\omega_m t} \tag{11.7-7}$$

The factor $(1 - e^{-i\omega_m\tau_d})/i\omega_m\tau_d$ represents the phase lag as well as the reduction in signal current due to the finite drift time τ_d. If the drift time is short compared to the modulation period, so $\omega_m \tau_d \ll 1$, it has its maximum value of unity, and the signal is maximum. This factor is plotted in Figure 11-17 as a function of the transit phase angle $\omega_m \tau_d$.

Numerical example. As an estimate of the limitation on the modulation frequency due to the transit time, consider the case of a photodiode with the following characteristics:

material: p-type germanium

$\bar{v}$ (from Figure 11.16) $= 6 \times 10^6$ cm/s

$d = 10^{-4}$ cm

transit time: $\tau_d = \dfrac{d}{\bar{v}} \simeq 1.67 \times 10^{-11}$ second

We choose, somewhat arbitrarily, the highest useful modulation frequency as that for which $\omega_m \tau_d = 1$. This is the point where, according to Figure 11-17, the detector photocurrent is down to 95 percent of its maximum value. Using the foregoing calculated value of τ_d, we obtain

$$(f_m)_{\max} \simeq 10^{10} \text{ Hz}$$

for the highest modulation frequency as allowed by transit time.

Returning to (11.7-7) we can determine the value of the constant a by requiring that (11.7-7) agree with the experimental observation according to which in the absence of modulation, $m = 0$, each incident photon will create η carriers. Thus the dc (average) current is

$$\bar{I} = \frac{Pe\eta}{h\nu} \tag{11.7-8}$$

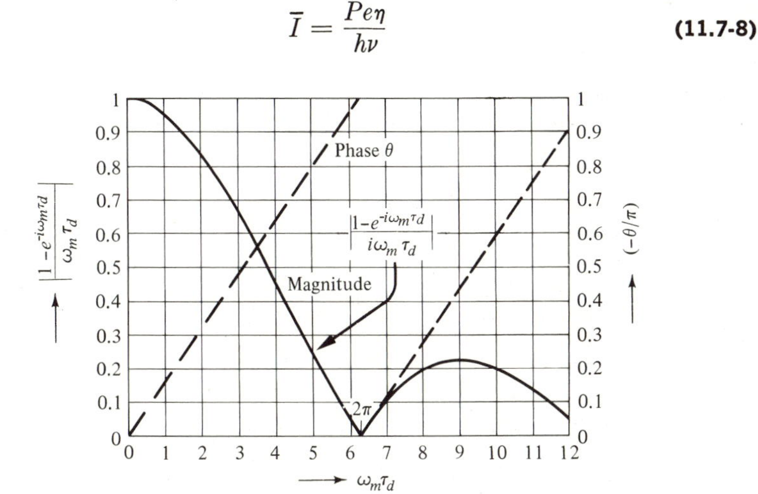

Figure 11-17 Phase and magnitude of the transit-time reduction factor $(1 - e^{-i\omega_m\tau_d})/i\omega_m\tau_d$.

where P is the optical (signal) power when $m = 0$. Using (11.7-8), we can rewrite (11.7-7) as

$$i(t) = \frac{Pe\eta}{h\nu}\left(1 + \frac{m^2}{2}\right) + \frac{Pe\eta}{h\nu}2m\left(\frac{i - e^{-i\omega_m\tau_d}}{i\omega_m\tau_d}\right)e^{i\omega_m t} \tag{11.7-9}$$

Detection sensitivity of photodiodes. We assume that the modulation frequency of the light to be detected is low enough that the transit time factor is unity and that the condition

$$\omega_m \ll \frac{1}{R_e C_d} \tag{11.7-10}$$

is fulfilled and, therefore, according to (11.7-1), the shunting of signal current by the diode capacitance C_d can be neglected. The diode current is given by (11.7-9) as

$$i(t) = \frac{Pe\eta}{h\nu}\left(1 + \frac{m^2}{2}\right) + \frac{Pe\eta}{h\nu}2me^{i\omega_m t} \tag{11.7-11}$$

The noise equivalent circuit of a diode connected to a load resistance R_L is shown in Figure 11-18. The signal power is proportional to the mean-square value of the sinusoidal current component, which, for $m = 1$, is

$$\overline{i_s^2} = 2\left(\frac{Pe\eta}{h\nu}\right)^2 \tag{11.7-12}$$

Two noise sources are shown. The first is the shot noise associated with the random generation of carriers. Using (10.4-9), this is represented by a noise generator $\overline{i_{N1}^2} = 2e\bar{I}\Delta\nu$, where $\bar{I}$ is the average current as given by the first term on the right side of (11.7-11). Taking $m = 1$, we obtain

$$\overline{i_{N1}^2} = \frac{3e^2(P + P_B)\eta\Delta\nu}{h\nu} + 2ei_d\Delta\nu \tag{11.7-13}$$

where P_B is the background optical power entering the detector (in addition to the signal power) and i_d is the "dark" direct current that exists even when $P_s = P_b = 0$. The second noise contribution is the thermal (Johnson

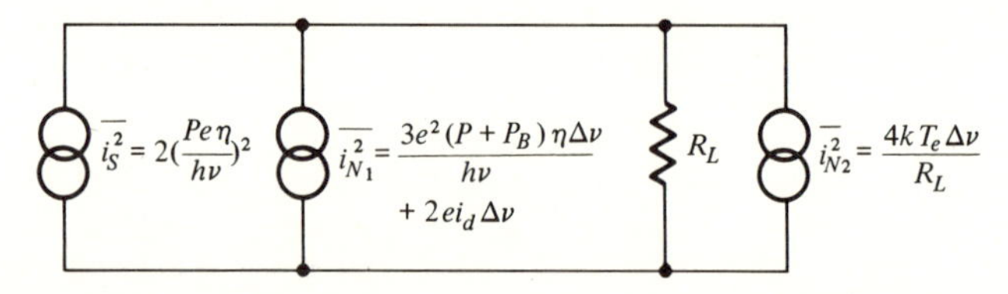

Figure 11-18 Equivalent circuit of a photodiode operating in the direct (video) mode. The modulation index m is taken as unity, and it is assumed that the modulation frequency is low enough that the junction capacitance and transit-time effects can be neglected. The resistance R_L is assumed to be much smaller than the shunt resistance R_d of the diode, so the latter is neglected. Also neglected is the series diode resistance, which is assumed small compared with R_L.

noise) generated by the output load, which, using (10.5-9), is given by

$$\overline{i_{N2}^2} = \frac{4kT_e\Delta\nu}{R_L} \tag{11.7-14}$$

where T_e is chosen to include the equivalent input noise power of the amplifier following the diode.[15] The signal-to-noise power ratio at the amplifier output is thus

$$\frac{S}{N} = \frac{\overline{i_s^2}}{\overline{i_{N1}^2} + \overline{i_{N2}^2}} = \frac{2(Pe\eta/h\nu)^2}{[3e^2(P + P_B)\eta\Delta\nu/h\nu] + 2ei_d\Delta\nu + (4kT_e\Delta\nu/R_L)} \tag{11.7-15}$$

In most practical systems the need to satisfy Equation (11.7-10) forces one to use small values of load resistance R_L. Under these conditions and for values of P that are near the detectability limit ($S/N = 1$), the noise term (11.7-14) is much larger than the shot noise (11.7-13) and the detector is consequently not operating near its quantum limit. Under these conditions we have

$$\frac{S}{N} \simeq \frac{2(Pe\eta/h\nu)^2}{4kT_e\Delta\nu/R_L} \tag{11.7-16}$$

The "minimum detectable optical power" is by definition that yielding $S/N = 1$ and is, from (11.7-16),

$$(P)_{\min} = \frac{h\nu}{e\eta}\sqrt{\frac{2kT_e\Delta\nu}{R_L}} \tag{11.7-17}$$

which is to be compared to the theoretical limit of $h\nu\Delta\nu/\eta$, which, according to (11.3-11), obtains when the signal shot-noise term predominates. In practice, the value of R_L is related to the desired modulation bandwidth $\Delta\nu$ and the junction capacitance C_d by

$$\Delta\nu \simeq \frac{1}{2\pi R_L C_d} \tag{11.7-18}$$

which, when used in (11.7-16), gives

$$P_{\min} \simeq 2\sqrt{\pi}\,\frac{h\nu\Delta\nu}{e\eta}\sqrt{kT_eC_d} \tag{11.7-19}$$

This shows that sensitive detection requires the use of small-area junctions so that C_d will be minimum.

---

[15] In practice it is imperative that the signal-to-noise ratio take account of the noise power contributed by the amplifier. This is done by characterizing the "noisiness" of the amplifier by an effective input noise "temperature" T_A. The amplifier noise output power is thus given by $GkT_A\Delta\nu$, where G is the power gain. This power can be referred to the input by dividing by G, thus becoming $kT_A\Delta\nu$. The total effective noise power at the amplifier input is the sum of this power and the Johnson noise $kT\Delta\nu$ due to the diode load resistance; that is, $k(T + T_A)\Delta\nu \equiv kT_e\Delta\nu$. The amplifier noise temperature T_A is related to its noise figure F by

$$F = 1 + \frac{T_A}{290}$$

Numerical example. Assume a typical Ge photodiode operating at $\lambda = 1.4\,\mu\text{m}$ with $C_d = 1$ pF, $\Delta\nu = 1$ GHz, and $\eta = 50$ percent. Let the amplifier following the diode have an effective noise temperature $T_e = 1200 + 290 = 1490$ °K[15] [14] [15]. Substitution in (11.7-19) gives

$$P_{\min} \simeq 3.34 \times 10^{-7} \text{ watt}$$

for the minimum detectable signal power.

11.8 The Avalanche Photodiode

By increasing the reverse bias across a *p-n* junction, the field in the depletion layer can increase to a point at which carriers (electrons or holes) that are accelerated across the depletion layer can gain enough kinetic energy to "kick" new electrons from the valence to the conduction band, while still traversing the layer. This process, illustrated in Figure 11-19, is referred to as avalanche multiplication. An absorbed photon (A) creates an electron–hole pair. The electron is accelerated until at point C it has gained sufficient energy to excite an electron from the valence to the conduction band, thus creating a new electron–hole pair. The newly generated carriers drift in turn in opposite directions. The hole (F) can also cause carrier multiplication as in G. The result is a dramatic increase (avalanche) in junction current that sets in when the electric field becomes high enough. This effect, discovered first in gaseous plasmas and more recently in *p-n* junctions (References [15] and [16]) gives rise to a multiplication of the current over its value in an ordinary (nonavalanching) photodiode. An

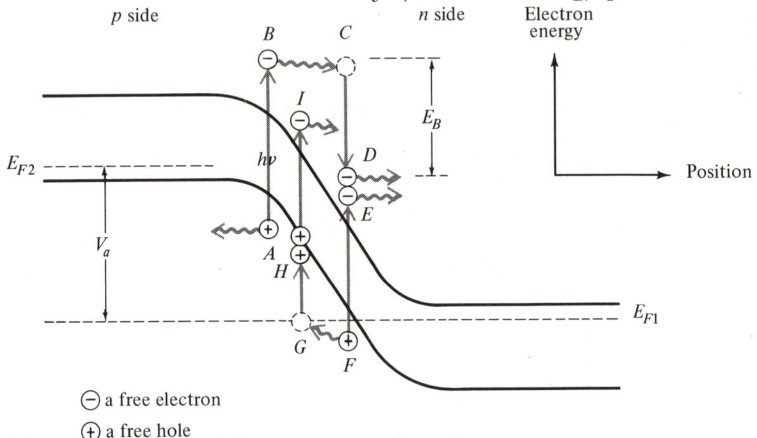

Figure 11-19 Energy-position diagram showing the carrier multiplication following a photon absorption in a reverse-biased avalanche photodiode.

[16] If the probability that a photo-excited electron-hole pair will create another pair during its drift is denoted by p, the current multiplication is

$$M = (1 + p + p^2 + p^3 + \cdots) = \frac{1}{1-p}$$

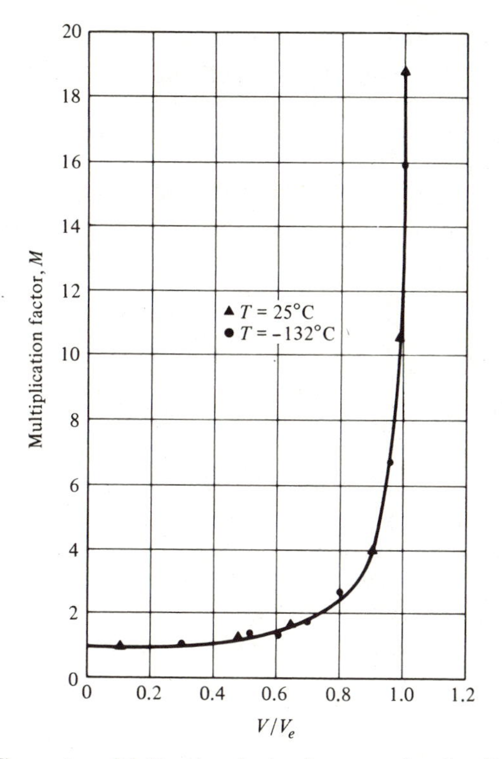

Figure 11-20 Current multiplication factor in an avalanche diode as a function of the electric field. (After Reference [16].)

experimental plot of the current gain M^{16} as a function of the junction field is shown in Figure 11-20.

Avalanche photodiodes are similar in their construction to ordinary photodiodes except that, because of the steep dependence of M on the applied field in the avalanche region, special care must be exercised to obtain very uniform junctions. A sketch of an avalanche photodiode is shown in Figure 11-21.

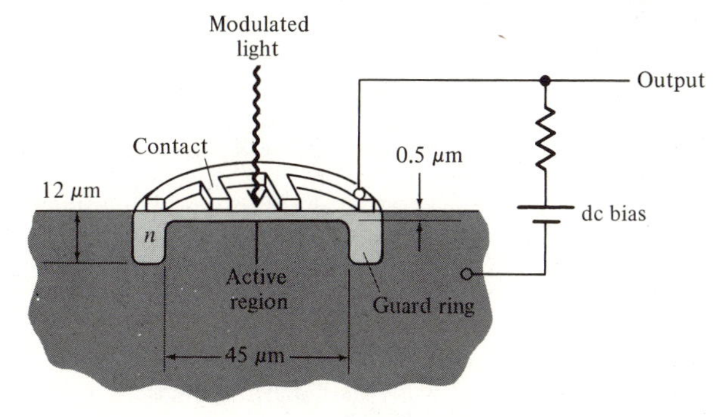

Figure 11-21 Planar avalanche photodiode. (After Reference [13].)

Since an avalanche photodiode is basically similar to a photodiode, its equivalent circuit elements are given by expressions similar to those given above for the photodiode. Its frequency response is similarly limited by diffusion, drift across the depletion layer, and capacitive loading, as discussed in Section 11.7.

A multiplication by a factor M of the photocurrent leads to an increase by M^2 of the signal power S over that which is available from a photodiode so that, using (11.7-12), we get

$$S \propto \overline{i_s^{\,2}} = 2M^2 \left(\frac{Pe\eta}{h\nu}\right)^2 \tag{11.8-1}$$

where P is the optical power incident on the diode. This result is reminiscent of the signal power from a photomultiplier as given by the numerator of (11.3-9), where the avalanche gain M plays the role of the secondary electron multiplication gain G. We may expect that, similarly, the shot noise power will also increase by M^2. The shot noise, however, is observed to increase as M^n, where $2 < n < 3$.[17] Experimental observation of a near ideal $M^{2.1}$ behavior is shown in Figure 11-22.

The signal-to-noise power ratio at the output of the diode is thus given,

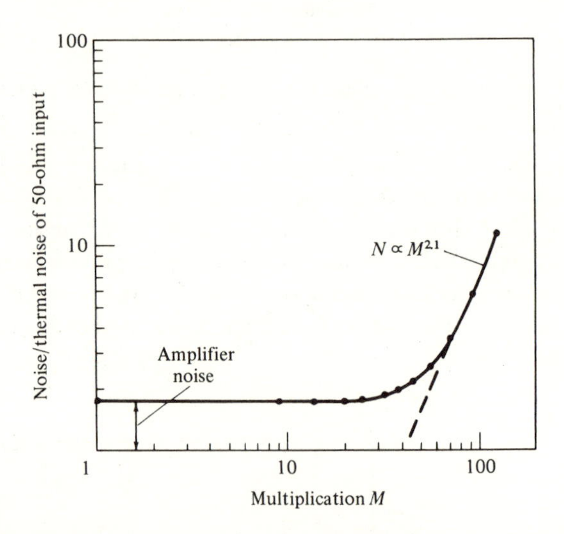

Figure 11-22 Noise power (measured at 30 MHz) as a function of photocurrent multiplication for an avalanche Schottky-barrier photodiode. (After Reference [18].)

---

[17] A theoretical study by McIntyre [17] predicts that if the multiplication is due to either holes or electrons, $n = 2$, whereas if both carriers are equally effective in producing electron–hole pairs, $n = 3$.

following (11.7-15), by

$$\left(\frac{S}{N}\right) = \frac{2M^2(Pe\eta/h\nu)^2}{[3e^2(P + P_B)\eta\Delta\nu/h\nu]M^n + 2ei_d\Delta\nu M^n + (4kT_e\Delta\nu/R_L)} \qquad \textbf{(11.8-2)}$$

The advantage of using an avalanche photodiode over an ordinary photodiode is now apparent. When $M = 1$, the situation is identical to that at the photodiode as described by (11.7-15). Under these conditions the thermal term $4kT_e\Delta\nu/R_L$ in the denominator of (11.8-2) is typically much larger than the shot-noise terms. This causes S/N to increase with M. This improvement continues until the shot-noise terms become comparable with $4kT_e\Delta\nu/R_L$. Further increases in M result in a reduction of S/N since $n > 2$, and the denominator of (11.8-2) grows faster than the numerator. If we assume that M is adjusted optimally so that the denominator of (11.8-2) is equal to twice the thermal term $4kT_e\Delta\nu/R_L$, we can solve for the minimum detectable power (that is, the power input for which $S/N = 1$) obtaining

$$P_{\min} = \frac{2h\nu}{M'e\eta}\sqrt{\frac{kT_e\Delta\nu}{R_L}} \qquad \textbf{(11.8-3)}$$

where M' is the optimum value of M as discussed previously. The improvement in sensitivity over the photodiode result (11.7-16) is thus approximately M'. Values of M' between 30 and 100 are commonly employed, so the use of avalanche photodiodes affords considerable improvement in sensitivity over that available from photodiodes.

11.9 A design of an optical fiber link for transmitting digital data

As an example of the use of some of the basic concepts involving optical detection and noise, let us consider the problem of designing an optical communication link using an optical fiber for the transmission of binary digital data over a distance of 5 km. The system requirements are:

1. Data rate—5×10^8 bits/second.
2. Error probability after amplification of detected signal at the receiver must be less than 10^{-9}.

The proposed system is shown in Figure 11-23. It consists of a CW diode laser emitting at 1.04 μm [19]. The laser output passes through an electrooptic waveguide modulator driven by the digital data stream. The electrooptic modulator impresses, in the manner described in Section 9.3, the digital data onto the optical beam whose intensity modulation becomes a replica of the digital input to the modulator. The modulated laser beam is coupled into a graded index optical fiber with a loss of 4 db/km.

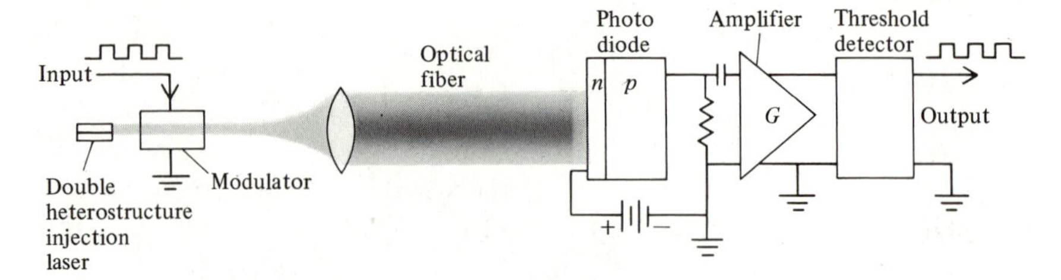

Figure 11-23 An optical fiber link for transmitting binary data.

The received signal is detected by a fast silicon *p-n* diode with a quantum efficiency of 0.5 at $\lambda = 1.04\ \mu$m and a total (junction plus package) capacitance of $3 \times 10^{-12} f$.

The detected signal is next amplified by a field-effect transistor amplifier with a noise figure of 6 db. The amplified signal is fed into a threshold detector which regenerates the original data stream to within the allowed error probability (per pulse) of $P_e < 10^{-9}$.

In order to meet the requirement $P_e < 10^{-9}$, we need, according to Figure 10-13, a signal-to-noise current ratio at the input to the threshold detector of

$$\frac{i_S}{\langle i_N \rangle} = 11.89 \text{ (that is, 21.5 db)}$$

where, we recall (see Section 10.11), that $\langle i_N \rangle \equiv (\overline{i_N^2})^{1/2}$.

The signal-to-noise power ratio of a *p-n* diode detector is given by (11.7-16) in the case where the dominant contributions to the noise power are the amplifier noise and the Johnson (thermal) noise of the load resistance R_L in the diode output circuit. The mean square noise current is then

$$\overline{i_N^2} \approx \frac{4kT_e \Delta\nu}{R_L} \tag{11.9-1}$$

The signal peak current is given by (11.7-8) as

$$i_S = \frac{P_S e \eta}{h\nu} \tag{11.9-2}$$

where P_s is the peak pulsed optical power incident on the detector. The signal-to-noise current ratio at the amplifier output[15] is thus

$$\frac{i_S}{\langle i_N \rangle} = \frac{(P_s e\eta / h\nu)}{(4kT_e \Delta\nu / R_L)^{1/2}} \tag{11.9-3}$$

Our next problem is that of finding the minimum value of the signal power P_s so that $i_S / \langle i_N \rangle$ in (11.9-3) exceeds the needed value of 11.89. We thus need to know T_e, R_L, and $\Delta\nu$. T_e is obtained from the given value of the amplifier noise figure (F = 6 db). Taking T = 290 °K, we obtain, using footnote 15, $T_e = 290 + (4 - 1)290 = 1160$. The bandwidth $\Delta\nu$ is taken, conservatively, as $\Delta\nu = 2/(\pi\tau)$ where $\tau = 2 \times 10^{-9}s$ is the pulse period. The result is $\Delta\nu = 3.18 \times 10^8$ Hz. In order to achieve this bandwidth, the load resistance R_L must not exceed (see 11.7-18) the value

$$R_L = \frac{1}{2\pi\Delta\nu C} \tag{11.9-4}$$

where C is the total output capacitance given as $3 \times 10^{-12}f$. Using the above value of $\Delta\nu$ and C, we obtain

$$R_L \leq 167\Omega$$

We return now to (11.9-3), which, using $\eta = 0.5$, $\lambda = 1.04$ μm, $i_S/\langle i_N \rangle = 11.89$, yields

$$P_S \cong 10^{-5} \text{ watts}$$

for the minimum power input to the photodiode.

The total transmission loss in the 5 km fiber is 20 db. We will assume that an additional 4 db loss is caused by coupling the laser output to the fiber and at the fiber output so that the total loss is 24 db (that is, 251). The laser power output must thus exceed

$$P_{\text{laser}} = 10^{-5} \times 251 = 2.51 \text{ milliwatts}$$

which is a reasonable power level for CW diode lasers.

If the fiber had been substantially lossier than in the above example, we could still have met our design specifications by using an avalanche photodiode. The avalanche photodiode was found (see Section 11.8) to be considerably more sensitive than the junction diode (that is, less noisy) and is thus capable of achieving a given signal-to-noise power ratio at lower signal input levels.

We also need to insure that group velocity dispersion does not broaden the output pulses during their 5 km journey by more than their average separation of $2 \times 10^{-9}s$. This can be done by using the results of Section 3.6 once the specific dispersion data of the fiber is given. If the group velocity dispersion of the fiber at $\lambda = 1.04$ μm exceeds the limits imposed by the fiber length (5 km in our example) and the pulse length, the detection, and regeneration will have to be performed at shorter distances. The regenerated pulse train will then serve as the input signal to the next segment of the fiber link.

PROBLEMS

11-1 Show that the total output shot noise power in a photomultiplier including that originating in the dynodes is given by

$$\overline{(i_{N^2})} = G^2 2e(\bar{i}_c + i_d)\Delta\nu \frac{1 - \delta^{-N}}{1 - \delta^{-1}}$$

where δ is the secondary-emission multiplication factor and N is the number of stages.

11-2 Calculate the minimum power that can be detected by a photoconductor in the presence of a strong optical background power P_B. *Answer:*

$$(P_s)_{\min} = 2\left(\frac{P_B h\nu\Delta\nu}{\eta}\right)^{1/2}$$

11-3 Derive the expression for the minimum detectable power using a photoconductor in the video mode (that is, no local-oscillator power) and assuming that the main noise contribution is the generation–recombination noise. The optical field is given by $e(t) = E(1 + \cos\omega_m t)\cos\omega t$ and the signal is taken as the component of the photocurrent at ω_m.

11-4 Derive the minimum detectable power of a Ge:Hg detector with characteristics similar to those described in Section 11-7 when the average current is due mostly to blackbody radiation incident on the photocathode. Assume $T = 295°\text{K}$, an acceptance solid angle $\Omega = \pi$ and a photocathode area of 1 mm^2. Assume that the quantum yield η for blackbody radiation at $\lambda < 14\,\mu\text{m}$ is unity and that for $\lambda > 14\,\mu\text{m}$, $\eta = 0$. *Hint:* Find the flux of photons with wavelengths $14\,\mu\text{m} > \lambda > 0$ using blackbody radiation formulas or, more easily, tables or a blackbody "slide rule.")

11-5 Find the minimum detectable power in Problem 11-4 when the input field of view is at $T = 4.2°\text{K}$.

11-6 Derive Equations (11.6-15) and (11.6-16).

11-7 Show that the transit time reduction factor $(1 - e^{-i\omega_m\tau_d})/i\omega_m\tau_d$ in Equation (11.7-6) can be written as

$$\alpha - i\beta$$

where

$$\alpha = \frac{\sin\omega_m\tau_d}{\omega_m\tau_d} \qquad \beta = \frac{1 - \cos\omega_m\tau_d}{\omega_m\tau_d}$$

Plot α and β as a function of $\omega_m\tau_d$.

11-8 Derive the minimum detectable power for a photodiode operated in the heterodyne mode. *Answer:* $P_{\min} = h\nu\Delta\nu/\eta$

11-9 Discuss the limiting sensitivity of an avalanche photodiode in which the noise increases as M^2. Compare it with that of a photomultiplier. What is the minimum detectable power in the limit of $M \gg 1$, and of zero background radiation and no dark current.

11-10 Derive an expression for the magnitude of the output current in a heterodyne detection scheme as a function of the angle θ between the signal and local-oscillator propagation directions. Taking the aperture diameter (see Figure 11-6) as D, show that if the output is to remain near its maximum ($\theta = 0°$) value, θ should not exceed λ/D. *Hint:* You may replace the lens in Figure 11-6 by the photoemissive surface. Show that instead of Equation (11.4-4) the current from an element $dx\, dy$ of the detector is

$$di(x, t) = \frac{P_L e\eta}{h\nu(\pi D^2/4)}\left[1 + 2\sqrt{\frac{P_s}{P_L}}\cos\,(\omega t + kx \sin\theta)\right] dx\, dy$$

The propagation directions lie in the z-x plane. The contribution of $dx\, dy$ to the (complex) signal current is thus

$$dI_s(x, t) = \frac{2\sqrt{P_s P_L}}{h\nu(\pi D^2/4)}\, e^{ikx \sin\theta}\, dx\, dy$$

11-11 Show that for a Poisson distribution (footnote 9) $\overline{(\Delta N)^2} = \bar{N}$.

REFERENCES

[1] Yariv, A., *Quantum Electronics*. New York: Wiley, 2d Ed. 1975, p. 54.

[2] Engstrom, R.W., " Multiplier phototube characteristics: Application to low light levels," *J. Opt. Soc. Am.*, vol. 37, p. 420, 1947.

[3] Sommer, A. H., *Photo-Emissive Materials*. New York: Wiley, 1968.

[4] Forrester, A. T., "Photoelectric mixing as a spectroscopic tool," *J. Opt. Soc. Am.*, vol. 51, p. 253, 1961.

[5] Siegman, A. E., S. E. Harris, and B. J. McMurtry, "Optical heterodyning and optical demodulation at microwave frequencies," in *Optical Masers*, J. Fox, ed. New York: Wiley, 1963, p. 511.

[6] Mandel, L., "Heterodyne detection of a weak light beam," *J. Opt. Soc. Am.*, vol. 56, p. 1200, 1966.

[7] Chapman, R. A., and W. G. Hutchinson, "Excitation spectra and photoionization of neutral mercury centers in germanium," *Phys. Rev.*, vol. 157, p. 615, 1967.

[8] Teich, M. C., "Infrared heterodyne detection," *Proc. IEEE*, vol. 56, p. 37, 1968.

[9] Buczek, C., and G. Picus, "Heterodyne performance of mercury doped germanium," *Appl. Phys. Letters*, vol. 11, p. 125, 1967.

[10] Lucovsky, G., M. E. Lasser, and R. B. Emmons, "Coherent light detection in solid-state photodiodes," *Proc. IEEE*, vol. 51, p. 166, 1963.

[11] Riesz, R. P., "High speed semiconductor photodiodes," *Rev. Sci. Instr.*, vol. 33, p. 994, 1962.

[12] Anderson, L. K., and B. J. McMurtry, "High speed photodetectors," *Appl. Opt.*, vol. 5, p. 1573, 1966.

[13] D'Asaro, L. A., and L. K. Anderson, "At the end of the laser beam, a more sensitive photodiode," *Electronics*, May 30, 1966, p. 94.

[14] Shockley, W., "Hot electrons in germanium and Ohm's law," *Bell System Tech. J.*, vol. 30, p. 990, 1951.

[15] McKay, K. G., and K. B. McAfee, "Electron multiplication in silicon and germanium," *Phys. Rev.*, vol. 91, p. 1079, 1953.

[16] McKay, K. G., "Avalanche breakdown in silicon," *Phys. Rev.*, vol. 94, p. 877, 1954.

[17] McIntyre, R., "Multiplication noise in uniform avalanche diodes," *IEEE Trans. Electron Devices*, vol. ED-13, p. 164, 1966.

[18] Lindley, W. T., R. J. Phelan, C. M. Wolfe, and A. J. Foyt, "GaAs Schottky barrier avalanche photodiodes," *Appl. Phys. Letters*, vol. 14, p. 197, 1969.

[19] Nahory, R. E., M. A. Pollack, E. D. Beebe, and J. C. DeWinter, "Continuous operation of a 1.0 μm wavelength $GaAs_{1-x}Sb_x/Al_yGa_{1-y}As_{1-x}Sb_x$ double-heterostructure injection laser," *Appl. Phys. Letters*, vol. 28, p. 19, 1976.

CHAPTER

12

Interaction of Light and Sound

12.0 Introduction

Diffraction of light by sound[1] waves was predicted by Brillouin in 1922 [1] and demonstrated experimentally some ten years later [2]. Recent developments in high frequency acoustics [3] and in lasers caused a renewed interest in this field. This is due to the fact that the scattering of light from sound affords a convenient means of controlling the frequency, intensity, and direction of an optical beam. This type of control makes possible a large number of applications involving the transmission, display, and processing of intelligence [4].

12.1 Scattering of Light by Sound

A sound wave consists of a sinusoidal perturbation of the density of the material, or strain, which travels at the sound velocity v_s, as shown in Figure 12-1. A change in the density of the medium causes a change in its

[1] In this chapter we use the word *sound* to describe acoustic waves with frequencies which in practice may range through the microwave region ($f \simeq 10^{10}$ Hz).

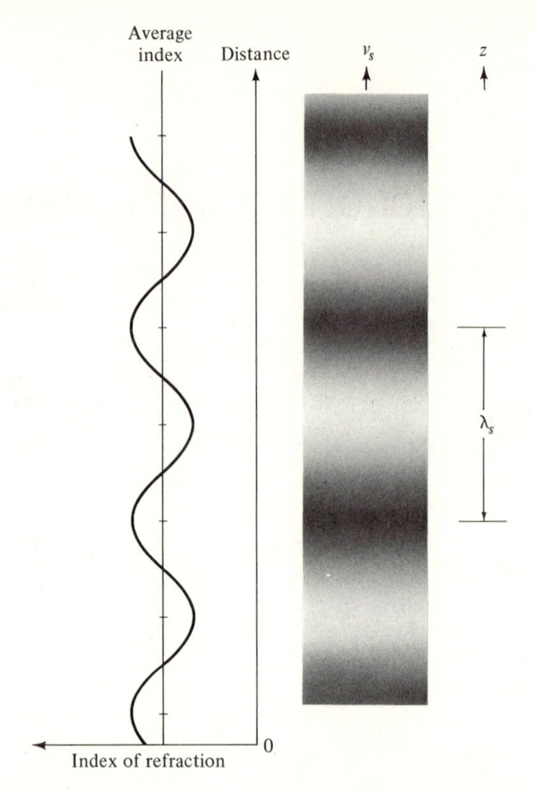

Figure 12-1 Traveling sound wave "frozen" at some instant of time. It consists of alternating regions of compressions (dark) and rarefaction (white), which travel at the sound velocity v_s. Also shown is the instantaneous spatial variation of the index of refraction that accompanies the sound wave.

index of refraction, which, to first order, is proportional to it.[2] We can, consequently, represent the sound wave shown in Figure 12.1 by

$$\Delta n(z, t) = \Delta n \sin (\omega_s t - k_s z) \tag{12.1-1}$$

where $\omega_s/k_s = v_s$.

Next consider an optical beam incident on a sound wave at an angle θ_i as in Figure 12-2. For the purpose of the immediate discussion we can characterize the sound wave as a series of partially reflecting mirrors,[3] separated by the sound wavelength λ_s, which are moving at a velocity v_s. Ignoring, for the moment, the motion of the mirrors, let us consider the diffracted wave and take the diffraction angle as θ_r. A necessary condition

---

[2] This is easily understood in the case where each atom (or molecule) contributes a constant amount to the index n, so the latter is proportional to the material density.

[3] This is due to the fact that the index of refraction is higher in the compressed portions of the sound wave and lower in the rarefied regions. Since a change in index causes reflection the mirrors' analogy follows.

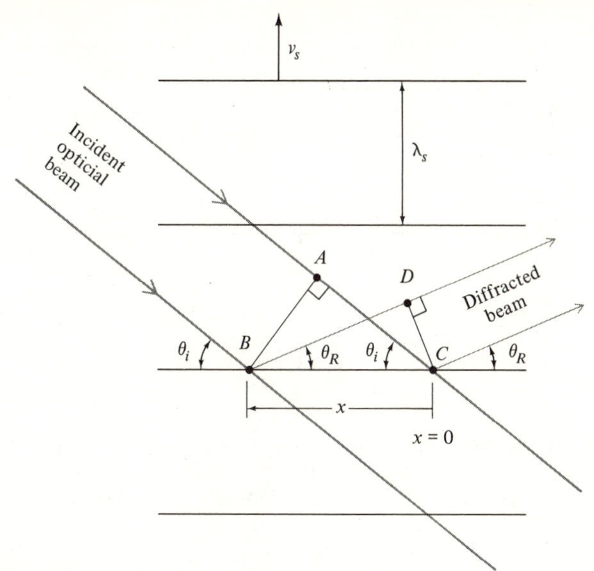

Figure 12-2 Diffraction of an incident optical beam from an array of equispaced reflectors.

for diffraction in a given direction is that *all the points on a given mirror contribute in phase* to the diffraction along this direction. Considering the diffraction from two points, such as C and B in Figure 12-2, it is then necessary that the optical path difference $AC - BD$ be some multiple of the optical wavelength λ/n for diffraction along θ_r to occur. This condition takes the form

$$x(\cos \theta_i - \cos \theta_r) = m\lambda/n \tag{12.1-2}$$

where $m = 0, \pm 1, \pm 2, \cdots$. The only way in which (12.1-2) can be satisfied simultaneously *for all points* x along a *given* reflector is if $m = 0$, from which it follows that

$$\theta_i = \theta_r \tag{12.1-3}$$

In addition to the requirement that the different parts of a given acoustic phase front interfere constructively, which leads to (12.1-3), we require that the diffraction from any two acoustic phase fronts add up in phase along the direction of the reflected beam. The path difference, $AO + OB$ shown in Figure 12-3, of a given optical wavefront resulting from reflection from two equivalent acoustic wavefronts (that is, planes separated by λ_s) must thus be equal to the optical wavelength λ. Using (12.1-3) and Figure 12-3 we find that this condition can be written as[4]

$$2\lambda_s \sin \theta = \lambda/n \tag{12.1-4}$$

where $\theta_i = \theta_r = \theta$.

---

[4] The reader may justly wonder why path differences of $2\lambda/n$, $3\lambda/n$, and so on, do not lead to maximum diffraction as well as a path difference of λ/n. This point is considered in Problem 12-6.

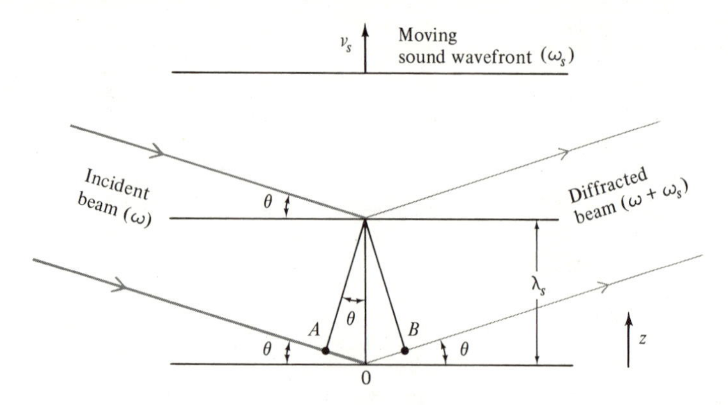

Figure 12-3 The reflections from two equivalent planes in the sound beam (that is, planes separated by the sound wavelength λ_s), which add up in phase along the direction θ if the optical path difference $AO + OB$ is equal to one optical wavelength.

The diffraction of light that satisfies (12.1-4) is known as Bragg diffraction after a similar law applying in X-ray diffraction from crystals. To get an idea of the order of magnitude of the angle θ, consider the case of diffraction of light with $\lambda = 0.5\ \mu$m from a 500-MHz sound wave. Taking the sound velocity as $v_s = 3 \times 10^5$ cm we have $\lambda_s = v_s/\nu_s = 6 \times 10^{-4}$ cm and, from (12.1-4),

$$\theta \simeq 4 \times 10^{-2} \text{ rad} \simeq 3.5°$$

12.2 Particle Picture of Bragg Diffraction of Light by Sound

Many of the features of Bragg diffraction of light by sound can be deduced if we take advantage of the dual particle-wave nature of light and of sound. According to this picture a light beam with a propagation vector $\mathbf{k}$[5] and frequency ω can be considered to consist of a stream of particles (photons) with a momentum $\hbar\mathbf{k}$ and energy $\hbar\omega$. The sound wave, likewise, can be thought of as made up of particles (phonons) with momentum $\hbar\mathbf{k}_s$ and energy $\hbar\omega_s$. The diffraction of light by an *approaching* sound beam illustrated in Figure 12-3 can be described as a series of collisions, each of which involves an annihilation of *one* incident photon at ω_i and *one* phonon and a simultaneous creation of a new (diffracted) photon at a frequency $\omega_d = \omega_s + \omega$, which propagates along the direction of the scattered beam.

[5] The beam is of the form $\cos(\omega t - \mathbf{k} \cdot \mathbf{r})$, so it propagates in a direction parallel to $\mathbf{k}$ with a wavelength $2\pi/k$.

The conservation of momentum requires that the momentum $\hbar(\mathbf{k}_s + \mathbf{k}_i)$ of the colliding particles is equal to the momentum $\hbar\mathbf{k}_d$ of the scattered photon, so

$$\mathbf{k}_d = \mathbf{k}_s + \mathbf{k}_i \tag{12.2-1}$$

The conservation of energy takes the form

$$\omega_d = \omega_i + \omega_s \tag{12.2-2}$$

From (12.2-2) we learn that the diffracted beam is shifted in frequency by an amount equal to the sound frequency. Since the interaction involves the annihilation of a phonon, conservation of energy decrees that the shift in frequency is such that $\omega_d > \omega_i$ and the phonon energy is *added* to that of the annihilated photon to form a new photon. Using this argument it follows that if the direction of the sound beam in Figure 12-3 were reversed so that it was receding from the incident optical wave, the scattering process could be considered as one in which a new photon (diffracted photon) and a *new* phonon are generated while the incident photon is annihilated. In this case, the conservation-of-energy principle yields

$$\omega_d = \omega_i - \omega_s$$

The relation between the sign of the frequency change and the sound propagation direction will become clearer using Doppler-shift arguments, as is done at the end of this section.

The conservation-of-momentum condition (12.2-1) is equivalent to the Bragg condition (12.1-4). To show why this is true, consider Figure 12-4.

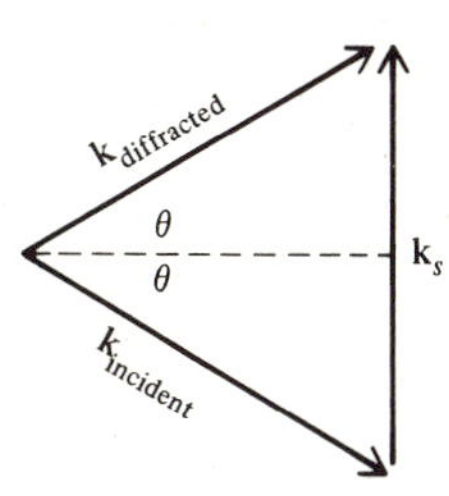

Figure 12-4 The momentum-conservation relation, Equation (12.2-1), used to derive the Bragg condition $2\lambda_s \sin\theta = \lambda/n$, for an optical beam that is diffracted by an approaching sound wave. θ is the angle between the incident or diffracted beam and the acoustic wavefront.

Since the sound frequencies of interest are below 10^{10} Hz and those of the optical beam are usually above 10^{13} Hz, we have

$$\omega_d = \omega_i + \omega_s \simeq \omega_i, \quad \text{so} \quad k_d \simeq k_i$$

and the magnitude of the two optical wave vectors is taken as k (see also Problem 12-4). The magnitude of the sound wave vector is thus

$$k_s = 2k \sin\theta \tag{12.2-3}$$

Using $k_s = 2\pi/\lambda_s$, this equation becomes

$$2\lambda_s \sin\theta = \lambda/n \tag{12.2-4}$$

which is the same as the Bragg-diffraction condition (12.1-4).

Doppler derivation of the frequency shift. The frequency-shift condition (12.2-2) can also be derived by considering the Doppler shift exercised by an optical beam incident on a mirror moving at the sound velocity v_s at an angle satisfying the Bragg condition (12.1-4). The formula for the Doppler frequency shift of a wave reflected from a moving object is

$$\Delta\omega = 2\omega \frac{v}{c/n}$$

where ω is the optical frequency and v is the component of the object velocity that is parallel to the wave propagation direction. From Figure 12-3 we have $v = v_s \sin\theta$, and thus

$$\Delta\omega = 2\omega \frac{v_s \sin\theta}{c/n} \tag{12.2-5}$$

Using (12.1-4) for $\sin\theta$ we obtain

$$\Delta\omega \equiv \frac{2\pi v_s}{\lambda_s} = \omega_s \tag{12.2-6}$$

and, therefore, $\omega_d = \omega + \omega_s$.

If the direction of propagation of the sound beam is reversed so that, in Figure 12-3, the sound recedes from the optical beam, the Doppler shift changes sign and the diffracted beam has a frequency $\omega - \omega_s$.

12.3 Bragg Diffraction of Light by Acoustic Waves—Analysis

In treating the diffraction of light by acoustic waves, we assume a long interaction path so that higher diffraction orders [5] are missing and the only two waves coupled by the sound are the incident wave at ω_i and a diffracted wave at $\omega_d = \omega_i + \omega_s$ or at $\omega_i - \omega_s$, depending on the direction of the Doppler shift as discussed in Section 12.2.

According to the discussion in Section 12.1, the sound wave causes a traveling modulation of the index of refraction given by

$$\Delta n(\mathbf{r}, t) = \Delta n \cos(\omega_s t - \mathbf{k}_s \cdot \mathbf{r}) \tag{12.3-1}$$

This modulation interacts with the fields at ω_i and ω_d to give rise to additional electric polarization in the medium which is given by[6]

$$\Delta\mathbf{p}(\mathbf{r}, t) = 2\sqrt{\epsilon\epsilon_0}\, \Delta n(\mathbf{r}, t)\mathbf{e}(\mathbf{r}, t) \tag{12.3-2}$$

---

[6] Equation (12.3-2) can be derived using the relations $\mathbf{d} = \epsilon_0\mathbf{e} + \mathbf{p} = \epsilon\mathbf{e}$, $\mathbf{p} \equiv \epsilon_0\chi\mathbf{e}$, and $n^2 \equiv \epsilon/\epsilon_0$. This leads to $\mathbf{p} = \epsilon_0(n^2 - 1)\mathbf{e}$ so that (12.3-2) follows.

where $\mathbf{e}(\mathbf{r}, t)$ is the sum of the fields at ω_i and ω_d. The polarization term $\Delta n\mathbf{e}$ in (12.3-2) will be shown, in what follows, to cause exchange of power between the fields at ω_i and ω_d.

We start with the wave equation (8.2-5)

$$\nabla^2\mathbf{e}(\mathbf{r}, t) = \mu\epsilon \frac{\partial^2 \mathbf{e}}{\partial t^2} + \mu \frac{\partial^2}{\partial t^2} \mathbf{p}_{NL}(\mathbf{r}, t) \tag{8.2-5}$$

where the medium is assumed lossless so that $\sigma = 0$. $\mathbf{p}_{NL}(\mathbf{r}, t)$ is given in our case by $\Delta\mathbf{p}(\mathbf{r}, t)$. Equation (8.2-5) must be satisfied separately for the fields at ω_i and ω_d. Writing it for the former case and assuming that both the incident and diffracted fields are linearly polarized result in

$$\nabla^2 e_i = \mu\epsilon \frac{\partial^2 e_i}{\partial t^2} + \mu \frac{\partial^2}{\partial t^2} (\Delta p)_i \tag{12.3-3}$$

where e_i is the magnitude of the vector $\mathbf{e}_i$ and $(\Delta p)_i$ is the component of $\Delta\mathbf{p}(\mathbf{r}, t)$ parallel to $\mathbf{e}_i$ which oscillates at a frequency ω_i. The polarization components oscillating at other frequencies are nonsynchronous and their contribution to e_i averages out to zero. The total field $\mathbf{e}(\mathbf{r}, t)$ is taken as the sum of two traveling waves

$$\begin{aligned} e_i(\mathbf{r}, t) &= \tfrac{1}{2}E_i(r_i)e^{i(\omega_i t - \mathbf{k}_i\cdot\mathbf{r})} + \text{c.c.} \\ e_d(\mathbf{r}, t) &= \tfrac{1}{2}E_d(r_d)e^{i(\omega_d t - \mathbf{k}_d\cdot\mathbf{r})} + \text{c.c.} \end{aligned} \tag{12.3-4}$$

where $\mathbf{k}_i$ and $\mathbf{k}_d$ are parallel to the direction of propagation of the incident and diffracted waves, respectively. Two differentiations of (12.3-4) lead to

$$\nabla^2 e_i(\mathbf{r}, t) = -\tfrac{1}{2}\left[k_i^2 E_i + 2ik_i \frac{dE_i}{dr_i} + \nabla^2 E_i\right] e^{i(\omega_i t - \mathbf{k}_i\cdot\mathbf{r})}$$

Assuming "slow" variation of $E_i(r_i)$ so that $\nabla^2 E_i \ll k_i\, dE_i/dr_i$, we combine (12.3-3) with the last equation and recalling that $k_i^2 = \omega_i^2\mu\epsilon$ obtain

$$k_i \frac{dE_i}{dr_i} = i\mu \left[\frac{\partial^2}{\partial t^2} (\Delta p)_i\right] e^{-i(\omega_i t - \mathbf{k}_i\cdot\mathbf{r})} \tag{12.3-5}$$

Using the relation $\Delta\mathbf{p} = 2\sqrt{\epsilon\epsilon_0}\, \Delta n(\mathbf{r}, t) \cdot [\mathbf{e}_i(\mathbf{r}, t) + \mathbf{e}_d(\mathbf{r}, t)]$ $(\Delta p)_i$ is given by

$$[\Delta p(\mathbf{r}, t)]_i = \tfrac{1}{2}\sqrt{\epsilon\epsilon_0}\Delta n E_d\{e^{i[(\omega_s+\omega_d)t - (\mathbf{k}_s+\mathbf{k}_d)\cdot\mathbf{r}]}\} + \text{c.c.} \tag{12.3-6}$$

Note that in taking the product $\Delta n(\mathbf{r}, t)e(\mathbf{r}, t)$ we assumed that $\omega_i = \omega_s + \omega_d$ and therefore neglected nonsynchronous terms with frequencies $\omega_d - \omega_s$ and $\omega_i \pm \omega_s$. Substituting (12.3-6) for $(\Delta p)_i$ in (12.3-5) leads to

$$\frac{dE_i}{dr_i} = -i\eta_i E_d e^{i(\mathbf{k}_i - \mathbf{k}_s - \mathbf{k}_d)\cdot\mathbf{r}} \tag{12.3-7}$$

and similarly

$$\frac{dE_d}{dr_d} = -i\eta_d E_i e^{-i(\mathbf{k}_i - \mathbf{k}_s - \mathbf{k}_d)\cdot\mathbf{r}}$$

with

$$\eta_{i,d} = \tfrac{1}{2}\omega_{i,d}\sqrt{\mu\epsilon_0}\Delta n = \frac{\omega_{i,d}\Delta n}{2c_0} \tag{12.3-8}$$

An inspection of (12.3-7) reveals that a prerequisite for continuous cumulative interaction between the incident field (E_i) and the diffracted field (E_d) is that

$$\mathbf{k}_i = \mathbf{k}_s + \mathbf{k}_d \tag{12.3-9}$$

otherwise, it follows from (12.3-7) that contributions to E_i, as an example, from different path elements do not add in phase and no sustained spatial growth of E_i is possible.

Equation (12.3-9) is, as shown in Section 12.2, the Bragg condition for scattering of light by sound. The difference between (12.3-9) and (12.2-1) is due to the fact that the latter was derived for the case of diffraction from an approaching sound beam so that $\omega_d = \omega_i + \omega_s$ resulting in a "momentum" condition $\mathbf{k}_d = \mathbf{k}_i + \mathbf{k}_s$, while in the treatment leading to (12.3-9) we recall that the sound wave is taken as receding from the incident field so that $\omega_d = \omega_i - \omega_s$.

Assuming that the Bragg condition (12.3-9) is satisfied, (12.3-7) becomes

$$\begin{aligned} \frac{dE_i}{dr_i} &= -i\eta E_d \\ \frac{dE_d}{dr_d} &= -i\eta E_i \end{aligned} \tag{12.3-10}$$

where, since $\omega_i \approx \omega_d$, we took $\eta_i = \eta_d \equiv \eta$.

Equations (12.3-10) are our main result. An apparent difficulty in solving (12.3-10) is the fact that they involve two different spatial coordinates r_i and r_d measured along the two respective ray directions. This difficulty can be resolved by transforming to a coordinate ζ measured along the bisector of the angle formed between $\mathbf{k}_i$ and $\mathbf{k}_d$ as shown in Figure 12-5. Defining the values of r_d and r_i, which correspond to a given ζ as the respective projections of ζ along $\mathbf{k}_d$ and $\mathbf{k}_i$, we have

$$r_i = \zeta \cos\theta \qquad r_d = \zeta \cos\theta \tag{12.3-11}$$

so that (12.3-10) become

$$\begin{aligned} \frac{dE_i}{d\zeta} &= \frac{dE_i}{dr_i}\cos\theta = -i\eta E_d \cos\theta \\ \frac{dE_d}{d\zeta} &= -i\eta E_i \cos\theta \end{aligned} \tag{12.3-12}$$

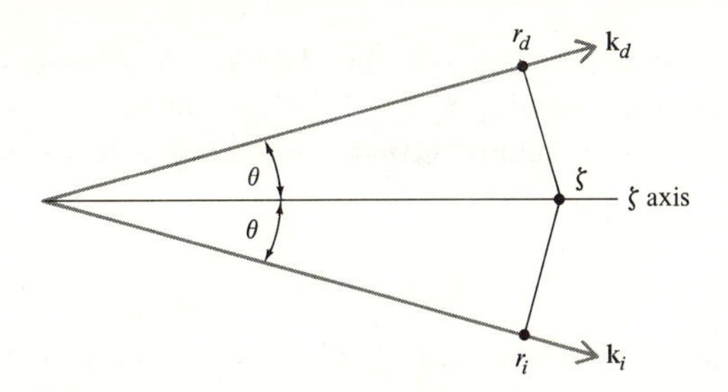

Figure 12-5 The directions and angles appearing in the diffraction equations (12.3–12).

whose solutions are

$$E_i(\zeta) = E_i(0) \cos (\eta\zeta \cos \theta) - iE_d(0) \sin (\eta\zeta \cos \theta)$$

$$E_d(\zeta) = E_d(0) \cos (\eta\zeta \cos \theta) - iE_i(0) \sin (\eta\zeta \cos \theta)$$

Using the correspondence between ζ, r_i, and r_d defined above, we can rewrite the solutions as

$$\begin{aligned} E_i(r_i) &= E_i(0) \cos (\eta r_i) - iE_d(0) \sin (\eta r_i) \\ E_d(r_d) &= E_d(0) \cos (\eta r_d) - iE_i(0) \sin (\eta r_d) \end{aligned} \qquad \textbf{(12.3-13)}$$

which is the desired result. It is of sufficient generality to describe the interaction between two input fields at ω_i and ω_d with arbitrary phases [$E_i(0)$ and $E_d(0)$ are complex] and arbitrary amplitudes as long as the Bragg condition (12.3-9) and the frequency condition $\omega_i = \omega_s + \omega_d$ are fulfilled. In the special case of a single frequency input at ω_i, $E_d(0) = 0$, and

$$\begin{aligned} E_i(r_i) &= E_i(0) \cos (\eta r_i) \\ E_d(r_d) &= -iE_i(0) \sin (\eta r_d) \end{aligned} \qquad \textbf{(12.3-14)}$$

we note that

$$|E_i(r_i)|^2 + |E_d(r_d = r_i)|^2 = |E_i(0)|^2 \qquad \textbf{(12.3-15)}$$

so that the total optical power carried by both waves is conserved.

If the interaction distance between the two beams is such that $\eta r_i = \eta r_d = \pi/2$, the total power of the incident beam is transferred into the diffracted beam. Since this process is used in a large number of technological and scientific applications, it may be worthwhile to gain some appreciation for the diffraction efficiencies possible using known acoustic media and conveniently available acoustic power levels.

The fraction of the power of the incident beam transferred in a distance l into the diffracted beam is given, using (12.3-8) and (12.3-14) by

$$\frac{I_{\text{diffracted}}}{I_{\text{incident}}} = \frac{E^2_{\text{diffracted}}}{E_i^2(0)} = \sin^2 \left(\frac{\omega l}{2c} \Delta n\right) \qquad \textbf{(12.3-16)}$$

It is advantageous to express the diffraction efficiency (12.3-16) in terms of the acoustic intensity I_{acoustic} (W/m^2) in the diffraction medium. First we relate the index change Δn to the strain s (see Section 12.1) by by [4]–[5]

$$\Delta n \equiv -\frac{n^3 p}{2} s \tag{12.3-17}$$

where p, the photoelastic constant of the medium,[7] is defined by (12.3-17). The strain s is related to the acoustic intensity I_{acoustic} by[8]

$$s = \sqrt{\frac{2I_{\text{acoustic}}}{\rho v_s^3}} \tag{12.3-18}$$

where v_s is the velocity of sound in the medium and ρ is the mass density (kg/m^3). Combining (12.3-7) and (12.3-18) in (12.3-16) we obtain

$$\frac{I_{\text{diffracted}}}{I_{\text{incident}}} = \sin^2\left[\frac{\pi l}{\sqrt{2}\lambda}\sqrt{\frac{n^6 p^2}{\rho v_s^3} I_{\text{acoustic}}}\right] \tag{12.3-19}$$

and using the following definition for the diffraction figure of merit

$$M \equiv \frac{n^6 p^2}{\rho v_s^3} \tag{12.3-20}$$

(12.3-19) becomes

$$\frac{I_{\text{diffracted}}}{I_{\text{incident}}} = \sin^2\left(\frac{\pi l}{\sqrt{2}\lambda}\sqrt{M I_{\text{acoustic}}}\right) \tag{12.3-21}$$

---

[7] In the case of interactions using crystals, (12.3-17) becomes a tensor relation and p becomes a fourth rank tensor. In this case we can often simplify the problem in such a way that only one tensor element is important so that (12.3-17) can be used.

[8] The (elastic) potential energy per unit volume due to an instantaneous strain $s(t)$ is

$$\tfrac{1}{2}Ts^2(t)$$

where T is the bulk modulus (elastic stiffness constant). The time averaged energy per unit volume due to the propagation of a sound wave with a strain amplitude s is the sum of the (equal) average potential and kinetic energy densities

$$\mathcal{E}/\text{vol} = 2(\tfrac{1}{2})T\overline{s^2(t)} = \tfrac{1}{2}Ts^2$$

since $\overline{s^2(t)} = \frac{1}{2}s^2$, the bar denoting time-averaging. Using the relation $I_{\text{acoustic}} = v_s\mathcal{E}/\text{vol}$ and $T/\rho = v_s^2$ where ρ is the mass density and v_s the velocity of sound, we get

$$I_{\text{acoustic}} = \tfrac{1}{2}\rho v_s^3 s^2$$

or

$$s = \sqrt{\frac{2I_{\text{acoustic}}}{\rho v_s^3}}$$

which is the result stated in (12.3-18).

Taking water as an example, an optical wavelength of $\lambda = 0.6328\ \mu\text{m}$, and the constants (taken from Table 12-1)

$$n = 1.33$$

$$p = 0.31$$

$$v_s = 1.5 \times 10^3\ \text{m/s}$$

$$\rho = 1000\ \text{kg/m}^3$$

Equation (12.3-21) gives

$$\left(\frac{I_{\text{diffracted}}}{I_{\text{incident}}}\right)_{\substack{\text{H}_2\text{O}\\ \text{at}\ \lambda = 0.6328\ \mu\text{m}}} = \sin^2 (1.4 l \sqrt{I_{\text{acoustic}}}) \qquad \textbf{(12.3-22)}$$

For other materials and at other wavelengths we can combine the last two equations to obtain a convenient working formula

$$\frac{I_{\text{diffracted}}}{I_{\text{incident}}} = \sin^2 \left(1.4 \frac{0.6328}{\lambda\ \mu\text{m}} l \sqrt{M_\omega I_{\text{acoustic}}}\right) \qquad \textbf{(12.3-23)}$$

where $M_\omega = M_{\text{material}}/M_{\text{H}_2\text{O}}$ is the diffraction figure of merit of the material relative to water. Values of M and M_ω for some common materials are listed in Tables 12-1 and 12-2.

Table 12-1 A LIST OF SOME MATERIALS COMMONLY USED IN THE DIFFRACTION OF LIGHT BY SOUND AND SOME OF THEIR RELEVANT PROPERTIES. ρ IS THE DENSITY, v_s THE VELOCITY OF SOUND, n THE INDEX OF REFRACTION, p THE PHOTOELASTIC CONSTANT AS DEFINED BY EQUATION (12.3-9), AND M_ω IS THE RELATIVE DIFFRACTION CONSTANT DEFINED ABOVE (AFTER REFERENCE [4]).

MATERIAL	ρ, (mg/m³)	v_s (km/s)	n	p	M_ω
Water	1.0	1.5	1.33	0.31	1.0
Extra-dense flint glass	6.3	3.1	1.92	0.25	0.12
Fused quartz (SiO_2)	2.2	5.97	1.46	0.20	0.006
Polystyrene	1.06	2.35	1.59	0.31	0.8
KRS-5	7.4	2.11	2.60	0.21	1.6
Lithium niobate ($LiNbO_3$)	4.7	7.40	2.25	0.15	0.012
Lithium fluoride (LiF)	2.6	6.00	1.39	0.13	0.001
Rutile (TiO_2)	4.26	10.30	2.60	0.05	0.001
Sapphire (Al_2O_3)	4.0	11.00	1.76	0.17	0.001
Lead molybdate ($PbMO_4$)	6.95	3.75	2.30	0.28	0.22
Alpha iodic acid (HIO_3)	4.63	2.44	1.90	0.41	0.5
Tellurium dioxide (TeO_2) (Slow shear wave)	5.99	0.617	2.35	0.09	5.0

Table 12-2 A LIST OF MATERIALS COMMONLY USED IN ACOUSTOOPTIC INTERACTIONS AND SOME OF THEIR RELEVANT PROPERTIES. $M = n^6p^2/\rho v_s^3$ IS THE FIGURE OF MERIT, DEFINED BY EQUATION (12.3-20) AND IS GIVEN IN MKS UNITS (AFTER REFERENCE [6]).

MATERIAL	$\lambda(\mu m)$	n	$\rho(g/cm^3)$	ACOUSTIC WAVE POLARIZATION AND DIRECTION	$v_s(10^5$ cm/s)	OPT. WAVE POLARIZATION AND DIRECTION[a]	$M = n^6p^2/\rho v_s^3$
Fused quartz	0.63	1.46	2.2	long.	5.95	$\perp$	1.51×10^{-15}
Fused quartz	0.63			trans.	3.76	$\parallel$ or $\perp$	0.467
GaP	0.63	3.31	4.13	long. in [110]	6.32	$\parallel$	44.6
GaP	0.63			trans. in [100]	4.13	$\parallel$ or $\perp$ in [010]	24.1
GaAs	1.15	3.37	5.34	long. in [110]	5.15	$\parallel$	104
GaAs	1.15			trans. in [100]	3.32	$\parallel$ or $\perp$ in [010]	46.3
TiO_2	0.63	2.58	4.6	long. in [11–20]	7.86	$\perp$ in [001]	3.93
$LiNbO_3$	0.63	2.20	4.7	long. in [11–20]	6.57	(b)	6.99
YAG	0.63	1.83	4.2	long. in [100]	8.53	$\parallel$	0.012
YAG	0.63			long. in [110]	8.60	$\perp$	0.073
YIG	1.15	2.22	5.17	long. in [100]	7.21	$\perp$	0.33
$LiTaO_3$	0.63	2.18	7.45	long. in [001]	6.19	$\parallel$	1.37
As_2S_3	0.63	2.61	3.20	long.	2.6	$\perp$	433
As_2S_3	1.15	2.46		long.		$\parallel$	347
SF-4	0.63	1.616	3.59	long	3.63	$\perp$	4.51
β-ZnS	0.63	2.35	4.10	long. in [110]	5.51	$\parallel$ in [001]	3.41
β-ZnS	0.63			trans. in [110]	2.165	$\parallel$ or $\perp$ in [001]	0.57
α-Al_2O_3	0.63	1.76	4.0	long. in [001]	11.15	$\parallel$ in [11–20]	0.34
CdS	0.63	2.44	4.82	long. in [11–20]	4.17	$\parallel$	12.1
ADP	0.63	1.58	1.803	long. in [100]	6.15	$\parallel$ in [010]	2.78
ADP	0.63			trans. in [100]	1.83	$\parallel$ or $\perp$ in [001]	6.43
KDP	0.63	1.51	2.34	long. in [100]	5.50	$\parallel$ in [010]	1.91
KDP	0.63			trans. in [100]		$\parallel$ or $\perp$ in [001]	3.83
H_2O	0.63	1.33	1.0	long.	1.5		160
Te	10.6	4.8	6.24	long. in [11–20]	2.2	$\parallel$ in [0001]	4400
$PbMO_4$[14]	0.63	2.4		long. $\parallel$ c axis	3.75	$\parallel$ or $\perp$	73

[a] The optical-beam direction actually differs from that indicated by the magnitude of the Bragg angle. The polarization is defined as parallel or perpendicular to the scattering plane formed by the acoustic and optical k vectors.

According to (12.3-19) at small diffraction efficiencies, the diffracted light intensity is proportional to the acoustic intensity. This fact is used in acoustic modulation of optical radiation. The information signal is used to modulate the intensity of the acoustic beam. This modulation is then transferred, according to (12.3-19), as intensity modulation onto the diffracted optical beam.

Numerical example—scattering in fused quartz. Calculate the fraction of 0.633 μm light which is diffracted under Bragg conditions from a sound wave in $PbMO_4$ with the following characteristics

$$\text{acoustic power} = 1 \text{ watt}$$
$$\text{acoustic beam cross section} = 1 \text{ mm} \times 1 \text{ mm}$$
$$l = \text{optical path in acoustic beam} = 1 \text{ mm}$$
$$M_\omega \text{ (from Table 12-1)} = 0.22$$

Substituting these data into (12.3-23) yields

$$\frac{I_{\text{diffracted}}}{I_{\text{incident}}} \simeq 40\%$$

12.4 Deflection of Light by Sound

One of the most important applications of acoustooptic interactions is in the deflection of optical beams. This can be achieved by changing the sound frequency while operating near the Bragg-diffraction condition. The situation is depicted in Figure 12-6 and can be understood using

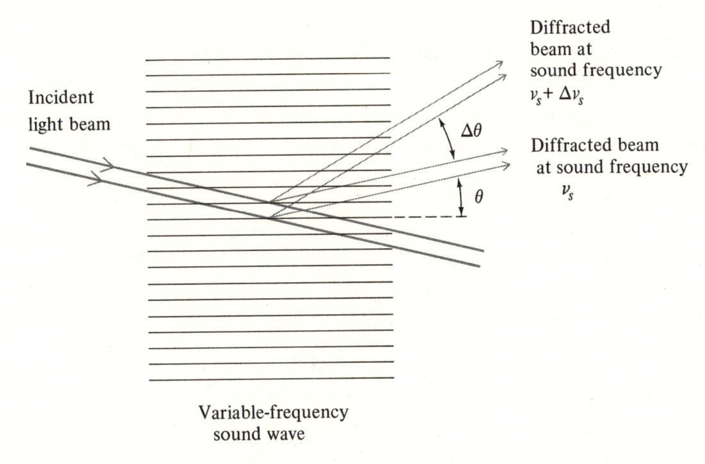

Figure 12-6 A change of frequency of the sound wave from ν_s to $\nu_s + \Delta\nu_s$ causes a change $\Delta\theta$ in the direction of the diffracted beam, according to Equation (12.4-1).

Figure 12-7. Let us assume first that the Bragg condition (12.1-4) is satisfied. The momentum vector diagram originally introduced in Figure 12-4 is closed and the beam is diffracted along the direction θ as given by (12.1-4). Now let the sound frequency change from ν_s to $\nu_s + \Delta\nu_s$. Since $k_s = 2\pi\nu_s/v_s$, this causes a change of $\Delta k_s = 2\pi(\Delta\nu_s)/v_s$ in the magnitude of the sound wave vector as shown. Since the angle of incidence remains θ and the magnitude of the diffracted k vector is unchanged,[9] so its tip is constrained to the circle locus shown in Figure 12-7, we can no longer close the momentum diagram and thus momentum is no longer strictly conserved. The beam will be diffracted along the direction that least violates the momentum conservation.[10] This takes place along the direction OB, causing a deflection of the beam by $\Delta\theta$. Recalling that the angles θ and $\Delta\theta$ are all small and that $k_s = 2\pi\nu_s/v_s$, we obtain

$$\Delta\theta = \frac{\Delta k_s}{k} = \frac{\lambda}{n v_s}\Delta\nu_s \qquad \textbf{(12.4-1)}$$

so that the deflection angle is proportional to the change of the sound frequency.

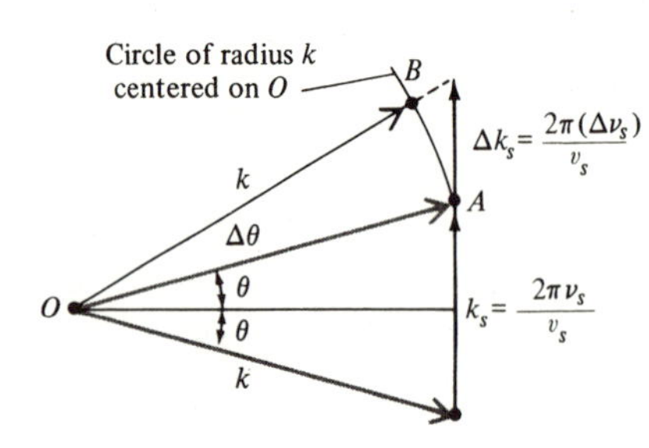

Figure 12-7 Momentum diagram, illustrating how the change in sound frequency from ν_s to $\nu_s + \Delta\nu_s$ deflects the diffracted light beam from θ to $\theta + \Delta\theta$.

As in the case of electrooptic deflection, we are not interested so much in the absolute deflection $\Delta\theta$ as we are in the number of resolvable spots—that is, the factor by which $\Delta\theta$ exceeds the beam divergence angle. If we take the diffraction angle as $\sim\lambda/nD$, where D is the beam diameter,[11] the

---

[9] The small change in the diffracted wave vector that is attributable to the frequency change is typically about $\Delta k/k \simeq 10^{-7}$ and is neglected. See Problem 12-4.

[10] The violation of momentum conservation is equivalent to destructive interference in the diffracted beam, so the beam intensity will be less than under Bragg conditions, where momentum is conserved. The diffracted beam will thus have its maximum value along the direction in which the destructive interference is smallest. This corresponds to the direction that minimizes the momentum mismatch, as shown in Figure 12-7.

[11] According to (3.2-18), $\theta_{\text{beam}} = \lambda/\pi n\omega_0$ is the half-apex diffraction angle, so the full beam diffraction angle can be taken as

$$\theta_{\text{diffraction}} = 2\theta_{\text{beam}} = \frac{4\lambda}{\pi n D}$$

where $D = 2\omega_0$ is the Gaussian spot diameter.

number of resolvable spots is

$$N = \frac{\Delta\theta}{\theta_{\text{diffraction}}} = \left(\frac{\lambda}{v_s}\right)\frac{\Delta\nu_s}{\lambda/D}$$
$$= \Delta\nu_s\left(\frac{D}{v_s}\right) = \Delta\nu_s\tau \tag{12.4-2}$$

where $\tau = D/v_s$ is the time it takes the sound to cross the optical-beam diameter.

Numerical example—beam deflection. Consider a deflection system using flint glass and a sound beam that can be varied in frequency from 80 MHz to 120 MHz; thus, $\Delta\nu_s = 40$ MHz. Let the optical beam diameter be $D = 1$ cm. From Table 12-1 we obtain $v_s = 3.1 \times 10^5$ cm/s; therefore, $\tau = D/v_s = 3.23 \times 10^{-6}$ second and the number of resolvable spots is $N = \Delta\nu_s\tau \simeq 130$.

Bragg interactions have recently been demonstrated [15] between surface acoustic waves and optical modes confined in thin film dielectric waveguides. Since the modulation efficiency depends, according to (12.3-19), on the acoustic intensity, the confinement of the acoustic power near the surface (to a distance $\sim\lambda_s$) leads to low modulation or switching power.

Figure 12-8 shows an experimental setup in which both the acoustic surface wave and the optical wave are guided in a single crystal of $LiNbO_3$. The dielectric waveguide is produced by out-diffusion from a layer of $\sim$10 μm near the surface, which raises the index of refraction.

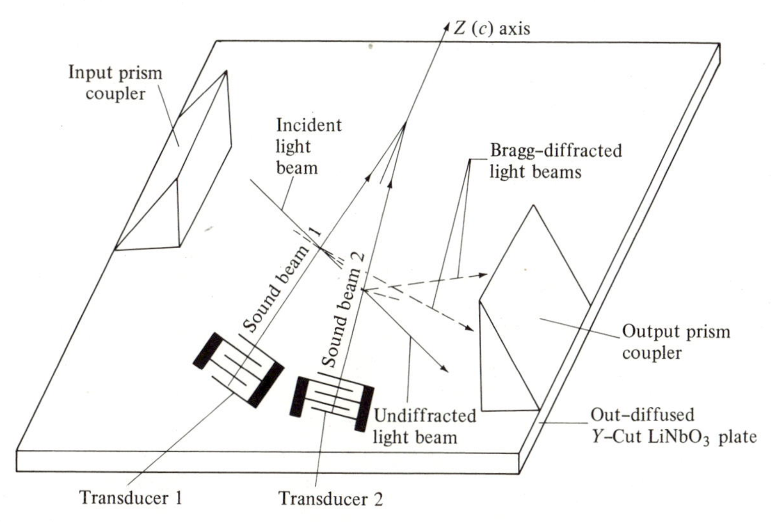

Figure 12-8 Guided-wave acoustooptic Bragg diffraction from two tilted surface acoustic waves. (Reference [16].)

PROBLEMS

12-1 Derive the expression of the frequency shift, under Bragg conditions, from a receding sound wave.

12-2 Design an acoustic modulation system for transferring the output of a magnetic-cartridge phonograph onto an optical beam with $\lambda_0 = 0.6328\,\mu\text{m}$ and $I_{\text{incident}} = 10^{-3}$ watt. Specify the power levels involved and the essential characteristics of all the key components. (*Hint:* Use the audio output of the cartridge to modulate a high-frequency (100 MHz, say) carrier, which is then used to transduce an acoustic beam.)

12-3 What happens in Bragg diffraction of light from a standing sound wave? Describe the frequency shifts and direction of diffraction.

12-4 Show that under Bragg conditions the change in wave vector of the diffracted wave is

$$\frac{k_{\text{diffracted}} - k}{k} = 2\sin\theta\,\frac{v_s}{c}$$

12-5 Consult the literature (see References [4] and [5], for example) and describe the difference between Bragg diffraction and Debye-Sears diffraction. Under what conditions is each observed?

12-6 Bragg's law for diffraction of X-rays in crystals is [7]

$$2d\sin\theta = m\frac{\lambda}{n} \qquad m = 1, 2, 3, \cdots$$

where d is the distance between equivalent atomic planes, θ is the angle of incidence, and λ/n is the wavelength of the diffracted radiation. Bragg diffraction of light from sound (see Eq. (12.1-4)) takes place when

$$2\lambda_s\sin\theta = \frac{\lambda}{n}$$

Thus, if we compare it to the X-ray result and take $\lambda_s = d$, only the case of $m = 1$ is allowed. Explain the difference. Why don't we get light diffracted along directions θ corresponding to $m = 2, 3, \cdots$? (*Hint:* The diffraction of X-rays takes place at discrete atomic planes, which can be idealized as infinitely thin sheets, whereas the sound wave is continuous in z; see Figure 12-3.)

12-7 Design an acoustic deflection system using $LiTaO_3$ to be used in scanning an optical beam in a manner compatible with that of commercial television receivers.

REFERENCES

[1] Brillouin, L., "Diffusion de la lumière et des rayons X par un corps transparent homgène," *Ann. Physique,* vol. 17, p. 88, 1922.

[2] Debye, P., and F. W. Sears, "On the scattering of light by supersonic waves," *Proc. Nat. Acad. Sci. U.S.*, vol. 18, p. 409, 1932.

[3] Dransfeld, K., "Kilomegacycle ultrasonics," *Sci. Am.*, vol. 208, p. 60, 1963.

[4] See, for example, Robert Adler, "Interaction between light and sound," *IEEE Spectrum*, vol. 4, May 1967, p. 42.

[5] Born, M., and E. Wolf, *Principles of Optics.* New York: Pergamon, 1965, Chap. 12.

[6] Dixon, R. W., "Photoelastic properties of selected materials and their relevance for applications to acoustic light modulators and scanners," *J. Appl. Phys.*, vol. 38, p. 5149, 1967.

[7] Kittel, C., *Introduction to Solid State Physics*, 3d Ed. New York: Wiley, 1967, p. 38.

[8] Quate, C. F., C. D. W. Wilkinson, and D. K. Winslow, "Interactions of light and microwave sound," *Proc. IEEE*, vol. 53, p. 1604, 1965.

[9] Cohen, M. G., and E. I. Gordon, "Acoustic beam probing using optical frequencies," *Bell System Tech. J.*, vol. 44, p. 693, 1965.

[10] Cummings, H. Z., and N. Knable, "Single sideband modulation of coherent light by Bragg reflection from acoustical waves," *Proc. IEEE*, vol. 51, p. 1246, 1963.

[11] Yariv, A., *Quantum Electronics.* New York: Wiley, 1967, Eq. (25.4-14).

[12] Gordon, E. I., "A review of acousto-optical deflection and modulation devices," *Proc. IEEE*, vol. 54, p. 1391, 1966.

[13] Gordon, E. I., "Measurement of light-sound interaction efficiencies in solids," *IEEE J. Quantum Electron.*, vol. QE-1, p. 283, 1965.

[14] D. A. Pinnow, L. G. Van Uitert, A. W. Warner, and W. A. Bonner. "$PbMO_4$: A melt grown crystal with a high figure of merit for acoustooptic device applications" *Appl. Phys. Lett.*, vol. 15, 83 1969.

[15] Kuhn, L. M., L. Dakss, F. P. Heidrich, and B. A. Scott, "Deflection of optical guided waves by a surface acoustic wave." *Appl. Phys. Lett.*, vol. 17, p. 265 (1970).

[16] C. S. Tsai, Le T. Nguyen, S. K. Yao, and M. H. Alhaider, "High performance acousto-optic guided light beam device using two tilting surface acoustic waves," *Appl. Phys. Lett.*, vol. 26, p. 140, 1975.

CHAPTER

13

Propagation, Modulation, and Oscillation in Optical Dielectric Waveguides

13.0 Introduction

In this chapter we discuss a number of topics that involve propagation of optical modes in dielectric films with thicknesses comparable to the wavelength.

The ability to generate, guide, modulate, and detect light in such thin film configurations ([1], [2], and [3]) opens up new possibilities for monolithic "optical circuits" [4]—an endeavor going under the name of integrated optics [5].

We will first consider the basic problem of TE and TM mode propagation in slab dielectric waveguides. A coupled mode formalism is then developed to describe situations involving exchange of power between modes. These include (a) periodic (corrugated) optical waveguides and filters, (b) distributed feedback lasers, (c) electrooptic mode coupling, and (d) directional couplers.

13.1 Waveguide Modes—A General Discussion

A prerequisite to an understanding of guided wave interactions is a knowledge of the properties of the guided modes. A mode of a dielectric waveguide at a (radian) frequency ω is a solution of the wave equation

$$\nabla^2\mathbf{E}(\mathbf{r}) + k_0^2n^2(\mathbf{r})\mathbf{E}(\mathbf{r}) = 0 \tag{13.1-1}$$

where $k_0^2 \equiv \omega^2\mu\epsilon_0 = (2\pi/\lambda)^2$ and n is tbe index of refraction. The solutions are subject to the continuity of the tangential components of $\mathbf{E}$ and $\mathbf{H}$ at the dielectric interfaces. In (13.1-1) the form of the field is taken as

$$\mathbf{E}(\mathbf{r}, t) = \mathbf{E}(x, y)e^{i(\omega t-\beta z)} \tag{13.1-2}$$

so that (13.1-1) becomes

$$\left(\frac{\partial^2}{\partial x^2} + \frac{\partial^2}{\partial y^2}\right)\mathbf{E}(x, y) + [k_0^2n^2(\mathbf{r}) - \beta^2]\mathbf{E}(x, y) = 0 \tag{13.1-3}$$

The basic features of the behavior of dielectric waveguides can be elucidated with the help of a slab (planar) model in which no variation exists in one (for example, y) dimension. Channel waveguides, in which the waveguide dimensions are finite in both the x and y directions, approach the behavior of the planar guide when one dimension is considerably larger than the other ([6] and [7]). Even when this is not the case, most of the phenomena of interest are only modified in a simple quantitative way when going from a planar to a channel waveguide. Because of the immense mathematical simplification which results, we will limit most of the following treatment to planar waveguides such as the one shown in Figure 13-1.

Putting $\partial/\partial y = 0$ in (13.1-3) and writing it separately for regions I, II, and III yields

Region I
$$\frac{\partial^2}{\partial x^2}E(x, y) + (k_0^2n_1^2 - \beta^2)E(x, y) = 0 \tag{13.1-4[a]}$$

Region II
$$\frac{\partial^2}{\partial x^2}E(x, y) + (k_0^2n_2^2 - \beta^2)E(x, y) = 0 \tag{13.1-4[b]}$$

Region III
$$\frac{\partial^2}{\partial x^2}E(x, y) + (k_0^2n_3^2 - \beta^2)E(x, y) = 0 \tag{13.1-4[c]}$$

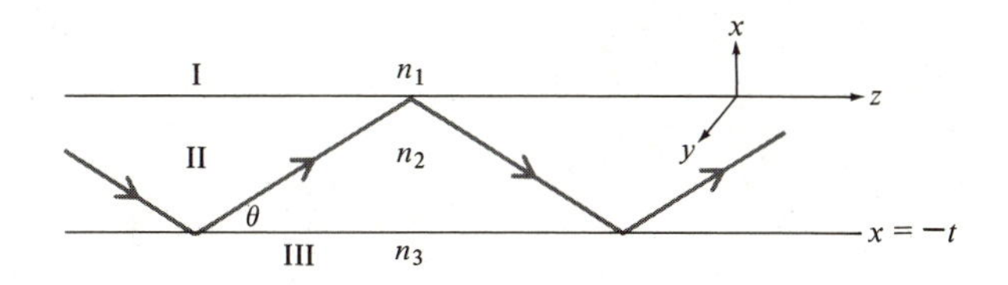

Figure 13-1 A slab ($\partial/\partial y = 0$) dielectric waveguide.

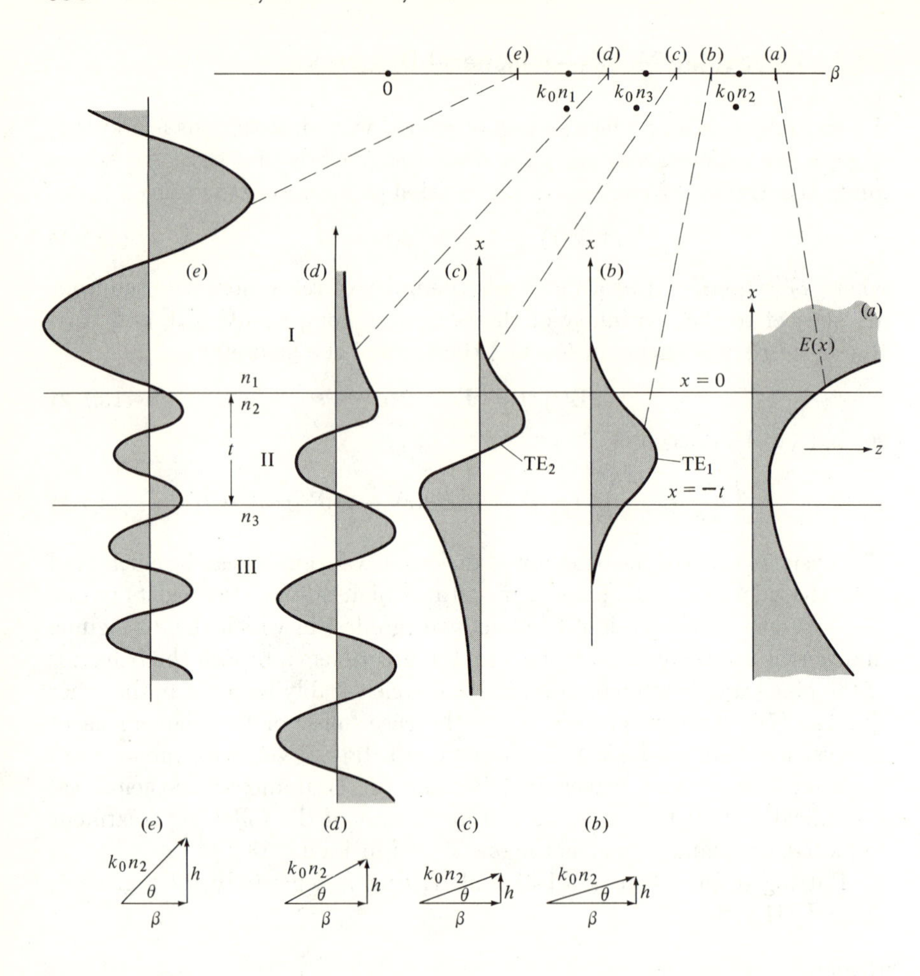

Figure 13-2 (*top*) The different regimes (a, b, c, d, e) of the propagation constant, β, of the waveguide shown in Figure 13.1. (*middle*) The field distributions corresponding to the different value of β. (*bottom*) The propagation triangles corresponding to the different propagation regimes.

where $E(x, y)$ is a Cartesian component of $\mathbf{E}(x, y)$. Before embarking on a formal solution of (13.1-4) we may learn a great deal about the physical nature of the solutions by simple arguments. Let us consider the nature of the solutions as a function of the propagation constant β at a *fixed* frequency ω. Let us assume that $n_2 > n_3 > n_1$. For $\beta > k_0n_2$ [that is, regime (a) in Figure 13-2] it follows directly from (13.1-4) that $(1/E)(\partial^2E/\partial x^2) > 0$ everywhere, and $E(x)$ is exponential in all three layers (I, II, III) of the waveguides. Because of the need to match both $E(x)$ and its derivatives at the two interfaces, the resulting field distribution is as shown in Figure 13-2(a). The field increases without bound away from the waveguide so that the solution is not *physically realizable* and thus does not correspond to a real wave.

For $k_0n_3 < \beta < k_0n_2$, as in points (b) and (c), it follows from (13.1-4) that the solution is sinusoidal in region II, since $(1/E)(\partial^2E/\partial x^2) < 0$, but is exponential in regions I and III. This makes it possible to have a solution $E(x)$ that satisfies the boundary conditions while *decaying* exponentially in regions I and III. Two such solutions are shown in Figure 13-2(b) and (c). The energy carried by these modes is confined to the vicinity of the guiding layer II, and we will, consequently, refer to them as confined, or guided, modes. From the above discussion it follows that a necessary condition for their existence is that $k_0n_1, k_0n_3 < \beta < k_0n_2$ so that confined modes are possible only when $n_2 > n_1, n_3$; that is, the inner layer possesses the highest index of refraction.

Mode solutions for $k_0n_1 < \beta < k_0n_3$, regime (d), correspond according to (13.1-4) to exponential behavior in region I and to sinusoidal behavior in regions II and III as illustrated in Figure 13-2(d). We will refer to these modes as substrate radiation modes. For $0 < \beta < k_0n_1$, as in (e), the solution for $E(x)$ becomes sinusoidal in all three regions. These are the so-called radiation modes of the waveguides.

A solution of (13.1-4) subject to the boundary conditions at the interfaces given in the next section shows that while in regimes (d) and (e) β is a continuous variable, the values of allowed β in the propagation regime $k_0n_3 < \beta < k_0n_2$ are *discrete.* The number of confined modes depends on the width, t, the frequency, and the indices of refraction n_1, n_2, n_3. At a given wavelength the number of confined modes increases from 0 with increasing t. At some t, the mode TE_1 becomes confined. Further increases in t will allow TE_2 to exist as well, and so on.

A useful point of view is one of considering the wave propagation in the inner layer 2 as that of a plane wave propagating at some angle θ to the horizontal axis and undergoing a series of total internal reflections at the interfaces II–I and II–III. This is based on [13.1-4(b)]. Assuming a solution in the form of $E \propto \sin(hx + \alpha) \exp(-i\beta z)$, we obtain

$$\beta^2 + h^2 = k_0^2n_2^2 \tag{13.1-5}$$

The resulting right-angle triangles with sides β, h, and k_0n_2 are shown in Figure 13-2. Note that since the frequency is constant, $k_0n_2 \equiv (\omega/c)n_2$ is the same for cases (b), (c), (d), and (e). The propagation can thus be considered formally as that of a plane wave along the direction of the hypotenuse with a *constant* propagation constant k_0n_2. As β decreases, θ increases until, at $\beta = k_0n_3$, the wave ceases to be totally internally reflected at the interface III–II. This follows from the fact that the guiding condition $\beta > k_0n_3$ leads, using $\beta = k_0n_2 \cos\theta$, to $\theta < \cos^{-1}(n_3/n_2) = \theta_c$, where θ_c is the total internal reflection angle at the interface between layers II–III. Since $n_3 > n_1$, total reflection at the II–III interface guarantees total internal reflection at the I–II interface.

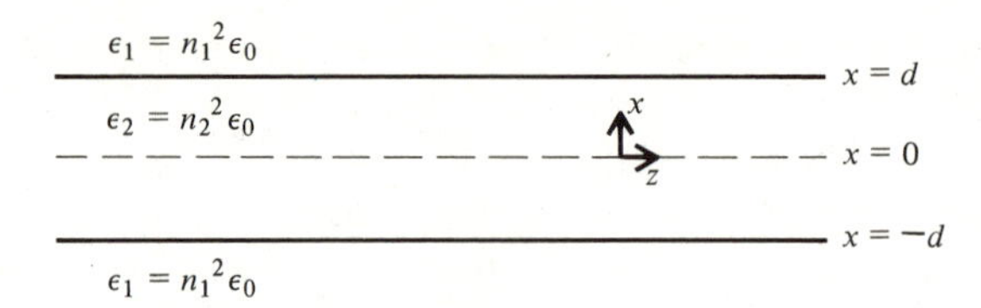

Figure 13-3 A symmetric slab waveguide.

Confined modes in a symmetric slab waveguide. Before considering the more general, and difficult, case of asymmetric waveguides, we will solve in some detail the case of the symmetric slab waveguide where $n_1 = n_3$. The guiding layer of index $n_2 > n_1$ occupies the region $-d < x < d$, as in Figure 13-3.

We consider the case of harmonic time behavior in the form of exp $(i\omega t)$ and an infinite slab geometry so that there is no variation in the y directions $(\partial/\partial y = 0)$. Maxwell's equations (1.2-1) and (1.2-2) become

$$\frac{\partial E_y}{\partial z} = i\omega\mu H_x \qquad \textbf{(13.1-6[a])}$$

$$\frac{\partial E_x}{\partial z} - \frac{\partial E_z}{\partial x} = -\, i\omega\mu H_y \qquad \textbf{(13.1-6[b])}$$

$$\frac{\partial E_y}{\partial x} = -i\omega\mu H_z \qquad \textbf{(13.1-6[c])}$$

$$\frac{\partial H_y}{\partial z} = -i\omega\epsilon E_x \qquad \textbf{(13.1-6[d])}$$

$$\frac{\partial H_x}{\partial z} - \frac{\partial H_z}{\partial x} = i\omega\epsilon E_y \qquad \textbf{(13.1-6[e])}$$

$$\frac{\partial H_y}{\partial x} = i\omega\epsilon E_z \qquad \textbf{(13.1-6[f])}$$

Next we assume that the modes propagate in the z direction with the z dependence in the form of exp $(-i\beta z)$ so that in (13.1-6) we can replace $\partial/\partial z$ by $-i\beta$. An inspection of (13.1-6) reveals that we may obtain two self-consistent types of solutions. The first contains only E_y, H_x, and H_z and is referred to as transverse electric (TE) modes, since the electric field (E_y) is restricted to the transverse (that is, normal to the direction of propagation) plane. Maxwell's equations (13.1-6) for the TE modes reduce to

$$E_y = -\frac{\omega\mu}{\beta} H_x \qquad \textbf{(13.1-7[a])}$$

$$\frac{\partial E_y}{\partial x} = -i\omega\mu H_z \qquad \textbf{(13.1-7[b])}$$

The second type of mode is transverse magnetic (TM) and involves H_y, E_x, and E_z. These are related, according to (13.1-6), by

$$H_y = \frac{\omega\epsilon}{\beta} E_x \tag{13.1-8[a]}$$

$$E_z = -\frac{i}{\omega\epsilon} \frac{\partial H_y}{\partial x} \tag{13.1-8[b]}$$

The solutions for the TE and TM modes are basically similar, so that in order to be specific we will consider the case of TE modes. Since the waveguide is symmetric about the plane $x = 0$, the mode solutions must be either even or odd in x, that is,

$$E_y(x, z, t) = E_y(-x, z, t)$$

in the case of even modes and

$$E_y(x, z, t) = -E_y(-x, z, t)$$

for the odd modes. The solution for the even modes is taken in the form

$$E_y = A \exp[-p(|x| - d) - i\beta z] \qquad |x| \geq d \tag{13.1-9}$$

and

$$E_y = B \cos(hx) \exp(-i\beta z) \qquad |x| \leq d \tag{13.1-10}$$

where p and h are positive real constants to be determined. From (13.1-7[b]) we obtain

$$H_z = \mp \frac{ipA}{\omega\mu} \exp[-p(|x| - d) - i\beta z] \qquad |x| \geq d \tag{13.1-11}$$

the $(-)$ sign is used with $x \geqslant d$ and $(+)$ for $x \leqslant -d$

$$H_z = -\frac{ihB}{\omega\mu} \sin(hx) \exp(-i\beta z) \qquad |x| \leq d \tag{13.1-12}$$

Next we require that the tangential field components E_y and H_z be continuous across the interfaces.[1] The continuity of E_y at $x = \pm d$ leads according to (13.1-9) and (13.1-10) to

$$A = B \cos(hd) \tag{13.1-13}$$

while the continuity of H_z results in

$$pA = hB \sin(hd) \tag{13.1-14}$$

From (13.1-13) and (13.1-14) it follows that

$$pd = hd \tan(hd) \tag{13.1-15}$$

---

[1] The reasons for these conditions are discussed in any elementary text on electromagnetic theory.

Since the field solutions (13.1-9) and (13.1-10) must satisfy the wave equation (13.1-4[a], [b]), the following relations are obeyed

$$\beta^2 = k_0^2 n_2^2 - h^2$$
$$\beta^2 = k_0^2 n_1^2 + p^2 \qquad \textbf{(13.1-16)}$$

The last two equations can be combined to give

$$(pd)^2 + (hd)^2 = (n_2^2 - n_1^2)k_0^2 d^2 \qquad \textbf{(13.1-17)}$$

The propagation constants p and h of a given mode need to satisfy, simultaneously, (13.1-15) and (13.1-17). A straightforward graphical solution is illustrated by Figure 13-4 and consists of finding the intersections in the pd-hd plane of the circle $(pd)^2 + (hd)^2 = (n_2^2 - n_1^2)k_0^2 d^2$ with the curve $pd = hd \tan(hd)$. Each intersection with a $p > 0$ corresponds to a confined mode. The propagation constant β of a given mode can be obtained, once p and h are given, from (13.1-16).

To appreciate the nature of the solutions, let us consider what happens in a given waveguide (that is, fixed n_1, n_2, and d) as the frequency increases gradually from zero. Since $k_0 = \omega/c$, the effect of increasing the frequency is to increase the radius of the circle $(pd)^2 + (hd)^2 = (n_2^2 - n_1^2)k_0^2 d^2$. At low frequencies such that

$$0 < \sqrt{n_2^2 - n_1^2}\, k_0 d < \pi \qquad \textbf{(13.1-18)}$$

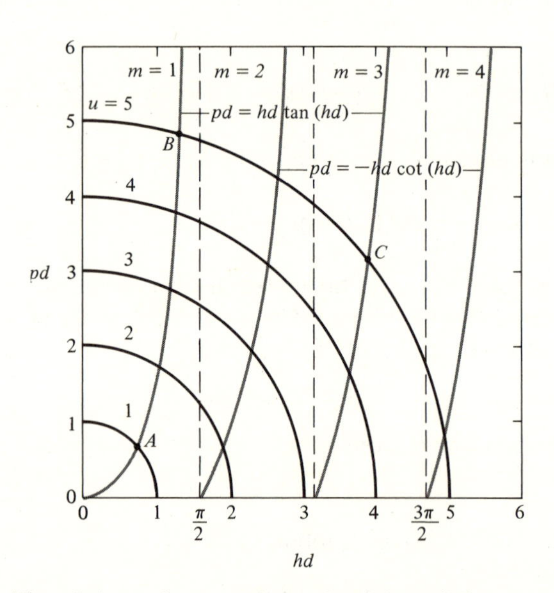

Figure 13-4 Plot of eigenvalue equations $pd = hd \tan(hd)$ for even TE modes, $pd = -hd \cot(hd)$ for odd TE modes, and the supplementary relationship $(pd)^2 + (hd)^2 = (n_2^2 - n_1^2)(k_0 d)^2 \equiv u^2$. (After Reference [8].)

only one intersection (point A) exists between the circle and the curve $pd = hd \tan(hd)$ with $p > 0$. This is evident from an inspection of Figure 13-4. The mode is designated as TE_1 and has a transverse h parameter falling within the range

$$0 < h_1 d < \frac{\pi}{2} \qquad \textbf{(13.1-19)}$$

so that it has no zero crossings in the interior of the slab $|x| \leq d$.

When the parameter $u \equiv \sqrt{n_2^2 - n_1^2} k_0 d$ falls within the range

$$\pi < \sqrt{n_2^2 - n_1^2} k_0 d < 2\pi \qquad \textbf{(13.1-20)}$$

we obtain two intersections with $p > 0$. One (point B) corresponds to a value of $hd < \pi/2$ and is thus that of the lowest order TE_1 mode. In the second mode (point C)

$$\pi < h_3 d < \frac{3\pi}{2} \qquad \textbf{(13.1-21)}$$

and consequently has two zero crossings (that is, points where $E_y = 0$) in the region $|x| < d$. This is the so-called TE_3 mode ($m = 3$ in Figure 13-4). Both of these modes correspond to the same frequency and can thus be excited simultaneously by the same input field. We notice, however, that the TE_1 mode has a larger value of p (that is, $p_1 > p_3$) and is therefore more highly confined to the interior slab. It also follows from (13.1-16) that $\beta_1 > \beta_3$, so that the phase velocity $v_1 = \omega/\beta_1$ of the TE_1 mode is smaller than that of the TE_3 mode.

From (13.1-19) and (13.1-21) one would conclude that no mode exists with $\pi/2 < hd < \pi$. This is due to the fact that up to this point, we considered only modes with even x symmetry as in (13.1-10). Another family of modes—the odd TE modes—exists and is described by

$$E_y = A \exp[-p(|x| - d) - i\beta z] \qquad |x| \geq d \qquad \textbf{(13.1-22)}$$

$$E_y = B \sin(hx) \exp(-i\beta z) \qquad |x| \leq d \qquad \textbf{(13.1-23)}$$

Applying the continuity conditions at $|x| = d$ leads to

$$pd = -hd \cot(hd) \qquad \textbf{(13.1-24)}$$

instead of (13.1-15). The mode solutions correspond to the intersection of (13.1-24) with the circle (13.1-17). Reference to Figure 13-4 shows that the corresponding values of h do indeed fill the gaps "avoided" by the even TE modes.

The lowest order odd TE mode is designated TE_2 ($m = 2$), since its h parameter h_2 satisfies

$$\frac{\pi}{2} < h_2 d < \pi \qquad \textbf{(13.1-25)}$$

thus falling between h_1 (of TE_1) and h_3. We can now generalize and state that the mth (TE or TM) mode satisfies

$$\left(m - 1\right)\frac{\pi}{2} < h_m d < m\frac{\pi}{2} \tag{13.1-26}$$

and has $m - 1$ zero crossings in the internal region $|x| \leq d$. The modes 1, 3, 5, . . . are even symmetric, while those with $m = 2, 4, 6$ . . . are odd. We note that all the modes except the fundamental ($m = 1$) can exist (that is, are confined) only above a "cutoff" frequency. The higher the mode index, the higher its cutoff frequency. The fundamental mode can exist at any frequency, as is evident from Figure 13-4. If the dielectric waveguide is asymmetric ($n_1 \neq n_3$), the lowest order ($m = 1$) modes also possess a cutoff frequency. This case will be taken up in the next section.

The general features of TM modes are similar to those of TE modes except that the corresponding values of p are somewhat smaller, indicating a lesser degree of confinement. A larger fraction of the total TM mode power thus propagates in the outer media compared to a TE mode of the same order. This point is taken up in Problem 13.7.

13.2 TE and TM Modes in an Asymmetric Slab Waveguide

In this section we will derive the mode solutions for the general asymmetric ($n_1 \neq n_3$) slab waveguide shown in Figure 13-1. We limit the derivation to the guided modes which according to Figure 13-2 have propagation constants β

$$k_0 n_3 < \beta < k_0 n_2$$

where $n_3 > n_1$.

TE modes. The field component E_y of the TE mode obeys the wave equation

$$\nabla^2 E_{yi}(x, y, z) + \omega^2 \mu \epsilon_0 n_i^2 E_{yi} = 0 \qquad i = 1, 2, 3 \tag{13.2-1}$$

where i refers to the layer and the (real) electric field is given by

$$E_{yi}(x, y, z, t) = \text{Re}\,[E_{yi}(x, y, z)e^{i\omega t}]$$

For waves propagating along the z direction and for $\partial/\partial y = 0$ we have

$$E_{yi}(x, y, z) = \mathcal{E}_{yi}(x)e^{-i\beta z} \tag{13.2-2}$$

The transverse function $\mathcal{E}_{yi}(x)$ is taken as

$$\mathcal{E}_y = \begin{cases} C \exp(-qx) & 0 \leq x < \infty \\ C[\cos(hx) - (q/h)\sin(hx)] & -t \leq x \leq 0 \\ C[\cos(ht) + (q/h)\sin(ht)]\exp[p(x+t)] & -\infty < x \leq -t \end{cases} \tag{13.2-3}$$

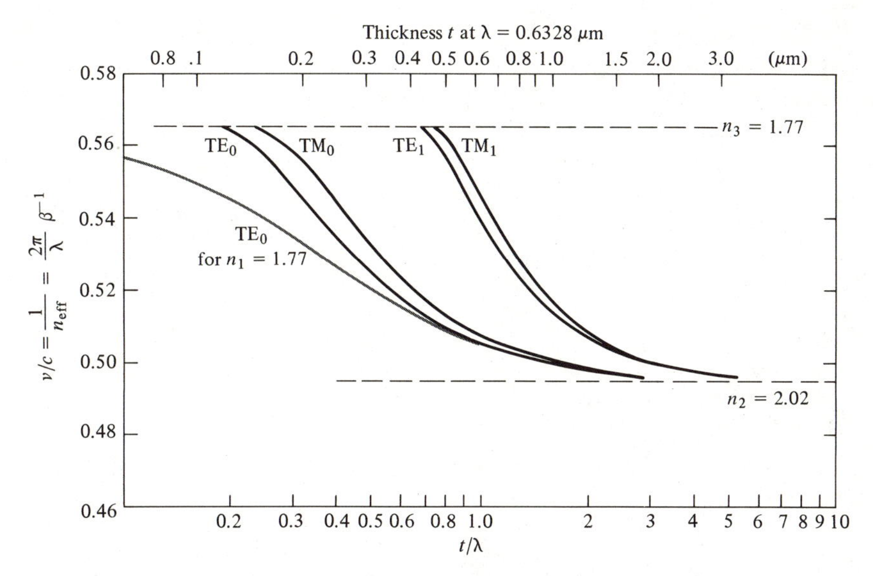

Figure 13-5 Dispersion curves for the confined modes of *Zn*O on sapphire waveguide. $n_1 = 1$. (After Reference [10].)

Applying (13.2-1) to (13.2-3) results in

$$\begin{aligned} h &= (n_2^2k_0^2 - \beta^2)^{1/2} \\ q &= (\beta^2 - n_1^2k_0^2)^{1/2} \\ p &= (\beta^2 - n_3^2k_0^2)^{1/2} \\ k_0 &\equiv \omega/c \end{aligned} \qquad \textbf{(13.2-4)}$$

The acceptable solutions for $\mathcal{E}_y$ and $\mathcal{H}_z = (i/\omega\mu)(\partial\mathcal{E}_y/\partial x)$ should be continuous at both $x = 0$ and $x = -t$. The choice of coefficients in (13.2-3) is such as to make $\mathcal{E}_y$ continuous at both interfaces as well as $(\partial\mathcal{E}_y/\partial x)$ at $x = 0$. By imposing the continuity requirement on $\partial\mathcal{E}_y/\partial x$ at $x = -t$, we get from (13.2-3)

$$h \sin(ht) - q \cos(ht) = p[\cos(ht) + (q/h) \sin(ht)]$$

or

$$\tan(ht) = \frac{p + q}{h(1 - pq/h^2)} \qquad \textbf{(13.2-5)}$$

In the symmetric case ($n_1 = n_3$) the field (13.2-3) must be odd or even about the mid-plane $x = -t/2$. This special case was treated in Section 13.1 and leads to the eigenvalue equation (13.1-15). The last equation in conjunction with (13.2-4) is used to obtain the eigenvalues β for the confined TE modes. An example of such a solution is shown in Figure 13.5.

The constant, C, appearing in (13.2-3) is arbitrary, yet for many applications, especially those in which propagation and exchange of power involve more than one mode, it is advantageous to define C in such a way that it is simply related to total power in the mode. This point will become clear in Section 13.3. We choose C so that the field $\mathcal{E}_y(x)$ in (13.2-3) corresponds to a power flow of one watt (per unit width in y direction) in the mode. A mode for which $E_y = A\mathcal{E}_y(x)$ will thus correspond to a power flow of $|A|^2$ watts/m. The normalization condition becomes

$$-\frac{1}{2}\int_{-\infty}^{\infty} E_y H_x^* \, dx = \frac{\beta_m}{2\omega\mu}\int_{-\infty}^{\infty} [\mathcal{E}_y^{(m)}(x)]^2 \, dx = 1 \qquad \textbf{(13.2-6)}$$

where the symbol m denotes the mth confined TE mode (corresponding to the mth eigenvalue of (13.2-5)) and $H_x = -i(\omega\mu)^{-1}\,\partial E_y/\partial z$.

Using (13.2-3) in (13.2-6) leads, after substantial but straightforward calculation, to

$$C_m = 2h_m\left[\frac{\omega\mu}{|\beta_m|\,[t + (1/q_m) + (1/p_m)](h_m^2 + q_m^2)}\right]^{1/2} \qquad \textbf{(13.2-7)}$$

Since the modes $\mathcal{E}_y^{(m)}$ are orthogonal (see problem 13.6) we have

$$\int_{-\infty}^{\infty} \mathcal{E}_y^{(l)}\mathcal{E}_y^{(m)} \, dx = \frac{2\omega\mu}{\beta_m}\,\delta_{l,m} \qquad \textbf{(13.2-8)}$$

TM modes. The derivation of the confined TM modes is similar in principle to that of the TE modes. Using (13.1-6) the field components are

$$\begin{aligned} H_y(x, z, t) &= \mathcal{H}_y(x)e^{i(\omega t - i\beta z)} \\ E_x(x, z, t) &= \frac{i}{\omega\epsilon}\frac{\partial H_y}{\partial z} = \frac{\beta}{\omega\epsilon}\mathcal{H}_y(x)e^{i(\omega t - \beta z)} \\ E_z(x, z, t) &= -\frac{i}{\omega\epsilon}\frac{\partial H_y}{\partial x} \end{aligned} \qquad \textbf{(13.2-9)}$$

The transverse function, $\mathcal{H}_y(x)$, is taken as

$$\mathcal{H}_y(x) = \begin{cases} -C\left[\dfrac{h}{\bar{q}}\cos(ht) + \sin(ht)\right]e^{p(x+t)} & x < -t \\ C\left[-\dfrac{h}{\bar{q}}\cos(hx) + \sin(hx)\right] & -t < x < 0 \\ -\dfrac{h}{\bar{q}}Ce^{-qx} & x > 0 \end{cases} \qquad \textbf{(13.2-10)}$$

The continuity of H_y and E_z at the two interfaces leads, in a manner similar to (13.2-5), to the eigenvalue equation

$$\tan(ht) = \frac{h(\bar{p} + \bar{q})}{h^2 - \bar{p}\bar{q}} \tag{13.2-11}$$

where

$$\bar{p} \equiv \frac{n_2^2}{n_3^2} p \qquad \bar{q} = \frac{n_2^2}{n_1^2} q$$

The normalization constant, C, is chosen so that the field represented by (13.2-9) and (13.2-10) carries *one* watt per unit width in the y direction

$$\frac{1}{2} \int_{-\infty}^{\infty} H_y E_x^* \, dx = \frac{\beta}{2\omega} \int_{-\infty}^{\infty} \frac{\mathcal{H}_y^2(x)}{\epsilon(x)} \, dx = 1$$

or, using $n_i^2 \equiv \epsilon_i/\epsilon_0$,

$$\int_{-\infty}^{\infty} \frac{[\mathcal{H}_y^{(m)}(x)]^2}{n^2(x)} \, dx = \frac{2\omega\epsilon_0}{\beta_m} \tag{13.2-12}$$

Carrying out the integration using (13.2-10) gives

$$\begin{aligned} C_m &= 2\sqrt{\frac{\omega\epsilon_0}{\beta_m t_{\text{eff}}}} \\ t_{\text{eff}} &\equiv \frac{\bar{q}^2 + h^2}{\bar{q}^2} \left(\frac{t}{n_2^2} + \frac{q^2 + h^2}{\bar{q}^2 + h^2} \frac{1}{n_1^2 q} + \frac{p^2 + h^2}{\bar{p}^2 + h^2} \frac{1}{n_3^2 p} \right) \end{aligned} \tag{13.2-13}$$

The general properties of the TE and TM mode solutions are illustrated in Figure 13-5. In general a mode becomes confined above a certain (cutoff) value of t/λ. At the cutoff value $p = 0$, and the mode extends to $x = -\infty$. For increasing values of t/λ, $p > 0$, and the mode becomes increasingly confined to layer 2. This is reflected in the effective mode index $\beta\lambda/2\pi$ that, at cutoff, is equal to n_3, and which, for large t/λ, approaches n_2. In a symmetric waveguide ($n_1 = n_3$) the lowest order modes TE_0 and TM_0 have no cutoff and are confined for all values of t/λ. The selective excitation of waveguide modes by means of prism couplers and a determination of their propagation constants β_m are described in Reference [11].

13.3 Theory of Coupled Modes

In Section 13.2 we obtained solutions for the confined modes supported by a slab dielectric waveguide such as that shown in Figure 13-1. An increasingly large number of experiments and devices involve coupling between such modes ([12], [13], and [17]). Two typical examples involve TM-to-TE mode conversion by the electrooptic or acoustooptic effect [12] or coupling of forward-to-backward modes by means of a corrugation in one of the waveguides interfaces [15], [16]. In this section we will develop a formalism for describing such coupling.

We start with the Maxwell wave equation in the form

$$\nabla^2\mathbf{E}(\mathbf{r}, t) = \mu\epsilon_0 \frac{\partial^2\mathbf{E}(\mathbf{r}, t)}{\partial t^2} + \mu \frac{\partial^2}{\partial t^2}\mathbf{P}(\mathbf{r}, t) \tag{13.3-1}$$

The total medium polarization can be taken as the sum

$$\mathbf{P}(\mathbf{r}, t) = \mathbf{P}_0(\mathbf{r}, t) + \mathbf{P}_{\text{pert}}(\mathbf{r}, t) \tag{13.3-2}$$

where

$$\mathbf{P}_0(\mathbf{r}, t) = [\epsilon(\mathbf{r}) - \epsilon_0]\mathbf{E}(\mathbf{r}, t) \tag{13.3-3}$$

is the polarization induced by $\mathbf{E}(\mathbf{r}, t)$ in the *unperturbed* waveguide whose dielectric constant is $\epsilon(\mathbf{r})$. The perturbation polarization $\mathbf{P}_{\text{pert}}(\mathbf{r}, t)$ is then defined by (13.3-2) and represents any deviation of the polarization from that of the unperturbed waveguide. Using (13.3-2) and (13.3-3) in (13.3-1) gives

$$\nabla^2 E_y - \mu\epsilon(\mathbf{r}) \frac{\partial^2 E_y}{\partial t^2} = \mu \frac{\partial^2}{\partial t^2}[P_{\text{pert}}(\mathbf{r}, t)]_y \tag{13.3-4}$$

and similar expressions for E_x and E_z.

Ignoring the possibility of coupling to the continuum of radiation modes, regimes *d* and *e* in Figure 13-2, we expand the total field in the "perturbed" waveguide as a superposition of confined modes

$$E_y(\mathbf{r}, t) = \tfrac{1}{2}\sum_m A_m(z)\mathcal{E}_y^{(m)}(x)e^{i(\omega t - \beta_m z)} + \text{c.c.} \tag{13.3-5}$$

where m indicates the mth discrete eigenmode of (13.2-5), which satisfies

$$\left(\frac{\partial^2}{\partial x^2} - \beta_m{}^2\right)\mathcal{E}_y^{(m)}(\mathbf{r}) + \omega^2\mu\epsilon(\mathbf{r})\mathcal{E}_y^{(m)}(\mathbf{r}) = 0 \tag{13.3-6}$$

where $\epsilon(\mathbf{r}) = \epsilon_0 n^2(\mathbf{r})$.

Substitution of (13.3-5) in (13.3-4) leads to

$$\begin{aligned} e^{i\omega t}\sum_m \Bigg[\frac{A_m}{2}\left(-\beta_m{}^2\mathcal{E}_y^{(m)} + \frac{\partial^2\mathcal{E}_y^{(m)}}{\partial x^2} + \omega^2\mu\epsilon(\mathbf{r})\mathcal{E}_y^{(m)}\right)e^{-i\beta_m z} \\ + \frac{1}{2}\left(-2i\beta_m \frac{dA_m}{dz} + \frac{d^2A_m}{dz^2}\right)\mathcal{E}_y^{(m)}e^{-i\beta_m z}\Bigg] + \text{c. c.} \\ = \mu \frac{\partial^2}{\partial t^2}[P_{\text{pert}}(\mathbf{r}, t)]_y \end{aligned} \tag{13.3-7}$$

First we note that in view of (13.3-6) the sum of the first three terms in (13.3-7) is zero. We assume "slow" variation so that

$$\left|\frac{d^2A_m}{dz^2}\right| \ll \beta_m \left|\frac{dA_m}{dz}\right|$$

and obtain from (13.3-7)

$$\sum_m -i\beta_m \frac{dA_m}{dz} \mathcal{E}_y^{(m)} e^{i(\omega t - \beta_m z)} + \text{c. c.} = \mu \frac{\partial^2}{\partial t^2} [P_{\text{pert}}(\mathbf{r}, t)]_y \qquad \textbf{(13.3-8)}$$

We take the product of (13.3-8) with $\mathcal{E}_y^{(s)}(x)$ and integrate from $-\infty$ to ∞. The result, using (13.2-8), is

$$\frac{dA_s^{(-)}}{dz} e^{i(\omega t + \beta_s z)} - \frac{dA_s^{(+)}}{dz} e^{i(\omega t - \beta_s z)} - \text{c. c.} = -\frac{i}{2\omega} \frac{\partial^2}{\partial t^2} \int_{-\infty}^{\infty} [P_{\text{pert}}(\mathbf{r}, t)]_y \mathcal{E}_y^{(s)}(x)\, dx \qquad \textbf{(13.3-9)}$$

where we recall that the summation over m in (13.3-8) contains two terms involving $\mathcal{E}_y^{(m)}(x)$ for each value of m—one, designated as $(-)$, traveling in the $-z$ direction, and the other $(+)$, traveling in the $+z$ direction.

Equation 13.3-9 can be used to treat a large variety of mode interactions [12]. Some important examples are considered in the following sections.

13.4 Periodic Waveguide

Consider a periodic dielectric waveguide in which the periodicity is due to a corrugation of one of the interfaces as shown in Figure 13-6. Such periodic waveguides are used for optical filtering [16] as well as in the distributed

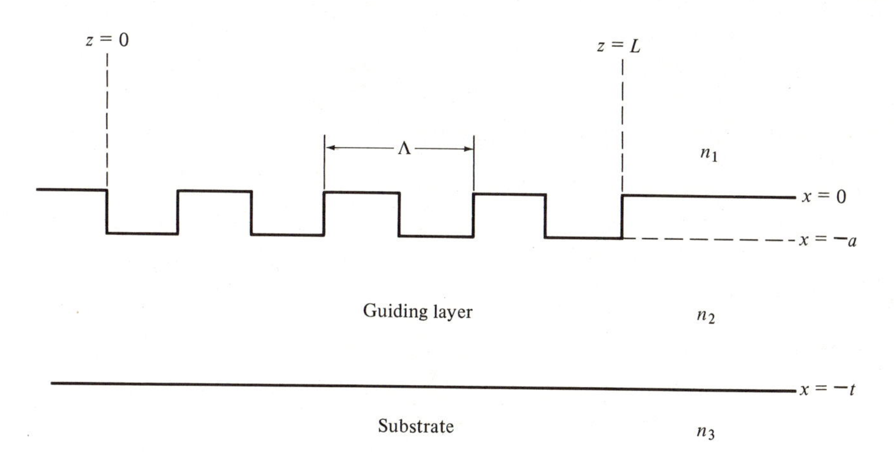

Figure 13-6 A corrugated periodic waveguide.

feedback laser ([17], [18], [19]). These two applications will be described further below.

The corrugation is described by the dielectric perturbation $\Delta\epsilon(\mathbf{r}) \equiv \epsilon_0 \Delta n^2(\mathbf{r})$ such that the total dielectric constant is

$$\epsilon'(\mathbf{r}) = \epsilon(\mathbf{r}) + \Delta\epsilon(\mathbf{r})$$

The perturbation polarization is from (13.3-2) and (13.3-3)

$$\mathbf{P}_{\text{pert}}(\mathbf{r}, t) = \Delta\epsilon(\mathbf{r})\mathbf{E}(\mathbf{r}, t) = \Delta n^2(\mathbf{r})\epsilon_0\mathbf{E}(\mathbf{r}, t) \quad \textbf{(13.4-1)}$$

Since $\Delta n^2(\mathbf{r})$ is a scalar, it follows, from (13.3-4), that the corrugation couples only TE to TE modes and TM to TM, but not TE to TM.

To be specific consider TE mode propagation. Using (13.3-5) in (13.4-1) gives

$$[P_{\text{pert}}(\mathbf{r}, t)]_y = \frac{\Delta n^2(\mathbf{r})\epsilon_0}{2} \sum_m [A_m \mathcal{E}_y^{(m)}(x) e^{i(\omega t - \beta_m z)} + \text{c. c.}] \quad \textbf{(13.4-2)}$$

which, when used in (13.3-9), leads to

$$\frac{dA_s^{(-)}}{dz} e^{i(\omega t + \beta_s z)} - \frac{dA_s^{(+)}}{dz} e^{i(\omega t - \beta_s z)} - \text{c. c.}$$

$$= -\frac{i\epsilon_0}{4\omega} \frac{\partial^2}{\partial t^2} \sum_m \left[A_m \int_{-\infty}^{\infty} \Delta n^2(x, z) \mathcal{E}_y^{(m)}(x) \mathcal{E}_y^{(s)}(x)\, dx e^{i(\omega t - \beta_m z)} + \text{c. c.} \right] \quad \textbf{(13.4-3)}$$

We may consider the right side of (13.4-3) as a source term driving the forward wave $A_s^+ \exp[i(\omega t - \beta_s z)]$ and the backward wave $A_s^- \exp[i(\omega t + \beta_s z)]$. In order for a wave to be driven by a source it is necessary that both waves have the same frequency so that the interaction will not average out to zero over a long time (long compared to a period of their difference frequency). Equally important: Both source and wave need have nearly the same phase dependence $\exp(i\beta z)$ so that the interaction does not average out to zero with distance of propagation z. If, for example, it is desired that the forward wave $A_s^{(+)} \exp[i(\omega t - \beta_s z)]$ be excited, it is necessary that at least one term on the right side of (13.4-3), say the lth one, vary as $\exp[i(\omega t - \beta z)]$ with $\beta \approx \beta_s$. If no other terms on the right side of (13.4-3) satisfy this condition, we simplify the equation by keeping only the forward wave on the left side and the lth on the right. We describe this situation by saying that the perturbation $\Delta n^2(x, z)$ couples the forward $(+s)$ mode to the lth mode and vice versa.

To be specific, let us assume that the period Λ of the perturbation $\Delta n^2(x, z)$ is so chosen that $l\pi/\Lambda \approx \beta_s$ for some integer l. We can expand $\Delta n^2(x, z)$ of a square wave perturbation as

$$\Delta n^2(x, z) = \Delta n^2(x) \sum_{-\infty}^{\infty} a_q e^{i(2q\pi/\Lambda)z} \quad \textbf{(13.4-4)}$$

The right side of (13.4-3) now contains a term ($q = l$, $m = s$) proportional to $A_s^{(+)} \exp [i(2l\pi/\Lambda - \beta_s)z]$. But

$$\frac{2l\pi}{\Lambda} - \beta_s \approx \beta_s$$

so that this term is capable of driving synchronously the amplitude $A_s^{(-)} \exp (i\beta_s z)$ on the left side of (13.4-3) with the result

$$\frac{dA_s^{(-)}}{dz} = \frac{i\omega\epsilon_0}{4} A_s^{(+)} \int_{-\infty}^{\infty} \Delta n^2(x)[\mathcal{E}_y^{(s)}(x)]^2 \, dx a_l e^{i[(2l\pi/\Lambda)-2\beta_s]z} \tag{13.4-5}$$

The coupling between the backward $A_s^{(-)}$ and the forward $A_s^{(+)}$ by the lth harmonic of $\Delta n^2(x, z)$ can thus be described by

$$\frac{dA_s^{(-)}}{dz} = \kappa A_s^{(+)} e^{-i2(\Delta\beta)z} \tag{13.4-6}$$

and reciprocally

$$\frac{dA_s^{(+)}}{dz} = \kappa^* A_s^{(-)} e^{i2(\Delta\beta)z}$$

where

$$\kappa = \frac{i\omega\epsilon_0 a_l}{4} \int_{-\infty}^{\infty} \Delta n^2(x)[\mathcal{E}_y^{(s)}(x)]^2 \, dx \tag{13.4-7}$$

$$\Delta\beta \equiv \beta_s - \frac{l\pi}{\Lambda} \equiv \beta_s - \beta_0 \tag{13.4-8}$$

We note that the total power carried by both modes is conserved, since

$$\frac{d}{dz} [|A_s^{(-)}|^2 - |A_s^{(+)}|^2] = 0 \tag{13.4-9}$$

Let us consider the specific "square wave" corrugation of Figure 13-6. In this case the periodicity (period $= \Lambda$) in the z direction is accounted for by taking

$$\begin{aligned} \Delta n^2(x, z) &= \Delta n^2(x) \left[\frac{1}{2} + \frac{2}{\pi}\left(\sin \eta z + \frac{1}{3} \sin 3\eta z + \cdots\right)\right] \\ &= \Delta n^2(x) \sum_l a_l e^{i\eta l z} \qquad l = 1, 3, 5, \cdots \end{aligned} \tag{13.4-10}$$

where

$$\Delta n^2(x) = \begin{cases} n_1^2 - n_2^2 & -a \le x \le 0 \\ 0 & \text{elsewhere} \end{cases} \tag{13.4-11}$$

$$\eta \equiv 2\pi/\Lambda$$

so that

$$a_l = \begin{cases} \dfrac{-i}{\pi l} & l \text{ odd} \\ 0 & l \text{ even} \\ a_0 = \frac{1}{2} & \end{cases}$$

and for l odd we obtain from (13.4-7) and (13.4-10)

$$\kappa = \frac{-\omega\epsilon_0}{4\pi l} \int_{-\infty}^{\infty} \Delta n^2(x)[\mathcal{E}_y^{(s)}(x)]^2\, dx \tag{13.4-12}$$

In practice the period Λ is chosen so that, for some particular l, $\Delta\beta \approx 0$. We note that for $\Delta\beta = 0$

$$\Lambda = l\frac{\lambda_g^{(s)}}{2} \tag{13.4-13}$$

where $\lambda_g^{(s)} = 2\pi/\beta_s$ is the guide wavelength of the sth mode.

We can now use the field expansion (13.2-3) plus (13.4-11) to perform the integration of (13.4-12).

$$\int_{-\infty}^{\infty} \Delta n^2(x)[\mathcal{E}_y^{(s)}(x)]^2\, dx = (n_1^2 - n_2^2) \int_{-a}^{0} [\mathcal{E}_y^{(s)}(x)]^2\, dx$$

$$= (n_1^2 - n_2^2)C_s^2 \int_{-a}^{0} \left[\cos(h_s x) - \frac{q_s}{h_s}\sin(h_s x)\right]^2 dx \tag{13.4-14}$$

Although the integral can be calculated exactly using (13.2-3) and (13.2-5), an especially simple result follows if we consider that operation is sufficiently above propagation cutoff, $t(n_2 - n_3)/s\lambda \gg 1$ so that from (13.2-4) and (13.2-5)

$$\beta_s \approx n_2 k_0$$

$$h_s \to \frac{\pi s}{t} \qquad s = 1, 2, \cdots = \text{transverse mode number}$$

$$\frac{q_s}{h_s} \approx (n_2^2 - n_1^2)^{1/2}\left(\frac{2t}{s\lambda}\right) \tag{13.4-15}$$

The results can be verified using (13.2-4) and (13.2-5). In addition since $q_s \gg h_s$, we have, from (13.2-7),

$$C_s^2 = \frac{4h_s^2\omega\mu}{\beta_s t q_s^2} \tag{13.4-16}$$

in the well-confined regime and for $h_s a \ll 1$ the integral of (13.4-14) becomes

$$(n_1^2 - n_2^2) \int_{-a}^{0} [\mathcal{E}_y^{(s)}(x)]^2 \, dx = (n_1^2 - n_2^2) \frac{4\pi^2 \omega\mu}{3n_2 k_0} \left(\frac{a}{t}\right)^3 \left(1 + \frac{3}{q_s a} + \frac{3}{q_s^2 a^2}\right)$$

and, using (13.4-15),

$$\kappa_s \approx -\frac{2\pi^2 s^2}{3l\lambda} \frac{(n_2^2 - n_1^2)}{n_2} \left(\frac{a}{t}\right)^3 \left[1 + \frac{3}{2\pi} \frac{\lambda/a}{(n_2^2 - n_1^2)^{1/2}} + \frac{3}{4\pi^2} \frac{(\lambda/a)^2}{(n_2^2 - n_1^2)}\right] \tag{13.4-17}$$

The problem has thus been reduced to a pair of coupled differential equations (13.4-6) and an expression (13.4-17) for the coupling constant.

13.5 Coupled-Mode Solutions

Let us return to the coupled-mode equations (13.4-6). For simplicity let us put $A_s^{(-)} \equiv A$, $A_s^{(+)} \equiv B$ and write them as

$$\begin{aligned} \frac{dA}{dz} &= \kappa_{ab} B e^{-i2(\Delta\beta)z} \\ \frac{dB}{dz} &= \kappa_{ab}^* A e^{+i2(\Delta\beta)z} \end{aligned} \tag{13.5-1}$$

Consider a waveguide with a corrugated section of length L as in Figure 13-6. A wave with an amplitude $B(0)$ is incident from the left on the corrugated section.

The solution of (13.5-1) for this case subject to $A(L) = 0$ is

$$\begin{aligned} A(z)e^{i\beta z} &= B(0) \frac{i\kappa_{ab} e^{i\beta_0 z}}{-\Delta\beta \sinh(SL) + iS\cosh(SL)} \sinh[S(z-L)] \\ B(z)e^{-i\beta z} &= B(0) \frac{e^{-i\beta_0 z}}{-\Delta\beta \sinh(SL) + iS\cosh(SL)} \\ &\quad \cdot \{\Delta\beta \sinh[S(z-L)] + iS\cosh[S(z-L)]\} \end{aligned} \tag{13.5-2}$$

where

$$\begin{aligned} S &= \sqrt{\kappa^2 - (\Delta\beta)^2} \\ \kappa &\equiv |\kappa_{ab}| \end{aligned} \tag{13.5-3}$$

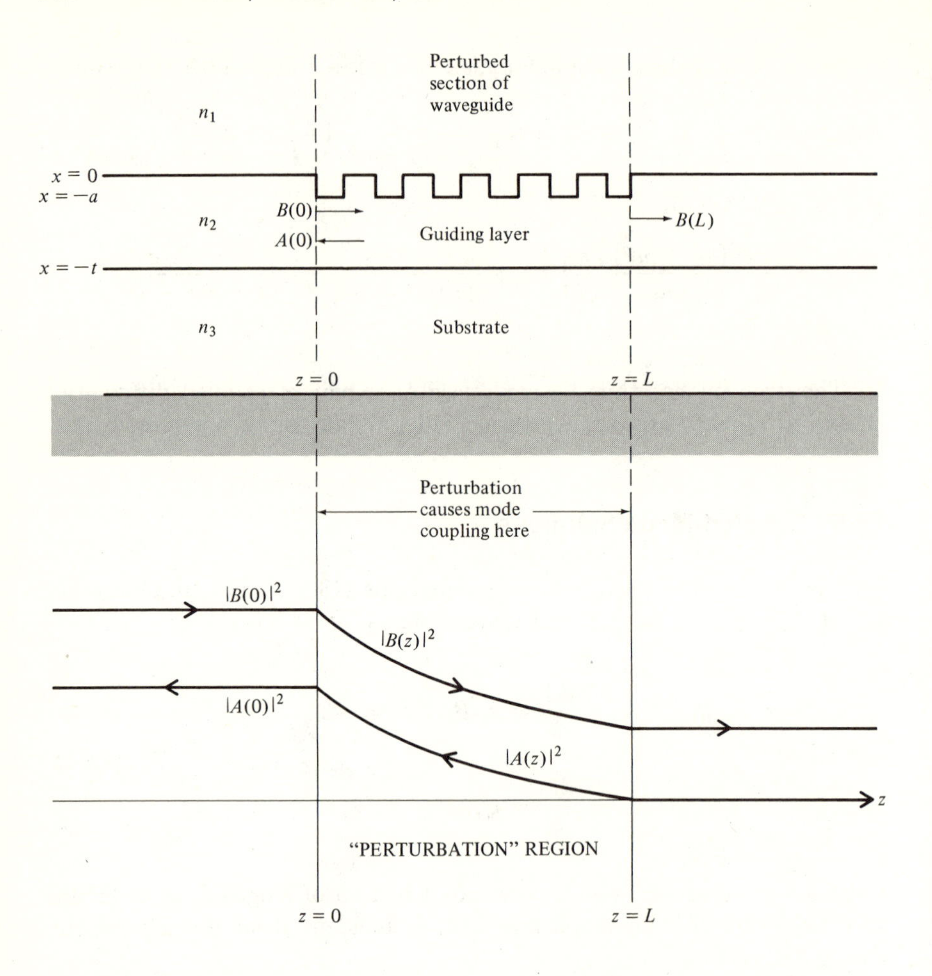

Figure 13-7 (*upper*) A corrugated section of a dielectric waveguide. (*lower*) The incident and reflected intensities inside the corrugated section.

Under phase matching conditions $\Delta\beta = 0$ we have

$$A(z) = B(0)\left(\frac{\kappa_{ab}}{\kappa}\right)\frac{\sinh\,[\kappa(z-L)]}{\cosh\,(\kappa L)}$$

$$B(z) = B(0)\,\frac{\cosh\,[\kappa(z-L)]}{\cosh\,(\kappa L)} \qquad \textbf{(13.5-4)}$$

A plot of the mode powers $|B(z)|^2$ and $|A(z)|^2$ for this case is shown in Figure 13-7. For sufficiently large arguments of the cosh and sinh functions in (13.5-4), the incident mode power drops off exponentially along the perturbation region. This behavior, however, is due not to absorption but to *reflection* of power into the backward traveling mode, A.

From (13.3-5) and (13.5-2) we find that the z dependent part of the wave solutions in the periodic waveguide are exponentials with propagation constants

$$\beta' = \beta_0 \pm iS = \frac{l\pi}{\Lambda} \pm i\sqrt{\kappa^2 - [\beta(\omega) - \beta_0]^2} \tag{13.5-5}$$

where we used $\Delta\beta \equiv \beta - \beta_0$, $\beta_0 \equiv \pi l/\Lambda$.

We note that for a range of frequencies such that $\Delta\beta(\omega) < \kappa$, β' has an imaginary part. This is the so-called "forbidden" region in which the evanescence behavior shown in Figure 13-7 occurs and which is formally analogous to the energy gap in semiconductors where the periodic crystal potential causes the electron propagation constants to become complex. Note that for each value of $l, l = 1, 2, 3 \ldots$, there exists a gap whose center frequency ω_{0l} satisfies $\beta(\omega_{0l}) = l\pi/\Lambda$. The exceptions are values of l for which κ is zero. Returning to (13.5-5) and approximating $\beta(\omega)$ near the Bragg value $(\pi l/\Lambda)$ by $\beta(\omega) \approx (\omega/c)n_{\text{eff}}$, where n_{eff} is an effective index of refraction. We have

$$\beta' \cong \frac{l\pi}{\Lambda} \pm i\left[\kappa^2 - \left(\frac{n_{\text{eff}}}{c}\right)^2 (\omega - \omega_0)^2\right]^{1/2} \tag{13.5-6}$$

where ω_0, the midgap frequency, is the value of ω for which the unperturbed β is equal to $\beta_0 \equiv l\pi/\Lambda$.

A plot of Re β' and Im β' (for $l = 1$) versus ω, based on (13.5-6), is shown in Figure 13-8. We note that the height of the "forbidden" fre-

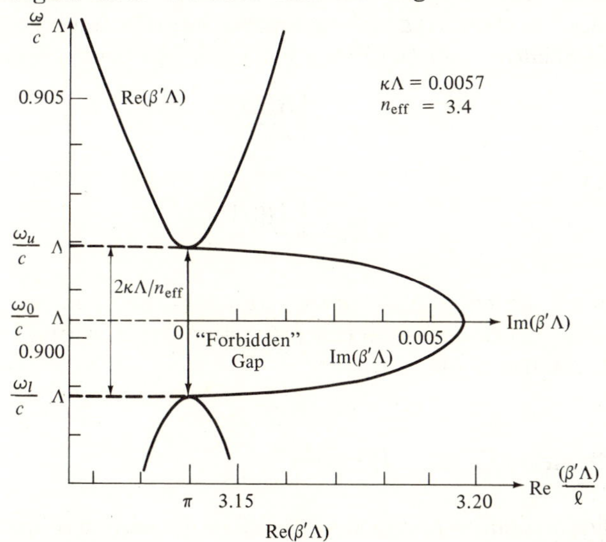

Figure 13-8 Dependence of the real and imaginary parts of the mode propagation constant, β', of the modes in a periodic waveguide. At frequencies $\omega_l < \omega < \omega_u$, Im$(\beta') \neq 0$ and the modes are evanescent. At these frequencies, Re $\beta' = l\pi/\Lambda$.

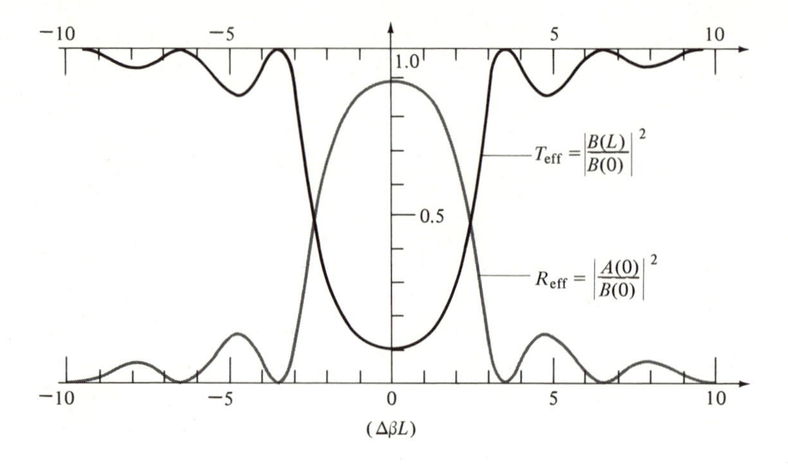

Figure 13-9 Transmission and reflection characteristics of a corrugated section of length L, as a function of the detuning $\Delta\beta L \approx [(\omega - \omega_0)L/c]n_{\text{eff}}(\kappa L = 1.84)$.

quency zone is

$$(\Delta\omega)_{\text{gap}} = \frac{2\kappa c}{n_{\text{eff}}} \tag{13.5-7}$$

where κ is according to (13.4-17) a function of the integer l. It follows from (13.5-6) that

$$(\text{Im}\,\beta')_{\text{max}} = \kappa \tag{13.5-8}$$

A short section of a corrugated waveguide thus acts as a high reflectivity mirror for frequencies near the Bragg value, ω_0. The transmission

$$T_{\text{eff}} = \left|\frac{B(L)}{B(0)}\right|^2$$

and reflection

$$R_{\text{eff}} = \left|\frac{A(0)}{B(0)}\right|^2$$

of such a filter are obtainable directly from (13.5-2) and are plotted in Figure 13-9. Actual transmission characteristics of a corrugated waveguide are shown in Figure 13-10.

13.6 Distributed Feedback Laser

If a periodic medium is provided with sufficient gain at frequencies near the Bragg frequency ω_0 where $(l\pi/\Lambda \approx \beta)$, oscillation can result without the benefit of end reflectors. The feedback is now provided by the continuous coherent backscattering from the periodic perturbation. In the

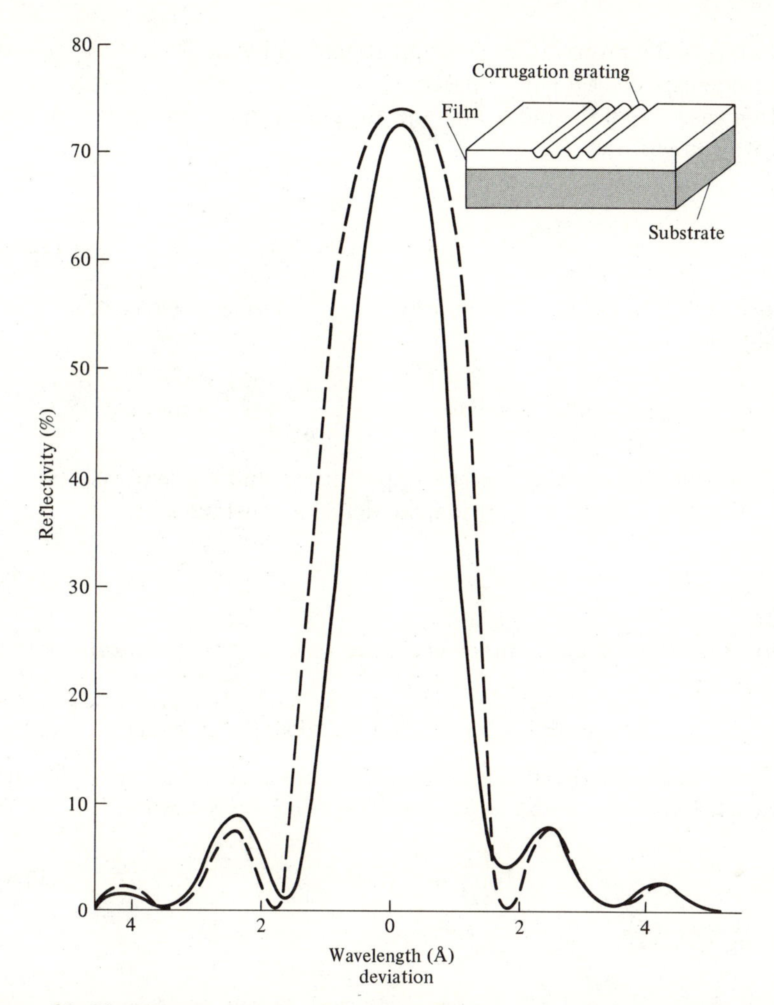

Figure 13-10 Illustration of a corrugation filter in a thin-film waveguide, plot (solid line) of reflectivity of filter versus wavelength deviation from the Bragg condition, and calculated response of filter (dotted line) $|A(0)/B(0)|^2$ using (13.5-2). (After Reference [16].)

following discussion we will consider two generic cases: (1) the bulk properties of a medium are perturbed periodically [17]; (2) the boundary of a waveguide laser is perturbed periodically [18]. Both cases will be found to lead to the same set of equations.

Bulk periodicity. Consider a medium with a complex dielectric constant so that the propagation constant, k, is given by

$$k^2 = \omega^2 \mu \epsilon = \omega^2 \mu (\epsilon_r + i\epsilon_i) = k_0^2 n^2(z) \left[1 + i \frac{2\lambda(z)}{k_0 n} \right] \tag{13.6-1}$$

so that k_0 is the propagation constant in vacuum and (for $\gamma \ll k_0$) γ is the amplitude exponential gain constant[2].

In a case where the index $n(z)$ and the gain $\gamma(z)$ are harmonic functions of z, we can write

$$n(z) = n + n_1 \cos 2\beta_0 z$$

$$\gamma(z) = \gamma + \gamma_1 \cos 2\beta_0 z \tag{13.6-2}$$

Using (13.6-2) in (13.6-1) and limiting ourselves to the case $n_1 \ll n$, $\gamma_1 \ll \gamma$ we obtain from (13.6-1)

$$k^2(z) = k_0^2 n^2 + i2k_0 n\gamma + 4k_0 n\left(\frac{\pi n_1}{\lambda} + i\frac{\gamma_1}{2}\right) \cos 2\beta_0 z$$

The propagation constant in the unperturbed and lossless case ($\gamma_1 = 0$, $n_1 = 0$) is $\beta = k_0 n$. If, in addition, we define a constant κ by

$$\kappa = \frac{\pi n_1}{\lambda} + i\frac{\gamma_1}{2} \tag{13.6-3}$$

where λ is the vacuum wavelength, we can rewrite the expression for $k^2(z)$ as

$$k^2(z) = \beta^2 + i2\beta\gamma + 4\beta\kappa \cos(2\beta_0 z), \qquad \beta \equiv k_0 n \tag{13.6-4}$$

For a small fractional variation of k^2 per wavelength it was shown in Section 3.1 that the scalar wave equation can be written as

$$\frac{d^2E}{dz^2} + k^2(z)E = 0 \tag{13.6-5}$$

or using (13.6-4)

$$\frac{d^2E}{dz^2} + [\beta^2 + i2\beta\gamma + 4\beta\kappa \cos(2\beta_0 z)]E = 0$$

In the discussion following (13.4-3) it was pointed out that a spatially modulated parameter varying as $\cos 2\beta_0 z$ can couple a forward traveling wave $\exp(-i\beta z)$ and a backward $\exp(i\beta z)$ wave of the same frequency provided $\beta_0 \cong \beta$. When this (Bragg) condition is nearly satisfied, it is impossible to describe the field $E(z)$ by a single traveling wave, but to a high degree of approximation we can take the total complex field amplitude as a linear superposition of both oppositely traveling waves

$$E(z) = A'(z)e^{i\beta' z} + B'(z)e^{-i\beta' z} \tag{13.6-6}$$

[2] γ used here is thus one-half of that appearing in (5.4-22).

where β' is the propagation constant of the uncoupled ($\kappa = 0$) waves

$$\beta'^2 = \beta^2 + i2\beta\gamma$$
$$(\beta' \approx \beta + i\gamma,\ \gamma \ll \beta) \tag{13.6-7}$$

Using (13.6-6) and

$$\frac{d^2}{dz^2}\left(A'(z)e^{i\beta' z}\right) = -\left(\beta'^2 A' - 2i\beta' \frac{dA'}{dz} - \frac{d^2A'}{dz^2}\right) e^{i\beta' z}$$

in (13.6-5), assuming "slow" variation so that $d^2A'/dz^2 \ll \beta'\, dA'/dz$, gives

$$i\beta' \frac{dA'}{dz} e^{i\beta' z} - i\beta' \frac{dB'}{dz} e^{-i\beta' z} = -\beta\kappa e^{i(2\beta_0-\beta')z}B' - \beta\kappa e^{-i(2\beta_0-\beta')z}A'$$
$$- \beta\kappa e^{i(2\beta_0+\beta')}A' - \beta\kappa e^{-i(2\beta_0+\beta')z}B' \tag{13.6-8}$$

For operation near the Bragg condition, $\beta_0 \approx \beta$, we can separately equate terms with nearly equal phase variation (that is, synchronous), thus ignoring the last two terms in (13.6-8). We obtain

$$\frac{dA'}{dz} = i\kappa B' e^{-i2(\beta'-\beta_0)z} = i\kappa B' e^{-i2(\Delta\beta+i\gamma)z}$$

$$\frac{dB'}{dz} = -i\kappa A' e^{+i2(\beta'-\beta_0)z} = -i\kappa A' e^{i(2\Delta\beta+i\gamma)z} \tag{13.6-9}$$

$$\Delta\beta \equiv \beta - \beta_0 = \frac{\omega}{c} n - \frac{\pi}{\Lambda} \qquad \Lambda = \text{period of spatial variation}$$

We will next derive a similar set of equations to describe a corrugated waveguide laser.

Corrugated waveguide laser. The case of a passive corrugated waveguide is described by (13.5-1). If the guiding medium possesses gain we simply need to modify these equations by adding gain terms so that when $\kappa = 0$ the two independent solutions, $A(z)$ and $B(z)$, correspond to exponentially growing waves along the $-z$ and the $+z$ directions, respectively. We thus replace (13.5-1) by

$$\frac{dA}{dz} = \kappa_{ab} B e^{-i2(\Delta\beta)z} - \gamma A$$
$$\frac{dB}{dz} = \kappa_{ab}^* A e^{i2(\Delta\beta)z} + \gamma B \tag{13.6-10}$$

where κ is given by (13.4-12) and γ is the exponential gain constant of the medium. Defining $A'(z)$ and $B'(z)$ by

$$\begin{aligned} A(z) &= A'(z)e^{-\gamma z} \\ B(z) &= B'(z)e^{\gamma z} \end{aligned} \tag{13.6-11}$$

Equations 13.6-10 become

$$\begin{aligned} \frac{dA'}{dz} &= \kappa_{ab}B'e^{-i2(\Delta\beta+i\gamma)z} \\ \frac{dB'}{dz} &= \kappa_{ab}^{*}A'e^{+i2(\Delta\beta+i\gamma)z} \end{aligned} \tag{13.6-12}$$

and are thus in a form identical to that of (13.6-9) derived for the case of a bulk periodic medium with index modulation.

Equations 13.6-12 become identical to (13.5-1) provided we replace

$$\Delta\beta \rightarrow \Delta\beta + i\gamma \tag{13.6-13}$$

With this substitution we can then use (13.5-2) to write directly the solutions for the incident wave, $E_i = B'(z) \exp[(-i\beta + \gamma)z]$, and the reflected wave $E_r = A(z)e^{i\beta z} = A'(z)^{[i\beta-\gamma]z}$, within a section of length L in the case of a single mode with amplitude $B(0)$ incident on the corrugated section at $z = 0$.

$$E_i(z) = B(0)\frac{e^{-i\beta_0 z}\{(\gamma - i\Delta\beta)\sinh[S(L-z)] - S\cosh[S(L-z)]\}}{(\gamma - i\Delta\beta)\sinh(SL) - S\cosh(SL)}$$

$$E_r(z) = B(0)\frac{\kappa_{ab}e^{i\beta_0 z}\sinh[S(L-z)]}{(\gamma - i\Delta\beta)\sinh(SL) - S\cosh(SL)} \tag{13.6-14}$$

where

$$S^2 = \kappa^2 + (\gamma - i\Delta\beta)^2 \tag{13.6-14[a]}$$

The fact that S now is complex makes for a qualitative difference between the behavior of the passive periodic guide (13.5-2) and the periodic guide with gain (13.6-14). To demonstrate this difference consider the case when the condition

$$(\gamma - i\Delta\beta)\sinh(SL) = S\cosh(SL) \tag{13.6-15}$$

is satisfied. It follows from (13.6-14) that both the reflectance, $E_r(0)/E_i(0)$, and the transmittance, $E_i(L)/E_i(0)$, become infinite. The device acts as an oscillator, since it yields finite output fields $E_r(0)$ and $E_i(L)$ with no input $[E_i(0) = 0]$. Condition (13.6-15) is thus the oscillation condition for a distributed feedback laser [17]. For the case of $\gamma = 0$ it follows, from (13.5-2), that $|E_i(L)/E_i(0)| < 1$ and $|E_r(0)/E_i(0)| < 1$ as appropriate to a passive device with no internal gain.

For frequencies very near the Bragg frequency $\omega_0(\Delta\beta \cong 0)$ and for sufficiently high gain constant γ so that (13.6-15) is nearly satisfied, the guide acts as a high gain amplifier. The amplified output is available either in reflection with a ("voltage") gain

$$\frac{E_r(0)}{E_i(0)} = \frac{\kappa_{ab} \sinh (SL)}{(\gamma - i\Delta\beta) \sinh (SL) - S \cosh (SL)} \tag{13.6-16}$$

or in transmission with a gain

$$\frac{E_i(L)}{E_i(0)} = \frac{-Se^{-i\beta_0 L}}{(\gamma - i\Delta\beta) \sinh (SL) - S \cosh (SL)} \tag{13.6-17}$$

The behavior of the incident and reflected field for a high gain case is sketched in Figure 13-11. Note the qualitative difference between this case and the passive one depicted in Figure 13-7.

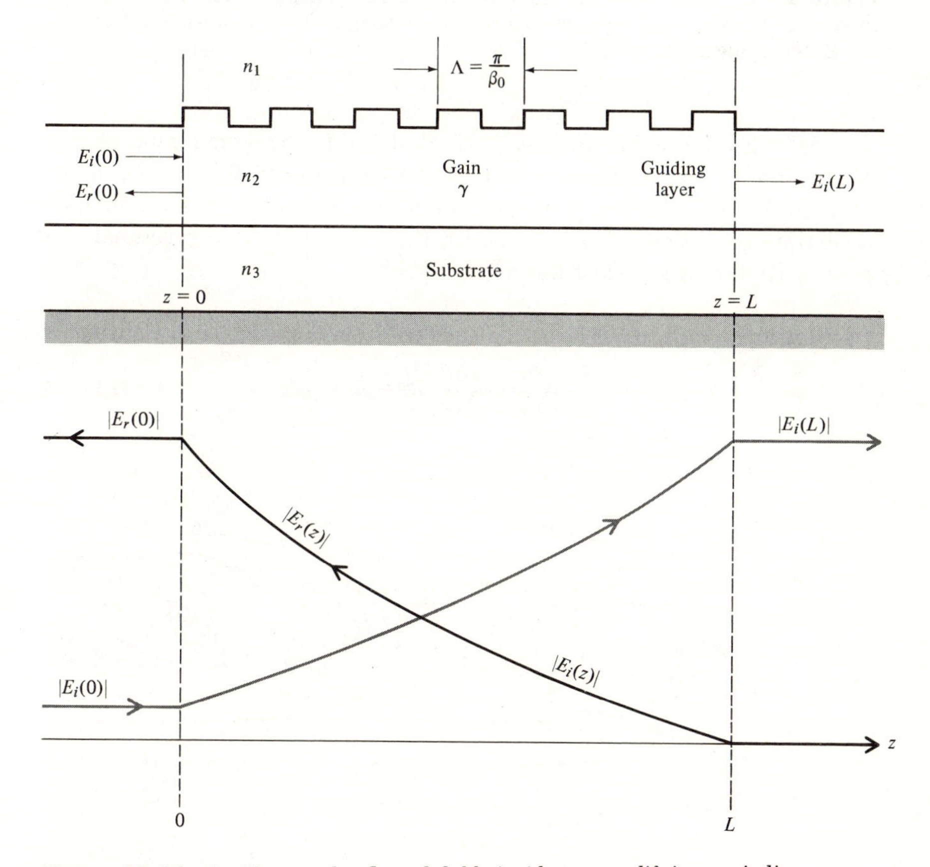

Figure 13-11 Incident and reflected fields inside an amplifying periodic waveguide.

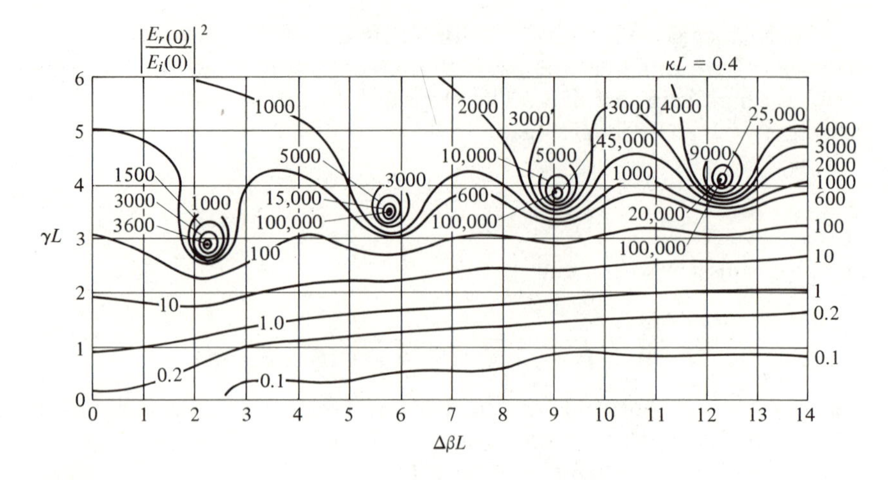

Figure 13-12 Reflection gain contours in the $\Delta\beta L$-γL plane. $\Delta\beta$ is defined following (13.6-9) and is proportional to the deviation of the frequency ω from the Bragg value $\omega_0 \equiv \pi c/\Lambda n$.

The reflection gain, $|E_r(0)/E_i(0)|^2$, and the transmission gain, $|E_i(L)/E_i(0)|^2$, are plotted in Figures 13-12 and 13-13, respectively, as a function of $\Delta\beta$ and γ. Each plot contains four infinite gain singularities at which the oscillation condition (13.6-15) is satisfied. These are four longitudinal laser modes. Higher orders exist but are not shown.

Oscillation condition. The oscillation condition (13.6-15) can be written as

$$\frac{S - (\gamma - i\Delta\beta)}{S + (\gamma - i\Delta\beta)} e^{2SL} = -1 \tag{13.6-18}$$

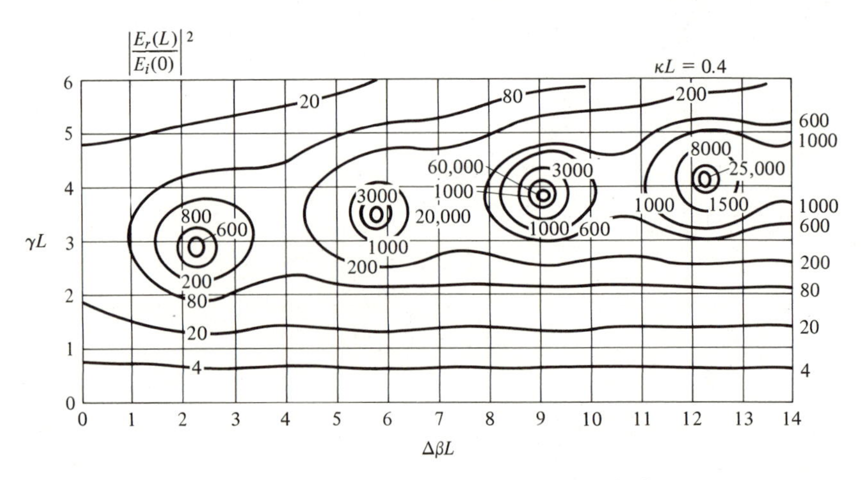

Figure 13-13 Transmission gain contours in the $\Delta\beta L$-γL plane.

In general, one has to resort to a numerical solution to obtain the threshold values of $\Delta\beta$ and γ for oscillation [17]. In some limiting cases, however, we can obtain approximate solutions. In the high-gain $\gamma \gg \kappa$ case we have from the definition of $s^2 = \kappa^2 + (\gamma - i\Delta\beta)^2$

$$S \approx -(\gamma - i\Delta\beta)\left[1 + \frac{\kappa^2}{2(\gamma - i\Delta\beta)^2}\right] \qquad \gamma \gg \kappa$$

so that

$$S - (\gamma - i\Delta\beta) \cong -2(\gamma - i\Delta\beta)$$

$$S + (\gamma - i\Delta\beta) \cong \frac{-\kappa^2}{2(\gamma - i\Delta\beta)}$$

and (13.6-18) becomes

$$\frac{4(\gamma - i\Delta\beta)^2}{\kappa^2} e^{2SL} = -1 \tag{13.6-19}$$

Equating the phases on both sides of (13.6-19) results in

$$2 \tan^{-1} \frac{(\Delta\beta)_m}{\gamma_m} - 2(\Delta\beta)_m L + \frac{(\Delta\beta)_m L \kappa^2}{\gamma_m^2 + (\Delta\beta)_m^2} = (2m + 1)\pi$$

$$m = 0, \pm 1, \pm 2, \cdots \tag{13.6-20}$$

In the limit $\gamma_m \gg (\Delta\beta)_m$, κ, the oscillating mode frequencies are given by

$$(\Delta\beta_m)L \cong -\left(m + \frac{1}{2}\right)\pi \tag{13.6-21}$$

and since $\Delta\beta \equiv \beta - \beta_0 \approx (\omega - \omega_0)n_{\text{eff}}/c$

$$\omega_m = \omega_0 - \left(m + \frac{1}{2}\right)\frac{\pi c}{n_{\text{eff}}L} \tag{13-6-22}$$

We note that no oscillation can take place exactly at the Bragg frequency ω_0. The mode frequency spacing is

$$\omega_{m-1} - \omega_m \simeq \frac{\pi c}{n_{\text{eff}}L} \tag{13.6-23}$$

and is approximately the same as in a two-reflector resonator of length L.

The threshold gain value γ_m is obtained from the amplitude equality in (13.6-19)

$$\frac{e^{2\gamma_m L}}{\gamma_m^2 + (\Delta\beta)_m^2} = \frac{4}{\kappa^2} \tag{13.6-24}$$

indicating an increase in threshold with increasing mode number m. This is also evident from the numerical gain plots (Figures 13-12 and 13-13).

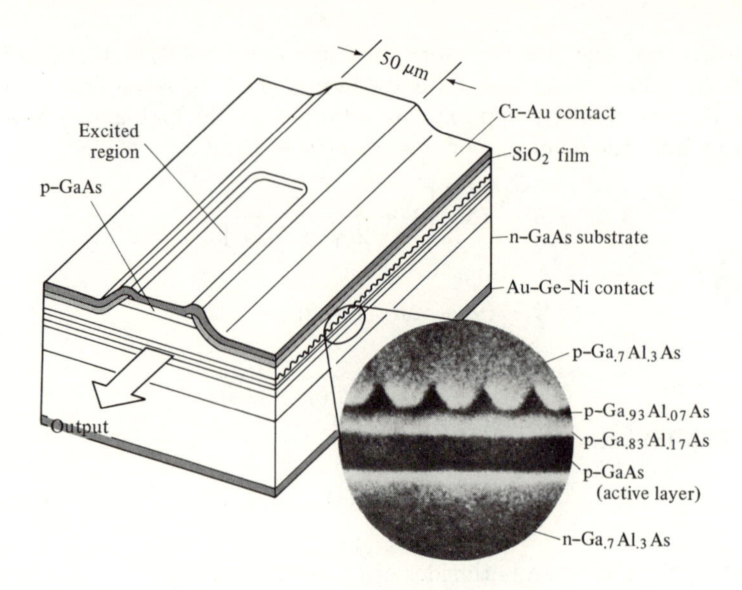

Figure 13-14 A GaAs-GaAlAs *cw* injection laser with a corrugated interface. The insert shows a scanning electron microscope photograph of the layered structure. The feedback is in third order ($l = 3$) and is provided by a corrugation with a period $\Lambda = 3\,\lambda g/2 = 0.345\ \mu$m. The thin (0.2 μm) p-$Ga_{.83}A_{.17}As$ layer provides a potential barrier which confines the injected electrons to the active (p-GaAs) layer, thus increasing the gain. (After Reference [19].)

A diagram of a distributed feedback laser using a GaAs–GaAlAs structure similar to that of Figure 7-26 is shown in Figure 13-14. The waveguiding layer as well as that providing the gain (active layer) is that of p-GaAs. The feedback is provided by corrugating the interface between the p-$Ga_{.93}Al_{.07}As$ and p-$Ga_{.7}Al_{.3}As$, where the main index discontinuity responsible for the guiding (see 7.8-12) occurs.

The increase in threshold gain with the longitudinal mode index m predicted by (13.6-24) and by the plots of Figures 13-12 and 13-13 manifests itself in a high degree of mode discrimination in the distributed feedback laser. This fact is displayed by the oscillation spectrum shown in Figure 13-15 in which a single ($m = 0$) mode is evident.

13.7 Electrooptic Modulation and Mode Coupling in Dielectric Waveguides

One of the most important applications of thin film waveguiding is in optical modulation and switching. The reason is twofold: (1) The confinement of the optical radiation to dimensions comparable to λ makes it

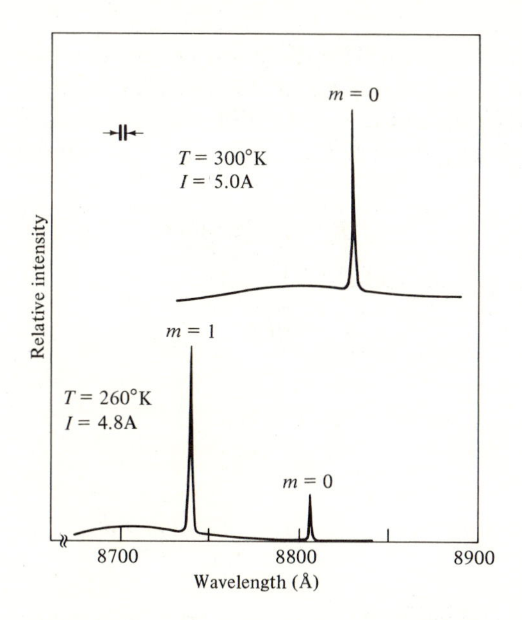

Figure 13-15 Oscillation spectrum of a distributed feedback laser in GaAs-GaAlAs double heterostructure diode. m is the transverse mode number. (After Reference [19].)

possible to achieve the magnitude of electric fields that is necessary for modulation (see Section 9.5) with relatively small applied voltages. This leads to smaller modulation powers. (2) The absence of diffraction in a guided optical beam makes it possible to use longer modulation paths (see discussion in Section 14.1).

The main principle of electrooptic modulation in dielectric waveguides involves the diversion of all or part of the power from an input TE (or TM) mode to an output TM (or TE) mode which is caused by an applied dc (that is, "low" frequency) field. To be specific we consider next the case of TM → TE mode conversion. This coupling is due to a perturbation polarization[3]

$$[P_{\text{pert}}(t)]_y \propto rE^{(0)}E_x^{(\omega)}(x)e^{i\omega t} \qquad \textbf{(13.7-1)}$$

caused by the TM mode at ω whose field is $E_x^{(\omega)}(x)$ in the presence of the dc field $E^{(0)}$. The symbol r is an appropriate linear combination of electro-

[3] This follows from the wave equation (13.3-1) or (13.3-9), according to which a TE mode with a field E_y can be excited by P_y. An input TM mode with a field E_x can thus excite the TE wave ($E_y \neq 0$) if the medium has an off-diagonal ϵ_{yx} dielectric tensor component which generates $P_y = \epsilon_{yx}E_x$. In the electrooptic case ϵ_{yx} is induced by and is proportional to the applied dc field $E^{(0)}$ as in (13.7-1).

optic coefficients, which will be discussed below. This polarization, acting as a source, can excite, according to (13.3-9), a TE wave $A_s^{(+)}$. The application of a dc field thus causes a TM → TE power transfer.

The complex amplitude of the y polarization at ω produced by the TM field

$$E_x^{(\omega)}(x)e^{i(\omega t-\beta_{\mathrm{TM}} z)} \tag{13.7-2}$$

in the presence of a dc field $E^{(0)}$ is[4]

$$[P_{\mathrm{pert}}^{(\omega)}]_y = -\frac{\epsilon^2 r E^{(0)}}{\epsilon_0} E_x^{(\omega)}(x)e^{-i\beta_{\mathrm{TM}} z} \tag{13.7-3}$$

Using (13.2-9) we take the TM input field $E_x^{(\omega)}(x) \exp [i(\omega t - \beta_{\mathrm{TM}} z)]$ as that of the lth mode.

$$E_x{}^l(\mathbf{r}, t) = \frac{\beta_l}{2\omega\epsilon(x)} B_l \mathcal{H}_y^{(l)}(x)e^{i(\omega t-\beta_l z)} + \text{c. c.} \tag{13.7-4}$$

where $\mathcal{H}_y^{(l)}(x)$ is given by (13.2-10) and $|B_l|^2$ is the mode power per unit width in the y direction. The polarization (13.-7-2) can thus be written as

$$[P_{\mathrm{pert}}(\mathbf{r}, t)]_y = \frac{\epsilon(x) r(x, z) E^{(0)}}{2\omega\epsilon_0} \beta_l B_l \mathcal{H}_y^{(l)}(x)e^{i(\omega t-\beta_l z)} + \text{c. c.} \tag{13.7-5}$$

---

[4] The origin of relation (13.7-3) is as follows: From the basic definition of the electrooptic tensor elements $(1/n^2)_{ij}$ in (1.4-1) and from (1.4-9) it follows that a change in the indicatrix constant $(1/n^2)_{ij}$ is related to the corresponding change in the elements of the dielectric tensor ϵ_{ij} by

$$\Delta\epsilon_{ij} = -\frac{\epsilon_{ii}\epsilon_{jj}}{\epsilon_0} \Delta\left(\frac{1}{n^2}\right)_{ij}$$

Using (9.1-3) we thus relate $\Delta\epsilon_{ij}$ to an applied dc field $E_k^{(0)}$ by

$$\Delta\epsilon_{ij} = -\frac{\epsilon_{ii}\epsilon_{jj}}{\epsilon_0} r_{ijk} E_k^{(0)}$$

where r_{ijk} is the electrooptic tensor and where we sum over repeated indices. From the relation $D_i = \epsilon_{ij}E_j + P_i$ we obtain

$$[P_{\mathrm{pert}}^{(\omega)}]_i = \Delta\epsilon_{ij}E_j^{(\omega)} = -\frac{\epsilon_{ii}\epsilon_{jj}}{\epsilon_0} r_{ijk} E_k^{(0)} E_j^{(\omega)}$$

for the change in the ith component of the complex amplitude of the polarization at ω induced by an optical field with amplitude $E_j^{(\omega)}$ at ω in the presence of a dc field $E_k^{(0)}$. This last relation appears above in the form of (13.7-3) where the z and x dependence of $E_j^{(\omega)}$ are expressed explicitly and where

$$\epsilon^2 r E^{(0)} \to \epsilon_{ii}\epsilon_{jj} r_{ijk} E_k^{(0)}.$$

Substitution of (13.7-5) into the wave equation (13.3-9) leads to

$$\frac{dA_m^{(+)}}{dz} \exp(-i\beta_m^{\text{TE}} z) - \frac{dA_m^{(-)}}{dz} \exp(i\beta_m^{\text{TE}} z)$$
$$= -\frac{i}{4} \int_{-\infty}^{\infty} \frac{\epsilon^2 r(x, z) E^{(0)}(x, z)}{\epsilon(x)\epsilon_0} \beta_l B_l \mathcal{H}_y^{(l)}(x) \mathcal{E}_y^{(m)}(x)\, dx \exp(-i\beta_l^{\text{TM}} z) \tag{13.7-6}$$

If $\beta_l^{\text{TM}} \approx \beta_m^{\text{TE}}$, the coupling excites only the $A_m^{(+)}$ wave, that is, it is codirectional. Dropping the plus and minus superscripts we can rewrite (13.7-6) as

$$\frac{dA_m}{dz} = -i\kappa_{ml}(z) B_l e^{-i(\beta_m^{\text{TE}} - \beta^{\text{TM}})z} \tag{13.7-7}$$

$$\kappa_{ml} = \frac{\beta_l}{4} \int_{-\infty}^{\infty} \frac{\epsilon(x) r(x, z) E^{(0)}(x, z)}{\epsilon_0} \mathcal{H}_y^{(l)}(x) \mathcal{E}_y^{(m)}(x)\, dx \tag{13.7-8}$$

Equation 13.7-8 is general enough to apply to a large variety of cases. The dependence of $E^{(0)}$ and $r(x, z)$ on x accounts for coupling by electrooptic material in the guiding or in the bounding layers. The z dependence allows for situations where $E^{(0)}$ or r depend on the longitudinal position. To be specific, we consider first the case where the guiding layer $-t < x < 0$ is uniformly electrooptic and where $E^{(0)}$ is uniform over the same region so that the integration in (13.7-8) is from $-t$ to 0. In that case, the overlap integral of (13.7-8) is maximum when the TE(m) and TM(l) modes are well confined and of the *same* order so that $l = m$. Under well-confined conditions, $p, q \gg h$ and the expressions (13.2-3), (13.2-7) for $\mathcal{E}_y^{(m)}(x)$, (13.2-10) and (13.2-13) for $\mathcal{H}_y^{(m)}(x)$ in the guiding layer become

$$\mathcal{E}_y^{(m)}(x) \to \left(\frac{4\omega\mu}{t\beta_m^{\text{TE}}}\right)^{1/2} \sin\frac{m\pi x}{t}$$

$$\mathcal{H}_y^{(m)}(x) \to \left(\frac{4\omega\epsilon_0 n_2^2}{t\beta_m^{\text{TM}}}\right)^{1/2} \sin\frac{m\pi x}{t}$$

where for well-confined mode $\beta_l^{\text{TM}} \simeq \beta_m^{\text{TE}} \equiv \beta \approx k_0 n_2$. In this case the overlap integral in (13.7-8) becomes

$$\int_{-t}^{0} \mathcal{H}_y^{(m)}(x) \mathcal{E}_y^{(m)}(x)\, dx = \frac{4\omega\sqrt{\mu\epsilon_2}}{t\beta} \int_{-t}^{0} \sin^2\frac{m\pi x}{t}\, dx = 2$$

and the coupling coefficient (13.7-8) achieves a maximum value of

$$\kappa \to \frac{n_2^3 k_0 r E^{(0)}}{2} \tag{13.7-9}$$

The coupling is thus described by

$$\frac{dA_m}{dz} = -i\kappa B_m e^{-i(\beta_m^{\text{TM}} - \beta_m^{\text{TE}})z}$$

and **(13.7-10)**

$$\frac{dB_m}{dz} = -i\kappa A_m e^{i(\beta_m^{\text{TM}} - \beta_m^{\text{TE}})z}$$

The second equation of (13.7-10) can be obtained by a process similar to that leading to the first equation or by invoking the conservation of total power [12], which shows that the above expression for dB_m/dz is needed to satisfy

$$\frac{d}{dz}(|A_m|^2 + |B_m|^2) = 0$$

For the phase-matched condition $\beta_m^{\text{TM}} = \beta_m^{\text{TE}}$ the solution of (13.7-10) in the case of a single input ($B_m(0) \equiv B_0$, $A_m(0) = 0$) is

$$B_m(z) = B_0 \cos(\kappa z)$$

$$A_m(z) = -iB_0 \sin(\kappa z) \qquad \textbf{(13.7-11)}$$

Using (13.7-9) we can show that the field length product $E^{(0)}L$ for which $\kappa L = \pi/2$, which is necessary to effect a complete TM $\leftrightarrow$ TE power transfer in a distance L, is the same as that needed to go from "on" to "off" in the bulk modulator shown in Figure 9-4. This result applies only in the limit of tight confinement. In general the coupling coefficient, κ, is smaller than the value given by (13.7-9), and the $E^{(0)}L$ product needed to achieve a complete power transfer is correspondingly larger.

When $\beta_m^{\text{TM}} \neq \beta_m^{\text{TE}}$, the solution of (13.7-10), subject to boundary conditions $B_m(0) = B_0$, $A_m(0) = 0$, is

$$B(z) = B_0 e^{i\delta z}\left\{\cos[(\kappa^2 + \delta^2)^{1/2}z] - i\frac{\delta}{(\kappa^2 + \delta^2)^{1/2}}\sin[(\kappa^2 + \delta^2)^{1/2}z]\right\}$$

(13.7-12)

$$A(z) = -iB_0 e^{-i\delta z}\frac{\kappa}{(\kappa^2 + \delta^2)^{1/2}}\sin[(\kappa^2 + \delta^2)^{1/2}z]$$

where

$$2\delta \equiv \beta_m^{\text{TM}} - \beta_m^{\text{TE}} \qquad \textbf{(13.7-13)}$$

In contrast to the phase-matched case (13.7-11), the maximum fraction of the power that can be coupled from the input mode, B, to A is

$$\text{Fraction of power exchanged} = \frac{\kappa^2}{\kappa^2 + \delta^2} \qquad \textbf{(13-7-14)}$$

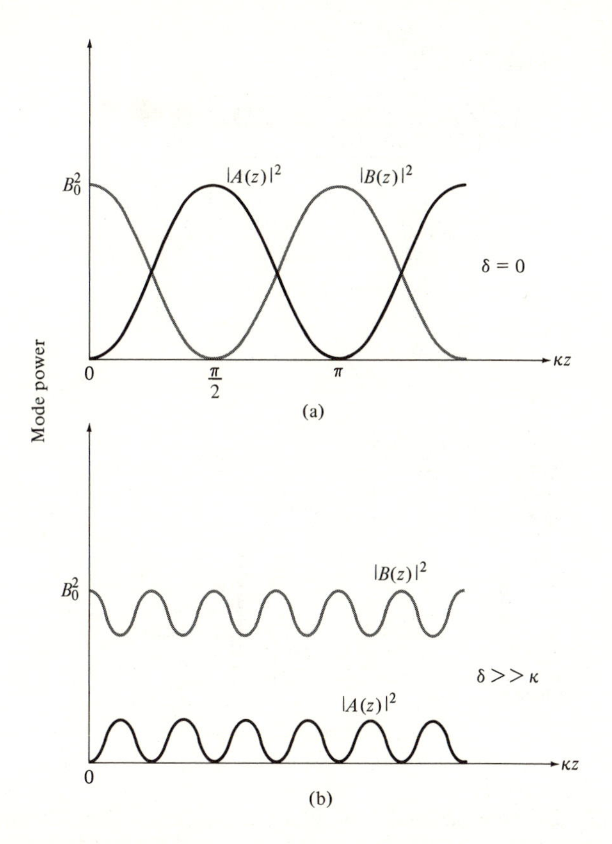

Figure 13-16 Power exchange between two coupled modes under (**a**) phase-matched conditions ($\beta_m^{\text{TM}} = \beta_m^{\text{TE}}$) as described by Equation (13.7-11); (**b**) $\beta_m^{\text{TM}} \neq \beta_m^{\text{TE}}$, Equation (13.7-12).

and becomes negligible once $\delta \gg \kappa$.

A plot of the mode power for the phase-matched ($\delta = 0$) and $\delta \neq 0$ case is shown in Figure 13-16.

A deliberate periodic variation of $E^{(0)}(z)$ or $r(z)$, in this case, with a period $2\pi/(\beta_m^{\text{TE}} - \beta_m^{\text{TM}})$ can be used, according to (13.7-8), to compensate for the mismatch factor $\exp[-i(\beta_m^{\text{TE}} - \beta_m^{\text{TM}})z]$ in (13.7-10), thus leading again to a phase-matched operation.

Example: GaAs thin-film modulator at $\lambda = 1\ \mu\text{m}$**.** To appreciate the order of magnitude of the coupling, consider a case where the guiding layer is GaAs and $\lambda = 1\ \mu$m. In this case (see Table 9-2)

$$n_2 \simeq 3.5 \qquad n_2{}^3 r = 59 \times 10^{-12}\ \frac{\text{m}}{\text{volt}}$$

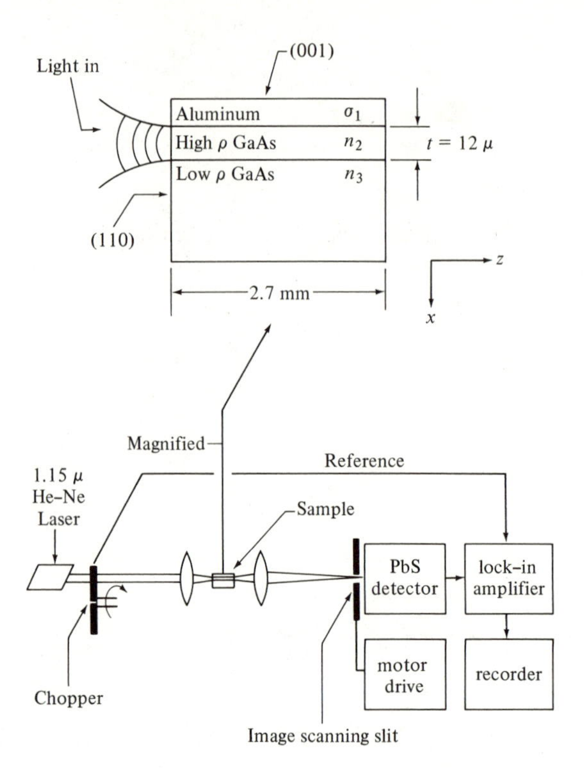

Figure 13-17 Electrooptic modulator in a GaAs epitaxial film. The modulation field is due to a reverse bias voltage applied to the metal semiconductor junction [3].

Taking an applied field $E^{(0)} = 10^6$ volt/m, we obtain, from (13.7-9),

$$\kappa = 1.85\ \text{cm}^{-1}$$

$$l \equiv \frac{\pi}{2\kappa} = 0.85\ \text{cm}$$

for the coupling constant and the power-exchange distance, respectively.

An experimental setup used in one of the earliest demonstrations [3] of electrooptic thin film modulation is depicted in Figure 13-17. The modulation scheme is identical to that illustrated in Figure (9-8) and depends on an electrooptic induced phase retardation (9.5-1)

$$\Gamma = (\beta_{\text{TM}} - \beta_{\text{TE}})l$$

$$\approx \frac{\pi n_2^3 r_{41} l}{\lambda t} V \tag{13.7-15}$$

where V is the applied voltage, t and l are the height and length of the waveguide, respectively.

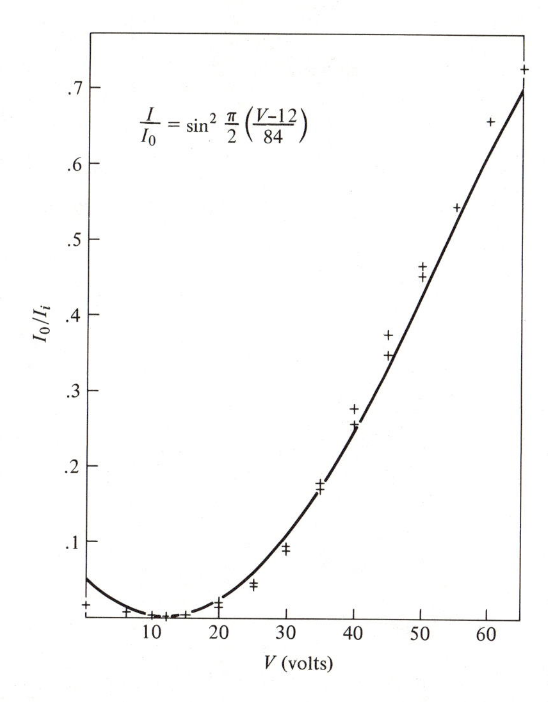

Figure 13-18 Transmittance of the waveguide, placed between crossed polarizers, as a function of the applied reverse voltage [3].

The ratio of the transmitted to the input intensities is given by (9.3-4), or, equivalently by (13.7-11), (note that Γ of (9.2-4) is the same as κ in (13.7-9)) as

$$\frac{I_0}{I_i} = \sin^2 \frac{\Gamma}{2} \qquad \textbf{(13.7-16)}$$

An experimental transmission plot is shown in Figure 13-18.

Another form of electrooptic thin film modulation utilizes two-dimensional Bragg diffraction of a waveguide mode from a spatially periodic modulation of the index of refraction. A periodic electric field set up by an interdigital electrode structure as in Figure 13-19 gives rise to the periodic variation of the index of refraction. The situation is equivalent formally to that of Bragg scattering from a sound wave, discussed in Chapter 12, where the index modulation was caused by the acoustic strain.

A necessary condition for diffraction from a mode with a propagation vector $\boldsymbol{\beta}_i$ into a mode with $\boldsymbol{\beta}_d$ is, in analogy to (12.2-1),

$$\boldsymbol{\beta}_d = \boldsymbol{\beta}_i + \mathbf{a}_G \frac{2\pi}{\Lambda} m \qquad m = 1, 2, 3 \cdots \qquad \textbf{(13.7-17)}$$

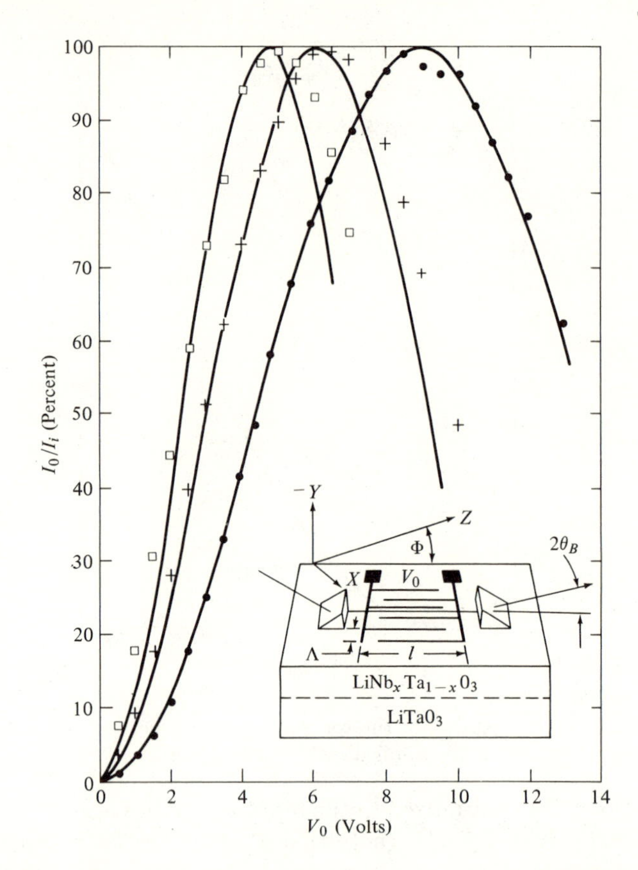

Figure 13-19 Schematic diagram of an electrooptic grating modulator in a $LiNb_xTa_{1-x}O_3$-$LiTaO_3$ waveguide. The input mode, which is coupled into the waveguide via a prism, is deflected through an angle $2\theta_B$ when a voltage is applied to the interdigital electrode structure. $\Lambda = 7.6\ \mu$m and $l = 0.3$ cm. The curves show percentages of light diffracted as a function of voltage. Open squares 4976 Å (He-Se laser), crosses 5598 Å (He-Se laser), and solid circles 6328 Å (He-Ne laser). The solid curves are plots of $\sin^2 (BV_0)$ normalized to the data at 75 percent. (After J. Hammer and W. Phillips, *Appl. Phys. Lett.*, vol. 24, p. 545, 1974.)

where Λ is the period of the spatial modulation and $\mathbf{a}_G$ is a unit vector normal to the equi-index lines (that is, normal to the electrodes in Figure 13-19). The vector $\boldsymbol{\beta}_i$ and $\boldsymbol{\beta}_d$ are in the plane of the waveguide.

The "momentum" diagram corresponding to (13.7-17) is shown in Figure 13-20 for the case of $m = 1$.

We take advantage of the formal equivalence of this case to that of Bragg diffraction from sound waves to write the diffraction efficiency, as in (12.3-16), in the form

$$\frac{I_d}{I_i} = \sin^2\left(\frac{\omega l}{2c}\Delta n\right) \qquad \textbf{(13.7-18)}$$

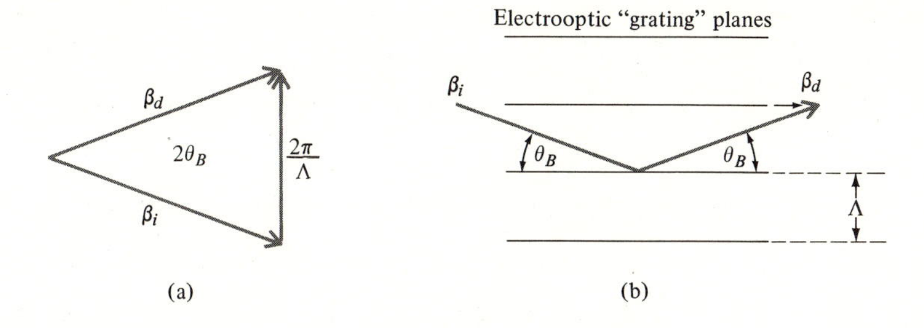

Figure 13-20 (**a**) A "momentum" diagram for diffraction of an input mode ($\boldsymbol{\beta}_i$) into the direction $\boldsymbol{\beta}_d$ by setting up (electrooptically) an index grating with a period Λ. (**b**) A top view of the waveguide plane showing the direction of the incident ($\boldsymbol{\beta}_i$), diffracted ($\boldsymbol{\beta}_d$) beams as well as the grating planes.

where Δn is the amplitude of the index modulation and is related to the appropriate Fourier component ($m = 1$ for first order Bragg diffraction), E_1 of the low frequency electric field by (9.1-10), that is

$$\Delta n = \frac{n^3}{2} r E_1 \tag{13.7-19}$$

where r is a suitable electrooptic tensor element which depends on the crystal orientation. Combining the last two equations leads to

$$\frac{I_d}{I_i} = \sin^2 \left(\frac{\pi n^3 r l}{2\lambda} E_1 \right) \tag{13.7-20}$$

Since E_1 is proportional to the applied voltage V_0 the diffraction efficiency, (13.7-20) can be written as

$$\frac{I_d}{I_i} = \sin^2 (B V_0) \tag{13.7-21}$$

with B inversely proportional to the optical wavelength λ.

The experimental transmission curves in Figure 13-19 are in agreement with (13.7-21).

13.8 Directional Coupling

Exchange of power between guided modes of adjacent waveguides is known as directional coupling. Waveguide directional couplers perform a number of useful functions in thin-film devices, including power division, modulation, switching, frequency selection, and polarization selection.

Waveguide coupling can be treated by the coupled-mode theory. Consider the case of the two planar waveguides illustrated in Figure 13-21. Refractive index distributions for the two guides in the absence of coupling are given by $n_a(x)$ and $n_b(x)$. The transverse electric field distribution for a particular guided mode of waveguide **a** alone and a particular

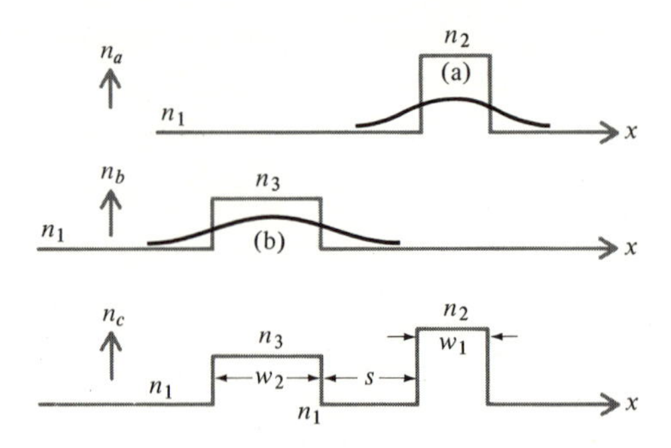

Figure 13-21 Spatial variation of the refractive index for uncoupled waveguides $n_a(x)$ and $n_b(x)$, and for a parallel waveguide structure $n_c(x)$.

mode of waveguide **b** alone will be denoted by $\mathcal{E}_y^{(a)}(x)$ and $\mathcal{E}_y^{(b)}(x)$, and their propagation constants by β_a and β_b. The field in the coupled-guide structure with an index $n_c(x)$ (for propagation in the positive z direction) is approximated by the sum of the unperturbed fields

$$E_y = A(z)\mathcal{E}_y^{(a)}(x)e^{i(\omega t-\beta_a z)} + B(z)\mathcal{E}_y^{(b)}(x)e^{i(\omega t-\beta_b z)} \tag{13.8-1}$$

In the absence of coupling—that is, if the distance between guides a and b were infinite—$A(z)$ and $B(z)$ do not depend on z and will be independent of each other, since each of the two terms on the right side of (13.8-1) satisfies the wave equation (13.3-1) separately.

The perturbation polarization responsible for the coupling is calculated by substituting (13.8-1) into (13.3-2) and (13.3-3). The result is

$$P_{\text{pert}} = e^{i\omega t}\epsilon_0[\mathcal{E}_y^{(a)}A(z)(n_c^2(x) - n_a^2(x))e^{-i\beta_a z} + \mathcal{E}_y^{(b)}B(z)(n_c^2(x) - n_b^2(x))e^{-i\beta_b z}] \tag{13.8-2}$$

where $n_c(x)$ is the index profile of the two-guide structure. Substituting (13.8-2) in (13.3-9) and integrating over x gives

$$\begin{aligned} \frac{dA}{dz} &= -i\kappa_{ab}Be^{-i(\beta_a-\beta_b)z} - iM_aA \\ \frac{dB}{dz} &= -i\kappa_{ba}Ae^{i(\beta_a-\beta_b)z} - iM_bB \end{aligned} \tag{13.8-3}$$

where

$$\kappa_{\substack{ab\\ba}} = \frac{\omega\epsilon_0}{4}\int_{-\infty}^{\infty}[n_c^2(x) - n_{\substack{b\\a}}^2(x)]\mathcal{E}_y^{(a)}\mathcal{E}_y^{(b)}\,dx \tag{13.8-4}$$

$$M_{(a,b)} = \frac{\omega\epsilon_0}{4}\int_{-\infty}^{\infty}[n_c^2(x) - n_{(a,b)}^2(x)](\mathcal{E}_y^{(a,b)})^2\,dx \tag{13.8-5}$$

The terms M_a and M_b represent a small correction to the propagation constants β_a and β_b, respectively, due to the presence of the second guide. So if we take the total field as

$$E_y = A(z)\mathcal{E}_y^{(a)}e^{i[\omega t-(\beta_a+M_a)z]} + B(z)\mathcal{E}_y^{(b)}e^{i[\omega t-(\beta_b+M_b)z]}$$

instead of (13.8-1), Equations (13.8-3) become

$$\frac{dA}{dz} = -i\kappa_{ab}Be^{-i2\delta z}$$
$$\frac{dB}{dz} = -i\kappa_{ba}Ae^{i2\delta z} \tag{13.8-6}$$

where

$$2\delta = (\beta_a + M_a) - (\beta_b + M_b)$$

The solution of (13.8-6) subject to a single input at guide b ($B(0) = B_0$, $A(0) = 0$) is given by (13.7-12). In terms of powers $P_a = AA^*$ and $P_b = BB^*$ in the two guides, the solution in the case $\kappa_{ab} = \kappa_{ba}$ becomes

$$P_a(z) = P_0\frac{\kappa^2}{\kappa^2+\delta^2}\sin^2[(\kappa^2+\delta^2)^{1/2}z]$$
$$P_b(z) = P_0 - P_a(z) \tag{13.8-7}$$

where $P_0 = |B(0)|^2$ is the input power to guide b. Complete power transfer occurs in a distance $L = \pi/2\kappa$ provided $\delta = 0$ (that is, equal phase velocities in both modes). For $\delta \neq 0$, the maximum fraction of power that can be transferred is from (13.8-7)

$$\frac{\kappa^2}{\kappa^2+\delta^2} \tag{13.8-8}$$

The coupling constant κ is given by (13.8-4). It can be evaluated straightforwardly using the field expressions (13.2-3) in the case of TE modes. In the special case of identical waveguides, $h_1 = h_2$ and $p_1 = p_2$ in Figure 13-21, one obtains

$$\kappa = \frac{2h^2pe^{-ps}}{\beta(w+2/p)(h^2+p^2)} \tag{13.8-9}$$

The extension to channel waveguide couplers which are confined in the y, as well as in the x, direction is simple and is discussed in Reference [20]. In the well-confined case $w \gg 2/p$ and (13.8-9) simplifies to [21]

$$\kappa = \frac{2h^2pe^{-ps}}{\beta w(h^2+p^2)} \tag{13.8-10}$$

A typical value of κ obtained at $\lambda \sim 1$ μm with w, $s \sim 3$ μm, and $\Delta n \sim 5 \times 10^{-3}$ is $\kappa \sim 5$ cm^{-1} so that coupling distances are of the order of magnitude of $\kappa^{-1} \approx 2$ mm.

A form of an electrooptic switch based on directional coupling [20] is as follows.

The length L of the coupler is chosen so that for $\delta = 0$ (that is, synchronous case) $\kappa L = \pi/2$. From (13.8-7) it follows that all the input power to guide b exits from guide a at $z = L$. The switching is achieved by applying an electric field to guide a (or b) in such a way as to change its propagation constant until

$$\delta L = \tfrac{1}{2}(\beta_a - \beta_b)L = \frac{\sqrt{3}}{2}\pi \tag{13.8-11}$$

that is, $\delta = \sqrt{3}\kappa$. It follows from (13.8-7) that at this value of δ

$$P_a = 0 \qquad P_b = P_0$$

that is, the power reappears at the output of guide b. A control of δ can thus be used to achieve any division of the powers between the outputs of guides a and b.

A power switch based on this principle has been demonstrated in GaAs [23]. Figure 13-22 shows the output powers P_a and P_b as a function of the applied voltage.

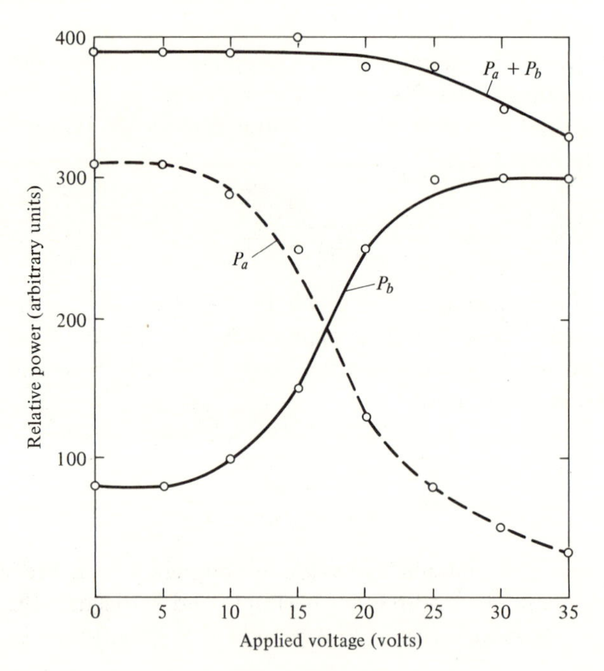

Figure 13-22 Variation of the guides' power in a directional coupler as a function of the applied field. (After Reference [23].)

A multiguide directional coupler such as that shown in Figure 13-15 is described by a set of equations

$$\frac{dA_n}{dz} = -i\kappa A_{n-1} - i\kappa A_{n+1} \qquad \textbf{(13.8-12)}$$

which is an obvious extension of (13.8-6) to the multimode synchronous case ($\delta = 0$) when only adjacent modes couple to each other. The solution of (13.8-12) in the case of a single input, that is, $A_n(0) = 1, n = 0, A_n(0) = 0$ $n \neq 0$ is [22]

$$A_n(z) = (-i)^n J_n(2\kappa z) \qquad \textbf{(13.8-13)}$$

where J_n is the Bessel function of order n.

A directional coupler based on this principle is shown in Figure 13-23(a). The predicted Bessel function distribution of the intensity at various propagation distances is shown in Figure 13-23(b).

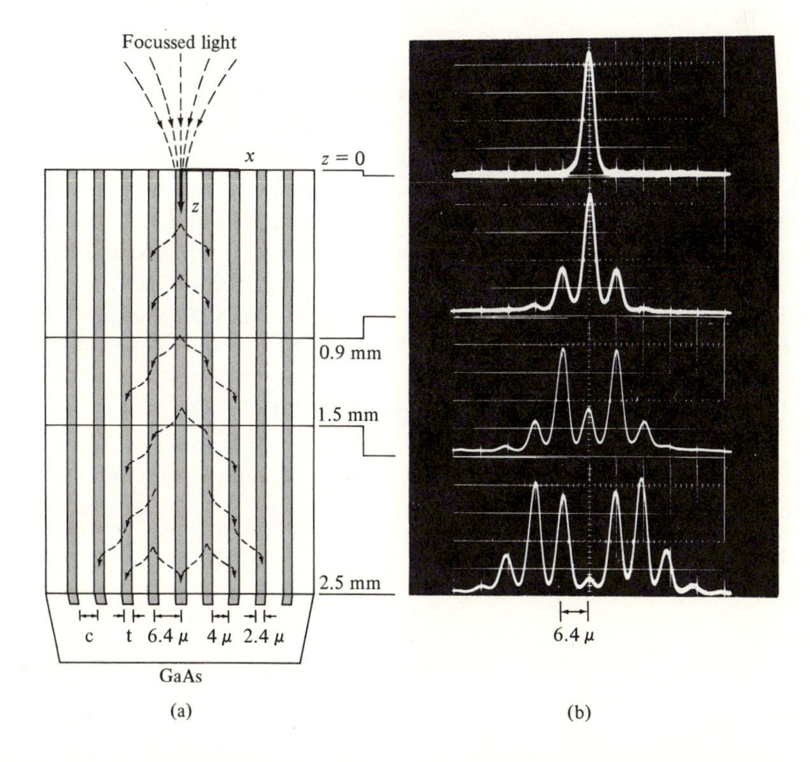

Figure 13-23 (**a**) Sketch of channel optical waveguide directional coupler showing flow of light energy into adjacent channels. (**b**) Measured guided-light intensity profiles at various lengths. The profiles have been displayed relative to the sketch at the proper value of z. Intensity scale is arbitrary. The guides were produced by proton implantation into p^+-GaAs crystal. (After Reference [22].)

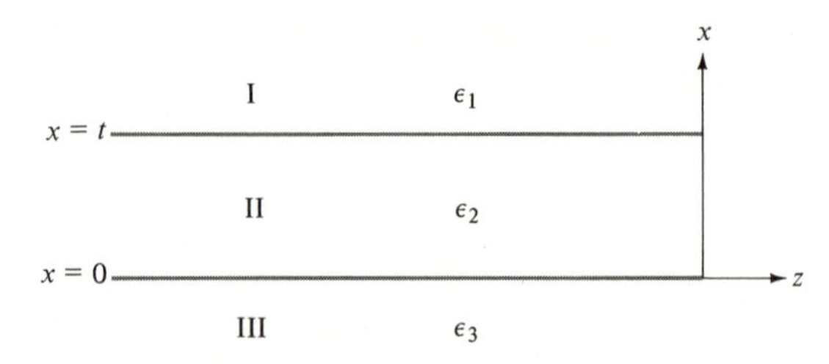

Figure 13-24 A "leaky" waveguide in which $\epsilon_3 > \epsilon_2 > \epsilon_1$.

13.9 Leaky Dielectric Waveguides

In Section 13.1 we derived the properties of the TE and TM modes propagating in thin dielectric waveguides. It was shown that a necessary condition for confined guiding is that (referring to Figure 13-1) the inner layer index, n_2, satisfy

$$n_3 < n_2 > n_1 \tag{13.9-1}$$

If (13.9-1) is not satisfied, the wave intensity increases exponentially with x in medium 1 or medium 2 or in both. A wave fed into such a waveguide will thus proceed to lose power by "leaking" to the continuum and will attenuate with z (the propagation direction).

A number of investigators [24] have recognized that the losses described above can, under certain circumstances, be quite small and the guides may thus be suitable for laser applications [25]. Recently, an increasing number of lasers—the so-called "waveguide lasers"—using "leaky" waveguides as the transmission media have been demonstrated [26], [27], and [28].

In the following discussion we will derive the TE mode characteristics and loss constants for such modes.

The model used is shown in Figure 13-24. We choose the case where $\epsilon_3 > \epsilon_2 > \epsilon_1$ ($\epsilon_i \equiv \epsilon_0 n_i^2$).

In the two-dimensional case ($\partial/\partial y = 0$) we assume a TE mode in the form

Region I
$$E_y = A \exp [i(h_1 x + \gamma z - \omega t)] \qquad x \geq t$$

Region II
$$E_y = [B \cos (h_2 x) + C \sin (h_2 x)] \exp [i(\gamma z - \omega t)] \qquad 0 \leq x \leq t \tag{13.9-2}$$

Region III
$$E_y = D \exp [i(-h_3 x + \gamma z - \omega t)] \qquad x \leq 0$$

The wave equation (13.1-1) can thus be written as

$$\left(\gamma^2 + \frac{\partial^2}{\partial x^2}\right) E_y + k_i^2 E_y = 0 \tag{13.9-3}$$

where

$$k_i^2 \equiv \omega^2\mu\epsilon_i = k_0^2 n_i^2 \qquad i = 1, 2, 3$$

Applying (13.9-3) to regions I, II, and III results in

$$\begin{aligned} h_1^2 &= k_2^2\left(\frac{\epsilon_1}{\epsilon_2}\right) - \gamma^2 \\ h_2^2 &= k_2^2 - \gamma^2 \\ h_3^2 &= \frac{\epsilon_3}{\epsilon_2} k_2^2 - \gamma^2 \end{aligned} \tag{13.9-4}$$

Matching E_y and $\partial E_y/\partial x$ at $x = 0$ and $x = t$ leads to

$$\begin{aligned} A \exp(ih_1t) - B\cos(h_2t) - C\sin(h_2t) &= 0 \\ h_1A\exp(ih_1t) - iBh_2\sin(h_2t) + iCh_2\cos(h_2t) &= 0 \\ B - D &= 0 \\ ih_2C - h_3D &= 0 \end{aligned} \tag{13.9-5}$$

Equating the determinant of the coefficients of A, B, C, and D in (13.9-5) to zero gives

$$\tan(h_2t) = -i\,\frac{h_1h_2 + h_3h_2}{h_2^2 + h_1h_3} \tag{13.9-6}$$

We note here that the determinantal equation (13.9-6) can be obtained directly from (13.2-5) by substituting $q \to -ih_1$, $h \to h_2$, $p \to -ih_3$. This correspondence between the two sets of constants follows from comparing the form of the confined mode solution (13.2-3) to the "leaky" mode solution (13.9-2).

Equation 13.9-6 can be solved together with (13.9-4) for h_1, h_2, h_3, and γ. We will obtain an approximate solution by assuming that the waveguide is "large" so that $k_2 \gg h_2$. In this case it follows from the second of (13.9-4) that

$$\gamma \to k_2$$

and

$$\begin{aligned} h_1 &\approx ik_2\left(1 - \frac{\epsilon_1}{\epsilon_2}\right)^{1/2} \\ h_3 &\approx k_2\left(\frac{\epsilon_3}{\epsilon_2} - 1\right)^{1/2} \end{aligned} \tag{13.9-7}$$

so that $h_3, h_1 \gg h_2$. We can thus approximate (13.9-6) by

$$h_2 t = \tan^{-1}\left[i\left(-\frac{h_2}{h_1} - \frac{h_2}{h_3}\right)\right] + n\pi \qquad n = \text{integer}$$

$$\simeq \tan^{-1}\left\{i\left[\frac{-h_2}{ik_2(1-\epsilon_1/\epsilon_2)^{1/2}} - \frac{h_2}{k_2(\epsilon_3/\epsilon_2 - 1)^{1/2}}\right]\right\} + n\pi$$

$$= \tan^{-1}\left(-\frac{h_2}{k_2} f_2 - i\frac{h_2}{k_2} f_3\right) + n\pi \tag{13.9-8}$$

with

$$f_2 \equiv \left(\frac{1}{1-\epsilon_1/\epsilon_2}\right)^{1/2} \qquad f_3 \equiv \left(\frac{1}{\epsilon_3/\epsilon_2 - 1}\right)^{1/2} = \left(\frac{1}{n_3^2/n_2^2 - 1}\right)^{1/2}$$

since f_2 and f_3 are o(1) and $k_2 \gg h_2$, (13.9-8) can be written as

$$h_2 t \left(1 + \frac{f_2}{k_2 t} + i\frac{f_3}{k_2 t}\right) = n\pi \qquad n = 1, 2, 3 \cdots$$

$$h_2 t \approx n\pi\left[1 - \frac{f_2}{k_2 t} - i\frac{f_3}{k_2 t}\right] \tag{13.9-9}$$

Substituting (13.9-9) into the second of (13.9-4) gives

$$\gamma_{\text{TE}} = k_2\left[1 - \frac{n^2\pi^2}{2k_2^2 t^2}\left(1 - \frac{2f_2}{k_2 t}\right)\right] + i\frac{n^2\pi^2 f_3}{k_2^2 t^3} \tag{13.9-10}$$

The exponential intensity decay coefficient is thus

$$\alpha_{\text{TE}} = 2\,\text{Im}\,\gamma_{\text{TE}} = \frac{2k_2 n^2\pi^2 f_3}{(k_2 t)^3} \qquad n = 1, 2, 3 \cdots \tag{13.9-11}$$

for $\Delta n \equiv n_3 - n_2 \ll n_2, n_3$

$$\alpha_{\text{TE}} = \frac{n^2}{2\lambda(t/\lambda)^3 n_2^2\sqrt{2\Delta n/n_2}} \tag{13.9-12}$$

This result was obtained by [29]. We thus find that the loss constant decreases as t^{-3}.

Some typical loss values for $n = 1$, $n_2 = 3.3$, $\Delta n = 0.1$, and $\lambda = 0.8\ \mu$m are

$t(\mu\text{m})$	0.8	2.4	4	6.4	8	10.4	50
$\alpha_{\text{TE}}(\text{cm}^{-1})$	2331	86	18	4.6	2.33	1.00	8.5×10^{-3}

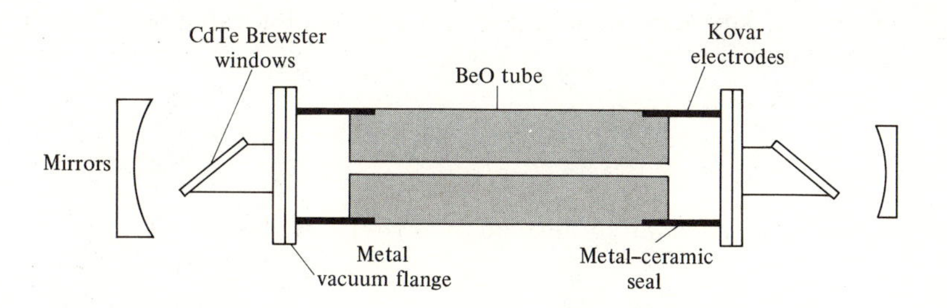

Figure 13-25 Construction details of a BeO capillary bore laser. (After Reference [28].)

We thus find that for $t > 10\lambda$, $\alpha_{\text{TE}}\lambda \lesssim 10^{-4}$. Such losses can easily be overcome by the gain of most laser media ([26], [28]).

A schematic diagram of a CO_2 10.6-μm waveguide laser, using the leaky mode described above, is shown in Figure 13-25. The waveguiding is accomplished in a BeO capillary tube.

PROBLEMS

13-1 Derive Equation 13.2-7.

13-2 Show that the form of Equation (13.7-10) is consistent with the conservation of the modes' power.

13-3 Derive the equations in (13.7-12).

13-4 Derive an expression for the modulation power of a transverse electrooptic waveguide modulator of length L and cross section $2\lambda \times 2\lambda$ (λ is the vacuum wavelength of the light). Compare to the bulk result ([9], p. 325). Estimate the power requirement for a $LiNbO_3$ modulator at $\lambda = 1\ \mu$m, $L = 5$ mm.

13-5 Derive the condition for distributed feedback laser oscillation for the case of gain perturbation, that is $\gamma_1 \neq 0$, $n_1 = 0$. Compare with the result of Reference 17.

13-6 Prove the orthogonality relation Equation (13.2-8).

13-7 Compare the ratio of the power propagating in the regions with an index n_1 to that of the total power for a TE_m and TM_m mode.

13-8 Show that in the case of electrooptic mode coupling in which $\beta_m^{\text{TM}} \neq \beta_m^{\text{TE}}$ (see Equation (13.7-10)), one can use a z periodic electrooptic constant or electric field to obtain phase matched operation. How would you accomplish this in practice? (Be bold and invent freely.)

13-9 Show that in the case of coupling between modes which carry power in opposite directions the conservation of total power condition takes the form of

$$\frac{d}{dz}(|A|^2 - |B|^2) = 0$$

which can be satisfied only if instead of (13.7-10) we have

$$\frac{dA}{dz} = \kappa_{ab} B e^{-i(\beta_B - \beta_A)z}$$

$$\frac{dB}{dz} = \kappa_{ab}^* A e^{i(\beta_B - \beta_A)z}$$

Compare these equations with the contradirectional coupling case described by Equations (13.5-1).

REFERENCES

[1] Yariv, A., and R. C. C. Leite, "Dielectric waveguide mode of light propagation in *p-n* junctions," *Appl. Phys. Lett.*, vol. 2, p. 55, 1963.

[2] Osterberg, H., and L. W. Smith, "Transmission of optical energy along surfaces," *J. Opt. Soc. Amer.*, vol. 54, p. 1073, 1964.

[3] Hall, D., A. Yariv, and E. Garmire, "Optical guiding and electro-optic modulation in GaAs epitaxial layers," *Opt. Comm.*, vol. 1, p. 403, 1970.

[4] Shubert, R. and J. H. Harris, "Optical surface waves on thin films and their application to integrated data processors," *IEEE Trans. Microwave Theory Tech.* (1968 Symp. issue), vol. MTT-16, pp. 1048–1054, Dec. 1968.

[5] Miller, S. E., "Integrated optics, an introduction," *Bell Syst. Tech. J.*, vol. 48, p. 2059, 1969.

[6] Goell, J. E., "A circular harmonic computer analysis for rectangular dielectric waveguides," *Bell Syst. Tech. J.*, vol. 48, p. 2133, 1968.

[7] Marcatili, E. A. J., "Dielectric rectangular waveguide and directional couplers for integrated optics," *Bell Syst. Tech. J.*, vol. 48, p. 2071, 1969.

[8] Lotspeich, J. F., "Explicit general eigenvalue solutions for dielectric slab waveguides," *Appl. Opt.*, vol. 14, p. 327, 1975.

[9] Chapter 14, this book.

[10] Hammer, J. M., D. J. Channin, and M. T. Duffy, "High speed electrooptic grating modulators," *RCA Technical Report*, unpublished.

[11] Tien, P. K., R. Ulrich, and R. J. Martin, "Modes of propagating light in thin deposited semiconductor films," *Appl. Phys. Lett.*, vol. 144, 1969.

[12] Yariv, A., "Coupled mode theory for guided wave optics," *IEEE J. of Quant. Elec.*, vol. 9, p. 919, 1973.

[13] Kuhn, L., M. L. Dakss, P. F. Heidrich, and B. A. Scott, "Deflection of optical guided waves by a surface acoustic wave," *Appl. Phys. Lett.*, vol. 17, p. 265, 1970.

[14] Dixon, R. W., "The photoelastic properties of selected materials and their relevance to acoustic light modulators and scanners," *J. Appl. Phys.*, vol. 38, p. 5149, 1967.

[15] Stoll, H., and A. Yariv, "Coupled mode analysis of periodic dielectric waveguides," *Opt. Commun.*, vol. 8, p. 5, 1973.

[16] Flanders, D. C., H. Kogelnik, R. V. Schmidt, and C. V. Shank, "Grating filters for thin film optical waveguides," *Appl. Phys. Lett.*, vol. 24, p. 194, 1974.

[17] Kogelnik, H., and C. V. Shank, "Coupled wave theory of distributed feedback lasers," *J. Appl. Phys.*, vol. 43, 2328, 1972.

[18] Nakamura, M., A. Yariv, H. W. Yen, S. Somekh, and H. L. Garvin, "Optically pumped GaAs surface laser with corrugation feedback," *Appl. Phys. Lett.*, vol. 22, p. 515, 1973.

[19] K. Aiki, M. Nakamura, J. Umeda, A. Yariv, A. Katzir, and H. W. Yen, "GaAs-GaAlAs distributed feedback laser with separate optical and carrier confinement," *Appl. Phys. Lett.*, vol. 27, p. 145, 1975.

[20] Somekh, S., Ph.D. Thesis, California Institute of Technology, 1973.

[21] Marcatili, E. A. J., "Dielectric rectangular waveguides and directional couplers for integrated optics," *Bell Syst. Tech. J.*, vol. 48, p. 2071, 1969.

[22] Somekh, S., E. Garmire, A. Yariv, H. L. Garvin, and R. G. Hunsperger, "Channel optical waveguide directional couplers," *Appl. Phys. Lett.*, vol. 22, p. 46, 1973.

[23] Campbell, J. C., F. A. Blum, D. W. Shaw, and K. L. Lawley, "GaAs directional coupler switch," *Appl. Phys. Lett.*, vol. 27, 1975.

[24] Marcatili, E. A. J., and R. A. Schmeltzer, "Hollow metallic and dielectric waveguides for long distance optical transmission and lasers," *Bell Syst. Tech. J.*, vol. 43, pp. 1783–1809, 1974.

[25] Steffen, H., and F. K. Kneubuhl, "Dielectric tube resonators for infrared and submillimeterwave lasers," *Phys. Lett.*, vol. 27A, pp. 612–613, 1968.

[26] Smith, P. W., "A waveguide gas laser," *Appl. Phys. Lett.*, vol. 19, p. 132, 1971.

[27] Bridges, T. J., E. G. Burkhardt, and P. W. Smith, "CO_2 waveguide lasers," *App. Phys. Lett.*, vol. 20, p. 403, 1972.

[28] Abrams, R. L., and W. B. Bridges, "Characteristics of sealed-off CO_2 waveguide lasers," *IEEE J. of Quant. Elec.*, vol. 9, p. 940, 1973.

[29] Hall, D. B., and C. Yeh, "Leaky waves in heteroepitaxial films," *J. of Appl. Phys.*, vol. 44, p. 2271, 1973.

SUPPLEMENTARY REFERENCES

[1] For a collection of reprints up to 1972 in integrated optics, *Integrated Optics*, D. Marcuse, ed. (New York: IEEE Press, 1972).

[2] Tien, P. K., "Light waves in thin films and integrated optics," *Appl. Opt.*, vol. 10, p. 2395, 1971.

[3] Tamir E., editor, *Integrated Optics*, Berlin: Springer-Verlag, 1975.

CHAPTER

14

Two Laser Applications

14.1 Design Considerations Involving an Optical Communication System

One potential area of application for lasers is that of communication between satellites. One reason is that in this case the problem of atmospheric absorption and distortion of laser beams is of no concern. In addition, the high directionality available with laser beams can be utilized effectively.

To be specific, we will consider a communication link between a system of three synchronous (24-hour orbit) satellites, as shown in Figure 14-1. Each satellite should be able to transmit and receive simultaneously.

Specifically, we agree on the following operating conditions:

1. The operating wavelength is $\lambda = 0.53\ \mu$m. This wavelength, which can be obtained by doubling the output of a Nd^{3+}:YAG laser (see Section 7.3), is chosen because of the high quantum efficiency of photomultiplier tubes at this wavelength (see Figure 11-4).

2. The detection will be performed by a photomultiplier tube operating in the video (that is, no local oscillator) mode, as described in Section 11.3.

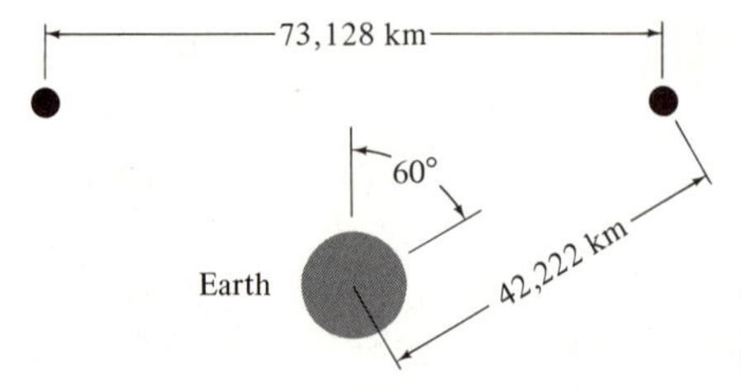

Figure 14-1 The disposition of three earth satellites with synchronous (24-hour) orbits.

3. The modulation signal will be impressed on the optical beam by an electrooptic modulator. The modulation signal will consist of a microwave subcarrier with a center frequency of $\nu_m = 3 \times 10^9$ Hz and sidebands[1] (caused by the information modulation) between $\nu_{\min} = 2.5 \times 10^9$ Hz and $\nu_{\max} = 3.5 \times 10^9$ Hz. The information bandwidth is thus $\Delta\nu = 10^9$ Hz.

4. The electrooptic crystal modulator will be used in the transverse mode (see Section 9.5) and will have an electrooptic coefficient of $r \simeq 4 \times 10^{-11}$ MKS) and a microwave dielectric constant of $\epsilon = 55\epsilon_0$. The peak modulation index is $\Gamma_m = \pi/3$.

5. The collimating lens and the receiving lens will have radii of 10 cm.

6. The signal-to-noise power ratio at the output of the amplifier following the receiver photomultiplier tube should be 10^3.

Our main concern is that of calculating the total primary (dc) power that the satellite must supply in order to meet the foregoing performance specifications. We will calculate first the optical power level of the transmitted beam and then the modulation power needed to meet these performance criteria.

Synchronous satellites. A synchronous satellite has an orbiting period of 24 hours, so its position relative to the earth is fixed. To find the distance from the earth to the satellite, we equate the centrifugal force caused

[1] The information signal $f(t)$, which may consist of, as an example, the video output of a vidicon television-camera tube, is impressed as modulation on the microwave signal. This modulation can take the form of AM, FM, PCM, or other types of modulation. The modulated microwave carrier is then applied to the electrooptic crystal to modulate the optical beam in one of the ways discussed in Chapter 9.

by the satellite's rotation to the gravitational attraction force

$$m\,\frac{v^2}{R_{E-S}} = mg\,\frac{(R_{\text{earth}})^2}{(R_{E-S})^2} \tag{14.1-1}$$

where v is the satellite's velocity, m its mass, g the gravitational acceleration at the earth's surface, R_{E-S} the distance from the center of the earth to the satellite, and R_{earth} the earth's radius. The synchronous orbit constraint (that is, a 24-hour period) is

$$\frac{v}{R_{E-S}} = \frac{2\pi}{24 \times 60 \times 60}$$

We use it in (13.1-1) to solve for R_{E-S}, obtaining $R_{E-S} = 42{,}222$ km.

We employ three satellites so as to obtain coverage of the earth's surface, as shown in Figure 14-1. The distance between two satellites is $R = 73{,}128$ km.

Calculation of the transmitted power. First we will derive an expression relating the received power to the transmitted power as a function of the transmitted beam diameter, the receiving aperture diameter, and the distance R between the transmitter and the receiver.

If the transmitted power P_T is beamed into a solid angle Ω_T and if the receiving aperture subtends a solid angle Ω_R at the transmitter, the power received is

$$P_R = P_T\,\frac{\Omega_R}{\Omega_T} \tag{14.1-2}$$

The transmitted beam diffracts with a half-apex angle θ_{beam}, as shown in Figure 14-2. This angle is related to the minimum beam radius ω_0 (spot size) by (3.2-18):

$$\theta_{\text{beam}} = \frac{\lambda}{\pi\omega_0} \tag{14.1-3}$$

The corresponding solid angle is $\Omega_T = \pi(\theta_{\text{beam}})^2$. If we choose ω_0 to be

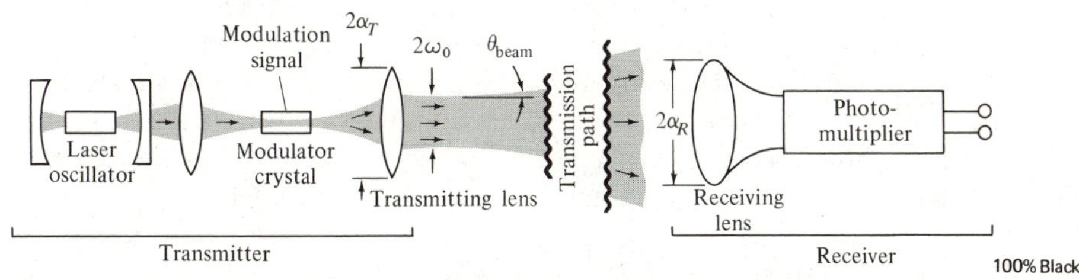

Figure 14-2 An optical communication link consisting of a laser oscillator, an electrooptic modulator, a collimating (transmitting) lens, a transmission medium, a receiving lens, and a receiver using a photomultiplier.

equal to the radius a_T of the transmitting lens, we obtain

$$\Omega_T = \frac{\lambda^2}{\pi a_T^2} \tag{14.1-4}$$

The receiving solid angle is

$$\Omega_R = \frac{\pi a_R^2}{R^2} \tag{14.1-5}$$

where a_R is the radius of the receiving lens and R is the distance between transmitter and receiver. Using (14.1-4) and (14.1-5) in (14.1-2) leads to

$$P_T = P_R \frac{\lambda^2 R^2}{\pi^2 a_T^2 a_R^2} \tag{14.1-6}$$

According to (11.3-5) and (11.3-9), the signal-to-noise power ratio at the output of a photomultiplier operating in the quantum-limited region (that is, the main noise contribution is the shot noise generated by the signal itself) is

$$\frac{S}{N} = \frac{2(P_R e\eta/h\nu)^2 G^2}{2G^2 e\bar{i}_e} = \frac{P_R \eta}{h\nu\Delta\nu}$$

where P_R is the optical power. Using $\lambda = 0.53\ \mu\text{m}$, $\eta = 0.2$, $\Delta\nu = 10^9$ Hz, and the required S/N value of 10^3, we obtain

$$P_R \simeq 2 \times 10^{-6} \text{ watt}$$

The required transmitted power is then calculated using (14.1-6), and $R = 7.31 \times 10^4$ meters, yielding

$$P_T \simeq 3 \text{ watts}$$

Calculation of the modulation power. In Section 9.6 we derived an expression for the power dissipated by an electrooptic modulator operated in the parallel RLC configuration shown in Figure 9-9. This power is given by

$$P = \frac{\Gamma_m^2 \lambda^2 d^2 \epsilon \Delta\nu}{4\pi l r^2 n^6} \tag{14.1-8}$$

where Γ_m is the peak electrooptic retardation, d the length of the side of the (square) crystal cross section, l the crystal length, r the appropriate electrooptic coefficient, $\Delta\nu$ the modulation bandwidth, and ϵ the dielectric constant of the crystal at the modulation frequency. To insure frequency-independent response of the crystal, the crystal length l is limited by transit-time considerations discussed in Section 9.6 to a length

$$l < \frac{c}{4\nu_{\max} n}$$

where n is the index of refraction and $\nu_{\max}$ is the highest modulation frequency. Using $\nu_{\max} = 3.5 \times 10^9$ Hz and $n = 2.2$, we obtain $l < 1$ cm.

We will consequently choose a crystal length of $l = 1$ cm. Having fixed l we may be tempted by (14.1-8) to use a crystal with a minimum thickness d. The choice of d is dictated, however, by the fact that we must be able to focus the laser beam into the crystal in such a way that its spread due to diffraction inside the crystal does not exceed the transverse dimension d. The situation is illustrated by Figure 14-3. If the beam is focused so that its waist occurs at the crystal mid-plane ($z = 0$), its radius at the two crystal faces ($z = -l/2$ and $l/2$) is given by (3.2-11) as

$$\omega\left(\frac{L}{2}\right) = \omega_0\left[1 + \left(\frac{l\lambda}{2\pi n}\right)^2 \omega_0^{-4}\right]^{1/2} \qquad (14.1\text{-}9)$$

Our problem is one of determining the value of ω_0 for which $\omega(z = l/2)$ is a minimum. The dimension d will then be chosen to be slightly larger than the minimum value of $2\omega(z = l/2)$. Setting the derivative of (14.1-9) (with respect to ω_0) equal to zero yields

$$(\omega_0)_{\min} = \sqrt{\frac{l\lambda}{2\pi n}}$$

and

$$\omega(z = l/2)_{\min} = \sqrt{\frac{l\lambda}{\pi n}}$$

so that choosing d_{optimum} as equal to $2\omega(z = l/2)_{\min}$

$$\left(\frac{d_{\text{optimum}}{}^2}{l}\right) = \frac{4\lambda}{\pi n} \qquad (14.1\text{-}10)$$

Substituting the last result for d^2/l in (14.1-8) yields an expression for the minimum modulation power

$$(P_{\min}) = \frac{\Gamma_m{}^2\lambda^3\epsilon\Delta\nu}{\pi^2 r^2 n_o{}^7} \qquad (14.1\text{-}11)$$

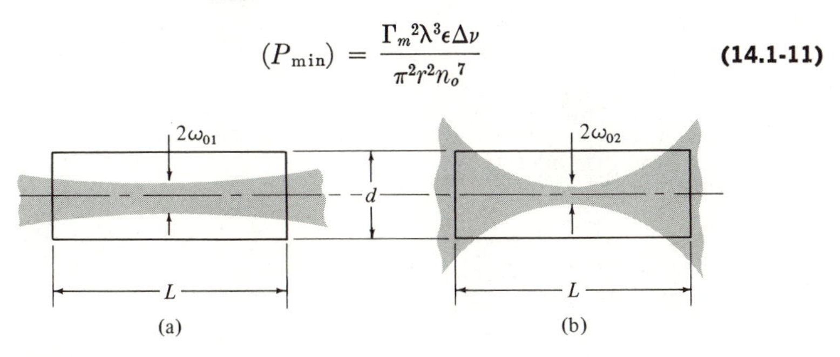

Figure 14-3 The problem of confining a fundamental Gaussian beam within a crystal of length l and height (and width) d. A decrease of the minimum beam radius from ω_{01} to ω_{02} causes the beam to expand faster and "escape" from the crystal.

Using $\Gamma_m = \pi/3$, $n = 2.2$, $r = 4 \times 10^{-11}$ (MKS), $\lambda = 0.53\ \mu$m, $\epsilon = 5$, and $\Delta\nu = 10^9$ yields

$$(P)_{\text{optimum}} = 0.0125 \text{ watts}$$

for the microwave modulation power under optimum focusing conditions.

We have thus determined that the laser power output should be approximately 3 watts and the modulation power around 0.0125 watt. If we assume that the efficiency of conversion of primary (dc) power to laser power is about 1 percent, each satellite will be required to supply approximately 300 watts of primary power.

13.2 Holography

One of the most important applications made practical by the availability of coherent laser radiation is holography, the science of producing images by wavefront reconstruction; see References [1]–[8]. Holography makes possible true reconstruction of three-dimensional images, magnified or reduced in size, in full color. It also makes possible the storage and retrieval of a large amount of optical information in a small volume.

Figure 14-4 illustrates the experimental setup used in making a simple hologram. A plane-parallel light beam illuminates the object whose hologram is desired. Part of the same beam is reflected from a mirror (at

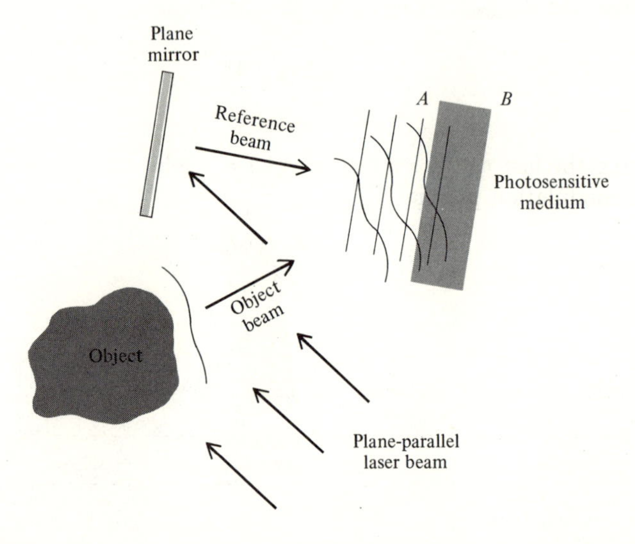

Figure 14-4 A hologram of an object can be made by exposing a photosensitive medium at the same time to coherent light, which is reflected diffusely from the object, and a plane-parallel reference beam, which is part of the same beam that is used to illuminate the object.

this point we refer to it as the reference beam) and is made to interfere within the *volume* of the photosensitive medium with the beam reflected diffusely from the object (object beam). The photosensitive medium is then developed and forms the hologram.

The image reconstruction process is illustrated in Figure 14-5. It is performed by illuminating the hologram with the same wavelength laser beam and in the same relative orientation that existed between the reference beam and the photosensitive medium when the hologram was made. An observer facing the far side (B) of the hologram will now see a three-dimensional image occupying the same spatial position as the original object. The image is, ideally, indistinguishable from the direct image of the laser-illuminated object.

The holographic process viewed as Bragg diffraction. To illustrate the basic process involved in holographic wavefront reconstruction, consider the simple case in which the two beams reaching the photosensitive medium in Figure 14-4 are plane waves. The situation is depicted in Figure 14-6. We choose the z axis as the direction of the bisector of the angle formed between the two propagation directions $\mathbf{k}_1$ and $\mathbf{k}_2$ of the reference and object plane waves inside the photosensitive layer. The x axis is contained in the plane of the paper. The electric fields of the two beams are taken as

$$\begin{aligned} e_{\text{object}}(\mathbf{r}, t) &= E_1 e^{i(\mathbf{k}_1 \cdot \mathbf{r} - \omega t)} \\ e_{\text{reference}}(\mathbf{r}, t) &= E_2 e^{i(\mathbf{k}_2 \cdot \mathbf{r} - \omega t)} \end{aligned} \tag{14.2-1}$$

From Figure 13-6 and the fact that $|\mathbf{k}_1| = |\mathbf{k}_2| = k$, we have

$$\begin{aligned} \mathbf{k}_1 &= \mathbf{a}_x k \sin \theta + \mathbf{a}_z k \cos \theta \\ \mathbf{k}_2 &= -\mathbf{a}_x k \sin \theta + \mathbf{a}_z k \cos \theta \end{aligned} \tag{14.2-2}$$

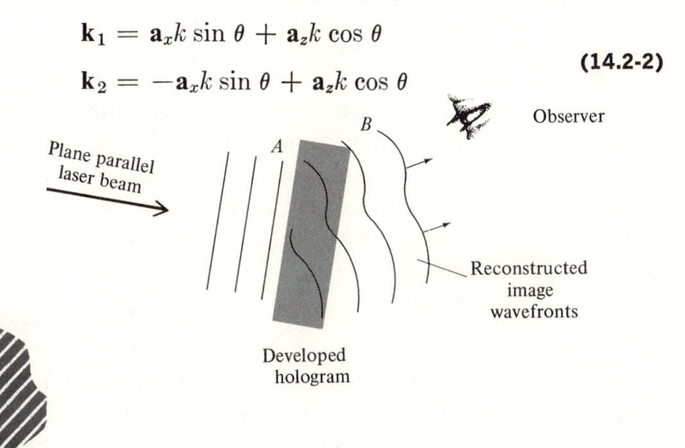

Figure 14-5 Wavefront reconstruction of the original image is usually achieved by illuminating the hologram with a laser beam of the same wavelength and relative orientation as the reference beam making it. An observer on the far side (B) sees a virtual image occupying the same space as the original subject.

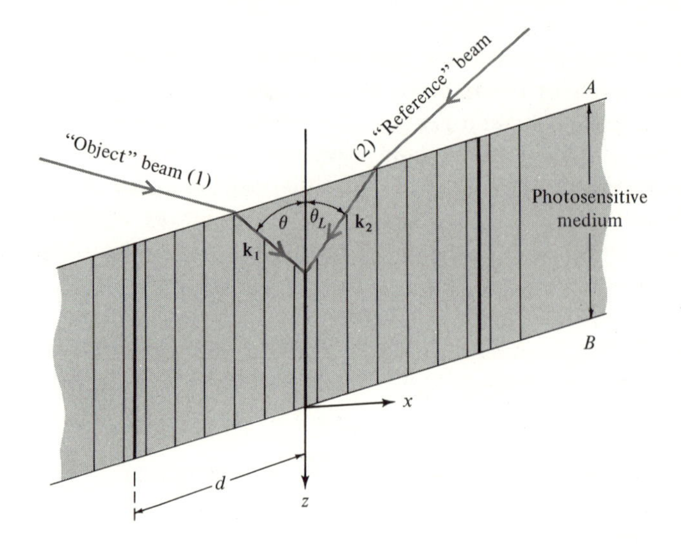

Figure 14-6 A sinusoidal "diffraction grating," produced by the interference of two plane waves inside a photographic emulsion. The density of black lines represents the exposure and hence the silver-atom density. The z direction is chosen as that of the bisector of the angle formed between the directions of propagation *inside* the photographic emulsion. It is not necessarily perpendicular to the surface of the hologram.

where $k = 2\pi/\lambda$, and $\mathbf{a}_x$ and $\mathbf{a}_z$ are unit vectors parallel to x and z, respectively.

The total complex field amplitude is the sum of the complex amplitudes of the two beams which, using (14.2-1) and (14.2-2) can be written as

$$E(x, z) = E_1 e^{ik(x \sin\theta + z \cos\theta)} + E_2 e^{ik(-x \sin\theta + z \cos\theta)} \tag{14.2-3}$$

If the photosensitive medium were a photographic emulsion, the exposure to the two beams and subsequent development would result in silver atoms developed out at each point in the emulsion in direct proportion to the time average of the square of the optical field. The density of silver in the developed hologram is thus proportional to $E(x, z)E^*(x, z)$, which, using (14.2-3), becomes

$$E(x, z)E^*(x, z) = E_1^2 + E_2^2 + 2E_1E_2 \sin(2kx \sin\theta) \tag{14.2-4}$$

The hologram is thus seen to consist of a sinusoidal modulation of the silver density. The planes $x =$ constant (that is, planes containing the bisector and normal to the plane of Figure 14-6) correspond to equidensity planes. The distance between two adjacent peaks of this spatial modulation pattern is, according to (14.2-4),

$$d = \frac{\pi}{k \sin\theta} = \frac{\lambda/n}{2 \sin\theta} \tag{14.2-5}$$

In the process of wavefront reconstruction the hologram is illuminated with a coherent laser beam. Since the hologram consists of a three-dimensional sinusoidal diffraction grating, the situation is directly analogous to the diffraction of light from sound waves, which was analyzed in Section 12.1. Applying the results of Bragg diffraction and denoting the wavelength of the light used in reconstruction (that is, in viewing the hologram) as λ_R, a diffracted beam exists *only* when the Bragg condition (12.1-4)

$$2d \sin \theta_B = \lambda_R / n_R \tag{14.2-6}$$

is fulfilled, where θ_B is the angle of incidence and of diffraction as shown in Figure 14-7. Substituting for d its value according to (14.2-5), we obtain

$$\sin \theta_B = \left(\frac{n}{n_R}\right) \frac{\lambda_R}{\lambda} \sin \theta \tag{14.2-7}$$

In the special case when $\lambda_R = \lambda$—that is to say, when the hologram is viewed with the same laser wavelength as that used in producing it—we have

$$\theta_B = \theta$$

so that wavefront reconstruction (that is, diffraction) results only when the beam used to view the hologram is incident on the diffracting planes at the same angle as the beam used to make the hologram. The diffracted beam emerges along the same direction ($\mathbf{k}_1$) as the original "object" beam, thus constituting a reconstruction of the latter.

We can view the complex beam reflected from the object toward the photographic emulsion when the hologram is made, as consisting of a "bundle" of plane waves each having a slightly different direction. Each one of these waves interferes with the reference beam, creating, after development, its own diffraction grating, which is displaced slightly in angle from that of the other gratings. During reconstruction the illuminating laser beam is chosen so as to nearly satisfy the Bragg condition

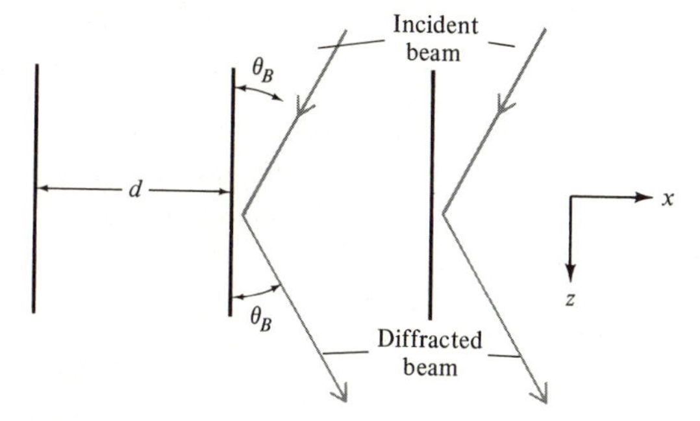

Figure 14-7 Bragg diffraction from a sinusoidal volume grating. The grating periodic distance d is the distance in which the grating structure repeats itself. In the case of a hologram we may consider the vertical lines in the figure as an edge-on view of planes of maximum silver density.

(14.2-6) for these gratings. Each grating gives rise to a diffracted beam along the same direction as that of the object plane wave that produced it, so the total field on the far side of the hologram (B) is identical to that of the object field.

Basic holography formalism. The point of view introduced above, according to which a hologram may be viewed as a volume diffraction grating is extremely useful in demonstrating the basic physical principles. A slightly different approach is to take the total field incident on the photosensitive medium as

$$A(\mathbf{r}) = A_1(\mathbf{r}) + A_2(\mathbf{r}) \tag{14.2-8}$$

where $A_1(\mathbf{r})$ may represent the complex amplitude of the diffusely reflected wave from the object while $A_2(\mathbf{r})$ is the complex amplitude of the reference beam. $A_2(\mathbf{r})$ is not necessarily limited to plane waves and may correspond to more complex wavefronts.

The intensity of the total radiation field can be taken, as in (14.2-4), to be proportional to

$$AA^* = A_1A_1^* + A_2A_2^* + A_1A_2^* + A_1^*A_2 \tag{14.2-9}$$

The first term $A_1A_1^*$ is the intensity I_1 of the light arriving from the object. If the object is a diffuse reflector, its unfocused intensity I_1 can be regarded as essentially uniform over the hologram's volume. $A_2A_2^*$ is the intensity I_2 of the reference beam. The change in the amplitude transmittance of the hologram ΔT can be taken as proportional to the exposure density so that

$$\Delta T \propto I_1 + I_2 + A_1A_2^* + A_1^*A_2$$

The reconstruction is performed by illuminating the hologram with the reference beam A_2 in the *same* relative orientation as that used during the exposure. Limiting ourselves to the portion of the transmitted wave modified by the exposure, we have

$$R = A_2\Delta T \propto (I_1 + I_2)A_2 + A_1^*A_2A_2 + I_2A_1 \tag{14.2-10}$$

The first term corresponds to a wavefront proportional to the reference beam. The second term, not being proportional to A_1, may be regarded as undesirable "noise." Since I_2 is a constant, the third term I_2A_1 corresponds to a transmitted wave that is proportional to A_1 and is thus a reconstruction of the object wavefront.

Some additional aspects of holography, which follow straight-forwardly from the formalism introduced above, are treated in the problems.

Holographic storage. The use of holography for the storage of a large number of images and for their retrieval can be best understood using the Bragg diffraction point of view.

We consider, for the sake of simplicity, the problem of recording (storing) holographically and then reconstructing two objects. The storage of a larger number of images is then accomplished by repeating the procedure used in recording the two images.

An exposure of the photosensitive medium is performed using the beam reflected from the first object and the reference beam as illustrated in Figure 14-4. Next, the first object is replaced by the second one, the photosensitive plate is rotated by a small angle $\Delta\theta$, and another exposure is taken. The plate is now developed and forms the hologram.

During exposure each object gave rise to a "diffraction grating" pattern. The two sets of diffraction planes are not parallel to each other, since the plate was rotated between the two exposures. A reconstruction of the first object is obtained when the hologram is illuminated with a laser beam in such a direction as to satisfy the Bragg condition with respect to first diffraction grating. If the same wavelength is used in making the hologram and in the image reconstruction, the image of the first object is reconstructed when the laser beam is incident on the hologram at the same angle as that of the reference beam (during exposure). A rotation $\Delta\theta$ of the hologram will cause the second set of diffraction planes (that due to the second object) to satisfy the Bragg condition with respect to the incident laser beam, thus giving rise to a reconstructed image of the second object.

PROBLEMS

14-1 Show that if a hologram is made using a wavelength λ but is reconstructed with a wavelength λ_R the reconstructed image is magnified by a factor of λ_R/λ with respect to the original object. [*Hint:* Consider the process of forming a real image by placing a lens on the output side (B) of the illuminated hologram and then determining the linear scale of the image in view of Equation (14.2-7).]

14-2 By considering a complex waveform as a superposition of plane waves show that a complex wave $A^*(r)$ is that obtained from A by making it retrace its path; that is, the wavefronts are identical but their direction of propagation is reversed. A^* is called the conjugate (waveform) of A.

14-3 **a.** Show that if the hologram is illuminated with a plane wave A_2^* instead of A_2 the reconstructed image is A_1^* instead of A_1.

b. Show that the reconstructed image A_1^* is real—that is, that A_1^* actually converges to an image. (*Hint:* Consider what happens to a bundle of rays originally emanating from a point on the object.)

c. Show that the reconstructed image A_1 observed when the hologram is illuminated by A_2 is virtual; that is, rays corresponding to a given image point do not cross unless imaged by a lens.

14-4 Consider the problem of making a hologram in which the reference and object beam are incident on the emulsion from two opposite sides. Draw the equidensity planes for the case where the beams are nearly antiparallel. Show that the viewing (reconstructing) of this beam is performed in the reflection mode; (that is, the viewer faces the side of the emulsion that is illuminated by the beam.

14-5 Show that in an infinitely thin hologram both virtual and real images can be reconstructed simultaneously. [*Hint:* Consider the problem of light scattering from a surface grating (as opposed to a volume grating).]

14-6 Calculate the reconstruction angle sensitivity $d\theta_B/d\lambda_R$ for transmission holograms (as described in the text) and in reflection holograms (as described in Problem 14-4). θ_B is the Bragg angle and λ_R is the wavelength used in reconstruction. Show that $d\theta_B/d\lambda_R$ is much larger in the case of the transmission hologram. Which hologram will yield better results when illuminated by white light?

■ REFERENCES

[1] Gabor, D., "Microscopy by reconstructed wavefronts," *Proc. Roy. Soc. (London)*, ser. A, vol. 197, p. 454, 1949.

[2] Leith, E. N., and J. Upatnieks, "Wavefront reconstruction with diffused illumination and three-dimensional objects," *J. Opt. Soc. Am.*, vol. 54, p. 1295, 1964.

[3] Collier, R. J., "Some current views on wavefront reconstruction," *IEEE Spectrum*, vol. 3., p. 67, July 1966.

[4] Stroke, G. W., *An Introduction to Coherent Optics and Holography*, 2d Ed. New York: Academic Press, 1969.

[5] DeVelis, J. B., and G. O. Reynolds, *Theory and Applications of Holography*. Reading, Mass.: Addison-Wesley, 1967.

[6] Smith, H. M., *Principles of Holography*. New York: Interscience, 1969.

[7] Goodman, J. W., *Introduction to Fourier Optics*. New York: McGraw-Hill, 1968.

[8] Yu, T. S. F., *Introduction to Diffraction Information Processing and Holography*. Cambridge, Mass: MIT Press, 1973.

APPENDIX

A Unstable Resonators—Electromagnetic Analysis

Unstable resonators are, according to (4.4-2), resonators operating in the region of the stability diagram where

$$\left(1-\frac{l}{R_1}\right)\left(1-\frac{l}{R_2}\right)>1, \quad \left(1-\frac{l}{R_1}\right)\left(1-\frac{l}{R_2}\right)<0$$

which causes the Gaussian beam radii to become infinite. Clearly, in this regime the end reflectors cannot be assumed to be infinite in extent, an assumption highly justified in the "stable" regime where the beam spot size at the mirror is typically very small (see numerical example in Section 4.3) compared to the mirror radius.

One way to account qualitatively for the finite extent of real-life mirrors is to simulate analytically the abrupt drop of the reflectivity to zero at the mirror's edge by some tapering of the reflectivity. If we choose a Gaussian tapering function, we find, not too surprisingly, that we can apply the self-consistent ABCD method of Section 4.5 to the case of unstable resonators.

Consider a Gaussian beam incident on a mirror with a radius of curvature R whose reflectivity is given by

$$\rho(r) = \rho_0 e^{-r^2/a^2} \tag{A-1}$$

where r is the radius measured from the center of the mirror. Let the incident beam possess a spot size ω_i and a radius of curvature R_i at the mirror position and let the medium through which it propagates have an index n. The complex beam parameter at the mirror is, according to (3.3-5),

$$\frac{1}{q_1} = \frac{1}{R_1} - i\frac{\lambda}{\pi\omega_i^2 n} \tag{A-2}$$

The reflected beam has a radius of curvature R_0 and spot size ω_0 where, using (3.3-8)

$$\frac{1}{R_0} = \frac{1}{R_i} - \frac{2}{R} \tag{A-3}$$

By multiplying the incident field distribution $\exp(-r^2/\omega_i^2)$ by $\rho(r)$ we find that the $1/e$ Gaussian spot size is modified from ω_i to ω_0, where

$$\frac{1}{\omega_0^2} = \frac{1}{\omega_i^2} + \frac{1}{a^2} \tag{A-4}$$

The Gaussian tapered mirror thus modifies not only the beam radius of curvature but the spot size ω_0 as well. In a uniform reflectivity mirror ($a^2 = \infty$) the mirror spot size does not change upon reflection.

The transformation properties (A-3) and (A-4) would follow from the ABCD law (3.3-9)

$$q_0 = \frac{Aq_i + B}{Cq_i + D}$$

provided we take the A, B, C, D, matrix as

$$\begin{vmatrix} A & B \\ C & D \end{vmatrix} = \begin{vmatrix} 1 & 0 \\ -\frac{2}{R} - i\frac{\lambda}{\pi a^2 n} & 1 \end{vmatrix} \equiv \begin{vmatrix} 1 & 0 \\ -\frac{2}{r} & 1 \end{vmatrix} \tag{A-5}$$

so that the tapered reflectivity mirror is characterized by a complex radius of curvature r where

$$\frac{1}{r} = \frac{1}{R} + i\frac{\lambda}{2\pi a^2 n} \tag{A-6}$$

To obtain the mode characteristics in this case we apply the self-consistent formalism of Section 4.5, replacing the ABCD parameters representing the mirrors by (A, B, C, D) matrices in the form of (A-5). The tapering function of the left mirror is represented by a_1^2 and that of the right mirror by a_2^2. The (ABCD) matrix relating the beam parameter q_1 following reflection from the left mirror (mirror 1) to the beam parameter at the same position one complete round trip "earlier," is obtained from (2.1-6) after replacing $f_{1,2}$ by $(1/2)r_{1,2}$. The result is

$$\begin{aligned} A &= 1 - \frac{2}{r_2}l & B &= 2l\left(1 - \frac{l}{r_2}\right) \\ C &= \frac{4l}{r_1 r_2} - \frac{2}{r_1} - \frac{2}{r_2} & D &= \left(1 - \frac{2l}{r_1}\right)\left(1 - \frac{2l}{r_2}\right) - \frac{2}{r_1}l \end{aligned} \tag{A-7}$$

The solution for the steady-state beam parameter $\underset{\rightarrow}{q_1}$ immediately following reflection from mirror 1 is given by (4.5-3) as

$$\frac{1}{\underset{\rightarrow}{q_1}} = \frac{D - A}{2B} + i\frac{\sin\theta}{B} \tag{A-8}$$

where the arrow denotes the direction of beam travel and

$$\cos\theta \equiv \cos(\alpha + i\beta) = \tfrac{1}{2}(D + A) \tag{A-9}$$

α and β are real.

Using (A-7) and (A-9) we have

$$\cos\theta = (2g_1g_2 - 2t_1t_2 - 1) - i(2g_1t_2 + 2g_2t_1) \tag{A-10}$$

so that

$$\frac{1}{\underset{\rightarrow}{q_1}} = \frac{1}{l}\left[-\frac{l}{R_1} - it_1 + i\frac{\sin\alpha\cosh\beta + i\cos\alpha\sinh\beta}{2(g_2 - it_2)}\right] \tag{A-11}$$

and

$$Im\left(\frac{l}{\underset{\rightarrow}{q_1}}\right) = -\frac{l\lambda}{\pi\underset{\rightarrow}{\omega_1}^2 n} = -t_1 + \frac{g_2\sin\alpha\cosh\beta - t_2\cos\alpha\sinh\beta}{2(g_2^2 + t_2^2)} \tag{A-12}$$

where

$$g_{1,2} \equiv 1 - \frac{l}{R_{1,2}} \tag{A-13}$$

$$t_{1,2} = \frac{l\lambda}{2\pi a_{1,2}^2 n} \tag{A-14}$$

According to (A-10) and (A-11) the parameters α and β are determined implicitly by means of the equations

$$\begin{aligned} \cos\alpha\cosh\beta &= 2g_1g_2 - 2t_1t_2 - 1 \\ \sin\alpha\sinh\beta &= 2g_1t_2 + 2g_2t_1 \end{aligned} \tag{A-15}$$

Once α and β are determined, we substitute them in (A-11) to determine the beam parameter $\underset{\rightarrow}{q_1}$. The resulting expression is extremely complicated so it may be advantageous to consider the special case where only one mirror, say 1, has a positive reflectivity taper ($t_1 > 0$), and to assume a uniform reflectivity for the second mirror ($t_2 = 0$). By putting $t_2 = 0$ in (A-15) and using the result in (A-13) we obtain

$$\frac{l\lambda}{\pi\underset{\rightarrow}{\omega_1}^2 n} = t_1(1 - \coth\beta) > 2t_1 \qquad \text{for } \beta < 0 \tag{A-16}$$

In a confined beam it is necessary that $\underset{\leftarrow}{\omega_1}^{-2}$ be finite and positive (that is, the left-going beam before reflection from mirror 1 has a finite spot size). Using (A-4) and (A-14) this last requirement translates into $l\lambda/\pi\underset{\rightarrow}{\omega_1}^2 n > 2t_1$. This condition is indeed satisfied, according to (A-16) by $\underset{\rightarrow}{\omega_1}^2$ provided $\beta < 0$. We are always free to choose $\beta < 0$, since for each (α, β) satisfying (A-9) (α and β real), the pair $(-\alpha, -\beta)$ is an equally valid solution. The solution with $\beta < 0$ is thus seen to lead to a confined beam *regardless*[1] of g_1 $(= 1 - L/R_1)$ and g_2 $(= 1 - L/R_2)$.

We note, in contrast, that the stability (confinement) condition (4.4-2) for uniform reflectivity mirrors ($t_1 = t_2 = 0$) can be written as

$$0 \lesseqgtr g_1 g_2 \lesseqgtr 1 \tag{A-17}$$

and is to be contrasted with the unconditional stability of the resonator with positive tapered ($t_{1,2} > 0$) mirrors.

It can be shown[2] that for $t_1 > 0$ and $t_2 > 0$ the beam modes are stable for any combination of g_1 and g_2. The beam parameters are then determined from (A-1). A plot using (A-12) and (A-15) of $\underset{\leftarrow}{\omega_1}$ versus g with t as a parameter in a symmetric ($g_1 = g_2$, $t_1 = t_2$) resonator is shown in Figure

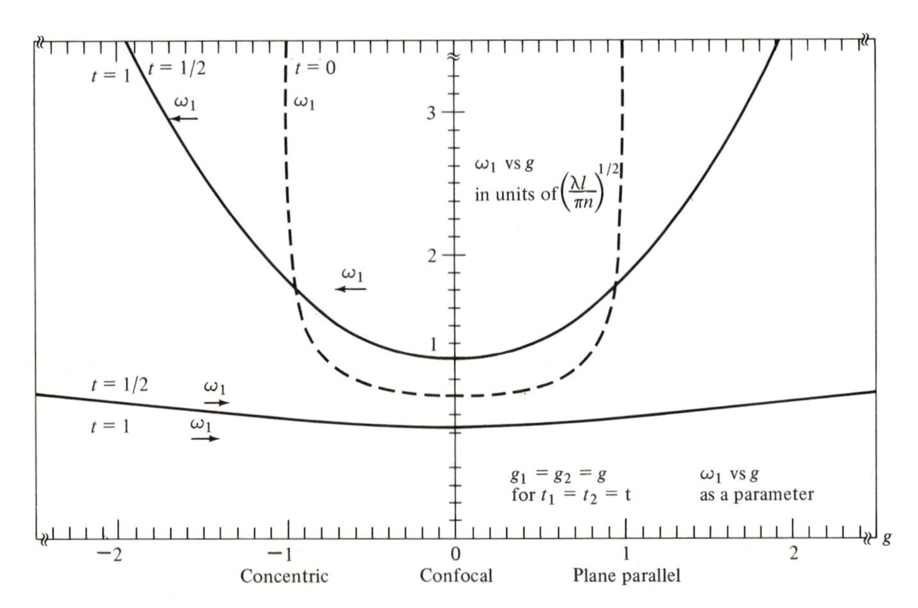

Figure A-1 Behavior of mirror spot size as a function of g in a symmetric resonator ($g_1 = g_2 = g$) for various degrees of reflectivity tapering (t).

---

[1] An exception is the case $g_2 = 0$ for which $\underset{\rightarrow}{\omega_1}^2 = 0$.

[2] A. Yariv, and P. Yeh, "Confinement and stability in optical resonators employing mirrors with Gaussian reflectivity tapers" *Optics Commun.*, vol. 13, pp. 370–374, April 1975.

A-1. We note that when $t = 0$, that is, both mirrors possess uniform reflectivity, ω_1 becomes infinite at $g = \pm 1$, while for $t > 0$, $\underleftarrow{\omega_1}$ is finite everywhere.

Perturbation Stability

The perturbation stability analysis of Section 4.5 can be used to derive the stability of the modes of the tapered mirror resonators. Using (A-9) in (4.5-11) leads to

$$\left.\frac{dq_{\text{out}}^{-1}}{dq_{\text{in}}^{-1}}\right|_{q_{\text{in}}=q_0} = e^{-2i\theta} = e^{-2i\alpha}e^{2\beta} \tag{A-18}$$

It was shown in the discussion following (A-16) that confined ($\underleftarrow{\omega_1}^{-2}$ finite and positive) modes solutions require that we choose the $\beta < 0$ branch of (A-9). With this choice it follows from (A-18) that

$$|\Delta q_{\text{out}}^{-1}| < |\Delta q_{\text{in}}^{-1}| \tag{A-19}$$

so that the beam perturbation decays progressively with each passage. The effect of positive reflectivity tapering ($a^2 > 0$) is thus to replace the neutral stability of (4.5-12) by the absolute stability of (A-19).

APPENDIX B

Mode Locking in Homogeneously Broadened Laser Systems

The analysis of mode locking in inhomogeneous laser systems in Section 6.6 assumed that the role of internal modulation was that of locking together the phases of modes that, in the absence of modulation, oscillate with random phases. In the case of homogeneous broadening, only one mode can normally oscillate. Experiments, however, reveal that mode locking leads to short pulses in a manner quite similar to that described in Section 6.6. One way to reconcile the two points of view and the experiments is to realize that *in the presence of internal modulation,* power is transferred continuously from the high gain mode to those of lower gain (that is, those which would not normally oscillate). This power can be viewed simply as that of the sidebands at $(\omega_0 \pm n\omega)$ of the mode at ω_0 created by a modulation at ω. Armed with this understanding we see that the physical phenomenon is not one of mode locking but one of mode generation. The net result, however, is that of a large number of oscillating modes with equal frequency spacing and fixed phases, as in the inhomogeneous case, leading to ultrashort pulses.

The analytical solution to this case ([1] and [2]) follows an approach used originally to analyze short pulses in traveling wave microwave oscillators [3].

Referring to Figure B-1, we consider an optical resonator with mirror reflectivities R_1 and R_2 that contains, in addition to the gain medium, a periodically modulated loss cell. The method of solution is to follow one pulse through a complete round trip through the resonator and to require that the pulse reproduce itself. The temporal pulse shape at each stage is assumed to be Gaussian.

Before proceeding, we need to characterize the effect of the gain medium and the loss cell on a traveling Gaussian pulse.

Transfer Function of the Gain Medium

Assume that an optical pulse with a field $E_{\text{in}}(t)$ is incident on an amplifying optical medium of length l. Taking the Fourier transform of $E_{\text{in}}(t)$ as $E_{\text{in}}(\omega)$, the amplifier can be characterized by a transfer function $g(\omega)$ where

$$E_{\text{out}}(\omega) = E_{\text{in}}(\omega)g(\omega) \quad \textbf{(B-1)}$$

is the Fourier transform of the output field. Equation (B-1) is a linear

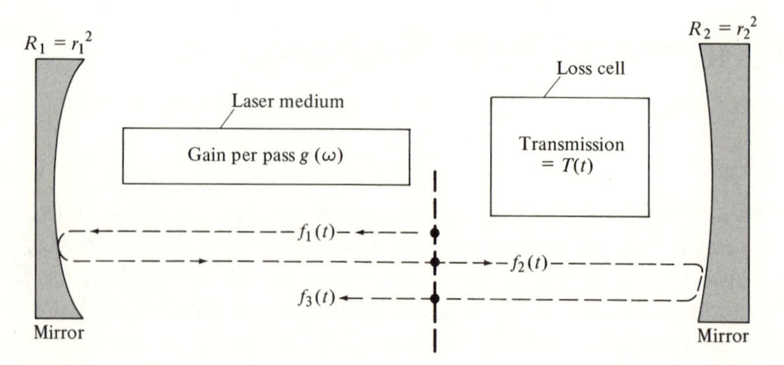

Figure B-1 The experimental arrangement assumed in the theoretical analysis of mode locking in homogeneously broadened lasers.

relationship and applies only in the limit of negligible saturation.

Using (6.1-1) and (5.5-1, 2) we have

$$g(\omega) = \exp\left\{-ikl\left[1 + \frac{1}{2n^2}(\chi' - i\chi'')\right]\right\}$$

$$= \exp\left\{-ikl - \left(\frac{kl}{2n^2}\right)\frac{\Delta N\lambda^3 T_2}{8\pi^2 n t_{\text{spont}}}\left[\frac{1}{1 + i(\omega - \omega_0)T_2}\right]\right\}$$

$$\simeq \exp\left\{-ikl + \frac{\gamma_{\max} l}{2}\left[1 - i(\omega - \omega_0)T_2 - (\omega - \omega_0)^2 T_2^2\right]\right\}$$

where we define $T_2 = (\pi\Delta\nu)^{-1}$, and the approximation is good for $(\omega - \omega_0)T_2 \ll 1$. We recall that $\Delta N_0 < 0$ for gain. Since the pulse is making two passes through the cell, we take

$$\frac{E_{\text{out}}(\omega)}{E_{\text{in}}(\omega)} = [g(\omega)]^2 = \exp\{-i2kl + \gamma_{\max} l[1 - i(\omega - \omega_0)T_2 - (\omega - \omega_0)^2 T_2^2]\}$$

The imaginary terms in the exponent correspond to a time delay (due to the finite group velocity of the pulse) of

$$\tau_d = \frac{2l}{c} + l\gamma_{\max} T_2$$

We are considering here only the effect on the pulse shape so that, ignoring the imaginary term,[1] we obtain

$$[g(\omega)]^2 = e^{\gamma_{\max} l[1-(\omega-\omega_0)^2 T_2^2]} \qquad \textbf{(B-1[a])}$$

Transfer Function of the Loss Cell

Here we need to express the effect of the cell on the pulse in the time domain.

Assume that the single pass amplitude transmission factor $T(t)$ of the loss cell is given by

$$E_{\text{out}}(t) = E_{\text{in}}(t)T(t) = E_{\text{in}}(t)\exp[-2\delta_l^2 \sin^2(\pi\Delta\nu_{\text{axial}} t)] \qquad \textbf{(B-2)}$$

[1] The finite propagation delay affects the round-trip pulse propagation time that must be equal to the period of the loss modulation.

where $\Delta\nu_{\text{axial}}$, the longitudinal mode spacing, is given by

$$\Delta\nu_{\text{axial}} = \frac{c}{2l_c}$$

where l_c is the effective optical length of the resonator. The transmission peaks are thus separated by $2l_c/c$ sec so that a mode-locked pulse can pass through the cell on successive trips with minimum loss. Since the pulses pass through the cell centered on the point of maximum transmission, we approximate (B-2) by

$$E_{\text{out}}(t) = E_{\text{in}}(t)T(t) = E_{\text{in}}(t) \exp\left[-2\delta_l^2(\pi\Delta\nu_{\text{axial}}t)^2\right] \tag{B-3}$$

We can view the form of (B-3) as the prescribed transmission function of the cell. The form, however, is suggested by physical considerations. In the case of an electrooptic shutter with a retardation (see Section 9.3) $\Gamma(t) = \Gamma_m \sin \omega_m t$, the transmission factor is $T(t) = \cos^2(\Gamma(t)/2)$. Near the transmission peaks $\Gamma(t) \ll 1$ and $T(t)$ is given by

$$T(t) \simeq \exp\left[-\tfrac{1}{4}(\Gamma_m{}^2\omega_m{}^2t^2)\right] = \exp\left[-2\delta_l^2(\pi\Delta\nu_{\text{axial}}t)^2\right]$$

where $\omega_m = \pi\Delta\nu_{\text{axial}}$ and $\Gamma_m = 2\sqrt{2}\,\delta_l$.

We now return to the main analysis. The starting pulse $f_1(t)$ in Figure B-1 is taken as

$$f_1(t) = Ae^{-\alpha_1 t^2}e^{i(\omega_0 t + \beta_1 t^2)} \tag{B-4}$$

corresponding to a "chirped" frequency

$$\omega(t) = \omega_0 + 2\beta_1 t \tag{B-5}$$

Its Fourier transform is

$$\begin{aligned} F_1(\omega) &= \frac{1}{2\pi}\int_{-\infty}^{\infty} f_1(t)e^{-i\omega t}\,dt \\ &= \frac{A}{2}\sqrt{\frac{1}{\pi(\alpha_1 - i\beta_1)}} \exp\left[-(\omega-\omega_0)^2/4(\alpha_1 - i\beta_1)\right] \end{aligned} \tag{B-6}$$

A double pass through the amplifier and one mirror reflection (r_1) are accounted for by multiplying $F_1(\omega)$ by the transfer factor $[g(\omega)]^2 r_1$

$$\begin{aligned} F_2(\omega) &= F_1(\omega)[g(\omega)]^2 r_1 \\ &= \frac{r_1 A}{2}e^{g_0}\sqrt{\frac{1}{\pi(\alpha_1 - i\beta_1)}} \exp\left\{[-(\omega - \omega_0)^2]\left[\frac{1}{4(\alpha_1 - i\beta_1)} + g_0 T_2{}^2\right]\right\} \end{aligned} \tag{B-7}$$

where $g_0 \equiv \gamma_{max} l$ and $[g(\omega)]^2$ is given by (B-1[a]). Transforming back to the time domain

$$f_2(t) = \int_{-\infty}^{\infty} F_2(\omega) e^{i\omega t}\, d\omega$$

$$= \frac{r_1 A e^{g_0}}{2\pi} \sqrt{\frac{\pi}{\alpha_1 - i\beta_1}}\, e^{-\omega_0{}^2 Q} \sqrt{\frac{\pi}{Q}} \exp\left[-(2i\omega_0 Q - t)^2/4Q\right] \quad \textbf{(B-8)}$$

where

$$Q \equiv \frac{1}{4(\alpha_1 - i\beta_1)} + g_0 T_2{}^2 \quad \textbf{(B-9)}$$

A reflection from mirror 2 and a passage through the loss cell lead according to (B-3) to

$$f_3(t) = r_2 f_2(t) e^{-2\delta_l{}^2 \pi^2 (\Delta\nu_{\text{axial}})^2 t^2}$$

$$= \frac{r_1 r_2 A e^{g_0}}{2} \sqrt{\frac{1}{(\alpha_1 - i\beta_1)Q}}\, e^{i\omega_0 t} e^{-[2\delta_l{}^2(\pi\Delta\nu_{\text{axial}})^2 + (1/4Q)]\, t^2} \quad \textbf{(B-10)}$$

For self-consistency we require that $f_3(t)$ be a replica of $f_1(t)$. We thus equate the exponent of (B-10) to that of (B-4)

$$\alpha_1 = 2\delta_l{}^2(\pi\Delta\nu_{\text{axial}})^2 + \text{Re}\left(\frac{1}{4Q}\right)$$

$$\beta_1 = -\text{Im}\left(\frac{1}{4Q}\right) \quad \textbf{(B-11)}$$

Using (B-9), the second of (B-11) gives

$$\beta_1 = \frac{\beta_1}{(1 + 4g_0 T_2{}^2 \alpha_1)^2 + (4g_0 T_2{}^2 \beta_1)^2}$$

so that a self-consistent solution requires that

$$\beta_1 = 0$$

that is, no chirp. With $\beta_1 = 0$ the first of (B-11), becomes

$$2\delta_l{}^2(\pi\Delta\nu_{\text{axial}})^2 + \frac{\alpha_1}{(1 + 4g_0 T_2{}^2 \alpha_1)} = \alpha_1 \quad \textbf{(B-12)}$$

that, assuming

$$4g_0 T_2{}^2 \alpha_1 \ll 1 \quad \textbf{(B-13)}$$

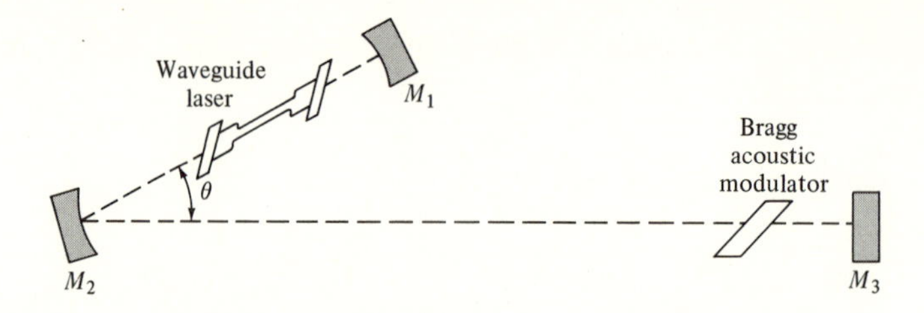

Figure B-2 A schematic drawing of the mode-locking experiment in a high pressure CO_2 laser. (After Reference [4].)

results in

$$\alpha_1 = \left(\frac{\delta_l^2}{2g_0}\right)^{1/2} \frac{\pi \Delta\nu_{\text{axial}}}{T_2}$$

The pulse width at the half intensity points is from (B-4)

$$\tau_p = (2 \ln 2)^{1/2} \alpha_1^{-1/2}$$

so that the self-consistent pulse has a width

$$\tau_p = \frac{(2 \ln 2)^{1/2}}{\pi} \left(\frac{2g_0}{\delta_l^2}\right)^{1/4} \left(\frac{1}{\Delta\nu_{\text{axial}}\Delta\nu}\right)^{1/2} \tag{B-14}$$

where $\Delta\nu \equiv (\pi T_2)^{-1}$. The condition (B-13) can now be interpreted as requiring that $\tau_p \gg 2\sqrt{g_0}\, T_2$, which is true in most cases.

An experimental setup demonstrating mode locking in a pressure broadened CO_2 laser is sketched in Figure B-2. The inverse square root dependence of τ_p on $\Delta\nu$ is displayed by the data of Figure B-3, while the dependence on the modulation parameter δ_l is shown in Figure B-4.

Mode Locking by Phase Modulation

Mode locking can be induced by internal phase, rather than loss, modulation. This is usually done by using an electrooptic crystal inside the resonator oriented in the basic manner of Figure (9-7) such that the passing wave undergoes a phase delay proportional to the instantaneous electric field across the crystal. The frequency of the modulating signal is equal, as in the loss modulation case, to the inverse of the round trip group delay time, that is, to the longitudinal intermode frequency separation.

We employ an analysis similar to that of the homogeneous case except

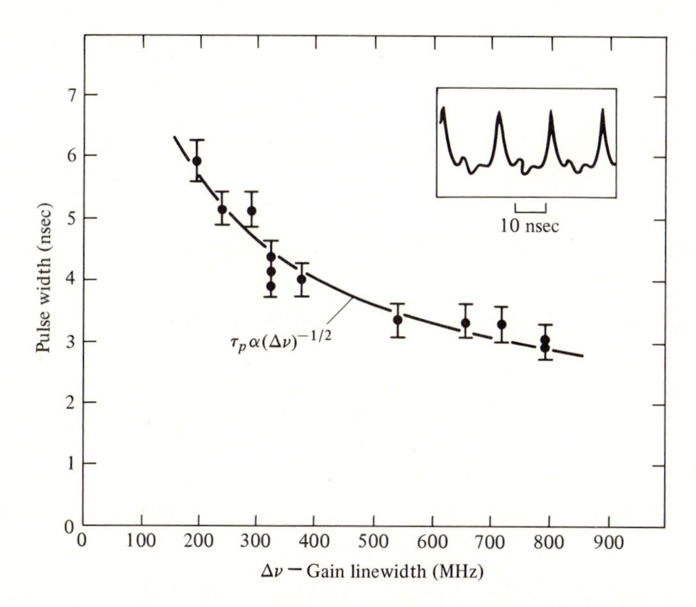

Figure B-3 The dependence of the pulse width on the gain linewidth, $\Delta\nu$, that is controlled by varying the pressure ($\Delta\nu = 8 \times 10^8$ at 150 torr). (After Reference [4].)

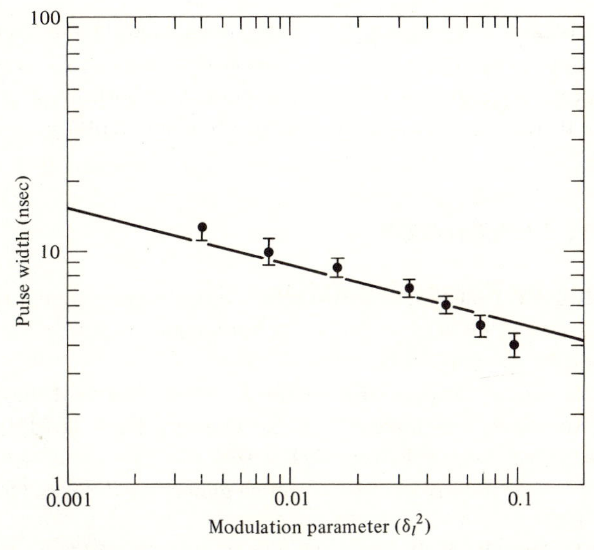

Figure B-4 The mode-locked pulse width as a function of the modulation parameter, δ_l^2. (After Reference [4].)

that the transfer function through the modulation cell is taken, instead of (B-2), as

$$E_{\text{out}}(t) = E_i(t) \exp\left(-i2\delta_\phi \cos 2\pi\Delta\nu_{\text{axial}}t\right) \quad \textbf{(B-15)}$$

For pulses passing near the extrema of the phase excursion we can approximate the last equation as

$$E_{\text{out}}(t) = E_i(t) \exp\left(\mp i2\delta_\phi \pm i\delta_\phi 4\pi^2\Delta\nu_{\text{axial}}^2 t^2\right) \quad \textbf{(B-16)}$$

An analysis identical to that leading to (B-14) yields

$$\tau_p = \frac{(2 \ln 2)^{1/2}}{\pi}\left(\frac{2g_0}{\delta_\phi}\right)^{1/4}\left(\frac{1}{\Delta\nu_{\text{axial}}\Delta\nu}\right)^{1/2} \quad \textbf{(B-17)}$$

In this case self-consistency leads to a chirped pulse with

$$\beta = \pm\alpha = \pm\pi^2\Delta\nu_{\text{axial}}\Delta\nu\sqrt{\frac{\delta_\phi}{2g_0}} \quad \textbf{(B-18)}$$

The upper and lower signs in (B-16) and (B-18) correspond to two possible pulse solutions, one passing through the cell near the maximum of the phase excursion and the other near its minimum.

We note that (B-17) is similar to the loss modulation result (B-14) except that δ_ϕ appears instead of δ_l^2. This difference can be traced into a difference between (B-2) and (B-5). The choice in both cases is such that δ *corresponds* to the *retardation* induced by the electrooptic crystal.

REFERENCES

[1] Siegman, A. E., and D. J. Kuizenga, "Simple analytic expressions for AM and FM mode locked pulses in homogeneous lasers," *Appl. Phys. Lett.*, vol. 14, p. 181, 1969.

[2] Kuizenga, D. J., and A. E. Siegman, "FM and AM mode locking of the homogeneous laser: Part I, Theory; Part II, Experiment," *J. Quant. Elect.*, vol. QE-6, p. 694, 1970.

[3] Cutler, C. C., "The regenerative pulse generator," *Proc. IRE*, vol. 43, p. 140, 1955.

[4] Smith, P. W., T. J. Bridges, E. G. Burkhardt, "Mode locked high pressure CO_2 laser," *Appl. Phys. Lett.*, vol. 21, p. 470, 1972.

APPENDIX

C

Electrooptic Effect in Cubic $\bar{4}3m$ Crystals

As an example of transverse modulation[1] and of the application of the electrooptic effect we consider the case of crystals of the $\bar{4}3m$ symmetry group. Examples of this group are: InAs, CuCl, GaAs, and CdTe. The last two are used for modulation in the infrared, since they remain transparent beyond 10 μm. These crystals are cubic and have axes of fourfold symmetry along the cube edges (⟨100⟩ directions), and threefold axes of symmetry along the cube diagonals ⟨111⟩.

To be specific, we apply the field in the ⟨111⟩ direction—that is, along a threefold-symmetry axis. Taking the field magnitude as E, we have

$$\mathbf{E} = \frac{E}{\sqrt{3}} (\mathbf{e}_1 + \mathbf{e}_2 + \mathbf{e}_3) \qquad \text{(C-1)}$$

where $\mathbf{e}_1$, $\mathbf{e}_2$, and $\mathbf{e}_3$ are unit vectors directed along the cube edges x, y, and z, respectively. The three nonvanishing electrooptic tensor elements are, according to Table 9-1 [see $\bar{4}3m$ tensor], r_{41}, $r_{52} = r_{41}$, and $r_{63} = r_{41}$. Thus, using Equations (9.1-2) through (9.1-4), with

$$\left(\frac{1}{n^2}\right)_1 = \left(\frac{1}{n^2}\right)_2 = \left(\frac{1}{n^2}\right)_3 \equiv \frac{1}{n_o^2}$$

[1] "Transverse modulation" is the term applied to the case when the field is applied normal to the direction of propagation.

we obtain

$$\frac{x^2 + y^2 + z^2}{n_o^2} + \frac{2r_{41}E}{\sqrt{3}}(xy + yz + xz) = 1 \tag{C-2}$$

as the equation of the index ellipsoid. One can proceed formally at this point to derive the new directions x', y', and z' of the principal axes of the ellipsoid. A little thought, however, will show that the $\langle 111 \rangle$ direction along which the field is applied will continue to remain a threefold-symmetry axis, whereas the remaining two orthogonal axes can be chosen *anywhere* in the plane normal to $\langle 111 \rangle$. Thus (C-2) is an equation of an ellipsoid of revolution about $\langle 111 \rangle$. To prove this we choose $\langle 111 \rangle$ as the z' axis, so

$$z' = \frac{1}{\sqrt{3}}x + \frac{1}{\sqrt{3}}y + \frac{1}{\sqrt{3}}z \tag{C-3}$$

and take

$$\begin{aligned} x' &= \frac{1}{\sqrt{2}}y - \frac{1}{\sqrt{2}}z \\ y' &= -\frac{2}{\sqrt{6}}x + \frac{1}{\sqrt{6}}y + \frac{1}{\sqrt{6}}z \end{aligned} \tag{C-4}$$

Therefore,

$$\begin{aligned} x &= -\frac{2}{\sqrt{6}}y' + \frac{1}{\sqrt{3}}z' \\ y &= \frac{1}{\sqrt{2}}x' + \frac{1}{\sqrt{6}}y' + \frac{1}{\sqrt{3}}z' \\ z &= -\frac{1}{\sqrt{2}}x' + \frac{1}{\sqrt{6}}y' + \frac{1}{\sqrt{3}}z' \end{aligned} \tag{C-5}$$

Substituting (C-5) in (C-2), we obtain the equation of the index ellipsoid in the x', y', z' coordinate system as

$$(x'^2 + y'^2)\left(\frac{1}{n_o^2} - \frac{r_{41}E}{\sqrt{3}}\right) + \left(\frac{1}{n_o^2} + \frac{2r_{41}}{\sqrt{3}}E\right)z'^2 = 1 \tag{C-6}$$

so the principal indices of refraction become

$$\begin{aligned} n_{y'} &= n_{x'} = n_o + \frac{n_o^3 r_{41}E}{2\sqrt{3}} \\ n_{z'} &= n_o - \frac{n_o^3 r_{41}E}{\sqrt{3}} \end{aligned} \tag{C-7}$$

It is clear from (C-6) that other choices of x' and y', as long as they are normal to z' and to each other, will work as well since x' and y' enter (C-6) as the combination $x'^2 + y'^2$, which is invariant to rotations about the z' axes. The principal axes of the index ellipsoid (C-6) are shown in Figure C-1.

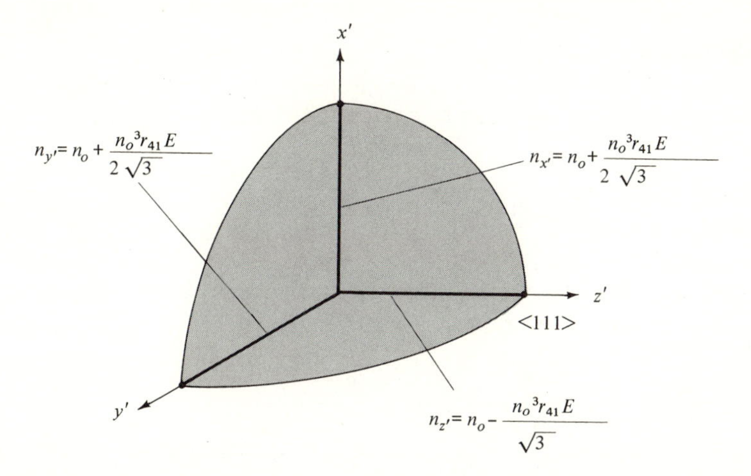

Figure C-1 The intersection of the index ellipsoid of $\bar{4}3m$ crystals (with E parallel to $\langle 111 \rangle$) with the planes $x' = 0$, $y' = 0$, $z' = 0$. The principal indices of refraction for this case are $n_{x'}$, $n_{y'}$, and $n_{z'}$.

An amplitude modulator based on the foregoing situation is shown in Figure C-2. The fractional intensity transmission is given by (9.3-4) as

$$\frac{I_o}{I_i} = \sin^2 \frac{\Gamma}{2}$$

where the retardation, using (A-7), is

$$\Gamma = \phi_{z'} - \phi_{y'} = \frac{(\sqrt{3}\pi) n_o^3 r_{41}}{\lambda_0} \left(\frac{Vl}{d} \right) \tag{C-8}$$

An important difference between this case where the electric field is applied normal to the direction of propagation and the longitudinal case (9.2-4) is that here Γ is proportional to the crystal length l.

A complete discussion of the electrooptic effect in $\bar{4}3m$ crystals is given in C. S. Namba, *J. Opt. Soc. Am.*, vol. 51, p. 76, 1961. A summary of his analysis is included in Table C-1.

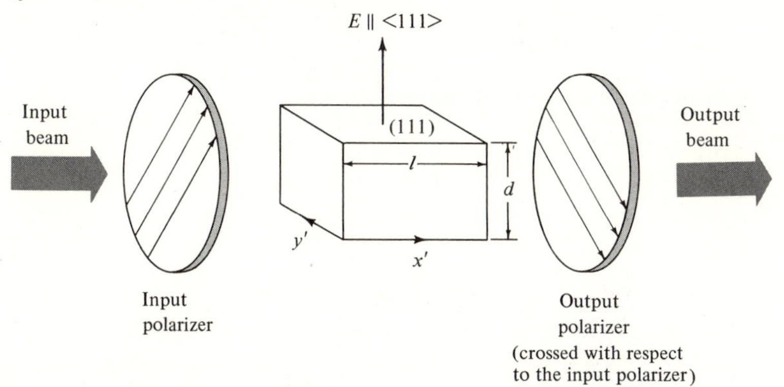

Figure C-2 A transverse electrooptic modulator using a zinc-blende type ($\bar{4}3m$) crystal with E parallel to the cube diagonal $\langle 111 \rangle$ direction.

Table C-1 ELECTROOPTICAL PROPERTIES AND RETARDATION IN $\bar{4}3m$ (ZINC BLENDE STRUCTURE) CRYSTALS FOR THREE DIRECTIONS OF APPLIED FIELD. AFTER C. S. NAMBA, JOUR. OPT. SOC. AM. VOL. 51, 76 (1961).

	$E \perp (001)$ plane $E_x = E_y = 0, E_z = E$	$E \perp (110)$ plane $E_x = E_y = \frac{E}{\sqrt{2}}, E_z = 0$	$E \perp (111)$ plane $E_x = E_y = E_z = \frac{E}{\sqrt{3}}$
Index ellipsoid	$\frac{x^2+y^2+z^2}{n_o^2} + 2 r_{41} E xy = 1$	$\frac{x^2+y^2+z^2}{n_o^2} + \sqrt{2} r_{41} E(yz + zx) = 1$	$\frac{x^2+y^2+z^2}{n_o^2} + \frac{2}{\sqrt{3}} r_{41} E(yz + zx + xy) = 1$
n_x'	$n_o + \frac{1}{2} n_o^3 r_{41} E$	$n_o + \frac{1}{2} n_o^3 r_{41} E$	$n_o + \frac{1}{2\sqrt{3}} n_o^3 r_{41} E$
n_y'	$n_o - \frac{1}{2} n_o^3 r_{41} E$	$n_o - \frac{1}{2} n_o^3 r_{41} E$	$n_o + \frac{1}{2\sqrt{3}} n_o^3 r_{41} E$
n_z'	n_o	n_o	$n_o - \frac{1}{\sqrt{3}} n_o^3 r_{41} E$
$x'y'z'$ coordinates	z,z′; y′; y; (001); x′; 45°; x	z; ($\bar{1}$10); y′; 45°; y; z′; (001); 45°; 45°; x′; x	z; x′; z′(111); y; y′; x
Directions of optical path and axes of crossed polarizer	E_{dc}; Γ_z; A(y); P(x); y; (001); (100); x; Γ = 0; Γ_{xy}; Γ = 0; A; E; P; 45°	Γ = 0; E_{dc}; (110); (001); ($\bar{1}$10); Γ = 0; z′; P(E); Γ_{max}; A	Γ = 0; E_{dc}; (111); d; E; Γ; Γ; 45°; P; A
Retardation phase difference $r(V = Ed)$	$\Gamma_z = \frac{2\pi}{\lambda} n_o^3 r_{41} V$ $\Gamma_{xy} = \frac{\pi}{\lambda} \frac{l}{d} n_o^3 r_{41} V$	$\Gamma_{max} = \frac{2\pi}{\lambda} \frac{l}{d} n_o^3 r_{41} V$	$\Gamma = \frac{\sqrt{3}\pi}{\lambda} \frac{l}{d} n_o^3 r_{41} V$

Author Index

Subject Index